CRISPR and Plant Functional Genomics

CRISPR is a crucial technology in plant physiology and molecular biology, resulting in more sustainable agricultural practices, including outcomes of better plant stress tolerance and crop improvement. *CRISPR and Plant Functional Genomics* explores ways to release the potential of plant functional genomics, one of the prevailing topics in plant biology and a critical technology for speed and precision crop breeding. This book presents achievements in plant functional genomics and features information on diverse applications using emerging CRISPR-based genome editing technologies producing high-yield, disease-resistant, and climate-smart crops. It also includes theories on organizing strategies for upgrading the CRISPR system to increase efficiency, avoid off-target effects, and produce transgene-free edited crops.

Features:

- Presents CRISPR-based technologies, releasing the potential of plant functional genomics
- Provides methods and applications of CRISPR/Cas-based plant genome editing technologies
- Summarizes achievements of speed and precision crop breeding using CRISPR-based technologies
- Illustrates strategies to upgrade the CRISPR system
- Supports the UN's sustainable development goals to develop future climate-resilient crops

CRISPR and Plant Functional Genomics provides extensive knowledge of CRISPR-based technologies and plant functional genomics and is an ideal reference for researchers, graduate students, and practitioners in the field of plant sciences as well as agronomy and agriculture.

CRISPR and Plant Functional Genomics

Edited by
Jen-Tsung Chen

CRC Press
Taylor & Francis Group
Boca Raton London New York

CRC Press is an imprint of the
Taylor & Francis Group, an **informa** business

First edition published 2024
by CRC Press
2385 NW Executive Center Drive, Suite 320, Boca Raton FL 33431

and by CRC Press
4 Park Square, Milton Park, Abingdon, Oxon, OX14 4RN

CRC Press is an imprint of Taylor & Francis Group, LLC

ISBN: 9781032469492 (hbk)
ISBN: 9781032480343 (pbk)
ISBN: 9781003387060 (ebk)

DOI: 10.1201/9781003387060

Typeset in Times
by codeMantra

Contents

Preface

For decades, functional genomics has been conducted intensively to explore plant biological networks and continues to identify critical components of regulatory and controlling machinery as well as their functions in sustaining growth and development. In the meantime, the resulting knowledge and technologies keep benefiting agricultural production through the creation of stress-tolerant, high-yield, and high-quality crops using multi-omics-assisted breeding strategies. In recent years, a goal that fits SDGs (Sustainable Development Goals) by the United Nations is to precisely breed climate-smart crops against global climate change, and therefore, definitely, functional genomics is the major technology for future sustainable agriculture.

The system of functional genomics evolves rapidly, and one of the game-changing tools is genome editing technology. Among them, an emerging one is based on CRISPR (Clustered Regularly Interspaced Short Palindromic Repeats) and the associated Cas proteins. CRISPR-based genome editing technology can specifically introduce targeted sequence alterations, and thus, it leads to a new era of plant functional genomics and, meanwhile, a practical system of precision molecular plant breeding. This book summarizes and comprehensively discusses the introduction and integration of CRISPR-based technology into plant functional genomics and precision breeding. It consists of 21 chapters to refine current knowledge from the literature and is organized by scientists and experts in this field.

Since the post-genomics era, bioinformatics has become an essential tool for dealing with huge amounts of genomic datasets and for the research field of systems biology, contributing to an array of research fields, including the construction of CRISPR systems. In this book, a part of the content pays attention to the design of guide RNA for CRISPR technology, which is critical to ensure efficient and precisely targeted modification of interest genes, and it can be customized by some tools belonging to bioinformatics. Another essential component of the CRISPR system is the association of Cas proteins, and a chapter was organized to illustrate the structures and functions of their varieties, which provide available types for developing appropriate systems in epigenetic manipulation.

To deal with the growing human population, the UN sets SDG goals to achieve zero hunger around the world. The issue of food security majorly depends on agricultural production, and it can be enhanced by crop improvement. Without a doubt, in the future, CRISPR-assisted precision plant breeding will be a promising tool to edit bases with the aid of advanced genomics tools, such as NGS (next-generation sequencing), single-cell omics, spatial transcriptomics, and so on. In modern times, another global threat is climate change, and therefore, scientists attempt to mitigate the effect by gaining future crops with the capacity of "climate-smart" or "climate-resilient" through cutting-edge breeding strategies. One of the effective ways is to apply the emerging CRISPR/Cas technology, particularly using Cas9, and some achievements have been reported, chiefly focusing on environmental/abiotic stress tolerance under heat, drought, flood, and so on, and disease resistance when faced with a changing climate that probably leads to severe damage on yield and quality of crops. For developing future crops, a revolutionary system must be taken into consideration, namely pan-genomics which explores a collection of genomic sequences for the entire species or population and can provide a rich resource for identifying critical genetic variations to advance plant breeding. In the future, crop breeders can perform CRISPR more efficiently based on the information and knowledge of the research in pan-genomes to optimize their programs for "speed breeding".

In agricultural production, a serious negative factor that impacts food security is the huge loss of crop yield and quality caused by pathogens and pests. Fortunately, plants have evolved an innate system of defense against viruses, bacteria, fungi, herbivores, nematodes, and so on. Theoretically, plant immunity has the potential to be released or improved, and it can be achieved by CRISPR-based technology through the understanding of the interactions between plants and microbes or pests and subsequently ensuring the editing of targeted genes for resistance. For plant pathologists, it's often challenging to monitor and control plant viral diseases, and some efforts have been made to use CRISPR-based technology against plant viruses. In this book, the current achievements using the emerging CRISPR/Cas13 system to successfully produce virus-resistant plants are systematically summarized.

Altogether, this book presents the current knowledge of the CRISPR technology based on refining literature with an emphasis on the fundamental technologies as well as upgraded tools for the exploration of functional genomics and then toward accelerated and precision plant breeding. The applications have been reported in some significant crops, such as rice, wheat, maize, soybean, oil palms, coconut, and ornamental plants. This book provides sufficient and updated knowledge of CRISPR technology and plant functional genomics and thus is an ideal reference for graduate students, teachers, researchers, and experts in the field of plant biology as well as agronomy and agriculture.

As the book editor, I'd like to thank all the authors of this book for their expertise and for their time and effort in organizing chapters. The friendly assistance and instructions from the staff of CRC Press/Taylor & Francis Group are very much appreciated.

Editor

Jen-Tsung Chen is a professor of cell biology at the National University of Kaohsiung in Taiwan. He also teaches genomics, proteomics, plant physiology, and plant biotechnology. His areas of research interest include bioactive compounds, chromatography techniques, plant molecular biology, plant biotechnology, bioinformatics, and systems pharmacology. He is an active editor of academic books and journals to advance the exploration of multidisciplinary knowledge involving plant physiology, plant biotechnology, nanotechnology, ethnopharmacology, and systems biology. He serves as an editorial board member in reputed journals, including *Plant Methods, GM Crops & Food, Plant Nano Biology, Biomolecules, International Journal of Molecular Sciences,* and a guest editor in *Frontiers in Plant Science, Frontiers in Pharmacology, Journal of Fungi,* and *Current Pharmaceutical Design.* Dr. Chen published books in collaboration with Springer Nature, CRC Press/Taylor & Francis Group, and In-TechOpen, and he is handling book projects for more international publishers on diverse topics such as drug discovery, herbal medicine, medicinal biotechnology, nanotechnology, bioengineering, plant functional genomics, plant speed breeding, and CRISPR-based plant genome editing.

Contributors

Malik Z. Abdin
Centre for Transgenic Plant Development, Department of Biotechnology
School of Chemical and Life Sciences
New Delhi, India

Siti Nor Akmar Abdullah
Department of Agriculture Technology, Faculty of Agriculture
Universiti Putra Malaysia
Selangor, Malaysia
and
Laboratory of Agronomy and Sustainable Crop Protection, Institute of Plantation Studies
Universiti Putra Malaysia
Selangor, Malaysia

Mahtab Ahmad
Department of Agricultural Biotechnology and Molecular Biology, College of Basic Sciences & Humanities
Dr. Rajendra Prasad Central Agricultural University
Pusa (Samastipur), India

Abdul Razak Ahmed
Molecular Mycology Laboratory, Plant Protection Department, Faculty of Agriculture
Akdeniz University
Antalya, Turkey

André Albuquerque
MED - Mediterranean Institute for Agriculture, Environment and Development & CHANGE - Global Change and Sustainability Institute, Institute for Advanced Studies and Research
Universidade de Évora, Pólo da Mitra
Évora, Portugal

Abdel Rahman Al-Tawaha
Department of Biological Sciences
Al-Hussein Bin Talal University
Maan, Jordon

Arumugam Vijaya Anand
Department of Human Genetics and Molecular Genetics
Bharathiar University
Coimbatore, India

Kumari Anjani
Department of Agricultural Biotechnology and Molecular Biology, College of Basic Sciences & Humanities
Dr. Rajendra Prasad Central Agricultural University
Pusa (Samastipur), India

Sinto Antoo
Division of Plant Physiology
ICAR-Indian Agricultural Research Institute
New Delhi, India

Anuradha
Department of Agricultural Biotechnology and Molecular Biology, College of Basic Sciences & Humanities
Dr. Rajendra Prasad Central Agricultural University
Pusa (Samastipur), India

Arpitha S R
Division of Biochemistry
ICAR-Indian Agricultural Research Institute
New Delhi, India

Boro Arthi
Department of Human Genetics and Molecular Genetics
Bharathiar University
Coimbatore, India

Muthukrishnan Arun
Department of Biotechnology
Bharathiar University
Coimbatore, India

Balaji Balamurugan
ICAR-National Institute for Plant Biotechnology
New Delhi, India

Mamta Bhattacharjee
Department of Agricultural Biotechnology
Assam Agriculture University
Jorhat, India

Sayan Bhattacharya
School of Ecology and Environment Studies
Nalanda University
Rajgir, India

Karma Landup Bhutia
Department of Agricultural Biotechnology and Molecular Biology, College of Basic Sciences & Humanities
Dr. Rajendra Prasad Central Agricultural University
Pusa (Samastipur), India

Nangsol Dolma Bhutia
Department of Vegetable Sciences, College of Horticulture and Forestry
Central Agricultural University (Imphal)
Pasighat, India

Bhavna Borah
Department of Agricultural Biotechnology and Molecular Biology, College of Basic Sciences & Humanities
Dr. Rajendra Prasad Central Agricultural University
Pusa (Samastipur), India

Ercan Bursal
Department of Biochemistry
Mus Alparslan University
Merkez/Muş, Turkey

Ozer Calis
Molecular Virology Laboratory, Plant Protection Department, Faculty of Agriculture
Akdeniz University
Antalya, Turkey

Maria Doroteia Campos
MED - Mediterranean Institute for Agriculture, Environment and Development & CHANGE - Global Change and Sustainability Institute, Institute for Advanced Studies and Research
Universidade de Évora, Pólo da Mitra
Évora, Portugal

Abira Chaudhuri
National Institute of Plant Genome Research
New Delhi, India

Channakeshavaiah Chikkaputtaiah
Biological Sciences and Technology Division
CSIR-North East Institute of Science and Technology (CSIR-NEIST)
Jorhat, India

Viswanathan Chinnusamy
Division of Plant Physiology
ICAR-Indian Agricultural Research Institute
New Delhi, India

Anamika Das
Department of Genetics and Plant Breeding
Bidhan Chandra Krishi Viswavidyalaya
Nadia, India
and
Department of Agriculture
Netaji Subhas University
Jamshedpur, India

Asis Datta
National Institute of Plant Genome Research
New Delhi, India

Johni Debbarma
Department of Botany
Assam Don Bosco University
Sonapur, India
and
Biological Sciences and Technology Division
CSIR-North East Institute of Science and Technology (CSIR-NEIST)
Jorhat, India

Sandip Debnath
Department of Genetics and Plant Breeding
Palli Siksha Bhavana (Institute of Agriculture), Visva-Bharati
Sriniketan, India

Barasha Rani Deka
Department of Agricultural Biotechnology and Molecular Biology, College of Basic Sciences & Humanities
Dr. Rajendra Prasad Central Agricultural University
Pusa (Samastipur), India

Abhijit Dey
Department of Life Sciences
Presidency University
Kolkata, India

Phat Tien Do
Institute of Biotechnology
Vietnam Academy of Science and Technology
Hanoi, Vietnam
and
Graduate University of Science and Technology
VietnamAcademy of Science and Technology
Hanoi, Vietnam

Maria do Rosário Félix
MED - Mediterranean Institute for Agriculture, Environment and Development & CHANGE - Global Change and Sustainability Institute, Departamento de Fitotecnia, Escola de Ciências e Tecnologia
Universidade de Évora, Pólo da Mitra
Évora, Portugal

Hakan Fidan
Molecular Mycology Laboratory, Plant Protection Department, Faculty of Agriculture
Akdeniz University
Antalya, Turkey

Nicolás Garrido
MED - Mediterranean Institute for Agriculture, Environment and Development & CHANGE - Global Change and Sustainability Institute, Institute for Advanced Studies and Research
Universidade de Évora, Pólo da Mitra
Évora, Portugal

Mimosa Ghorai
Department of Life Sciences
Presidency University
Kolkata, India

Gopika
Department of Human Genetics and Molecular Genetics
Bharathiar University
Coimbatore, India

Packiaraj Gurusaravanan
Plant Biotechnology Laboratory, Department of Botany
Bharathiar University
Coimbatore, India

Koushik Halder
National Institute of Plant Genome Research
New Delhi, India
and
Centre for Transgenic Plant Development, Department of Biotechnology
School of Chemical and Life Sciences
New Delhi, India

Muhammad Asyraf Md Hatta
Department of Agriculture Technology, Faculty of Agriculture
Universiti Putra Malaysia
Selangor, Malaysia

Akbar Hossain
Division of Soil Science
Bangladesh Wheat and Maize Research Institute
Dinajpur, Bangladesh

Hao Hu
Agriculture and Agri-Food Canada
Saskatoon Research and Development Centre
Saskatoon, SK, Canada

Ishfaq Majid Hurrah
Plant Biotechnology Division
CSIR-Indian Institute of Integrative Medicine
Srinagar, India

Tofazzal Islam
Institute of Biotechnology and Genetic Engineering (IBGE)
Bangabandhu Sheikh Mujibur Rahman Agricultural University
Gazipur, Bangladesh

M. James
School of Crop Improvement, College of Post Graduate Studies in Agricultural Sciences
Central Agricultural University (Imphal)
Imphal, India

Gayatri Jamwal
School of Biotechnology
Sher-e-Kashmir University of Agricultural Sciences and Technology of Jammu
Chatha, India

Senthil Kalaiselvi
Department of Biochemistry Biotechnology and Bioinformatics
Avinashilingam Institute for Home Science & Higher Education for Women
Coimbatore, India

Danish Mushtaq Khanday
Division of Plant Breeding and Genetics
Sher-e-Kashmir University of Agricultural Sciences and Technology of Jammu
Chatha, India

Anima Kisku
Department of Agricultural Biotechnology and Molecular Biology, College of Basic Sciences & Humanities
Dr. Rajendra Prasad Central Agricultural University
Pusa (Samastipur), India

Trishna Konwar
Department of Agricultural Biotechnology
Assam Agriculture University
Jorhat, India

Kumaresan Kowsalya
Department of Biotechnology
Bharathiar University
Coimbatore, India

Santhanu Krishnapriya
Department of Biochemistry Biotechnology and Bioinformatics
Avinashilingam Institute for Home Science & Higher Education for Women
Coimbatore, India

Sudhir Kumar
Division of Plant Physiology
ICAR-Indian Agricultural Research Institute
New Delhi, India

Sarita Kumari
Department of Agricultural Biotechnology and Molecular Biology, College of Basic Sciences & Humanities
Dr. Rajendra Prasad Central Agricultural University
Pusa (Samastipur), India

Bharti Lap
School of Crop Improvement, College of Post Graduate Studies in Agricultural Sciences
Central Agricultural University (Imphal)
Imphal, India

Luo-Yu Liang
Department of MOE Key Laboratory of Cell Activities and Stress Adaptations, School of Life Sciences
Lanzhou University
Lanzhou, China
and
Gansu Province Key Laboratory of Gene Editing for Breeding, School of Life Sciences
Lanzhou University
Lanzhou, China

Limasunep Longkumer
Department of Agricultural Biotechnology
Assam Agricultural University
Jorhat, India

Sayanti Mandal
Department of Biotechnology, Dr. D. Y. Patil Arts
Commerce & Science College (affiliated to Savitribai Phule Pune University)
Pune, India
and
Institute of Bioinformatics and Biotechnology
Savitribai Phule Pune University
Pune, India

Patrick Materatski
MED - Mediterranean Institute for Agriculture, Environment and Development & CHANGE - Global Change and Sustainability Institute, Institute for Advanced Studies and Research
Universidade de Évora, Pólo da Mitra
Évora, Portugal

Kanshouwa Modunshim
Department of Agricultural Biotechnology
Assam Agricultural University
Jorhat, India

Tabasum Mohiuddin
Govt. Degree College for Women
Baramulla, India

Tomás Monteiro
MED - Mediterranean Institute for Agriculture, Environment and Development & CHANGE - Global Change and Sustainability Institute, Institute for Advanced Studies and Research
Universidade de Évora, Pólo da Mitra
Évora, Portugal

Ponnuchamy Mugudeshwari
Indian Institute of Rice Research
Hyderabad, India

Israt Nadia
Mawlana Bhashani Science and Technology University
Tangail, Bangladesh

Cuong Xuan Nguyen
Institute of Biotechnology
Vietnam Academy of Science and Technology
Hanoi, Vietnam

Prabhakar Nishi
Department of Agricultural Biotechnology and Molecular Biology, College of Basic Sciences & Humanities
Dr. Rajendra Prasad Central Agricultural University
Pusa (Samastipur), India

Potshangbam Nongdam
Department of Biotechnology
Manipur University
Imphal, India

Devendra Kumar Pandey
Department of Biotechnology
Lovely Professional University
Punjab, India

Mariana Patanita
MED - Mediterranean Institute for Agriculture, Environment and Development & CHANGE - Global Change and Sustainability Institute, Institute for Advanced Studies and Research
Universidade de Évora, Pólo da Mitra
Évora, Portugal

Ayyadurai Pavithra
Department of Human Genetics and Molecular Genetics
Bharathiar University
Coimbatore, India

Gadpayale Durgeshwari Prabhakar
Division of Biochemistry
ICAR-Indian Agricultural Research Institute
New Delhi, India

Biswajit Pramanik
Department of Genetics and Plant Breeding
Palli Siksha Bhavana (Institute of Agriculture), Visva-Bharati
Sriniketan, India

S. M. Hisam Al Rabbi
Biotechnology Division
Bangladesh Rice Research Institute
Gazipur, Bangladesh

Gyanendra Kumar Rai
School of Biotechnology
Sher-e-Kashmir University of Agricultural Sciences and Technology of Jammu
Chatha, India

Bashyam Ramya
Department of Biochemistry
Holy Cross College (Autonomous)
Tiruchirappalli, India

Reena Rani
National Institute for Biotechnology and Genetic Engineering
Faisalabad, Pakistan
and
Constituent College Pakistan Institute of Engineering and Applied Sciences
Faisalabad, Pakistan

Ghulam Raza
National Institute for Biotechnology and Genetic Engineering
Faisalabad, Pakistan
and
Constituent College Pakistan Institute of Engineering and Applied Sciences
Faisalabad, Pakistan

Muhammad Khuram Razzaq
Soybean Research Institute & MARA National Center for Soybean Improvement & MARA Key Laboratory of Biology and Genetic Improvement of Soybean & National Key Laboratory for Crop Genetics and Germplasm Enhancement & Jiangsu Collaborative Innovation Center for Modern Crop Production
Nanjing Agricultural University
Nanjing, China

Joana A. Ribeiro
MED - Mediterranean Institute for Agriculture, Environment and Development & CHANGE - Global Change and Sustainability Institute, Institute for Advanced Studies and Research
Universidade de Évora, Pólo da Mitra
Évora, Portugal

Kattilaparambil Roshna
Plant Biotechnology Laboratory, Department of Botany
Bharathiar University
Coimbatore, India

Filipa Santos
MED - Mediterranean Institute for Agriculture, Environment and Development & CHANGE - Global Change and Sustainability Institute, Departamento de Fitotecnia, Escola de Ciências e Tecnologia
Universidade de Évora, Pólo da Mitra
Évora, Portugal

Chinnasamy Sashtika
Department of Biochemistry Biotechnology and Bioinformatics
Avinashilingam Institute for Home Science & Higher Education for Women
Coimbatore, India

Lekshmy Sathee
Division of Plant Physiology
ICAR-Indian Agricultural Research Institute
New Delhi, India

Natchiappan Senthilkumar
Division of Bioprospecting
Institute of Forest Genetics and Tree Breeding (IFGTB)
Coimbatore, India

Vinay Kumar Sharma
Department of Agricultural Biotechnology and Molecular Biology, College of Basic Sciences & Humanities
Dr. Rajendra Prasad Central Agricultural University
Pusa (Samastipur), India

Muhammad Basit Shehzad
College of Plant Protection
Nanjing Agricultural University
Nanjing, China

Mahipal S Shekhawat
Plant Biotechnology Unit
KM Government Institute for Postgraduate Studies and Research
Puducherry, India

Monika Singh
Department of Applied Sciences & Humanities
GL Bajaj Institute of Technology and Management
Greater Noida, India

Dhanawantari L. Singha
Department of Botany
Rabindranath Tagore University
Hojai, Assam

Annamalai Sivaranjini
Department of Biotechnology
Dwaraka Doss Goverdhan Doss Vaishnav College
Chennai, India

Rajalingam Amutha Sudhan
Department of Agricultural Biotechnology and Molecular Biology, College of Basic Sciences & Humanities
Dr. Rajendra Prasad Central Agricultural University
Pusa (Samastipur), India

Mumin Ibrahim Tek
Molecular Virology Laboratory, Plant Protection Department, Faculty of Agriculture
Akdeniz University
Antalya, Turkey

Mireia Uranga
Department of Plant Biotechnology and Bioinformatics
Ghent University
Ghent, Belgium
and
VIB Center for Plant Systems Biology
Ghent, Belgium
and
KU Leuven Plant Institute (LPI)
Heverlee, Belgium

Asif Ali Vadakkethil
Centre for Agricultural Bioinformatics
ICAR-Indian Agricultural Statistics Research Institute
New Delhi, India

Carla M. R. Varanda
ESAS, Instituto Politécnico de Santarém
Santarém, Portugal
and
MED - Mediterranean Institute for Agriculture, Environment and Development & CHANGE - Global Change and Sustainability Institute, Institute for Advanced Studies and Research
Universidade de Évora, Pólo da Mitra
Évora, Portugal

Nandakumar Vidya
Department of Botany
Bharathiar University
Coimbatore, India

Sathasivam Vinoth
Department of Biotechnology
Sona College of Arts and Science
Salem, India

Jun-Li Wang
Department of MOE Key Laboratory of Cell Activities and Stress Adaptations, School of Life Sciences
Lanzhou University
Lanzhou, China
and
Gansu Province Key Laboratory of Gene Editing for Breeding, School of Life Sciences
Lanzhou University
Lanzhou, China

Archana Watts
Division of Plant Physiology
ICAR-Indian Agricultural Research Institute
New Delhi, India

Lei Wu
Department of MOE Key Laboratory of Cell Activities and Stress Adaptations, School of Life Sciences
Lanzhou University
Lanzhou, China
and
Gansu Province Key Laboratory of Gene Editing for Breeding, School of Life Sciences
Lanzhou University
Lanzhou, China

Guangnan Xing
Soybean Research Institute & MARA National Center for Soybean Improvement & MARA Key Laboratory of Biology and Genetic Improvement of Soybean & National Key Laboratory for Crop Genetics and Germplasm Enhancement & Jiangsu Collaborative Innovation Center for Modern Crop Production
Nanjing Agricultural University
Nanjing, China

Fengqun Yu
Agriculture and Agri-Food Canada, Saskatoon Research and Development Centre
Saskatoon, SK, Canada

1 Advances in CRISPR/Cas-based Genome Editing

Break New Ground for Plant Functional Genomics

Hao Hu and Fengqun Yu

1.1 INTRODUCTION

Clustered Regularly Interspaced Short Palindromic Repeats-CRISPR associated protein (CRISPR/Cas) technology has revolutionized the field of genetic engineering and has become a prominent tool in genome editing. It is derived from the bacterial immune system and has been adapted for precise and efficient genome editing in a wide range of organisms, including plants. The discovery of CRISPR/Cas can be traced back to the early 1990s when unusual repetitive DNA sequences were identified in the genomes of bacteria and archaea.[1–4] However, it wasn't until 2007 that researchers recognized the potential role of these repetitive sequences as part of a prokaryotic immune system.[5] Further studies led to the understanding that CRISPR sequences serve as an adaptive immune defense mechanism against invading genetic elements such as viruses and plasmids.[6]

The CRISPR/Cas system consists of two main components: the guide RNA (gRNA) and the Cas protein. The gRNA is a synthetic RNA molecule that guides the Cas protein to the target DNA sequence of interest based on RNA-DNA recognition and binding. The Cas protein is an endonuclease that cleaves DNA at specific sites determined by the gRNA sequence. This cleavage, i.e., double-strand break (DSB), initiates DNA repair either by the error-prone non-homologous end joining (NHEJ) mechanism or by the donor-dependent homology-directed repair (HDR) mechanism, leading to various genetic modifications like sequence insertion/deletion (indel) and gene replacement.[6] The power and versatility of CRISPR/Cas technology lie in its flexibility to be easily reprogrammed by designing a specific gRNA sequence complementary to the target DNA region, which allows researchers to precisely edit genes and introduce designed genetic modifications. Compared to earlier genome editing techniques, such as Zinc-Finger Nucleases (ZFNs) and Transcription Activator-Like Effector Nucleases (TALENs), CRISPR/Cas is simpler, faster, and more cost-effective, making it widely adopted across various fields of research.[7]

Since its initial discovery, CRISPR/Cas technology has rapidly evolved and expanded, with different variants of Cas proteins, gRNA formats, and new effector proteins being discovered and engineered. Each Cas variant has unique characteristics, such as Protospacer Adjacent Motif (PAM) requirements, cleavage patterns, and target specificity. Together with other optimization strategies, especially by linking novel effectors (such as deaminase and reverse transcriptase) to Cas protein, current CRISPR/Cas genome editing in plants has been greatly improved in editing efficiency, specificity, targeting scopes, and the introduction of numerous types of precise editing. These advancements have deeply impacted fundamental biological research such as plant functional genomics and enabled the development of more sophisticated applications, such as multiplexed genome editing (simultaneously targeting multiple genes), base editing (precise nucleotide

DOI: 10.1201/9781003387060-1

substitutions), prime editing (precise search-and-replace genome editing to install indels and/or base substitutions), genome-wide screening, and epigenomic modifications. Since its first application in plants in 2013,[8–10] CRISPR/Cas has been extensively utilized in agriculture to improve yield, disease resistance, and other agronomically important traits.[11–13] Overall, the discovery and development of CRISPR/Cas technology have transformed the field of genetic engineering, providing plant scientists with an efficient and precise tool for understanding gene function, unraveling genetic networks, and driving innovations in various plant research areas including crop improvement.

1.2 PART I. ADVANCES IN CRISPR/CAS TECHNOLOGIES

1.2.1 Cas Variants

Cas proteins, the key components of the CRISPR/Cas system, exhibit remarkable diversity across different bacterial and archaeal species. CRISPR/Cas systems can be categorized into two classes based on their mechanism of action. In Class 1 systems, a multi-protein complex (multi-effector) is involved in the degradation of foreign nucleic acids. On the other hand, Class 2 systems accomplish the same function using a single, large Cas protein (single effector).[14–16] Due to its simplicity in application, Cas proteins from Class 2 systems are extensively studied and utilized. Cas9 is the most widely used Cas protein variant in Class 2. It was initially discovered in *Streptococcus pyogenes* and has become synonymous with CRISPR/Cas technology. Cas9 is an RNA-guided endonuclease that generates DSBs at target DNA sites. The two main domains, known as the RuvC and HNH nuclease domains, are responsible for cleaving the DNA strands.[17] These domains are guided to the target DNA sequence by a single guide RNA (sgRNA), which is a fusion of a CRISPR RNA (crRNA, carrying a sequence complementary to the target DNA) and a trans-activating crRNA (tracrRNA, helping in the processing and stability of the crRNA) and provides the necessary specificity by base-pairing with the complementary DNA sequence (namely protospacer). In addition, a short DNA sequence located immediately adjacent to the protospacer, i.e., PAM, is also essential for Cas9 binding and cleavage. The specific PAM sequence recognized by Cas9 varies depending on the Cas9 ortholog and species.[18–20] For the commonly used *S. pyogenes* Cas9 (SpCas9), the PAM sequence is NGG (where N can be any nucleotide). Over the years, researchers have identified, characterized, and even modified numerous Cas protein variants of Class 2[14,16] (Table 1.1). These Cas variants exhibit distinct features and functionalities, which greatly serve versatile applications and the ever-increasing need for editing specificity, efficiency, targeting range, etc.

1. **nCas9 (nickase Cas9):** nCas9 is a modified version of the Cas9 nuclease protein that retains its ability to bind to target DNA but has lost one of its two nuclease domain activities. Unlike the wild-type Cas9, which creates a DSB in the target DNA, nCas9 induces a single-strand break (SSB) or nick in one of the DNA strands within the protospacer region. The nicking activity of nCas9 is achieved through specific point mutations in the nuclease domains of the Cas9 protein. The most commonly used mutation is the D10A mutation, which disrupts the endonuclease activity of RuvC domain of Cas9. By using a pair of gRNAs, one targeting each DNA strand, nCas9 can be employed to create a pair of nicks on opposite DNA strands.[21] This generates a DSB at the desired genomic locus. The paired nicking strategy reduces off-target effects, as the nicks can be repaired by error-free mechanisms such as homologous recombination rather than relying on error-prone NHEJ repair pathways. The use of nCas9 offers several advantages over wild-type Cas9, including improved specificity and reduced potential for off-target effects. It allows for precise genome editing with minimal disruption to the DNA sequence, making it particularly useful in applications that require high accuracy and minimal impact on the target locus.[21]
2. **dCas9 (deactivated Cas9 or dead Cas9):** dCas9 is a modified version of the Cas9 protein that lacks both of its nuclease activities. Through point mutations, such as D10A and H840A,

TABLE 1.1
Popular Cas Variants in Class 2

Cas Type	Cas Variant	Target Type	PAM (5′-3′)	Characteristics	Applications	References
II	Cas9 (i.e., SpCas9 from *Streptococcus pyogenes*)	DNA	NGG	The activity of both RuvC and HNH nuclease domains create a DSB at the targeted location on dsDNA	Numerous applications involving dsDNA/genome editing such as gene knockout, replacement, etc.	14–17,23
	Cas9-VQR		NGA			19,27
	Cas9-EQR		NGAG			
	Cas9-VRER		NGCG			
	Cas9-NG		NGN			28,29
	Cas9-xCas9		NG/GAA/GAT			
	Cas9-SpG		NGN			30,31
	Cas9-SpRY		PAM-less			
	St1Cas9 (from *S. thermophilus*)		NNAGAA			18
	NmCas9 (from *Neisseria meningitidis*)		NNNNGATT			
	TdCas9 (from *Treponema denticola*)		NAAAAN			
	FnCas9 (from *Francisella novicida*)		NGG			32
	FnCas9-RHA		YG			
	nCas9		NGG	Lack one nuclease activity (usually RuvC) through mutation, creating a nick (single-strand break) at a specific location on dsDNA	Link with various effectors for precise editing (base editing, prime editing, etc.) or gene regulations	21,33,34
	dCas9		NGG	Lack both RuvC and HNH nuclease activities through mutations, no cleavage but can still bind to a specific location on dsDNA.	Link with various effectors for precise editing (prime editing, etc) or gene regulations	22,23,35,36
V	Cas12a	DNA	TTN, or TTTN	Staggered cutting	Auto-processing pre-crRNA activity for multiplex gene regulation	14–16,24,37,38
VI	Cas13	RNA	A, U, C (no G in PFS)	Non-specifically cleaves non-target RNA	mRNA knockdown and RNA editing	14–16,25,39
V	Cas14	DNA	None	ssDNA specific	ssDNA cutting and single nucleotide variant detection	14–16,26

which alter critical residues necessary for endonuclease activity, dCas9 is rendered inactive. It can still bind to target DNA sequences, guided by the associated sgRNA, but does not induce DNA cleavage. This binding alone is often sufficient to attenuate or block the transcription of the targeted gene, especially when the sgRNA positions dCas9 in a manner that obstructs the access of transcription factors and RNA polymerase to the DNA. Instead of introducing permanent genetic changes, dCas9 can be utilized for gene regulation purposes.[22] The modifiable regions of dCas9, typically located at the N- and C-terminus of the protein, can be utilized to attach transcriptional activators, repressors, or other effector domains. When dCas9 is fused with transcriptional activators, it is referred to as CRISPR activation (CRISPRa), enabling targeted upregulation of gene expression. Conversely, when dCas9 is fused with transcriptional repressors, it is known as CRISPR interference (CRISPRi), allowing targeted downregulation of gene expression. This approach offers a powerful tool for studying gene function, gene networks, and regulatory elements in plants.[22,23]

3. **Cas12a (previously known as Cpf1):** Cas12a is an alternative RNA-guided endonuclease that was first identified in the bacterium *Francisella novicida*. Cas12a exhibits a key distinction from Cas9 in its ability to create staggered breaks (i.e., sticky ends).[24] When Cas12a cleaves DNA, it produces overhangs of five base pairs, which proves advantageous for specific applications that necessitate the generation of single-stranded DNA, like HDR experiments. Moreover, Cas12a is well-suited for targeting genomic regions with a high adenine-thymine (A-T) content since its PAM does not rely on the presence of guanine-cytosine (G-C) pairs.[24]
4. **Cas13 (previously known as C2c2):** Cas13 is an RNA-guided nuclease that distinguishes itself from Cas9 by its ability to target RNA instead of DNA, and it has been found in different bacterial species. This unique characteristic makes Cas13 particularly valuable for generating transient modifications in signaling molecules, specifically RNA, as opposed to permanent alterations in the genome. Similar to Cas14, Cas13 exhibits non-specific cleavage of single-stranded RNA when bound to its RNA target sequence.[25] This RNA-targeting capability renders Cas13 highly effective in targeting mRNA and serves as a useful mechanism for temporarily suppressing gene expression. As a result, this system can be employed as an alternative to small interfering RNA (siRNA) or short hairpin RNA (shRNA) for efficient and multiplexable knockdown of RNA.
5. **Cas14 (or Cas12f):** In contrast to other Cas enzymes, Cas14 exhibits a preference for single-stranded DNA (ssDNA) rather than double-stranded DNA (dsDNA). It displays a remarkable ability to cleave ssDNA with exceptional accuracy, as even a single mismatch in the target sequence can be detected. Similar to Cas12, Cas14 is considerably smaller in size compared to the conventional Cas9 protein. Its small size offers advantages for packaging into viral vectors, potentially facilitating its delivery into cells for genome editing applications. In addition, akin to Cas12a, Cas14 demonstrates non-specific cleavage of non-complementary ssDNA upon engagement with its target sequence.[26]

1.2.2 Precise Editing Technologies

The efficacy of gene targeting technology largely depends on HDR for introducing desired sequence modifications, but the limited efficiency of HDR has hindered its application.[40,41] To address this constraint, alternative genome editing technologies have emerged, including base editing and prime editing. These CRISPR/Cas-based technologies offer precise sequence editing without requiring DSBs or donor DNA, making them more efficient than HDR in plants. Initially developed for human cells, cytosine base editors (CBEs) and adenine base editors (ABEs) laid the foundation for the advancement of dual base editors and precise DNA deletions specifically tailored for plants. In this section, we provide a brief overview of the recently developed CRISPR/Cas-based technologies used to achieve precise editing of plant genomes (Table 1.2).

TABLE 1.2
Summary of CRISPR/Cas-based Precise Editing Technologies in Plants

Editing Technology	Main Effectors	Cas Variant	Sequence Modification	Editing Window	Editing Efficiency (up to %)	References
APOBEC1-CBE2/CBE3/CBE4	Rat APOBEC1	nCas9 (D10A), dCas9 (D10A and H840A), or nCas9-NG	C:G>T:A	C4-C8, or C3-C9	75.00	34,43,45,47,82,83
hAID-CBE3	Human AID	nCas9, nSpCas9-NG, nScCas9, or nCas9-SpRY		C3-C8	97.92	31,33,84,85
APOBEC3A-CBE3/CBE4	Human APOBEC3A	nCas9		C1-C17	82.90	49
PmCDA1-CBE2/CBE3/CBE4	Petromyzon marinus CDA1	nCas9, dCas9, xCas9, nSpCas9-NG, nCas9-NG, or nScCas9++		C2-C5, or C1-C17	86.10	46,47,86–88
ABE7.10	TadA-TadA7.10	nCas9, nSpCas9-NG, nSaCas9, or nScCas9	A:T>G:C	A4-A8	94.12	50,53,54,85,89
ABE-P1S	TadA7.10	nCas9, or nSaCas9		A1-A12	96.30	55
ABE8e	TadA8e	nCas9, nCas9-NG, nCas9-SpG, or nCas9-SpRY		A4-A8, A3-A10, or A1-A14	100.00	31,90,91
ABE9	TadA9	nCas9, nSpCas9-NG, nCas9-SpRY, or nScCas9		A1-A12, A4-A10, A3-A10, or A4-A12	100.00	58
CGBE	Anc689(R33A)	nCas9	C:G>G:C	C4-C9	52.50	61
STEME	APOBEC3A-ecTadA-ecTad7.10	nCas9, or nCas9-NG	C:G>T:A and A:T>G:C	C1-C17, and A4-A8	15.10, 73.21	62
pDuBE1	eCDAL-TadA8e	nCas9	C:G>T:A and A:T>G:C	C2-C5, and A4-A8	87.60	92
AFID	APOBEC3A, APOBEC3Bctd	Cas9	Precise, predictable multinucleotide deletion	N/A	34.80	63
PE2	M-MLV-RT (H9Y/D200N/T306K/W313F/T330P/L603W)	nCas9 (H840A)	all kinds of base substitutions, precise insertions, deletions, and combinations of these sequence modifications	1-80	59.90	66,67,69,93–96
PE3					70.30	66,70,75,79,94,95,97–100
PE4					2.10	66
PE5					18.30	66
PEmax		nCas9 (R221K/N394K/H840A)			77.08	66,67

1. **CBE:** The initial version of CBE, known as CBE1, was created by linking a rat cytidine deaminase, rAPOBEC1, to a dCas9.[42] CBE1 induces the substitution of cytosine (C) to thymine (T) by deaminating C to uracil (U) on the non-target DNA strand, which is then converted to T during DNA repair and replication. However, the presence of uracil N-glycosylase (UNG) in the cellular base excision repair (BER) pathway leads to low editing efficiency of CBE1 due to the elimination of uracil.[42] To enhance efficiency, the second-generation CBE, CBE2, was developed by adding a uracil DNA glycosylase inhibitor (UGI) to the C-terminal of CBE1, preventing UNG activity.[42] CBE2 improves editing efficiency threefold and reduces unexpected indels.[43] To further enhance efficiency, the third-generation CBE, CBE3, was created by adding rAPOBEC1 and UGI to nCas9 (D10A).[42,44,45] Although CBE3 does not cleave dsDNA, it creates an incision in the target strand to initiate the repair process. The fourth-generation CBE, CBE4, improves deamination activity by linking two UGIs to nCas9, resulting in increased base editing efficiency and reduced incidents of undesired C to A or G transversions compared to CBE3.[46–48] In addition, the Mu Gam protein from bacteriophage was linked to CBE4, resulting in CBE4-Gam, which further enhances product purity and reduces indel occurrences.[48,49]
2. **ABE:** ABE7.10 was the first ABE created by linking nCas9 (D10A) to a dimer comprising a wild-type adenine deaminase called TadA and an evolved adenine deaminase called TadA7.10.[50] ABE7.10 allows the precise conversion of adenine (A) to inosine (I), which is then recognized as guanine (G) during DNA repair and replication processes.[51] The editing window of ABE7.10 is located at positions 4–8 nt in the protospacer region. To enhance editing efficiency, ABE7.10 underwent codon optimization to improve its performance. In addition, a nuclear localization sequence (NLS) was added to facilitate the transport of ABE7.10 into the cell nucleus.[52] ABEmax was developed by adding NLS sequences at both ends of ABE7.10. This modification further increased the editing efficiencies to less than 50% at most targets.[50,53,54] ABEmax demonstrated successful A-to-G conversions in various target genes in rice plants. A simplified version called ABE-P1S was developed (i.e., TadA7.10-nCas9 vs. commonly used TadA-TadA7.10-nCas9), and it showed significantly higher editing efficiency in rice, indicating its potential for precise editing in plants.[55] Later, by using another adenine deaminase variant TadA8e (V106W), the new ABE8e exhibited significantly improved A-to-G conversion and reduced off-target effects.[56,57] Further, TadA9, an improved adenosine deaminase by introducing V82S and Q154R mutations into TadA8e, was found to be compatible with various Cas variants. Thus, ABE9 with TadA9 should have quite relaxed PAM restrictions and demonstrate stronger editing capabilities with an expanded editing window.[58]
3. **CGBE (C-to-G base editor):** CBEs and ABEs are limited to inducing base transitions rather than base transversions. To overcome this limitation, a novel base editor known as CGBE has been developed. CGBE consists of a variant of the rAPOBEC1 cytidine deaminase (R33A), a nCas9 (D10A), and a UNG. Recent studies have demonstrated the successful use of CGBE in achieving efficient C-to-G transversion in bacteria and mammalian cells.[59,60] Further, a rice-specific base editor called OsCGBE03 was created through codon optimization on UNG, which showed efficient C-to-G editing in five rice genes.[61] The development of CGBE expands the repertoire of base editing tools, offering a powerful approach for generating diverse base substitution types in precise crop breeding and the creation of novel germplasm resources.
4. **Dual base editor:** A novel gene editing tool called the "saturated targeted endogenous mutagenesis editor" (STEME) has been developed for the simultaneous editing of cytosine and adenine bases in plants using a sgRNA.[62] This system consists of a cytidine deaminase (APOBEC3A), an adenosine deaminase (ecTadA–ecTadA7.10), a nCas9 (D10A), and a UGI. STEME enables the conversion of cytidines to uridines and adenosines to inosines within the editing window, which are then replicated and repaired by the plant's DNA

repair mechanisms, resulting in dual substitutions of C:G>T:A and A:T>G:C. To expand the editing capacity and target a wider range of sequences, an SpCas9–NG PAM variant is used in the STEME system, which recognizes NG PAM sequences.[28] This allows for the editing of a larger number of target sites. The ability to perform dual base editing with STEME opens up possibilities for the directed evolution of endogenous plant genes in their native context. In addition, STEME can be utilized for modifying cis-regulatory elements in gene regulatory regions and for high-throughput genome-wide screening in plants.

5. **Multinucleotide deletion:** APOBEC–Cas9 fusion-induced deletion systems (AFIDs) are a type of genome editing tool that combines the activity of APOBEC and Cas9 proteins to induce targeted deletions in DNA.[63] APOBEC proteins are naturally occurring enzymes that can induce DNA mutations, particularly cytosine-to-uracil changes. AFIDs utilize two cytidine deaminases, namely hAPOBEC3A and the C-terminal catalytic domain of hAPOBEC3B (hAPOBEC3Bctd), which could generate DNA deletions spanning from the targeted cytidine or a preferred TC motif to the DSB induced by Cas9, respectively.[63] These predictable deletions ensure more consistent editing outcomes. AFIDs could be applied to create predictable multinucleotide deletions and study DNA regulatory regions and protein domains.
6. **PE (prime editor):** PEs are a class of CRISPR/Cas-based genome editing tools that offer precise and versatile editing capabilities for targeted genetic modifications. They are designed to generate all 12 types of base substitutions, precise insertions, deletions, and combinations of these sequence modifications without the requirement for DSBs or the reliance on donor DNA templates. The prime editing system consists of a fusion protein composed of a nCas9 (H840A) (or dCas9), a Moloney murine leukemia virus reverse transcriptase (M-MLV-RT), and an engineered prime editing guide RNA (pegRNA). The pegRNA consists of a reverse transcriptase template (RTT, harboring designed edits) and a primer-binding site (PBS, initiating reverse transcription) at the 3′ end of a sgRNA (targeting the specific location). PBS pairs with the single-strand DNA nicked by nCas9, which allows reverse transcription and incorporation of the designed edits from the RTT into the genome. Subsequently, a series of processes, including equilibration, ligation, and repair, lead to the desired edit. Since its first introduction in 2019,[64] several generations of PEs (PE1, PE2, PE3, PE3b, and subsequent iterations) have been developed to expand their functionality and improve editing efficiency, and the engineering of core editor protein and the redesigning of pegRNA have played a crucial role in the gradual improvement. Initially, PE1 was created by adding a wild-type virus reverse transcriptase to nCas9 (H840A) to enable prime editing. To increase the editing efficiency, PE2 was developed by changing the wild-type reverse transcriptase in PE1 with a modified version containing six specific mutations. Subsequently, by incorporating an additional nicking sgRNA to induce another cut, PE3 was introduced to further increase editing efficiency. However, the PE3 system, inducing two SSBs on complementary DNA strands, could lead to higher indel frequencies through the NHEJ repair pathway. To mitigate undesired indels, PE3b employs a specific sgRNA to induce the second nick after the incorporation of the designed edit into the targeted genome locus is done. In addition, inhibiting critical components of the DNA mismatch repair (MMR) pathway, like MLH1, has been shown to effectively enhance prime editing efficiency.[65] PE4 and PE5 were generated by adding a dominant negative MMR effector (i.e., MLH1dn) to PE2 or PE3, respectively, which can help avoid MMR and thus enhance prime editing capacity and significantly increase editing efficiency by several folds. Furthermore, the ability to create nicks in the non-target strand is essential for efficient prime editing. By switching the nCas9 (H840A) of PE2 with the SpCas9max variant containing two additional mutations (i.e., R221K and N394K), PEmax has achieved significantly improved prime editing efficiency.[65–67] Prime editing has several advantages over conventional HDR strategies, including improved precise genome editing efficiency

and the ability to generate combinations of various sequence modifications at a relatively wide range of positions, which reduces constraints imposed by its PAM.[64,68–72] However, the editing efficiency of PEs in plants is still unsatisfying, even with the utilization of various strategies like alternative reverse transcriptase orthologs, ribozymes for precise pegRNA production, temperature optimization, enhanced sgRNA scaffold modifications, and selective markers for cell enrichment.[65,73–79] In addition, the capacity of PEs to induce larger genetic edits spanning hundreds of nucleotides and their level of specificity have yet to be verified in plants.[68,72,80,81]

1.2.3 Delivery Mechanisms

To effectively utilize CRISPR/Cas9 technology in plants, it is crucial to have a reliable and universally applicable method for delivering the necessary CRISPR/Cas components into plant cells. However, the two commonly used delivery strategies, i.e., *Agrobacterium*-mediated transformation and biolistics, are both inadequate to meet our needs. First of all, both strategies require tissue culture, which is time-consuming and laborious. The recipient plant species available for *Agrobacterium* infection is increasing, but the editing efficiency still varies significantly among different genotypes, especially in monocots.[101] Further, the intrinsic feature of random insertion of foreign DNA into the host genome stirs concerns over safety risks. Regarding biolistics, the mechanical force used for delivery may cause damage to the host genome, and the delivery efficiency is still unsatisfying. As a result, there is an urgent need for innovative delivery strategies to overcome these challenges and enhance the application of CRISPR/Cas9 in plant research.

De novo induction of meristem is a promising approach that utilizes morphogenetic regulators to facilitate CRISPR/Cas-mediated gene editing in plants. These regulators not only assist in transforming recalcitrant cultivars but can also trigger the formation of new meristems in plants, eliminating the need for tissue culture.[102] A recent study demonstrated the effectiveness of this approach by injecting morphogenetic regulators and sgRNA expression cassettes-loaded *Agrobacterium* into pruned sites of Cas9-overexpressing tobacco plants (*Nicotiana benthamiana*).[103] The resulting shoots directly gave rise to gene-edited plants with inheritable mutations. This methodology holds potential for various plant species and can greatly expedite plant research. Virus-assisted gene editing is another promising strategy that leverages the capabilities of plant viruses to generate gene-edited plants without relying on tissue culture. Certain plant viruses can efficiently deliver sgRNA for genome modifications. Several single-stranded RNA viruses, such as tobacco rattle virus and the single-stranded DNA cabbage leaf curl virus, have been utilized for this purpose, and the editing efficiencies could reach 80%.[104–109] Still, these viruses cannot deliver Cas9 and sgRNA expression cassettes together due to cargo limitations. To overcome this, researchers inserted Cas9 and sgRNA cassettes into the genomes of another two plant viruses with better capacity, which enabled systemic gene editing in tobacco plants.[110,111] To address the limitation of heritability, sgRNAs were fused with RNA mobile elements and incorporated into tobacco rattle virus RNA2. These mobile elements guided the sgRNAs into shoot apical meristem cells, resulting in high efficiencies of heritable mutations in the offsprings.[112]

1.3 PART II. APPLICATIONS OF CRISPR/CAS IN PLANT FUNCTIONAL GENOMICS

CRISPR/Cas technology has provided plant scientists with a precise, efficient, and versatile tool for studying gene function, unraveling genetic networks, and accelerating crop improvement efforts. It has significantly advanced our understanding of plant functional genomics and has the potential to revolutionize plant breeding and agriculture. Here are some examples of its vast applications in plant functional studies:

1. **Gene knockdown/activation/visualization:** Cas9 can effectively target and silence specific genes at the DNA level, a phenomenon observed in bacteria where the presence of Cas9 alone is sufficient to block transcription.[113,114] The use of dCas9 eliminates CRISPR's ability to cut DNA while retaining its capability to target specific sequences. When dCas9 is guided to the target gene by the sgRNA, it can effectively block gene transcription or interfere with RNA processing, leading to a decrease in the expression level of the target gene. This approach is known as CRISPRi or gene knockdown. CRISPRi operates similarly to RNA interference (RNAi) by targeting specific sites without cutting them, resulting in reversible gene silencing.[115] Further, various regulatory factors have been added to dCas9s by different research groups, enabling them to effectively control gene expression by turning genes on or off or adjusting their activity levels.[11,12,116,117] By fusing a repressor domain to dCas9, it becomes possible to enhance transcriptional repression by inducing heterochromatinization. One notable example is the fusion of the Krüppel-associated box (KRAB) domain with dCas9, which can effectively repress the transcription of the target gene by up to 99% in human cells.[114] On the other hand, CRISPRa enhances gene transcription. In this approach, dCas9 is fused with transcriptional activators or activation domains. When guided to the target gene's promoter by the sgRNA, dCas9 recruits transcriptional machinery to promote gene expression and enhance transcriptional activity.[116] CRISPRa enables researchers to study the effects of gene overexpression and gain-of-function phenotypes. Furthermore, CRISPR/Cas technology can be utilized for gene visualization or labeling within living cells. By fusing fluorescent proteins or other visual markers to dCas9, researchers can target specific genomic loci and visualize the dynamic behavior and spatial organization of genes in real time. This technique, known as CRISPR imaging or CRISPR Live-Cell Imaging, provides valuable insights into gene regulation, chromatin organization, and nuclear architecture.[118] Overall, these applications of CRISPR/Cas technology in gene regulations have greatly expanded our ability to investigate gene functions, regulatory networks, and cellular processes.
2. **RNA editing:** RNA editing using CRISPR/Cas technology has emerged as a powerful approach to precisely modify RNA sequences, opening up new possibilities for studying RNA functions at the RNA level without permanently changing the underlying DNA sequence. Several CRISPR/Cas systems targeting RNA, such as Cas13a, have recently been developed in plants.[12] These Cas proteins are engineered to be catalytically active and capable of performing targeted nucleotide conversions on the RNA molecule. This is achieved by incorporating an RNA-targeting domain into the Cas protein, enabling it to bind to the target RNA sequence with high specificity. Once bound, the Cas protein can introduce single-base changes, insertions, or deletions at specific positions within the RNA molecule. These systems can downregulate specific transcripts with higher specificity compared to RNAi.[119] Apart from direct RNA targeting, CRISPR/Cas systems can also modulate pre-mRNA splicing. By editing critical splicing motifs that adhere to the canonical GU-AG rule, splicing can be disrupted, resulting in changes to gene function.[89,120] Further, CRISPR/Cas systems can be utilized to modify micro-RNA and long-noncoding RNA in plants.[121] RNA editing allows for the investigation of RNA functions and regulatory mechanisms. By introducing specific modifications into RNA molecules, researchers can probe the consequences of these alterations on RNA stability, structure, splicing, translation, and protein interactions. This provides valuable insights into the roles of RNA in cellular processes and advances our understanding of RNA biology. However, it is important to note that RNA editing using CRISPR/Cas technology is still an evolving field, and there are challenges to overcome. Delivery of the editing components to specific tissues or cells remains a hurdle, as does the efficiency and specificity of RNA editing. Continued research and development in this area will further refine and expand the capabilities of RNA editing technologies.

3. **Epigenomic editing:** Epigenomic editing using CRISPR/Cas technology is a rapidly advancing field that allows precise modifications to be made to the epigenetic marks on DNA or chromatin, enabling the manipulation of gene expression and regulation. Epigenetic modifications, such as DNA methylation and histone modifications, play crucial roles in determining gene activity and cellular identity. Similar to the concept of utilizing the dCas9 fusion protein for gene knockdown/activation/visualization, the CRISPR/Cas system is adapted for epigenomic editing by fusing dCas9 with effector domains capable of modifying the epigenetic landscape (such as DNA methyltransferases, acetyltransferases, etc.).[35] By directing the fused protein to specific genomic loci using sgRNAs, targeted changes can be made to the epigenetic marks associated with those loci. For example, by fusing the catalytic domain of a DNA methyltransferase DNMT3a and a human acetyltransferase p300 with dCas9, dCas9-DNMT3a and dCas9-p300 have successfully accomplished DNA methylation and acetylation of the targeted region as specified by sgRNA, respectively.[122,123] However, challenges remain in optimizing the efficiency and specificity of epigenomic editing,[35] and the potential off-target effects and unintended alterations to the epigenome need to be carefully evaluated and minimized. To address this concern, a CRISPR/Cas-based epigenomic editing system, namely FIRE-Cas9, has been developed for rapid modification of targeted epigenomic marks, and it allows to reverse the changes made if something goes wrong.[124] Epigenomic editing using CRISPR/Cas technology provides a powerful means to investigate the functional impact of epigenetic modifications and offers potential therapeutic avenues for diseases linked to aberrant epigenetic regulation. Continued advances in this field will further refine and expand the scope of epigenomic editing, unlocking new insights into the complex interplay between the epigenome and gene expression.
4. **Multiplexed genome editing:** Multiplexed genome editing refers to the simultaneous targeting and modification of multiple genomic loci using CRISPR/Cas technology. It allows for the efficient and precise editing of multiple genes or DNA sequences within a single experiment. This approach is particularly useful for studying gene function, pathway engineering, and crop improvement. One of the main advantages of multiplexed genome editing is its efficiency and cost-effectiveness compared to traditional breeding or single-gene editing approaches, which could enable the generation of plant lines with multiple desired traits in a single generation, greatly accelerating the breeding process. Most studies of multiplexed genome editing in plants involve one kind of Cas protein and multiple sgRNAs. Various strategies have been developed to express and deliver multiple sgRNAs efficiently. These include using RNA polymerase III (Pol III)-driven systems, where multiple Pol III promoters (such as U3 and U6) are used to express multiple sgRNAs within a single construct.[125,126] Pol II-driven systems are also used, employing different strategies such as ribozyme sequences, polycistronic tRNA-sgRNA transcripts, or adding linkers to flank the sgRNAs.[127–129] These systems allow for the simultaneous expression of multiple sgRNAs, each targeting a specific genomic site. However, a single type of Cas protein cannot meet the needs of multiplexed orthogonal editing, which requires manipulation of the genome in a synthetic manner. By utilizing different Cas proteins or engineering the sgRNA sequences, each target site is modified independently without interfering with the editing activity at other sites. For example, one strategy utilizes a dCas9 combined with different single-chain RNAs (scRNAs) that carry RNA aptamers to recruit various transcription activators or repressors in mammalian cells.[130] Another strategy utilizes one sgRNA with a complete protospacer for gene knockout and another sgRNA with a partial protospacer for regulating another gene with Cas9, Cas12a-repressor, or Cas12a-activator systems.[131,132] These strategies have also been successfully implemented in plants. Furthermore, the development of a dual-function system, namely SWISS (simultaneous and wide editing induced by a single system), allows for concurrent base modifications and gene knockouts

in rice.[133] Another similar system combines a complete and a partial protospacer to modulate the activity of a modified CBE with improved specificity, enabling indels and C:G>T:A base transitions.[134] These multiplexed and/or orthogonal editing systems represent significant advances in genome manipulation, providing synthetic tools to precisely engineer the genome and explore diverse applications in plant research and biotechnology.

5. **Conditional genome editing:** Conditional CRISPR/Cas systems in plants refer to the use of regulatory elements and inducible promoters to precisely control the timing and location of gene editing events. These systems provide greater flexibility and specificity in targeting specific genes or tissues, allowing for more precise functional analysis and manipulation of plant genomes. One approach in conditional CRISPR/Cas systems is the use of tissue-specific promoters. By incorporating tissue-specific promoters into the CRISPR/Cas system, the Cas9 enzyme can only be expressed in certain types of cells or organs. This enables gene editing to be limited to particular tissues, preventing off-target effects and minimizing potential adverse effects in other parts of the plant. For example, a CRISPR/Cas system has been developed with tissue-specific promoters *CLV3* or *AP1* controlling the expression of Cas9, and inheritable editing results have been obtained as designed in *Arabidopsis thaliana* with superior editing efficiency.[135] Tissue-specific CRISPR/Cas systems have been employed to study gene function in various plant structures, such as stomata and lateral roots.[136] Inducible expression systems are another important component of conditional CRISPR/Cas systems. These systems allow researchers to control the timing and level of Cas9 expression by using exogenous inducers, such as chemical compounds or light. For example, researchers have developed CRISPR/Cas systems that are responsive to blue light repression or red light induction, as well as heat induction.[135,137] Furthermore, the combination of these elements allows for precise control over gene editing, limiting it to certain cells or organs and regulating it based on the timing of inducer application. This approach ensures that gene modifications occur only in the desired tissues and at the desired developmental stages. Overall, by utilizing tissue-specific promoters and/or inducible expression systems, conditional CRISPR/Cas systems provide researchers with the means to manipulate plant genomes with high specificity, minimal off-target effects, and temporal control. They enable a deeper understanding of gene functions, elucidation of gene regulatory networks, and potential applications in crop improvement and biotechnology.
6. **Functional genomics screening:** CRISPR screening is a powerful tool in functional genomics that allows for the systematic identification of genes involved in specific biological processes or phenotypes. It leverages the precision and efficiency of the CRISPR/Cas system to perturb the function of individual genes in a high-throughput manner, enabling the discovery of gene functions and genetic interactions on a genome-wide scale. The workflow of CRISPR screening typically involves the generation of a sgRNAs library targeting individual genes throughout the genome. These sgRNAs are then delivered to a population. Each sgRNA guides the Cas protein (e.g., Cas9) to a specific genomic locus, inducing targeted DNA cleavage or modulation of gene expression, depending on the experimental design. The pooled CRISPR library is subsequently subjected to selection or screening assays to identify individuals with specific phenotypes of interest. By analyzing the representation of each sgRNA in the selected population using high-throughput sequencing, the enrichment or depletion of specific sgRNAs provides insight into the functional relevance of the targeted genes. For example, over 14000 independent lines of rice with confirmed edits were developed using a library of 25604 pooled sgRNAs.[138] In maize, 1244 candidate loci were screened using high-throughput CRISPR/Cas editing, leading to the accurate mapping of genes related to agronomically important traits.[139] Similar approaches have also been applied to tomatoes and soybeans.[140,141] CRISPR screening enables the systematic interrogation of gene function and genetic interactions. It can identify essential

genes required for cell survival, genes involved in specific signaling pathways, or genes associated with disease phenotypes. Furthermore, CRISPR screens can uncover genetic interactions by assessing the effects of gene perturbations in combination, providing insights into complex biological networks and pathways. Despite its tremendous potential, CRISPR screening has certain challenges. Off-target effects of CRISPR/Cas systems and incomplete sgRNA representation in the library can introduce noise and false positives. Careful experimental design and rigorous bioinformatic analysis are essential to mitigate these issues and ensure reliable results.

7. **Directed evolution:** CRISPR/Cas-directed evolution is a cutting-edge technique that combines the precision of CRISPR/Cas genome editing with the power of directed evolution. Directed evolution aims to modify genes and proteins to acquire enhanced or novel properties through iterative rounds of mutagenesis and selection. In the context of CRISPR/Cas, this approach allows for the targeted and controlled evolution of specific genes or gene products. The traditional directed evolution methods, like error-prone PCR, generate random mutations throughout the entire gene or gene pool. In contrast, CRISPR/Cas-directed evolution offers a more targeted approach by utilizing the programmable nature of CRISPR/Cas systems to introduce precise mutations in specific regions of the genome.[12] The process of CRISPR/Cas-directed evolution typically involves the following steps: (1) Designing a library of sgRNAs to target the gene or genes of interest (GOIs), covering different regions or variations within the target sequences; (2) Introduction of mutations (indels, base substitutions, etc.) at the target sites within GOIs using CRISPR/Cas system guided by the sgRNA library; (3) Selection for desired traits using selection pressures (e.g., herbicides or antibiotics) or screening assays; (4) Iterative cycles of mutagenesis and selection to further enhance the desired traits. CRISPR/Cas-directed evolution has been successfully applied in various plant systems to modify genes and proteins with specific objectives. For example, it has been used to confer herbicide resistance in crops, enhance plant stress tolerance, improve enzyme efficiency, and modify metabolic pathways.[62,142,143] The ability to introduce precise mutations in a targeted manner greatly expands the possibilities of directed evolution in plants and enables the creation of customized genetic variants with desired properties. As the field progresses, ongoing research aims to develop more efficient and versatile selection methods, expand the range of target genes and traits that can be evolved, and improve the scalability and throughput of the process. CRISPR/Cas-directed evolution holds tremendous potential for applications in agriculture, biotechnology, and synthetic biology, as it allows for the rapid and controlled evolution of genes and proteins to meet specific needs and challenges.

1.4 CONCLUSION AND PROSPECTS

In this chapter, the topic of "CRISPR/Cas and plant functional genomics" was discussed by summarizing recent advancements in CRISPR/Cas technology and their applications in plant functional studies. CRISPR/Cas technology has revolutionized both basic and applied plant research by enabling precise genome manipulation. In addition to inducing indel mutations, numerous Cas variants have been developed to perform various precise modifications in the plant genome. Along with the development of Cas variants, various CRISPR/Cas-based plant biotechnologies have also been developed or improved, encompassing precise gene modulation techniques at different stages of gene expression, multiplexed and high-throughput methods for sequence modifications at multiple genomic sites, and novel delivery strategies for efficient gene editing in plants. The fast-evolving technology of CRISPR/Cas-based genome editing has enabled plant scientists, for the first time in history, to control the specific introduction of targeted sequence alterations in the plant, thus breaking new ground for studying gene functions in plants. Meanwhile, human society is facing unprecedented challenges in agriculture brought about by climate change and the booming population, which requires constant crop improvement.[144] CRISPR/Cas-based genome editing technologies have significant potential for accelerating the breeding process and ensuring the development of sustainable agriculture.[11–13] The new Cas variants, CRISPR/Cas-based biotechnologies, and the various applications in plant functional studies introduced in this chapter could also be used for crop improvement. For example, the strategy and workflow of CRISPR/Cas-directed evolution could be directly utilized in crop improvement if the aimed phenotype of directed evolution is set to be important agronomic traits such as biotic/abiotic stress tolerance. In fact, many CRISPR/Cas-based techniques developed for plant functional studies have been extensively utilized in the development of novel crops and plant breeding technologies. For instance, BEs and PEs have demonstrated the ability to introduce a substantial proportion of causative mutations (35% and 85%, respectively) in 225 important agronomic trait genes in rice.[71] Further, by combining CRISPR/Cas technology and conventional breeding methods, several innovative breeding methods have recently been developed. These novel breeding methods have utilized CRISPR/Cas to specifically target reproduction-related genes and successfully achieved valuable goals like inducing haploid lines, generating male sterile lines, fixing hybrid vigor, and manipulating self-incompatibility.[12]

Despite these exciting advancements, there are still unmet needs in CRISPR/Cas-based plant genome manipulation. Some agricultural traits are controlled by multiple genes, necessitating the development of efficient CRISPR/Cas-based technologies to stack desired alleles. To minimize fitness penalties caused by gene disruptions, further progress is required to improve the specificity of precise editing. Further, understanding the underlying mechanisms that influence the genome editing results will greatly enhance the precise editing of target genes in plants. Lastly, improvements in fundamental genetic research are also needed for discovering genes associated with agronomically important traits. This knowledge will aid in developing tailored genome editing strategies for targeted improvements in crops. Nevertheless, with ongoing efforts, CRISPR/Cas-based technologies are expected to become commonplace and adaptable tools for precise gene editing in fundamental plant research and crop enhancement in the coming years.

REFERENCES

1. Ishino Y, Shinagawa H, Makino K, Amemura M, Nakata A. Nucleotide sequence of the iap gene, responsible for alkaline phosphatase isozyme conversion in *Escherichia coli*, and identification of the gene product. *Journal of Bacteriology* 1987;169(12):5429–5433. DOI: 10.1128/jb.169.12.5429-5433
2. van Soolingen D, de Haas PE, Hermans PW, Groenen PM, van Embden JD. Comparison of various repetitive DNA elements as genetic markers for strain differentiation and epidemiology of *Mycobacterium tuberculosis*. *Journal of Clinical Microbiology* 1993;31(8):1987–1995. DOI: 10.1128/jcm.31.8.1987-1995
3. Groenen PM, Bunschoten AE, Soolingen Dv, Errtbden JD. Nature of DNA polymorphism in the direct repeat cluster of *Mycobacterium tuberculosis*; application for strain differentiation by a novel typing method. *Molecular Microbiology* 1993;10(5):1057–1065. DOI: 10.1111/j.1365-2958.1993.tb00976.x
4. Jansen R, Embden JD, Gaastra W, Schouls LM. Identification of genes that are associated with DNA repeats in prokaryotes. *Molecular Microbiology* 2002;43(6):1565–1575. DOI: 10.1046/j.1365-2958.2002.02839.x
5. Barrangou R, Fremaux C, Deveau H, et al. CRISPR provides acquired resistance against viruses in prokaryotes. *Science* 2007;315(5819):1709–1712. DOI: 10.1126/science.1138140
6. Jinek M, Chylinski K, Fonfara I, Hauer M, Doudna JA, Charpentier E. A programmable dual-RNA-guided DNA endonuclease in adaptive bacterial immunity. *Science* 2012;337(6096):816–821. DOI: 10.1126/science.1225829
7. Kumar K, Gambhir G, Dass A, et al. Genetically modified crops: Current status and future prospects. *Planta* 2020;251(4):91. DOI: 10.1007/s00425-020-03372-8
8. Shan Q, Wang Y, Li J, et al. Targeted genome modification of crop plants using a CRISPR/Cas system. *Nature Biotechnology* 2013;31(8):686–688. DOI: 10.1038/nbt.2650
9. Nekrasov V, Staskawicz B, Weigel D, Jones JD, Kamoun S. Targeted mutagenesis in the model plant *Nicotiana benthamiana* using Cas9 RNA-guided endonuclease. *Nature Biotechnology* 2013;31(8):691–693. DOI: 10.1038/nbt.2655

10. Li J-F, Norville JE, Aach J, et al. Multiplex and homologous recombination-mediated genome editing in *Arabidopsis* and *Nicotiana benthamiana* using guide RNA and Cas9. *Nature Biotechnology* 2013;31(8):688–691. DOI: 10.1038/nbt.2654
11. Li J, Zhang C, He Y, et al. Plant base editing and prime editing: The current status and future perspectives. *Journal of Integrative Plant Biology* 2023;65(2):444–467. DOI: 10.1111/jipb.13425
12. Zhu H, Li C, Gao C. Applications of CRISPR-Cas in agriculture and plant biotechnology. *Nature Reviews Molecular Cell Biology* 2020;21(11):661–677. DOI: 10.1038/s41580-020-00288-9.
13. Chen K, Wang Y, Zhang R, Zhang H, Gao C. CRISPR/Cas genome editing and precision plant breeding in agriculture. *Annual Review of Plant Biology* 2019;70:667–697. DOI: 10.1146/annurev-arplant-050718-100049
14. Makarova KS, Wolf YI, Iranzo J, et al. Evolutionary classification of CRISPR-Cas systems: A burst of class 2 and derived variants. *Nature Reviews Microbiology* 2020;18(2):67–83. DOI: 10.1038/s41579-019-0299-x
15. Shmakov S, Smargon A, Scott D, et al. Diversity and evolution of class 2 CRISPR-Cas systems. *Nature Reviews Microbiology* 2017;15(3):169–182. DOI: 10.1038/nrmicro.2016.184
16. Makarova KS, Zhang F, Koonin EV. SnapShot: Class 2 CRISPR/Cas systems. *Cell* 2017;168(1–2):328–328.e1. (In eng). DOI: 10.1016/j.cell.2016.12.038
17. Mali P, Esvelt KM, Church GM. Cas9 as a versatile tool for engineering biology. *Nature Methods* 2013;10(10):957–963. DOI: 10.1038/nmeth.2649
18. Esvelt KM, Mali P, Braff JL, Moosburner M, Yaung SJ, Church GM. Orthogonal Cas9 proteins for RNA-guided gene regulation and editing. *Nature Methods* 2013;10(11):1116–1121. DOI: 10.1038/nmeth.2681
19. Kleinstiver BP, Prew MS, Tsai SQ, et al. Engineered CRISPR/Cas9 nucleases with altered PAM specificities. *Nature* 2015;523(7561):481–485. DOI: 10.1038/nature14592
20. Anders C, Niewoehner O, Duerst A, Jinek M. Structural basis of PAM-dependent target DNA recognition by the Cas9 endonuclease. *Nature* 2014;513(7519):569–573. DOI: 10.1038/nature13579
21. Mali P, Aach J, Stranges PB, et al. CAS9 transcriptional activators for target specificity screening and paired nickases for cooperative genome engineering. *Nature Biotechnology* 2013;31(9):833–838. DOI: 10.1038/nbt.2675
22. Pinto BS, Saxena T, Oliveira R, et al. Impeding transcription of expanded microsatellite repeats by deactivated Cas9. *Molecular Cell* 2017;68(3):479–490.e5. DOI: 10.1016/j.molcel.2017.09.033
23. Barrangou R, Horvath P. A decade of discovery: CRISPR functions and applications. *Nature Microbiology* 2017;2(7):17092. DOI: 10.1038/nmicrobiol.2017.92
24. Paul B, Montoya G. CRISPR/Cas12a: Functional overview and applications. *Biomedical Journal* 2020;43(1):8–17. DOI: 10.1016/j.bj.2019.10.005
25. Cox DBT, Gootenberg JS, Abudayyeh OO, et al. RNA editing with CRISPR/Cas13. *Science* 2017;358(6366):1019–1027. DOI: 10.1126/science.aaq0180
26. Harrington LB, Burstein D, Chen JS, et al. Programmed DNA destruction by miniature CRISPR/Cas14 enzymes. *Science* 2018;362(6416):839–842. DOI: 10.1126/science.aav4294
27. Hirano S, Nishimasu H, Ishitani R, Nureki O. Structural basis for the altered PAM specificities of engineered CRISPR/Cas9. *Molecular Cell* 2016;61(6):886–894. DOI: 10.1016/j.molcel.2016.02.018
28. Nishimasu H, Shi X, Ishiguro S, et al. Engineered CRISPR/Cas9 nuclease with expanded targeting space. *Science* 2018;361(6408):1259–1262. DOI: 10.1126/science.aas9129
29. Kim N, Kim HK, Lee S, et al. Prediction of the sequence-specific cleavage activity of Cas9 variants. *Nature Biotechnology* 2020;38(11):1328–1336. DOI: 10.1038/s41587-020-0537-9
30. Liang F, Zhang Y, Li L, et al. SpG and SpRY variants expand the CRISPR toolbox for genome editing in zebrafish. *Nature Communications* 2022;13(1):3421. DOI: 10.1038/s41467-022-31034-8
31. Xu Z, Kuang Y, Ren B, et al. SpRY greatly expands the genome editing scope in rice with highly flexible PAM recognition. *Genome Biology* 2021;22(1):6. DOI: 10.1186/s13059-020-02231-9
32. Hirano H, Gootenberg Jonathan S, Horii T, et al. Structure and engineering of *Francisella novicida* Cas9. *Cell* 2016;164(5):950–961. DOI: 10.1016/j.cell.2016.01.039
33. Ren B, Yan F, Kuang Y, et al. Improved base editor for efficiently inducing genetic variations in rice with CRISPR/Cas9-guided hyperactive hAID mutant. *Molecular Plant* 2018;11(4):623–626. DOI: 10.1016/j.molp.2018.01.005
34. Lu Y, Zhu J-K. Precise editing of a target base in the rice genome using a modified CRISPR/Cas9 system. *Molecular Plant* 2017;10(3):523–525. DOI: 10.1016/j.molp.2016.11.013.
35. O'Geen H, Ren C, Nicolet CM, et al. dCas9-based epigenome editing suggests acquisition of histone methylation is not sufficient for target gene repression. *Nucleic Acids Research* 2017;45(17):9901–9916. DOI: 10.1093/nar/gkx578

36. Karlson CKS, Mohd-Noor SN, Nolte N, Tan BC. CRISPR/dCas9-based systems: Mechanisms and applications in plant sciences. *Plants* 2021;10(10) (In eng). DOI: 10.3390/plants10102055
37. Zetsche B, Gootenberg Jonathan S, Abudayyeh Omar O, et al. Cpf1 is a single RNA-guided endonuclease of a class 2 CRISPR/Cas system. *Cell* 2015;163(3):759–771. DOI: 10.1016/j.cell.2015.09.038
38. Fonfara I, Richter H, Bratovič M, Le Rhun A, Charpentier E. The CRISPR-associated DNA-cleaving enzyme Cpf1 also processes precursor CRISPR RNA. *Nature* 2016;532(7600):517–521. DOI: 10.1038/nature17945
39. Abudayyeh OO, Gootenberg JS, Konermann S, et al. C2c2 is a single-component programmable RNA-guided RNA-targeting CRISPR effector. *Science* 2016;353(6299):aaf5573. DOI: 10.1126/science.aaf5573
40. Atkins PAP, Voytas DF. Overcoming bottlenecks in plant gene editing. *Current Opinion in Plant Biology* 2020;54:79–84. DOI: 10.1016/j.pbi.2020.01.002
41. Huang T-K, Puchta H. CRISPR/Cas-mediated gene targeting in plants: Finally a turn for the better for homologous recombination. *Plant Cell Reports* 2019;38(4):443–453. DOI: 10.1007/s00299-019-02379-0
42. Komor AC, Kim YB, Packer MS, Zuris JA, Liu DR. Programmable editing of a target base in genomic DNA without double-stranded DNA cleavage. *Nature* 2016;533(7603):420–424. DOI: 10.1038/nature17946
43. Zong Y, Wang Y, Li C, et al. Precise base editing in rice, wheat and maize with a Cas9-cytidine deaminase fusion. *Nature Biotechnology* 2017;35(5):438–440. DOI: 10.1038/nbt.3811
44. Qin L, Li J, Wang Q, et al. High-efficient and precise base editing of CG to T•A in the allotetraploid cotton (*Gossypium hirsutum*) genome using a modified CRISPR/Cas9 system. *Plant Biotechnology Journal* 2020;18(1):45–56. DOI: 10.1111/pbi.13168
45. Zhang R, Liu J, Chai Z, et al. Generation of herbicide tolerance traits and a new selectable marker in wheat using base editing. *Nature Plants* 2019;5(5):480–485. DOI: 10.1038/s41477-019-0405-0
46. Liu T, Zeng D, Zheng Z, et al. The ScCas9++ variant expands the CRISPR toolbox for genome editing in plants. *Journal of Integrative Plant Biology* 2021;63(9):1611–1619. DOI: 10.1111/jipb.13164
47. Zeng D, Liu T, Tan J, et al. PhieCBEs: Plant high-efficiency cytidine base editors with expanded target range. *Molecular Plant* 2020;13(12):1666–1669. DOI: 10.1016/j.molp.2020.11.001
48. Komor AC, Zhao KT, Packer MS, et al. Improved base excision repair inhibition and bacteriophage Mu Gam protein yields C:G-to-T:A base editors with higher efficiency and product purity. *Science Advances* 2017;3(8):eaao4774. DOI:10.1126/sciadv.aao4774
49. Zong Y, Song Q, Li C, et al. Efficient C-to-T base editing in plants using a fusion of nCas9 and human APOBEC3A. *Nature Biotechnology* 2018;36(10):950–953. DOI: 10.1038/nbt.4261
50. Li C, Zong Y, Wang Y, et al. Expanded base editing in rice and wheat using a Cas9-adenosine deaminase fusion. *Genome Biology* 2018;19(1):59. DOI: 10.1186/s13059-018-1443-z
51. Gaudelli NM, Komor AC, Rees HA, et al. Programmable base editing of AT to G•C in genomic DNA without DNA cleavage. *Nature* 2017;551(7681):464–471. DOI: 10.1038/nature24644
52. Koblan LW, Doman JL, Wilson C, et al. Improving cytidine and adenine base editors by expression optimization and ancestral reconstruction. *Nature Biotechnology* 2018;36(9):843–846. (In eng). DOI: 10.1038/nbt.4172
53. Yan F, Kuang Y, Ren B, et al. Highly efficient AT to G•C base editing by Cas9n-guided tRNA adenosine deaminase in rice. *Molecular Plant* 2018;11(4):631–634. DOI: 10.1016/j.molp.2018.02.008
54. Hua K, Tao X, Yuan F, Wang D, Zhu J-K. Precise AT to G•C base editing in the rice genome. *Molecular Plant* 2018;11(4):627–630. DOI: 10.1016/j.molp.2018.02.007
55. Hua K, Tao X, Liang W, Zhang Z, Gou R, Zhu J-K. Simplified adenine base editors improve adenine base editing efficiency in rice. *Plant Biotechnology Journal* 2020;18(3):770–778. DOI: 10.1111/pbi.13244
56. Gaudelli NM, Lam DK, Rees HA, et al. Directed evolution of adenine base editors with increased activity and therapeutic application. *Nature Biotechnology* 2020;38(7):892–900. DOI: 10.1038/s41587-020-0491-6
57. Richter MF, Zhao KT, Eton E, et al. Phage-assisted evolution of an adenine base editor with improved Cas domain compatibility and activity. *Nature Biotechnology* 2020;38(7):883–891. DOI: 10.1038/s41587-020-0453-z
58. Yan D, Ren B, Liu L, et al. High-efficiency and multiplex adenine base editing in plants using new TadA variants. *Molecular Plant* 2021;14(5):722–731. DOI: 10.1016/j.molp.2021.02.007
59. Zhao D, Li J, Li S, et al. Glycosylase base editors enable C-to-A and C-to-G base changes. *Nature Biotechnology* 2021;39(1):35–40. DOI: 10.1038/s41587-020-0592-2
60. Kurt IC, Zhou R, Iyer S, et al. CRISPR C-to-G base editors for inducing targeted DNA transversions in human cells. *Nature Biotechnology* 2021;39(1):41–46. DOI: 10.1038/s41587-020-0609-x

61. Tian Y, Shen R, Li Z, et al. Efficient C-to-G editing in rice using an optimized base editor. *Plant Biotechnology Journal* 2022;20(7):1238–1240. DOI: 10.1111/pbi.13841
62. Li C, Zhang R, Meng X, et al. Targeted, random mutagenesis of plant genes with dual cytosine and adenine base editors. *Nature Biotechnology* 2020;38(7):875–882. (In eng). DOI: 10.1038/s41587-019-0393-7
63. Wang S, Zong Y, Lin Q, et al. Precise, predictable multi-nucleotide deletions in rice and wheat using APOBEC-Cas9. *Nature Biotechnology* 2020;38(12):1460–1465. DOI: 10.1038/s41587-020-0566-4
64. Anzalone AV, Randolph PB, Davis JR, et al. Search-and-replace genome editing without double-strand breaks or donor DNA. *Nature* 2019;576(7785):149–157. DOI: 10.1038/s41586-019-1711-4
65. Chen PJ, Hussmann JA, Yan J, et al. Enhanced prime editing systems by manipulating cellular determinants of editing outcomes. *Cell* 2021;184(22):5635–5652.e29. DOI: 10.1016/j.cell.2021.09.018
66. Jiang Y, Chai Y, Qiao D, et al. Optimized prime editing efficiently generates glyphosate-resistant rice plants carrying homozygous TAP-IVS mutation in EPSPS. *Molecular Plant* 2022;15(11):1646–1649. DOI: 10.1016/j.molp.2022.09.006
67. Li J, Chen L, Liang J, et al. Development of a highly efficient prime editor 2 system in plants. *Genome Biology* 2022;23(1):161. DOI: 10.1186/s13059-022-02730-x
68. Anzalone AV, Gao XD, Podracky CJ, et al. Programmable deletion, replacement, integration and inversion of large DNA sequences with twin prime editing. *Nature Biotechnology* 2022;40(5):731–740. DOI: 10.1038/s41587-021-01133-w
69. Jiang Y-Y, Chai Y-P, Lu M-H, et al. Prime editing efficiently generates W542L and S621I double mutations in two ALS genes in maize. *Genome Biology* 2020;21(1):257. DOI: 10.1186/s13059-020-02170-5
70. Li H, Li J, Chen J, Yan L, Xia L. Precise modifications of both exogenous and endogenous genes in rice by prime editing. *Molecular Plant* 2020;13(5):671–674. DOI: 10.1016/j.molp.2020.03.011
71. Hua K, Han P, Zhu J-K. Improvement of base editors and prime editors advances precision genome engineering in plants. *Plant Physiology* 2021;188(4):1795–1810. DOI: 10.1093/plphys/kiab591
72. Wang J, He Z, Wang G, et al. Efficient targeted insertion of large DNA fragments without DNA donors. *Nature Methods* 2022;19(3):331–340. DOI: 10.1038/s41592-022-01399-1
73. Lin Q, Jin S, Zong Y, et al. High-efficiency prime editing with optimized, paired pegRNAs in plants. *Nature Biotechnology* 2021;39(8):923–927. DOI: 10.1038/s41587-021-00868-w
74. Li X, Zhou L, Gao B-Q, et al. Highly efficient prime editing by introducing same-sense mutations in pegRNA or stabilizing its structure. *Nature Communications* 2022;13(1):1669. DOI: 10.1038/s41467-022-29339-9
75. Xu W, Yang Y, Yang B, et al. A design optimized prime editor with expanded scope and capability in plants. *Nature Plants* 2022;8(1):45–52. DOI: 10.1038/s41477-021-01043-4
76. Kim HK, Yu G, Park J, et al. Predicting the efficiency of prime editing guide RNAs in human cells. *Nature Biotechnology* 2021;39(2):198–206. DOI: 10.1038/s41587-020-0677-y
77. Liu Y, Yang G, Huang S, et al. Enhancing prime editing by Csy4-mediated processing of pegRNA. *Cell Research* 2021;31(10):1134–1136. DOI: 10.1038/s41422-021-00520-x
78. Nelson JW, Randolph PB, Shen SP, et al. Engineered pegRNAs improve prime editing efficiency. *Nature Biotechnology* 2022;40(3):402–410. DOI: 10.1038/s41587-021-01039-7
79. Zou J, Meng X, Liu Q, et al. Improving the efficiency of prime editing with epegRNAs and high-temperature treatment in rice. *Science China Life Sciences* 2022;65(11):2328–2331. DOI: 10.1007/s11427-022-2147-2
80. Jiang T, Zhang X-O, Weng Z, Xue W. Deletion and replacement of long genomic sequences using prime editing. *Nature Biotechnology* 2022;40(2):227–234. DOI: 10.1038/s41587-021-01026-y
81. Choi J, Chen W, Suiter CC, et al. Precise genomic deletions using paired prime editing. *Nature Biotechnology* 2022;40(2):218–226. DOI: 10.1038/s41587-021-01025-z
82. Li J, Sun Y, Du J, Zhao Y, Xia L. Generation of targeted point mutations in rice by a modified CRISPR/Cas9 system. *Molecular Plant* 2017;10(3):526–529. DOI: 10.1016/j.molp.2016.12.001
83. Ren B, Yan F, Kuang Y, et al. A CRISPR/Cas9 toolkit for efficient targeted base editing to induce genetic variations in rice. *Science China Life Sciences* 2017;60(5):516–519. DOI: 10.1007/s11427-016-0406-x
84. Ren B, Liu L, Li S, et al. Cas9-NG greatly expands the targeting scope of the genome-editing toolkit by recognizing NG and other atypical PAMs in rice. *Molecular Plant* 2019;12(7):1015–1026. DOI: 10.1016/j.molp.2019.03.010
85. Wang M, Xu Z, Gosavi G, et al. Targeted base editing in rice with CRISPR/ScCas9 system. *Plant Biotechnology Journal* 2020;18(8):1645–1647. DOI: 10.1111/pbi.13330
86. Shimatani Z, Kashojiya S, Takayama M, et al. Targeted base editing in rice and tomato using a CRISPR/Cas9 cytidine deaminase fusion. *Nature Biotechnology* 2017;35(5):441–443. DOI: 10.1038/nbt.3833

87. Zhong Z, Sretenovic S, Ren Q, et al. Improving plant genome editing with high-fidelity xCas9 and non-canonical PAM-targeting Cas9-NG. *Molecular Plant* 2019;12(7):1027–1036. DOI: 10.1016/j.molp.2019.03.011
88. Veillet F, Perrot L, Guyon-Debast A, et al. Expanding the CRISPR toolbox in *P. patens* using SpCas9-NG variant and application for gene and base editing in solanaceae crops. *International Journal of Molecular Sciences* 2020;21(3):1024. https://www.mdpi.com/1422-0067/21/3/1024
89. Kang B-C, Yun J-Y, Kim S-T, et al. Precision genome engineering through adenine base editing in plants. *Nature Plants* 2018;4(7):427–431. DOI: 10.1038/s41477-018-0178-x
90. Wei C, Wang C, Jia M, et al. Efficient generation of homozygous substitutions in rice in one generation utilizing an rABE8e base editor. *Journal of Integrative Plant Biology* 2021;63(9):1595–1599. DOI: 10.1111/jipb.13089
91. Tan J, Zeng D, Zhao Y, et al. PhieABEs: A PAM-less/free high-efficiency adenine base editor toolbox with wide target scope in plants. *Plant Biotechnology Journal* 2022;20(5):934–943. DOI: 10.1111/pbi.13774.
92. Xu R, Kong F, Qin R, Li J, Liu X, Wei P. Development of an efficient plant dual cytosine and adenine editor. *Journal of Integrative Plant Biology* 2021;63(9):1600–1605. DOI: 10.1111/jipb.13146.
93. Butt H, Rao GS, Sedeek K, Aman R, Kamel R, Mahfouz M. Engineering herbicide resistance via prime editing in rice. *Plant Biotechnology Journal* 2020;18(12):2370–2372. DOI: 10.1111/pbi.13399
94. Hua K, Jiang Y, Tao X, Zhu J-K. Precision genome engineering in rice using prime editing system. *Plant Biotechnology Journal* 2020;18(11):2167–2169. DOI: 10.1111/pbi.13395
95. Xu R, Li J, Liu X, Shan T, Qin R, Wei P. Development of plant prime-editing systems for precise genome editing. *Plant Communications* 2020;1(3):100043. DOI: 10.1016/j.xplc.2020.100043
96. Zong Y, Liu Y, Xue C, et al. An engineered prime editor with enhanced editing efficiency in plants. *Nature Biotechnology* 2022;40(9):1394–1402. DOI: 10.1038/s41587-022-01254-w
97. Lin Q, Zong Y, Xue C, et al. Prime genome editing in rice and wheat. *Nature Biotechnology* 2020;38(5):582–585. DOI: 10.1038/s41587-020-0455-x
98. Xu W, Zhang C, Yang Y, et al. Versatile nucleotides substitution in plant using an improved prime editing system. *Molecular Plant* 2020;13(5):675–678. DOI: 10.1016/j.molp.2020.03.012
99. Li H, Zhu Z, Li S, et al. Multiplex precision gene editing by a surrogate prime editor in rice. *Molecular Plant* 2022;15(7):1077–1080. DOI: 10.1016/j.molp.2022.05.009
100. Lu Y, Tian Y, Shen R, et al. Precise genome modification in tomato using an improved prime editing system. *Plant Biotechnology Journal* 2021;19(3):415–417. DOI: 10.1111/pbi.13497
101. Ran Y, Liang Z, Gao C. Current and future editing reagent delivery systems for plant genome editing. *Science China Life Sciences* 2017;60(5):490–505. DOI: 10.1007/s11427-017-9022-1
102. Lowe K, Wu E, Wang N, et al. Morphogenic regulators baby boom and wuschel improve monocot transformation. *The Plant Cell* 2016;28(9):1998–2015. DOI: 10.1105/tpc.16.00124
103. Maher MF, Nasti RA, Vollbrecht M, Starker CG, Clark MD, Voytas DF. Plant gene editing through de novo induction of meristems. *Nature Biotechnology* 2020;38(1):84–89. DOI: 10.1038/s41587-019-0337-2
104. Ali Z, Abul-Faraj A, Li L, et al. Efficient virus-mediated genome editing in plants using the CRISPR/Cas9 system. *Molecular Plant* 2015;8(8):1288–1291. DOI: 10.1016/j.molp.2015.02.011
105. Ali Z, Eid A, Ali S, Mahfouz MM. *Pea early-browning virus*-mediated genome editing via the CRISPR/Cas9 system in *Nicotiana benthamiana* and Arabidopsis. *Virus Research* 2018;244:333–337. DOI: 10.1016/j.virusres.2017.10.009
106. Cody WB, Scholthof HB, Mirkov TE. Multiplexed gene editing and protein overexpression using a *Tobacco mosaic virus* viral vector. *Plant Physiology* 2017;175(1):23–35. DOI: 10.1104/pp.17.00411
107. Hu J, Li S, Li Z, et al. A barley stripe mosaic virus-based guide RNA delivery system for targeted mutagenesis in wheat and maize. *Molecular Plant Pathology* 2019;20(10):1463–1474. DOI: 10.1111/mpp.12849
108. Jiang N, Zhang C, Liu J-Y, et al. Development of *Beet necrotic yellow vein virus*-based vectors for multiple-gene expression and guide RNA delivery in plant genome editing. *Plant Biotechnology Journal* 2019;17(7):1302–1315. DOI: 10.1111/pbi.13055
109. Yin K, Han T, Liu G, et al. A geminivirus-based guide RNA delivery system for CRISPR/Cas9 mediated plant genome editing. *Scientific Reports* 2015;5(1):14926. DOI: 10.1038/srep14926
110. Gao Q, Xu W-Y, Yan T, et al. Rescue of a plant cytorhabdovirus as versatile expression platforms for planthopper and cereal genomic studies. *New Phytologist* 2019;223(4):2120–2133. DOI: 10.1111/nph.15889
111. Ma X, Zhang X, Liu H, Li Z. Highly efficient DNA-free plant genome editing using virally delivered CRISPR-Cas9. *Nature Plants* 2020;6(7):773–779. DOI: 10.1038/s41477-020-0704-5

112. Ellison EE, Nagalakshmi U, Gamo ME, Huang P-J, Dinesh-Kumar S, Voytas DF. Multiplexed heritable gene editing using RNA viruses and mobile single guide RNAs. *Nature Plants* 2020;6(6):620–624. DOI: 10.1038/s41477-020-0670-y
113. Shariati SA, Dominguez A, Xie S, Wernig M, Qi LS, Skotheim JM. Reversible disruption of specific transcription factor-DNA interactions using CRISPR/Cas9. *Molecular Cell* 2019;74(3):622–633.e4. DOI: 10.1016/j.molcel.2019.04.011
114. Gilbert Luke A, Larson Matthew H, Morsut L, et al. CRISPR-mediated modular RNA-guided regulation of transcription in eukaryotes. *Cell* 2013;154(2):442–451. DOI: 10.1016/j.cell.2013.06.044
115. Dhamad AE, Lessner DJ. A CRISPRi-dCas9 system for archaea and its use to examine gene function during nitrogen fixation by *Methanosarcina acetivorans. Applied and Environmental Microbiology* 2020;86(21):e01402–20. DOI: 10.1128/AEM.01402-20
116. Dominguez AA, Lim WA, Qi LS. Beyond editing: Repurposing CRISPR-Cas9 for precision genome regulation and interrogation. *Nature Reviews Molecular Cell Biology* 2016;17(1):5–15. DOI: 10.1038/nrm.2015.2
117. Qi LS, Larson Matthew H, Gilbert Luke A, et al. Repurposing CRISPR as an RNA-guided platform for sequence-specific control of gene expression. *Cell* 2013;152(5):1173–1183. DOI: 10.1016/j.cell.2013.02.022
118. Chen B, Gilbert Luke A, Cimini Beth A, et al. Dynamic imaging of genomic loci in living human cells by an optimized CRISPR/Cas system. *Cell* 2013;155(7):1479–1491. DOI: 10.1016/j.cell.2013.12.001
119. Abudayyeh OO, Gootenberg JS, Essletzbichler P, et al. RNA targeting with CRISPR-Cas13. *Nature* 2017;550(7675):280–284. DOI: 10.1038/nature24049
120. Li Z, Xiong X, Wang F, Liang J, Li J-F. Gene disruption through base editing-induced messenger RNA missplicing in plants. *New Phytologist* 2019;222(2):1139–1148. DOI: 10.1111/nph.15647
121. Basak J, Nithin C. Targeting non-coding RNAs in plants with the CRISPR/Cas technology is a challenge yet worth accepting. *Frontiers in Plant Science* 2015;6 (Mini Review) (In eng). DOI: 10.3389/fpls.2015.01001
122. McDonald JI, Celik H, Rois LE, et al. Reprogrammable CRISPR/Cas9-based system for inducing site-specific DNA methylation. *Biology Open* 2016;5(6):866–874. DOI: 10.1242/bio.019067
123. Hilton IB, D'Ippolito AM, Vockley CM, et al. Epigenome editing by a CRISPR/Cas9-based acetyltransferase activates genes from promoters and enhancers. *Nature Biotechnology* 2015;33(5):510–517. DOI: 10.1038/nbt.3199
124. Braun SMG, Kirkland JG, Chory EJ, Husmann D, Calarco JP, Crabtree GR. Rapid and reversible epigenome editing by endogenous chromatin regulators. *Nature Communications* 2017;8(1):560. DOI: 10.1038/s41467-017-00644-y
125. Xing H-L, Dong L, Wang Z-P, et al. A CRISPR/Cas9 toolkit for multiplex genome editing in plants. *BMC Plant Biology* 2014;14(1):327. DOI: 10.1186/s12870-014-0327-y
126. Ma X, Liu Y-G. CRISPR/Cas9-based multiplex genome editing in monocot and dicot plants. *Current Protocols in Molecular Biology* 2016;115(1):31.6.1–31.6.21. DOI: 10.1002/cpmb.10
127. Gao Y, Zhao Y. Self-processing of ribozyme-flanked RNAs into guide RNAs in vitro and in vivo for CRISPR-mediated genome editing. *Journal of Integrative Plant Biology* 2014;56(4):343–349. DOI: 10.1111/jipb.12152
128. Ding D, Chen K, Chen Y, Li H, Xie K. Engineering introns to express RNA guides for Cas9- and Cpf1-mediated multiplex genome editing. *Molecular Plant* 2018;11(4):542–552. DOI: 10.1016/j.molp.2018.02.005
129. Mikami M, Toki S, Endo M. In planta processing of the SpCas9-gRNA complex. *Plant and Cell Physiology* 2017;58(11):1857–1867. DOI: 10.1093/pcp/pcx154
130. Zalatan JG, Lee ME, Almeida R, et al. engineering complex synthetic transcriptional programs with CRISPR RNA scaffolds. *Cell* 2015;160(1):339–350. DOI: 10.1016/j.cell.2014.11.052
131. Dahlman JE, Abudayyeh OO, Joung J, Gootenberg JS, Zhang F, Konermann S. Orthogonal gene knockout and activation with a catalytically active Cas9 nuclease. *Nature Biotechnology* 2015;33(11):1159–1161. DOI: 10.1038/nbt.3390
132. Campa CC, Weisbach NR, Santinha AJ, Incarnato D, Platt RJ. Multiplexed genome engineering by Cas12a and CRISPR arrays encoded on single transcripts. *Nature Methods* 2019;16(9):887–893. DOI: 10.1038/s41592-019-0508-6
133. Li C, Zong Y, Jin S, et al. SWISS: Multiplexed orthogonal genome editing in plants with a Cas9 nickase and engineered CRISPR RNA scaffolds. *Genome Biology* 2020;21(1):141. DOI: 10.1186/s13059-020-02051-x

134. Fan R, Chai Z, Xing S, et al. Shortening the sgRNA-DNA interface enables SpCas9 and eSpCas9(1.1) to nick the target DNA strand. *Science China Life Sciences* 2020;63(11):1619–1630. DOI: 10.1007/s11427-020-1722-0
135. Hu H, Yu F. A CRISPR/Cas9-based system with controllable auto-excision feature serving cisgenic plant breeding and beyond. *International Journal of Molecular Sciences* 2022;23(10):5597. DOI: 10.3390/ijms23105597
136. Decaestecker W, Buono RA, Pfeiffer ML, et al. CRISPR-TSKO: A technique for efficient mutagenesis in specific cell types, tissues, or organs in arabidopsis. *The Plant Cell* 2019;31(12):2868–2887. DOI: 10.1105/tpc.19.00454
137. Ochoa-Fernandez R, Abel NB, Wieland F-G, et al. Optogenetic control of gene expression in plants in the presence of ambient white light. *Nature Methods* 2020;17(7):717–725. DOI: 10.1038/s41592-020-0868-y
138. Meng X, Yu H, Zhang Y, et al. Construction of a genome-wide mutant library in rice using CRISPR/Cas9. *Molecular Plant* 2017;10(9):1238–1241. DOI: 10.1016/j.molp.2017.06.006
139. Liu H-J, Jian L, Xu J, et al. High-throughput CRISPR/Cas9 mutagenesis streamlines trait gene identification in maize[OPEN]. *The Plant Cell* 2020;32(5):1397–1413. DOI: 10.1105/tpc.19.00934
140. Jacobs TB, Zhang N, Patel D, Martin GB. Generation of a collection of mutant tomato lines using pooled CRISPR libraries. *Plant Physiology* 2017;174(4):2023–2037. DOI: 10.1104/pp.17.00489
141. Bai M, Yuan J, Kuang H, et al. Generation of a multiplex mutagenesis population via pooled CRISPR/Cas9 in soya bean. *Plant Biotechnology Journal* 2020;18(3):721–731. DOI: 10.1111/pbi.13239
142. Butt H, Eid A, Momin AA, et al. CRISPR directed evolution of the spliceosome for resistance to splicing inhibitors. *Genome Biology* 2019;20(1):73. DOI: 10.1186/s13059-019-1680-9
143. Liu X, Qin R, Li J, et al. A CRISPR/Cas9-mediated domain-specific base-editing screen enables functional assessment of ACCase variants in rice. *Plant Biotechnology Journal* 2020;18(9):1845–1847. DOI: 10.1111/pbi.13348
144. Tilman D, Balzer C, Hill J, Befort BL. Global food demand and the sustainable intensification of agriculture. *Proceedings of the National Academy of Sciences of the United States of America* 2011;108(50):20260–20264. DOI: 10.1073/pnas.1116437108

2 Functional Genome Analysis and Genome Editing in Plants
An Updated Overview

Karma Landup Bhutia, Sarita Kumari, Kumari Anjani, Anima Kisku, Bhavna Borah, Prabhakar Nishi, Vinay Kumar Sharma, Barasha Rani Deka, Kanshouwa Modunshim, Bharti Lap, M. James, Nangsol Dolma Bhutia, Limasunep Longkumer, and Akbar Hossain

2.1 INTRODUCTION

Since the initial publication of the Arabidopsis thaliana genome in 2000, more than 500 plant genomes have been sequenced, thanks to advancing sequencing technology and falling sequencing prices. While this has been happening, an increasing number of omics datasets have been made public using methods like genome re-sequencing, pan-genomics, RNA-Seq, metabolomics, and proteomics. We now have a fantastic chance and a wealth of data resources to investigate the evolution of plant genes, thanks to the publication of these databases. In addition to the bioinformatics field's explosive growth in recent years, a wide range of software and tools for comparative genomics analyses have also been accessible. But we still require more effective, potent, and user-friendly tools, software, pipelines, websites, and databases to handle the growing variety of omics data.

2.2 GENOME ANALYSIS

Genome analysis is the study of an organism's whole DNA sequence and its genetic makeup. It entails identifying genes and their functions, studying genetic variations, understanding the genetics of genes related to diseases, species evolution, development of new drugs, etc. All these considered, genome analysis provides a wide range of essential functions that can advance our understanding of genetics, disease, and evolution, as well as the creation of new medications and treatments for human health and crop improvement in agriculture. Some specific functions of genome analysis are as follows: (1) Identifying genes and their functions: Genome analysis assists in identifying the causative genes along with their possible functional effects through genetic variation detection or genome sequencing. By understanding the functions of these genes, researchers can develop new therapies or treatments. For example, researchers identified CIMMYT maize lines that are resistant to fall armyworm (*Spodoptera frugiperda*) (Kamweru et al., 2022) which provides useful information when formulating a resistance breeding plan. It also helped in the demonstration of the phylogenomic relationship of *Gossypium barbadense* in China (Zhao et al., 2022). It signifies the functions of candidate

 DOI: 10.1201/9781003387060-2

genes controlling various important features, including disease resistance, fiber length, fiber strength, and lint percentage, which were verified using gene expression studies and VIGS transgenic experiments, etc. (2) Study on evolution of species: genome analysis can be used to study evolution using phylogenetic analysis, comparative genomics, population genetics, etc. Combining these approaches, it provides a detailed understanding of the evolution of species, including the timing of major evolutionary events, the genetic changes that have occurred, and the processes that have driven the diversification and adaptation of different species. Therefore, it helped researchers understand the evolution process of different species by providing information about the genetic changes that have occurred over time. Researchers conducted experiments, for example, research on genome evolution and diversity of wild and cultivated potatoes (Tang et al., 2022) and when compared to closely comparable, seed-propagated solanaceous crops, it was shown that the potato genome significantly extended its repertoire of disease resistance genes. These findings are suggestive of the impact of tuber-based propagation tactics on the evolution of the potato genome. There is no evidence for a transcription factor that controls tuber identity and interacts with the mobile tuberization inductive signal SP6A. Additionally, 561,433 high-confidence structural variants have been found, and this information can be used to enhance inbred lines and prevent linkage drag that may be brought on by the carotenoid content of tubers. This will hasten the breeding of hybrid potatoes and deepen our knowledge of the biology and evolution of one of the world's most important crops. (3) Developing new drugs: the ideal therapeutic medicine should be free from side effects and efficient in combating disease. There have been several significant new medicine classes introduced in the past 25 years. Even the most successful and beneficial of them, nevertheless, only offer the best possible treatment to a portion of patients. A medicine may provide little or no benefit for some people with a certain ailment, while it may have negative side effects for other people. The high failure rates of novel drug candidates in clinical trials are caused by such individual variances in response to treatment. There are two causes for variation in individuals; the first of these might be the target molecule's structural variability. Not every individual may have the same receptor if a medicine blocks a certain receptor from working and the second cause may be differences in pharmacokinetics, or variations in how a certain medication is absorbed, distributed, metabolized, and eliminated by the body. A medication won't work if it isn't absorbed or metabolized too rapidly. To create novel medications that specifically target particular genes or disease-related pathways, genome analysis can be used to find possible therapeutic targets. A novel approach for analyzing changes in gene expression in the context of whole pathways is expression profiling of drug response from genes. Genome analysis may also be used to find current medications that could be repurposed for the treatment of a specific illness in the context of drug repurposing. This can be done by identifying genes or pathways that are similar to those targeted by existing drugs and testing whether these drugs are effective in treating the disease. Personalized medication that targets certain genetic alterations or pathways in particular individuals is often beneficial for this. It can increase efficacy and lessen the negative effects of therapy by adjusting the course of action to the unique genetic profile of each patient. Ultimately, genome (specifically pharmaco-genomics) analysis will be utilized to identify which pharmaceuticals should combat the disease and which drugs will be efficiently used for the treatment. Numerous modules of the analysis of expression data address various issues pertinent to drug response screening (Holleman et al., 2004). Finding genes whose expression differs across two or more conditions (for instance, between groups of patients who are susceptible to or resistant to a certain medicine) is one of the most crucial works to find typical gene expression patterns that can be used to categorize people and to find pertinent pathways that might explain the expression patterns.

2.2.1 Plant Functional Genomics

For a complete understanding, it's important to interpret molecular complexity and variety at various levels, including the genome, transcriptome, proteome, and metabolome. The collective term for data from several scales is "multi-omics". Insights into the flow of biological information at different levels are provided by multi-omics data gathered through a variety of methods. These insights can be used to identify the biological state of interest underlying mechanisms. Technology advancements in the last 10 years have shown a significant contribution to biological data, including DNA sequencing (Le Nguyen et al., 2019), transcriptomics analysis via RNA-seq (Mashaki et al., 2018), SWATH-based proteomics (Zhu et al., 2020), and metabolomics via UPLC-MS and GC-MS (Balcke et al., 2012). Genomic research, which examines entire genomes, is the first omics discipline to take off. Bhatta et al., (2019), Yoshida et al., (2022) and others have employed genomic research like QTL/association mapping to identify genomic areas related to agronomically significant traits and gave a basic framework for additional omics methods. Additionally, by employing transcriptomics investigations in a variety of crop plants, differentially expressed genes under various biotic and abiotic challenges were discovered (Chen et al., 2022; Pal et al., 2022). Crops like Pigeonpea (Pazhamala et al., 2017), chickpeas (Kudapa et al., 2018) and groundnuts (Sinha et al., 2020) all have gene expression atlases that offer insights into the subsets of genes expressed at various growth stages. The spatial transcriptomics method (Giacomello et al., 2017) combines histological imaging and RNA sequencing to provide high-throughput and spatially resolved transcriptomics in plant tissues. Proteomics is used to understand the functional analysis of translated portions of the genome, and metabolomics is used as a diagnostic tool to evaluate how well plants respond to various stimuli (Villate et al., 2021). To manage the data generated from various experiments and sequencing research, several repositories were created. The repositories contain databases for DNA, RNA, and protein sequences as well as specialized databases for particular types of data (Lai et al., 2012). Databases can be divided into four categories based on various omics data types: Nucleotide sequences or genomic sequences can be found in genomic databases, functional RNA sequences can be found in transcriptomics databases, amino acid sequences and protein structures can be found in proteomic databases, and metabolites and metabolic pathways can be found in metabolomics databases (Table 2.1).

TABLE 2.1
Databases Frequently Used in Plant Functional Genome Analysis

Databases	Key Features
PlantTFDB	Contains detailed information about the reported plant transcription factors
AtMAD	Contains high-quality Arabidopsis thaliana multi-omics data.
GoMapMan	Contains Plant Gene functional annotations
HapRice	Database for SNP-haplotype of rice.
PGDJ	Database for DNA markers and linkages of plants.
PIECE	Database for plant intron-exon comparison and evolution.
Plant rDNA	Database for plant rDNA.
PlantRNA	Database for tRNAs of photosynthetic eukaryotes.
PlnTFDB	Database for the prediction of plant transcription factors
PMRD	Database for plant microRNA.
PTGBase	Database for plant tandem duplicated genes.

2.3 ADVANCEMENTS IN GENOMICS AND EPIGENOMICS TOOLS

Genomics has played a key role in plant sciences since its advent in the 1970s. Today 1,000 plant genome sequences are publicly available including more than 600 angiosperms (Kress et al., 2022). The genomic tools provide opportunities and avenues that can help increase crop productivity and sustainability in the present scenario of climate change. The genomic studies can be used to utilize the genetic diversity of wild relatives for the improvement of cultivated species, to identify new crops for domestication, to identify genes having crucial roles in abiotic and biotic stress tolerance and many others. Recent advancement in genomics along with high-throughput phenotyping has helped in the identification of promising agronomic characters. Genome editing tools can be used to modify and manipulate the discovered genes to facilitate the development of climate-resilient crops with biotic and abiotic stress tolerance and increased nutritional value (Meena et al., 2022). With the increasing availability of improved plant genome assemblies, new research targets like long non-coding RNAs and cis-regulatory elements have been identified, which can help open further new avenues for crop improvement. The development of bioinformatics tools for genomic studies has helped to target huge data sets. Consequently, metagenomic and pan-genomics approaches using the help of bioinformatics tools can further expand the horizon of research.

The understanding of plant systems has been greatly improved by these advanced genomic tools, but one more aspect of plant functioning needs to be explored to have a complete understanding of the plant system. These are the epigenetic modifications, which play a crucial role in the interaction of the plants with the environment. Epigenomics aims at understanding the role of epigenetic marks such as DNA methylation, histone modification, chromatin remodeling, non-coding long RNA regulation, etc. in plant growth and development and genome function. Recent years have seen a surge in research for the exploration of the role of these aspects of the genome. The diversity of epigenomes and their role in plant adaptation to environmental factors have been identified. Advancement in technological tools has enabled Epigenome-Wide Association Studies or EWAS, similar to GWAS and large-scale projects like the human ENCyclopedia Of DNA Elements (ENCODE) project, the NIH Roadmap Epigenomics effort, the Human Epigenome Project and BLUEPRINT, to study the epigenome of different cells and tissues (Ballereau et al., 2013). The possibility to fuse genome editing techniques like dCas9 protein with DNA methylation and histone modification is also being explored to improve plant survival in changing climatic conditions. Some of the recent genomic and epigenomic tools which have revolutionized plant science research have been listed in the following sections:

2.3.1 Next-Generation Sequencing

The next-generation sequencing (NGS) platforms like Illumina/Solexa, Roche 454, Ion Torrent, SOLiD sequencing, Pac Bio and others have reduced the cost of genome sequencing by many folds and have provided unparalleled information on genome, transcriptome and epigenome (Ray and Satya, 2014). NGS in combination with comprehensive and fast computational pipelines allows the discovery of refined information on a large scale. The NGS methods, in combination with bioinformatic tools, have enabled the easy and cost-effective discovery of high-throughput markers such as SSR, SNP, etc., which can be used in molecular breeding programs (Stapley et al., 2010). These methods have also helped in the construction of high-density, high-resolution genetic maps, which further enable the detection of numerous DNA sequence polymorphisms. The methodologies based on this technique for crop breeding include microarray-based genomic selection (Okou et al., 2007), molecular inversion probe (Porreca et al., 2007), reduced-representation libraries (Ray and Satya, 2014), solution hybrid selection (Gnirke et al., 2009), complexity reduction of polymorphic sequences (Mammadov et al., 2010), exome sequencing (Teer and Mullikin, 2010) and sequencing of the genomic region associated with particular trait (Teer et al., 2010), multiplexed shotgun genotyping (Andolfatto et al., 2011), genotyping-by-sequencing (GBS) (Elshire et al., 2011), restriction-site associated DNA sequencing (RAD-seq), sequence-based polymorphic marker technology (Sahu et al., 2012).

2.3.2 Genome-Wide Approaches for Functional Annotation

The improved sequencing methodologies are generating voluminous information at a low cost and in a short frame of time. This information can be processed to answer all the unsolved mysteries in plant sciences including stress tolerance mechanisms. The information can help in accurate gene prediction and candidate gene identification through genome-wide searches and homology studies. The bioinformatics tools can help in the functional annotation of the available sequences, which can further help to understand the functioning of the genome as a whole. The functional annotation can be done by GO study, homology modeling, pathway analysis, structure prediction analysis, and many other methods. This helps in understanding the functional role of the sequences and ultimately gene identification (Shi et al., 2023).

2.3.3 Discovery of Non-Coding Part of the Genome

The large amounts of available information and computational platforms have helped to venture into the depths of the genome, which were practically inaccessible earlier. New genomic regions having critical regulatory roles in the genome have been identified. The discovery of non-coding roles of genomic loci like long non-coding transcripts, cis-regulatory elements, microRNA, etc. has improved our understanding of the complexity of plant transcriptomes (Van Bel et al., 2019). The discovery of these elements has helped in the understanding of the regulatory role of the genome. These regions can be a target of genome editing and other advanced breeding approaches for crop improvement,

2.3.4 Pan-Genome Approach

The plant genomes are highly variable, and the study of a single reference genome cannot help in the understanding of the complexity of genome function. The pan-genome represents the collection of all the genomic sequences and gene content found in a species (Shi et al., 2023). Soybean was the first crop for which pan-genome assembly was first attempted (Li et al., 2014). Pan-genomes of key crop species, such as rice, bread wheat, maize, barley, cotton, tomato, potato, citrus, and oilseed rape, are now available (Gao et al., 2019; Shi et al., 2023). The assembly of these genomes has revealed that huge variations are present even in closely related crops. Thus, there is an increasing need for using pan-genome for mapping, GWAS, gene expression quantification and variant identification instead of single reference genome (Tian et al., 2019). This new development has changed molecular breeding approaches. The adoption of the pan-genome reference will also allow the inclusion of variants beyond SNPs in GWAS.

2.3.5 Advances in DNA Methylation Studies

The commonly used method for studying histone modification is the sodium bisulfite method in which the unmethylated cytosines are converted to uracil to convert the epigenetic code to genetic code, which can be detected by microarray. However, the method has limited hybridization specificity. The NGS methods can be linked to the classical bisulfite method to improve detection. One method known as Reduced Representation Bisulfite Sequencing (RRBS) combines these two for specific detection. The whole-genome bisulfite sequencing, MethylC-seq involving NGS of BS treated DNA makes the detection high throughput. Alternatively, nanopore sequencing to sequence mC directly, without BS treatment can increase (Branton et al., 2008) coverage. Another commonly used method for DNA methylation detection is the use of methylation-dependent restriction enzymes (MDRE) like HpaII and MspI, which recognize and digest the methylated DNA. Recently, methods such as microarray-based methylation assessment of single samples (MMASS), comprehensive high-throughput array for relative methylation (CHARM), and NGS sequencing of DNA enriched for CpG-containing regions (Methyl-seq) have been used for better detection results (Laird, 2010).

2.3.6 Advances in Histone Modification and Chromatin Studies

The affinity-based tool, i.e., chromatin immunoprecipitation (ChIP), is used to detect histone modifications in the genome sequences. In this method, histone is allowed to interact with targeted antibodies and recovered by heating DNA–protein complexes. It is detected by quantitative PCR or NGS (ChIP-seq) or microarray (ChIP-chip). The modifications of this method include MeDIP-chip and MeDIP-seq (Methylated DNA Immunoprecipitation), a monoclonal antibody-based technique which is used against methylated cytosine to enrich single-strand methylated DNA and MDB-seq in which high-affinity binding ability of a Methyl-CpG Binding Domain (MBD) protein complex to double-strand methylated DNA is used for detection. High-throughput DNAse I hypersensitivity assay (DNAseI-seq aka DHS-seq), formaldehyde-assisted isolation of regulatory elements followed by sequencing (FAIRE–seq) and Sono-seq are used for identification of low nucleosomal content and open chromatin structure. The long-range chromosomal interaction is identified with chromosomal conformation capture (3C), 3C on chip (4C), 3C carbon copy (5C) and coupled with NGS using Hi-Cand ChIA-PET. Nucleosome positioning and remodeling are studied with CATCH-IT and MNase-seq while haploChIP identifies allele-specific chromatin profiles, including SNPS that affect gene expression (Ballereau et al., 2013).

2.3.7 Genome Editing and Epigenomics

The possibility of genome editing tools to modify transcriptional regulation, histone modification, DNA methylation, and non-coding RNA has been explored (Qi et al., 2023). One methodology known as CRISPRi targets transcriptional regulation by fusion of transcriptional repressor domain (TRD) to dCas9 protein without affecting DNA sequence. The method has been successfully used in common wheat, *N. Benthamiana* and *Nicotiana tabacum* (Zhou et al., 2022; Karlson et al., 2022 and Xu et al., 2023). CRISPR/dCas can be used to carry out specific epigenetic changes to the target site by deploying different epigenetic effector domains. One example is the targeting of CpG methylation by the fusion of dCas9 endonuclease with the DNMT3a catalytic domain. This system has been used in human embryonic kidney cells to increase the methylation level of specific promoters (Vojta et al., 2016). In *Arabidopsis*, the fusion of the catalytic domain of human demethylase ten-eleven translocation 1 (TET1cd) with an artificial zinc finger (ZF) has been used to promote the demethylation of the *FWA* promoter (Papikian et al., 2019). Similar strategies have been used in *Arabidopsis* for modulating flowering time, drought tolerance, and other responses (Hilton et al., 2015). The CRISPR/Cas system can be used for lncRNA targeting, knockout, knockdown, overexpression, and imaging.

2.4 ADVANCES IN TRANSCRIPTOMICS

The total set of RNAs produced by the genome from a particular cell, tissue, or organism at a certain developmental stage in response to a particular physiological condition or environmental cues is known as the transcriptome. Transcriptomics, on the other hand, is the study of the transcriptome, which is created in a particular cell or under specific circumstances at a certain time (Yang et al., 2021). The transcriptome is made up of both coding (mRNA) and non-coding (ncRNA) RNAs, such as rRNA, tRNA, siRNA, miRNA, snoRNA, snRNA, and piwi-interacting RNA. Charles Auffray coined the word "transcriptome" in 1953 (McGettigan, 2013). Transcriptomic research enhances breeding selection and cultivation practices by identifying transcripts, transcript abundance, gene structure, expression profiles, functional genes, and regulatory mechanisms in plants (Guo et al., 2021). The use of transcriptomics in plant biology is significant because it reveals the roles of genes, their connections, and the pathways involved in plant growth, disease resistance, and stress responses. Additionally, they aid in the discovery of secondary metabolites, genetic diversity, and the mechanisms underlying epigenetic alterations.

Transcriptomics can be studied under three categories: first-generation sequencing, second-generation sequencing, and third-generation sequencing, which is frequently used in the fields of genetics, plant breeding, developmental biology, molecular biology, and biomedical research (Figure 2.1). Transcriptomics offers insights into the gene expression patterns of an organism. Transcriptomic data are generated and interpreted using a variety of experimental approaches, data analytic strategies, and visualization tools. Real-time quantitative polymerase chain reaction (RT-qPCR), microarray, and RNA sequencing (RNA-seq) are examples of experimental procedures. The data obtained from transcriptomics experiments is processed and analyzed using data analysis techniques. These include tools for quality assurance, alignment, and quantification, including Bowtie, HISAT2, STAR, GMAP, Salmon, StringTie2, and Salmon, for aligning RNA-seq reads to a reference genome, and eXpress, RSEM, for measuring gene expression levels. Differential expression analysis tools, such as DESeq2, edgeR, and limma+voom, are some additional tools for analysis (Chung et al., 2021). These techniques are used to find genes that are expressed differently in various datasets. Heatmaps, volcano plots, and principle component analysis (PCA) plots are some of the visualization tools used to analyze and visualize transcriptomics data. These tools help to spot patterns and linkages in the data. During the early 1980s, first-generation sequencing relied on total RNA isolation from certain tissues or cells, cDNA library creation, and the sequencing of arbitrary individual transcripts using the low-throughput Sanger sequencing method, which was referred to as expressed sequence tags (ESTs). The sequence of the organism is not necessary to use ESTs. Following that, work on transcript quantification was carried out leveraging northern blotting,

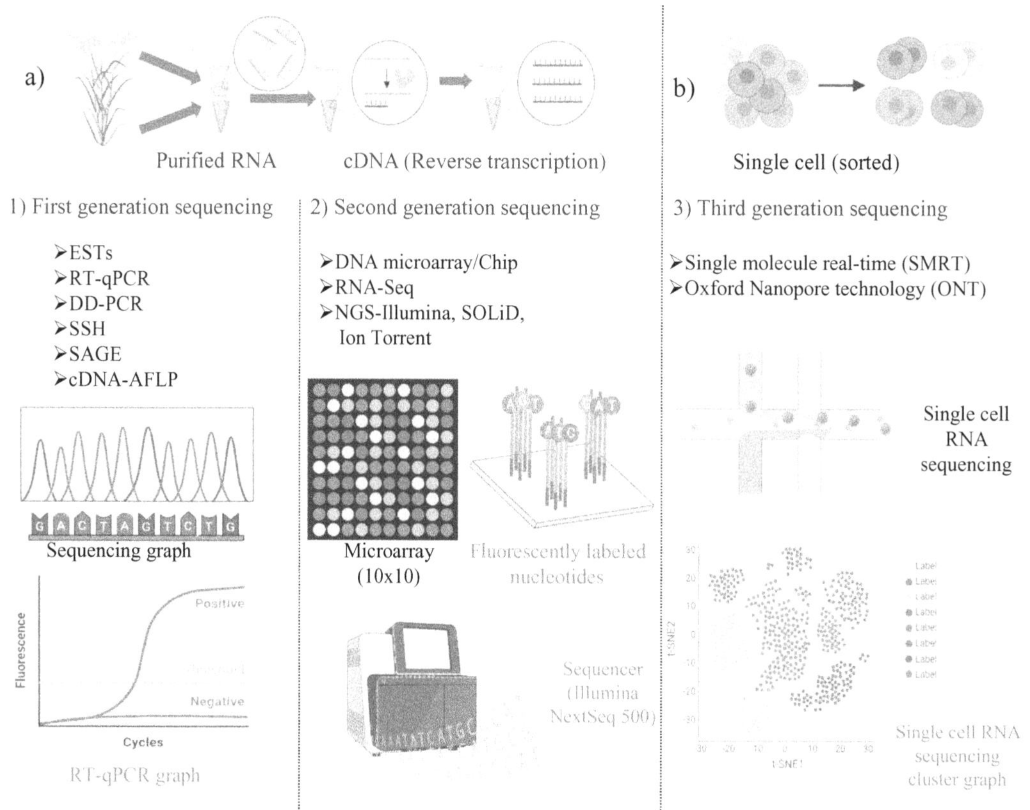

FIGURE 2.1 Different sequencing technologies used for transcriptomic studies. Raw material for first and second-generation sequencing are the (a) complementary DNA sequences, whereas for the third-generation sequencing (b) single cells are used.

nylon membrane arrays, and reverse transcriptase quantitative PCR (RT-qPCR). Other techniques included differential display-PCR (DD-PCR), cDNAs-AFLP, suppression subtractive hybridization (SSH), and serial analysis of gene expression (SAGE). SSH, SAGE, and DD-PCR all depended on Sanger sequencing to determine the cDNA sequence. However, these techniques were expensive, time-consuming, labor-intensive, had low throughput, and could only record a tiny portion of the transcriptome (Wang et al., 2019; Yang et al., 2021). The introduction of DNA microarray and RNA-seq, which were hybridization or sequencing-based approaches using NGS platforms, saw the emergence of second-generation sequencing of whole transcripts. When using the DNA microarray/Chip approach, fluorescently tagged transcripts (ds-cDNA) are hybridized to an array of known short nucleotide oligomers, or probes, that are arranged in an organized fashion on a solid substrate. A reference genome or previous sequencing data is required for the generation of probes (Wang et al., 2019). Sequence data was frequently provided by EST libraries for early microarray designs. Transcriptomics microarrays can be either high-density short probe arrays or low-density spotted arrays. For instance, the NimbleGen Assay uses 45–85 mer probes while the AffimetrixGeneChip array has 25 mer probes.

In massively parallel signature sequencing (MPSS), a flow cell is made up of millions of DNA-tagged microbeads that are aligned and examined using a fluorescence-based sequencing technique (Bhattacharjee, 2021). MPSS is a mix of microarray technology and sequence sampling. The direct sequencing of RNAs is possible using RNA-Seq, a high-throughput, high-sensitivity, and high-resolution approach, without the need for probes. It distinguishes and measures the expression of isoforms and unidentified transcripts, and it has a high level of data repeatability. When compared to microarrays (microgram amount), RNA-seq requires substantially less RNA input (nanogram quantity). To more accurately characterize the transcriptome, RNA-seq uses NGS techniques (Mahmood et al., 2022).

NGS relies on synthesis- or ligation-based sequencing rather than Sanger sequencing. To examine plant–virus interactions, RNA-seq variants such as sRNA-seq and Degradome-seq have been created (Zanardo et al., 2019). The read-length efficiency of NGS technologies has increased over time. NGS is one of the most popular technologies for transcriptomes because it is low cost, quick, high throughput, and deep coverage (Guo et al., 2021). The mRNA and small RNA in the transcriptome are predominantly sequenced and analyzed using NGS. NGS platforms like the Roche 454, Applied Biosynthesis SOLiD, Ion Torrent/PGM/Chef systems, IlluminSolexa, and PacBio RSII system are some of the more popular ones. SOLiD employs sequencing by ligation, whereas other methods use sequencing by synthesis. Pyrosequencing is the foundation of the Roche 454 platform, while the Solexa platform is built on the notion of sequencing by synthesis, which involves randomly attaching DNA fragments to the flow cell (Guo et al., 2021). The PacBio technology is the only one that uses single molecules as sequencing templates. Both single-end and paired-end sequencing options are available for the transcript molecules (reads) created for these NGS platforms. For gene annotation and the finding of transcript isoforms, paired-end sequencing generates more reliable alignments and/or assemblies. These technologies yield reads with a range of lengths, including 75 bp (paired-end), 300 bp (overlapping paired-end), 400 bp (bidirectional), 700 bp (paired-end), and as much as 8,500 bp, produced by the SOLiD, Illumina, Ion Torrent, 454 and PacBio systems, respectively. Additionally, depending on the platforms, the number of reads per NGS run varies from 1 million to 5,000 million (Jiang et al., 2015; Zanardo et al., 2019). Using software tools like Trimmomatic or Cutadapt, the raw readings are trimmed and then subjected to quality control using FASTQC, FASTX-toolkit, or NGS QC. Then, these reads are aligned using either de novo assembly or assembly based on reference sequences. Trinity, SOAPdenovo-Trans, rnaSPAdes, Trans-ABYSS, and Oases are tools for de novo assembly, and Cufflinks, StringTie, CLASS2, Ryuto, Transcomb, Scripture, Strawberry, Bayessembler, and Scallop are tools for genome sequence-based assembly (Chung et al., 2021). Tools like limma, EdgeR, DESeq2 edgeR, Cuffdiff/Cuffdiff2, DEGseq, baySeq, SAMseq, sleuth, and NOIseq, which are useful for gene annotation and transcript isoform finding, are used to further analyze differentially expressed genes (Liu et al., 2021). The two main tools used for gene annotation are the Gene Ontology (GO) function and the Kyoto Encyclopaedia of Genes and Genomes (KEGG) function. Although NGS has high accuracy, its biggest drawback is that it does not consistently cover the entire gene area, which leads to bias in gene length and transcript composition (Guo et al., 2021).

The two main third-generation sequencing technologies are Oxford Nanopore technology (ONT) and single-molecule real-time (SMRT) sequencing technology from Pacific Biosciences (PacBio) (Mikheenko et al., 2022). They are also referred to as long-read sequencing approaches because they deliver high-quality long-read sequencing data that can span whole transcripts or even full-length isoforms. These can reliably assess transcript isoform expression, unravel complex gene architectures, and locate instances of alternative splicing. The detection of alternative splicing (AS), alternative polyadenylation, and alternative transcription start sites and improvement in genome assemblies in repetitive sections have been made possible by these methods (Daszkowska-Golec et al., 2023). The nanopore single-molecule sequencing technology is a cutting-edge SMRT sequencing technique that primarily employs changes in electrical impulses to estimate the base composition. The SMRT sequencing technology is based on the principle of sequencing by synthesis. These sequencing techniques are favorable due to their high throughput, extended read length, and low cost (Guo et al., 2021). Single-cell RNA sequencing (scRNA-seq) and spatially resolved transcriptomics (SRT) are two further cutting-edge methods. Tang et al. (2009) conducted the first single-cell transcriptome study in mice. While scRNA-seq cannot offer spatial information, it can enable the investigation of gene expression in individual cells, which can uncover previously unnoticed heterogeneity in cell populations. Additionally, scRNA-seq offers insights into cell differentiation and development and aids in the identification of unusual cell types. Because of variables including cell size and cell cycle stage, some cell types may significantly vary during scRNA-seq, which can lead to less accurate results (Du et al., 2023). Single-cell RNA-seq has the distinct capacity to examine and take advantage of regulatory processes in complex tissues based on single-cell behavior. There are other scRNA-seq methodologies, including SMART-seq and Drop-Seq (Rich-Griffin et al., 2020). Numerous high-throughput methods for analyzing in situ gene expression, including DbiT-seq (deterministic barcoding in tissue-sequencing), Slide-seq, Stereo-seq (spatial enhanced resolution omics-sequencing), and 10x Genomics Visium platform, offer fresh perspectives on the identification of cell types and the cell-type specificity of complex traits. The temporal and spatial reconstruction of cell state and cell type in tissue is made possible by scRNA-seq when used in conjunction with imaging technology (Kaur et al., 2023). Likewise, SRT identifies the patterns of gene expression in particular areas of tissues and organs by combining spatial information with RNA sequencing. Understanding how cells and tissues are arranged spatially can be aided by spatial transcriptomics. The 10x Genomics Visium, NanostringGeoMx DSP, and VizgenMERSCOPE are frequently employed for spatial transcriptomics (Du et al., 2023). SRT supports the analysis of cell-cell interactions, the identification of the cell-type architecture of tissues, and the observation of molecular interactions between tissue constituents. It enables the spatial resolution acquisition of gene expression data from intact tissue sections in the original physiological context, provides biological insights into the microenvironment and intercellular communication, and aids in the clarification of molecular interactions between tissue constituents (Tian et al., 2023). The two types of SRT technologies are imaging-based and sequencing-based approaches, both of which can capture the target molecule objectively and without the need for any prior knowledge. The fields of single-cell, spatial, and sequencing transcriptomics, as well as many other bioinformatics techniques, are all developing quickly. Transcriptomics in conjunction with omic technologies, such as genomics, proteomics, and metabolomics, can offer a thorough understanding of gene expression and its control in the context of plant biology. A deeper knowledge of gene functions could be achieved by fusing transcriptomic data with other forms of data.

2.5 ADVANCES IN PROTEOMICS

In a broad sense, proteomics is the quantitative and/or qualitative study of all the proteins that are expressed in any given cell, tissue, organ, or organism in spatial and temporal bases (Mehta et al., 2019). The proteome (sum of all proteins in an organism) is dynamic, and it is expected to change in different tissues under varied conditions. Proteomics complements both transcriptome and metabolomic studies, and its role in crop improvement is inevitable (Majumdar and Keller, 2021). In general, in a proteomics study, the proteins are isolated, quantified, and fractionated by methods like gel electrophoresis or liquid chromatography (LC). Later the separated proteins are enzymatically cleaved and the peptides are studied in mass spectrometry (MS). The results of the MS were analyzed in silico in databases for final protein identification and to deduce their functional roles in the respective organism (Yan et al., 2022).

2.5.1 Technological Interventions in Proteomics

One cannot imagine the field of proteomics without the first generation's gel-based electrophoresis techniques, especially two-dimensional gel electrophoresis (2DE), which was and is still in use. In the second generation, isotopic/isobaric labeling was used, followed by shotgun, gel, and label-free techniques in the third generation. In the current fourth-generation targeted, mass-western, and many MS-based methods were used (Jorrin-Novo et al., 2019). For the identification of protein sequences, interactions, and post-translation modification (PTM) studies, MS is widely used (Yan et al., 2022). Each technique had its own advantages and limitations, but recently, chromatography techniques coupled with MS, i.e., liquid chromatography-tandem mass spectrometry (LC-MS/MS), have become the most sought-after technique in proteomic studies because of their sensitivity, accuracy, and wider proteome coverage (Mergner and Kuster, 2022). Another recent technique called isobaric tags for relative and absolute quantification systems (iTRAQ) was also capable of identifying around 4,000 proteins per sample.

2.5.2 Study of Sub-Cellular Proteomics and PTM

Sub-cellular components like stomata, chloroplast, plasma membrane, nucleus, ribosomes, mucilage substances, and their related proteins are generally studied using 2DE, LC-MS, and MS-based methods (Hochholdinger et al., 2018, Bhattacharya et al., 2020). Apart from these, other specialized techniques were also used to study individual sub-cellular components. Vacuum infiltration-based centrifugation was used to study the apoplastic fluids in the roots of cotton (Han et al., 2019). Proteins isolated from Arabidopsis seedlings were processed with two-phase partitioning combined with free-flow electrophoresis to study plasma membrane proteins (de Michele et al., 2016). Mitochondrial proteins of potato tubers were studied by centrifugation combined with one-dimensional blue native PAGE (polyacrylamide gel electrophoresis) (Senkler et al., 2017). PTMs like glycosylation, phosphorylation, ubiquitination, s-nitrosylation, lysine malonylation, lysine acetylation, in leaves were studied using the LC-MS method in various crops (Kausar et al., 2022). Among them, half of the PTMs were N-glycosylation type, which was mostly involved in signaling and molecular trafficking under abiotic stresses (Strasser et al., 2021). Other PTMs like phosphorylation and acetylation play an important regulatory role in the nitrogen fixation of legume crops (Marx et al., 2016).

2.5.3 Proteomics of the Plants under Various Conditions

Proteins are denoted as workhorses of a cell and differential proteins are expressed under various stress conditions in plants. A total of 29% of amino acids were reduced during various stresses in plants like Arabidopsis, which were likely to be used inside the plant system to develop alternative substrates and secondary metabolites that can alleviate the stress (Hildebrandt, 2018). Starch, lignin, fatty acid synthesis, reactive oxygen species (ROS) scavenging proteins, chaperons, photosynthesis, and signal transduction-related proteins were known to be differentially expressed under various biotic and abiotic stresses in different crop plants (Mergner and Kuster, 2022). There is also cross-talk between different stresses and compounds like phytohormones, plant growth regulators, and signaling molecules, which are known to involve multiple stress tolerance (Aftab and Roychoudhury, 2021).

2.5.4 Tools and Databases Used in Proteomics Studies

As mentioned earlier, many researchers use MS in their proteomics study and an application called PRIDE Inspector can be used to visualize and assess the raw data directly from MS (Li et al., 2019) and the peptide retention time can be predicted in the DeepRT tool, which functions based on deep learning methods (Ma et al., 2017). Ionbot tool uses machine learning for peptide identification (Degroeve et al., 2021) and AlphaViz can validate proteins and peptides at the raw data level (Voytik et al., 2022). Another tool, pathway matrix, can be used to visualize the binary relations between proteins (Dang et al., 2015). The data generated through these tools is enormous, which emphasizes the need to deposit them in databases so that they can be studied, reviewed, and retrieved by other researchers. The Proteomics Identifications (PRIDE) database (https://www.ebi.ac.uk/pride; accessed on 09 July 2023) is the largest repository of proteomics data obtained from MS methods (Perez-Riverol et al., 2022). PSP database provides information on a plant stress-related protein (Kumar et al., 2014). In the PlantMP database, we can find single polypeptides that can perform more than two biological functions (Su et al., 2019). qPTM plants database can be used to find details of PTMs identified in plants to date (Xue et al., 2022). The physiochemical, and structural properties of plant proteomes can be obtained from Plant-PrAS data (Kurotani et al., 2015). Apart from this, there are many other tools and databases which were well discussed by Misra (2018).

2.5.5 Proteomics Studies in Genome-Edited Crops

CRISPR/Cas9-based gene editing was employed in rice to generate targeted mutagenesis for Pi21 gene (*Magnoporthae oryzae*) and the proteomic profiling was studied using iTRAQ-based proteomic analysis. The differentially expressed proteins (DEP) were related to secondary metabolites biosynthesis, metabolic, ribosomal, and carbon metabolism pathway (Nawaz et al., 2020). In a similar study, targeted mutagenesis for GS3 (Grain size) gene was performed in rice, and it increased the grain length. The iTRAQ method identified 30 DEPs, which were cysteine proteinase inhibitors and ubiquitin-related proteins (Usman et al., 2021). Gene editing by CRISPR/Cas9 and RNAi was used to silence selected amylase/trypsin-inhibitors (ATIs) genes in wheat. Different methods were employed to find the content of ATIs in selected lines. Labeled isotope-based stable isotope dilution assay (SIDA) and iTRAQ were superior to the other methods, as they were able to detect all the ATIs even in small doses (Geisslitz et al., 2022). CRISPR/Cas9 mediated mutation in GmVPS8a (Vacuolar Protein Sorting) gene resulted in compact plant architecture in soybeans and tobacco. Transcriptome and proteomic data of the mutant plants were analyzed together, which identified that dysfunction of this gene affects sugar and lipid metabolism, transport, and auxin signal transduction pathways (Kong et al., 2023).

In the current era, technologies like LC-MS/MS, iTRAQ and SIDA are used in proteomic studies, and various tools and databases are available, which can be utilized to identify the protein and its function. In spite of the advent of new technologies and tools, there are still many proteins which are poorly understood and some are underutilized. Most of the studies conducted by researchers focus mainly on well-known proteins whose functions and pathways have already been reported. Hence, there is ample scope for finding new or understudied proteins and deducing their molecular function. An understudied protein initiative can also be taken to annotate the uncharacterized proteins, which will lay the groundwork for futuristic researchers (Kustatscher et al., 2022).

2.6 ADVANCEMENTS IN GENOME EDITING OF CROP PLANTS

Genome editing implies the site-specific addition, deletion, and modification of the genome. The editing services of the genome provide in vivo modification of the target site by transforming the set of nucleotides and proteins exogenously. Recent decades have provided several genome editing tools with different applications in crop plants as follows: Mega nucleases (MegNs), zinc finger nucleases (ZFNs), transcription activator-like effector nucleases (TALENs), CRISPR/Cas9 and CRISPR/Cpf1 system (Mohanta et al., 2017; Khalil, 2020; Safari et al., 2019). The comparisons of different tools are given in Table 2.2. The major apprehensions with the genome editing tools are the target specificity, accuracy, off-target effect, removal of residual components after target editing, and feasibility of customization of effector molecules for performing genome editing experiments. Thus, a gradual improvement is required for genome editing tools to achieve the convenient mode with technically feasible site-specific modification in the genome. MegNs are a homing endonuclease having two domains: DNA binding and DNA cleavage domain. DNA binding domain recognizes the longer DNA sequences of 14–40 bp and the DNA cleavage domain cleaves double-stranded DNA. The MegNs are the first endonucleases that initiate the avenue for genome editing technology (Silva et al., 2011). It acts in homodimer form to cleave the target. As the recognition motif of the DNA binding domain recognizes longer sequences, prior protein engineering is always required to recognize the novel target in the genome (Khalil, 2020). The extensive protein engineering adds time, cost, and labor to the genome editing tools. Thus, the routine application of MegNs as a genome editing tool is limited. ZFNs are an engineered genome editing system having a DNA binding domain from ZF and a DNA cleavage domain from *Fok*I (Carroll, 2011). ZF recognition sites are 3–4 bp in length. Each ZF protein contains two-finger modules, each of which recognizes 3 bp. Thus, a set of three zinc finger proteins (ZFPs) recognizes 18 bp. Cleavage requires two sets of ZFN that make heterodimers of FokI at the site of cleavage and produce a double-strand break (Khalil, 2020). The ZFN may be a preferred choice over MegNs, but it has certain shortfalls. The protein engineering is required to add different finger modules to recognize the longer sequence to increase the specificity. The availability of finger module for each of the 64 variable nucleotide triplet is limited and requires extensive customization. The high risk of an off-target effect is also associated with ZFNs. In addition to this, TALEN is a genome editing tool that works on the same principle of DNA–protein interactions as MegNs and ZFNs, but it has high specificity, is cheaper, safer, and negligible off-target effect (Zhang et al., 2011). As each module of the TALEN protein binds to a single nucleotide, it provides ease of customization for engineering DNA binding motifs (Khan, 2019). A major breakthrough came in the genome editing tools with the identification of RNA-DNA-based CRISPR/Cas9 system (Cong et al., 2013; Mali et al., 2013). The source of the system was Streptococcus pyogenes, which works for adaptive immunity against viruses and then continuously evolved to be identified as a potent genome editing system (Kumari et al., 2021). The target customization is highly feasible for DNA-RNA-based tools in comparison to the DNA – protein-based genome editing tools. The target region is recognized by the guide RNA, which is the duplex of crRNA and tracrRNA. The double-stranded cleavage of the target is implied with the Cas 9 nucleases that cleave the protospacer adjacent motif (PAM) region present 12 bp downstream of seed sequences (Bortesi and Fischer, 2015). The ease of customization of guide RNA for target recognition, multiple targeting, and high cleavage efficiency make the CRISPR/Cas9 a potent technique for genome editing, but off–the target is a major apprehension. The specificity of cleavage depends on the designing of guide RNA and the availability of PAM sequences. The guide RNA can tolerate the mismatch of up to 8bp (Anders et al., 2014). The length of PAM sequences is three bp (5′ –NGG– 3′), which may often be present in the vicinity of mismatched guide RNA regions. Thus, the chances of off-target with CRISPR/Cas9 are high in comparison to the ZFN and TALEN. The off-target cleavage leads to malignancies and death of the organism that is being edited.

The advancement in genome editing tools has been achieved to reduce the off-target cleavage and improve the specificity and efficiency of the existing system. The various improvements of existing technologies called CRISPR/Cas 9 variants system and novel systems based on other than Cas9 are being tried by researchers. CRISPR/Cas9 variant systems have been implied to combat the off-target cleavage by CRISPR/Cas 9 system are: double-nicking method, FokI-dCas9 fusion protein method, the truncatedsgRNA method, and CRISPR-SKIP. The double-nicking method is used to facilitate the homology-directed repair (HDR) using mutant Cas9 as Cas9 nickase (nCas9). Here, a pair of CRISPR/Cas9 – nickase is used for nick at two strands to produce staggered fragments and promote the HDR system. It was observed that hundred-fold more specificity was observed with mutated Cas9 nickase in the double-nicking method compared to wild-type Ca9 (Ran et al., 2013). The off-target site cleavage is often associated with the mismatch pairing of guide RNA, in addition to the on-target site pairing. The co-administered dead-truncated guide RNA with active Cas9 guide RNA reduced off-target cleavage and mutation. The dead guide RNA is a perfect complementarity to the off-target site that guides the Cas9 binding to the site but not cleavage. The perfect complementarity of dead RNA to the off-target site out-competes the active guide RNA as per their imperfect complementarity (Rose et al., 2020). The efficiency of off-target binding can be improvised by 40%. The cleavage specificity has been also increasing with the FokI-dCas9 fusion protein method. In this method, a pair of FokI-dCas9 fusion proteins target the region and is cleaved by the FokI heterodimer. The specificity is enhanced as a pair of guide RNAs needed for localizing the target region in comparison to a single guide RNA-Cas9 system (Saifaldeen et al., 2020). The malignancies because of a double-strand break at off-target have been reduced with CRISPR-SKIP approaches. It includes a base editing methodology instead of a double-strand break by implying the CRISPR/Cas9 system (Gapinske et al., 2018). Several bioinformatics tools are frequently utilized in genome editing processes; some of the important tools are listed in Table 2.2.

Although the above-mentioned modified strategies of CRISPR/Cas9-based systems have been implied, they are still associated with certain shortfalls. Thus, to overcome the obstacles

TABLE 2.2
Principles and Apprehensions of Different Genome Editing Tools

Genome Editing Tools	Principle	Target	Apprehensions	References
Mega nucleases (MegNs)	DNA – Protein interactions	DNA	• Engineering of complex protein • Mismatch and off-target rate is high	Silva et al. (2011)
Zinc Finger Nucleases (ZFN)	do-	do-	• Complex design for each triplet recognition motif • Off-target	Carroll (2011)
Transcription Activator-Like Effector Nucleases (TALENs)	do-	do-	• Difficult to customize because it requires multiple identical repeat sequences	Zhang et al. (2011)
CRISPR/Cas 9 system	do-	do-	• The presence of redundancy in PAM site affects target specificity • Require transacting RNA	Cong et al. (2013)
CRISPR/Cas9 variant system	do-	DNA and RNA	• Frameshift mutation can be achieved mostly	Khalil (2020)
CRISPR/Cpf1	do-	do-	• High target efficiency, and ease of customization but need similar to Cas9 variant system to enhance specificity	Zetsche et al. (2015)

TABLE 2.3
Comparative Studies of Cas9 and Cpf1 for CRISPR-Based Genome Editing Methodology

Features	Cas9	Cpf1
Source	Streptococcuspyogenes	Prevotellafrancisellal
Classification	Class II and type II system	Class II and type V system
Nucleases domain	HNH and RuvC-like domain	RuvC-like domain only
PAM sequences	• 3 bp • G – rich (5′–NGG– 3′)	• 4 bp • T – rich (5′–TTTN– 3′)
Guide RNA	• crRNA • tracrRNA	• crRNA only
Length of guide RNA	75 bp	44 bp
Cleavage	• The proximal end of PAM • Blunt	• The distal end of PAM • Staggered

of off-target effect, an identification of a novel system is required. One such identification is the CRISPR/Cpf1 from *Prevotella francisella*l (Zetsche et al., 2015; Bin Moon et al., 2018). The comparative study of Cas9 and Cpf1 shows that Cpf1 is highly efficient in target recognition with minimum off-target effect (Table 2.3). Cpf1 cleavage does not require a crRNA-tracrRNA hybrid which affects the ex-situ binding of Cas9 and ultimately affects the cleavage efficiency. While a single crRNA required for Cpf1 that introduced the staggered cutting in DNA which leads to HDR reduces the apprehensions of off-target double-stranded cleavage. Thus, Cpf1-based CRISPR technology has high efficiency with limited risk and is being used as a plant genome editing tool (Alok et al., 2020).

2.7 BIOINFORMATICS TOOLS FOR FUNCTIONAL GENOME ANALYSIS AND GENOME EDITING IN CROP PLANTS

Functional genomics is a complex field that utilizes genome-wide approaches to understand the function of the entire genome. To decipher this difficult puzzle, the cell is regarded as a system, and the relationship between different entities is studied. The approach aims to understand the function of individual genes, pathways, networks, and the genome on the whole. This creates a huge amount of information that has to be simultaneously processed to gain meaningful information. This cumbersome task requires the aid of many bioinformatic tools to ease the process. There are uncountable bioinformatics aids that can be employed in functional genomic studies.

Robust and reliable genetic modification methods like CRISPR/Cas9 are excellent examples of genome editing technologies. These technologies have proven to be incredibly powerful tools, both in the field of plant breeding and in basic science. Because they do away with the limitations of conventional breeding methods, genome editing technologies (such as ZFN, TALEN, CRISPR/Cas9, etc.) have received a lot of interest (Matres et al., 2021). The time it takes to acquire plants with desired features for the creation of new crop varieties is significantly reduced by these approaches, which enable accurate and efficient targeted genome changes.

Small guide RNA and sequence-specific nucleases are the essential elements of a CRISPR-based gene editing strategy that produces precise alteration. Off-target effects and on-target efficiency remain two major challenges for the CRISPR/Cas system, which is still under development (Liu et al., 2020). In order to solve these problems, small guide RNA optimization using efficient computer approaches is crucial (Hassan et al., 2021). The target sequence's nucleotide composition is one of the main elements influencing gRNA efficiency. For greater effectiveness, the PAM sequence and its neighboring nucleotide play a vital role (Liu et al., 2020).

Thymines are not favored within four nucleotides up/downstream of the PAM sequence, however, guanines are preferred at the first and second nucleotide positions before the PAM region. Furthermore, sequences downstream of PAMs can alter gRNA efficiency, whereas sequences upstream of PAMs have no discernible influence (Doench et al., 2014). Cytosine is favored at the cleavage site, and GC content enhances the high efficiency of gRNA downstream of the PAM sequence. There are many efficiency prediction models available that were created using this crucial data. Based on these models, a variety of tools have been created to design gRNA using alignment-based, hypothesis-driven, or learning-based models (Konstantakos et al., 2022). Alignment-based models do not perform as well as hypothesis-driven and learning model-based technologies.

To predict the gRNA with high target efficiency, a number of tools have been created, namely, CHOPCHOP, E-CRISP, CRISPR-FOCUS, CRISPETa, PROTOSPACER, CLD, and CRISPOR. A web-based bioinformatics tool called WheatCRISPR is frequently used to create target-specific gRNA in wheat (Cram et al., 2019). The first open-source bioinformatics tool, CROPSR, also assists in designing genome-wide guide RNA for CRISPR-based genome editing at a fast rate of speed, which lessens the difficulties of complicated crop genomes (Müller Paul et al., 2022).

2.8 CONCLUSION

For several years, functional genomics analysis has been performed utilizing a variety of techniques and instruments. However, only lately have high-throughput methods varying from traditional real-time polymerase chain reactions to more complex systems like NGS or MS have undergone significant progress. Furthermore, laboratory research alone is insufficient for reliable bioinformatics analysis and sound scientific results. Genomics, epigenomics, proteomics, and interactomics are just a few of the research areas that may be precisely and thoroughly analyzed using these methodologies to fill the knowledge gaps about dynamic biological processes at the cellular and organismal levels.

REFERENCES

Aftab T, Roychoudhury A. 2021. Crosstalk among plant growth regulators and signaling molecules during biotic and abiotic stresses: Molecular responses and signaling pathways. *Plant Cell Reports*, 40: 2017–2019.

Alok A, Sandhya D, Jogam P, Rodrigues V, Bhati KK, Sharma H, Kumar J. 2020. The rise of the CRISPR/Cpf1 system for efficient genome editing in plants. *Frontiers in Plant Science*. 11: 264. doi: 10.3389/fpls.2020.00264

Anders C, Niewoehner O, Duerst A, Jinek M. 2014. Structural basis of PAM dependent target DNA recognition by the Cas9 endonuclease. *Nature*, 513: 569–573.

Andolfatto P, Davison D, Erezyilmaz D, Hu TT, Mast J, Sunayama-Morita T, Stern DL. 2011. Multiplexed shotgun genotyping for rapid and efficient genetic mapping. *Genome Research*. 21(4): 610–617.

Balcke GU, Handrick V, Bergau N, Fichtner M, Henning A, Stellmach H, ... Frolov A. 2012. An UPLC-MS/MS method for highly sensitive high-throughput analysis of phytohormones in plant tissues. *Plant Methods*. 8(1): 1–11.

Ballereau S, Glaab E, Kolodkin A, Chaiboonchoe A, Biryukov M, Vlassis N, ... Auffray C. 2013. Functional Genomics, Proteomics, Metabolomics and Bioinformatics for Systems Biology. In: Prokop A, Csukás B. (eds) Systems Biology. Springer, Dordrecht. https://doi.org/10.1007/978-94-007-6803-1_1.

Bhatta M, Morgounov A, Belamkar V, Wegulo SN, Dababat AA, Erginbas-Orakci G, ... Demir L. 2019. Genome-wide association study for multiple biotic stress resistance in synthetic hexaploid wheat. *International Journal of Molecular Sciences*. 20(15): 3667.

Bhattacharjee S. 2021. Advances of transcriptomics in crop improvement: A Review. *Journal of Emerging Technologies and Innovative Research*. 8(6): 22–37

Bhattacharya O, Ortiz I, Walling LL. 2020. Methodology: An optimized, high-yield tomato leaf chloroplast isolation and stroma extraction protocol for proteomics analyses and identification of chloroplast co-localizing proteins. *Plant Methods*. 16: 131.

Bin Moon S, Lee JM, Kang JG, Lee NE, Ha DI, Kim DY, … Kim YS. 2018. Highly efficient genome editing by CRISPR-Cpf1 using CRISPR RNA with a uridinylate-rich 3′-overhang. *Nature Communications*. 9: 3651.

Bortesi L, Fischer R. 2015. The CRISPR/Cas9 system for plant genome editing and beyond. *Biotechnology Advances*. 33: 41–52.

Branton D, Deamer DW, Marziali A, Bayley H, Benner SA, Butler T, … Schloss JA. 2008. The potential and challenges of nanopore sequencing. *Nature Biotechnology*. 26(10): 1146–1153.

Carroll D. 2011. Genome engineering with zinc-finger nucleases. *Genetics*. 188: 773–782.

Chen C, Shang X, Sun M, Tang S, Khan A, Zhang D, … Xie Q. 2022. Comparative transcriptome analysis of two sweet sorghum genotypes with different salt tolerance abilities to reveal the mechanism of salt tolerance. *International Journal of Molecular Sciences*. 23(4): 2272.

Chung M, Bruno VM, Rasko DA, Cuomo CA, Muñoz JF, Livny J, … Dunning Hotopp JC. 2021. Best practices on the differential expression analysis of multi-species RNA-seq. *Genome Biology*. 22(1): 1–23.

Cong L, Ran FA, Cox D, Lin S, Barretto R, Habib N, … Zhang F. 2013. Multiplex genome engineering using CRISPR/Cas systems. *Science*. 339: 819–823.

Cram D, Kulkarni M, Buchwaldt M, Rajagopalan N, Bhowmik P, Rozwadowski K, ... Kagale S. 2019. Wheat CRISPR: a web-based guide RNA design tool for CRISPR/Cas9-mediated genome editing in wheat. BMC Plant Biology. 19: 1–8.

Dang TN, Murray P, Forbes AG. 2015. PathwayMatrix: Visualizing binary relationships between proteins in biological pathways. BMC proceedings. 9(6):1–13.

Daszkowska-Golec A, Mascher M, Zhang R. 2023. Applications of long-read sequencing in plant genomics and transcriptomics. *Frontiers in Plant Science*. 14: 1141429.

de Michele R, McFarlane HE, Parsons HT, Meents MJ, Lao JM, FernandezNino SMG, … Heazlewood JL. 2016. Free-fow electrophoresis of plasma membrane vesicles enriched by two-phase partitioning enhances the quality of the proteome from *Arabidopsis* seedlings. *Journal of Proteome Research*. 15(3): 900–13.

Degroeve S, Gabriels R, Velghe K, Bouwmeester R, Tichshenko N, Martens L. 2021. Ionbot: A novel, innovative and sensitive machine learning approach to LC-MS/MS peptide identification. *BioRxiv*, 2021-07.

Doench JG, Hartenian E, Graham DB, Tothova Z, Hegde M, Smith I, ... & Root DE. 2014. Rational design of highly active sgRNAs for CRISPR-Cas9–mediated gene inactivation. *Nature Biotechnology*. 32(12): 1262–1267.

Du J, Yang YC, An ZJ, Zhang MH, Fu XH, Huang ZF, Yuan Y, Hou, J. 2023. Advances in spatial transcriptomics and related data analysis strategies. *Journal of Translational Medicine*. 21(1): 1–21.

Elshire RJ, Glaubitz JC, Sun Q, Poland JA, Kawamoto K, Buckler ES, Mitchell SE. 2011. A robust, simple genotyping-by-sequencing (GBS) approach for high diversity species. *PLoS One*. 6(5): e19379.

Gao L, Gonda I, Sun H, Ma Q, Bao K, Tieman DM, … Fei Z. 2019. The tomato pan-genome uncovers new genes and a rare allele regulating fruit flavor. *Nature Genetics*. 51(6): 1044–1051.

Gapinske M, Luu A, Winter J, Woods WS, Kostan KA, Shiva N, Song JS, Perez-Pinera P. 2018. CRISPR-SKIP: Programmable gene splicing with single base editors. *Genome Biology*. 19(1): 107. doi:10.1186/s13059-018-1482-5.

Geisslitz S, Islam S, Buck L, Grunwald-Gruber C, Sestili F, Camerlengo F., … D'Amico S. 2022. Absolute and relative quantitation of amylase/trypsin-inhibitors by LC-MS/MS from wheat lines obtained by CRISPR-Cas9 and RNAi. *Frontiers in Plant Science*. 13: 974881.

Giacomello S, Salmen F, Terebieniec BK, Vickovic S, Navarro JF, Alexeyenko A, … Stahl PL. 2017. Spatially resolved transcriptome profiling in model plant species. *Nature Plants*. 3(6): 1–11.

Gnirke A, Melnikov A, Maguire J, Rogov P, LeProust EM, Brockman W, … Nusbaum C. 2009. Solution hybrid selection with ultra-long oligonucleotides for massively parallel targeted sequencing. *Nature Biotechnology*. 27(2): 182–189.

Guo J, Huang Z, Sun J, Cui X, Liu Y. 2021. Research progress and future development trends in medicinal plant transcriptomics. *Frontiers in Plant Science*. 12: 691838.

Han LB, Li YB, Wang FX, Wang WY, Liu J, Wu JH, … Xia GX. 2019. The cotton apoplastic protein CRR1 stabilizes chitinase 28 to facilitate defense against the fungal pathogen *Verticillium dahliae*. *The Plant Cell*. 31(2): 520–536.

Hassan MM, Chowdhury AK, Islam T. 2021. *In silico* analysis of gRNA secondary structure to predict its efficacy for plant genome editing. *CRISPR-Cas Methods*. 2: 15–22.

Hildebrandt TM. 2018. Synthesis versus degradation: Directions of amino acid metabolism during Arabidopsis abiotic stress response. *Plant Molecular Biology*. 98: 121–135.

Hilton IB, D'ippolito AM, Vockley CM, Thakore PI, Crawford GE, Reddy TE, Gersbach CA. 2015. Epigenome editing by a CRISPR-Cas9-based acetyltransferase activates genes from promoters and enhancers. *Nature Biotechnology*. 33(5): 510–517.

Hochholdinger F, Marcon C, Baldauf JA, Yu P, Frey FP. 2018. Proteomics of maize root development. *Frontiers in Plant Science*. 9: 143.

Holleman A, Cheok MH, denBoer ML, Yang W, Veerman AJ, Kazemier KM, ... Evans WE. 2004. Gene-expression patterns in drug-resistant acute lymphoblastic leukemia cells and response to treatment. *New England Journal of Medicine*. 351: 533–542.

Jiang Z, Zhou X, Li R, Michal JJ, Zhang S, Dodson MV, Zhang Z, Harland RM. 2015. Whole transcriptome analysis with sequencing: Methods, challenges and potential solutions. *Cellular and Molecular Life Sciences*. 72: 3425–3439.

Jorrin-Novo JV, Komatsu S, Sanchez-Lucas R, de Francisco LER. 2019. Gel electrophoresis-based plant proteomics: Past, present, and future. Happy 10th anniversary Journal of Proteomics! *Journal of Proteomics*. 198: 1–10.

Kamweru I, Anani BY, Beyene Y, Makumbi D, Adetimirin VO, Prasanna BM, Gowda M. 2022. Genomic analysis of resistance to fall armyworm (*Spodoptera frugiperda*) in CIMMYT maize lines. *Genes*. 13(2): 251

Karlson CKS, Mohd Noor SN, Khalid N, Tan BC. 2022. CRISPRi-mediated down-regulation of the cinnamate-4-hydroxylase (C4H) gene enhances the flavonoid biosynthesis in *Nicotiana tabacum*. *Biology*. 11(8): 1127.

Kaur H, Jha P, Ochatt SJ, Kumar V. 2023. Single-cell transcriptomics is revolutionizing the improvement of plant biotechnology research: Recent advances and future opportunities. *Critical Reviews in Biotechnology*. 11: 1–16.

Kausar R, Wang X, Komatsu S. 2022. Crop proteomics under abiotic stress: From data to insights. *Plants*. 11: 2877. doi:10.3390/plants11212877

Khalil AM. 2020. The genome editing revolution: Review. *Journal of Genetic Engineering and Biotechnology*. 18: 68. doi:10.1186/s43141-020-00078-y.

Khan SH. 2019. Genome-editing technologies: Concept, pros, and cons of various genome-editing techniques and bioethical concerns for clinical application. *Molecular Therapy-Nucleic Acids*. 16:326–334.

Kong K, Xu M, Xu Z, Lv W, Lv P, Begum N, ... Zhao T. 2023. Dysfunction of GmVPS8a causes compact plant architecture in soybean. *Plant Science*. 331: 111677.

Kress WJ, Soltis DE, Kersey PJ, Wegrzyn JL, Leebens-Mack JH, Gostel MR, ... Soltis PS. 2022. Green plant genomes: What we know in an era of rapidly expanding opportunities. *Proceedings of the National Academy of Sciences*. 119(4): e2115640118.

Konstantakos V, Nentidis A, Krithara A, Paliouras G. 2022. CRISPR–Cas9 gRNA efficiency prediction: an overview of predictive tools and the role of deep learning. *Nucleic Acids Research*. 50(7): 3616–3637.

Kudapa H, Garg V, Chitikineni A, Varshney RK. 2018. The RNA-Seq-based high resolution gene expression atlas of chickpea (*Cicer arietinum* L.) reveals dynamic spatio-temporal changes associated with growth and development. *Plant Cell & Environment*. 41(9): 2209–2225.

Kumar AS, Kumari HP, Sundararajan VS, Suravajhala P, Kanagasabai R, Kavi Kishor PB. 2014. PSPDB: Plant stress protein database. *Plant Molecular Biology Reporter*. 32: 940–942.

Kumari S, Singh SK, Sharma VK, Kumar R, Mathur M, Upadhyay TK, Prajapat RK. 2021. CRISPR-Cas: A continuously evolving technology. *Indian Journal of Agricultural Sciences*. 91(9): 10–15.

Kurotani A, Yamada Y, Shinozaki K, Kuroda Y, Sakurai T. 2015. Plant-PrAS: A database of physicochemical and structural properties and novel functional regions in plant proteomes. *Plant and Cell Physiology*. 56(1): e11.

Kustatscher G, Collins T, Gingras AC, Guo T, Hermjakob H, Ideker T, ... Rappsilber J. 2022. Understudied proteins: Opportunities and challenges for functional proteomics. *Nature Methods*. 19(7): 774–779.

Lai K, Lorenc MT, Edwards D. 2012. Genomic databases for crop improvement. *Agronomy*. 2(1): 62–73.

Laird PW. 2010. Principles and challenges of genome-wide DNA methylation analysis. *Nature Reviews Genetics*. 11(3): 191–203

Le Nguyen K, Grondin A, Courtois B, Gantet P. 2019. Next-generation sequencing accelerates crop gene discovery. *Trends in Plant Science*, 24(3): 263–274

Li K, Vaudel M, Zhang B, Ren Y, Wen B. 2019. PDV: An integrative proteomics data viewer. *Bioinformatics*. 35(7): 1249–1251.

Li YH, Zhou G, Ma J, Jiang W, Jin LG, Zhang Z, ... Qiu LJ. 2014. De novo assembly of soybean wild relatives for pan-genome analysis of diversity and agronomic traits. *Nature Biotechnology*. 32(10): 1045–1052.

Liu G, Zhang Y, Zhang T. 2020. Computational approaches for effective CRISPR guide RNA design and evaluation. *Computational and Structural Biotechnology Journal*. 18: 35–44.
Liu S, Wang Z, Zhu R, Wang F, Cheng Y, Liu Y. 2021. Three differential expression analysis methods for RNA sequencing: Limma, EdgeR, DESeq2. *Journal of Visualized Experiments*. 175: e62528.
Ma C, Zhu Z, Ye J, Yang J, Pei J, Xu S, … Liu, S. 2017. DeepRT: Deep learning for peptide retention time prediction in proteomics. arXiv preprint arXiv:1705.05368.
Mahmood U, Li X, Fan Y, Chang W, Niu Y, Li J, Qu C, Lu K. 2022. Multi-omics revolution to promote plant breeding efficiency. *Frontiers in Plant Science*. 13: 1062952.
Majumdar S, Keller AA. 2021. Omics to address the opportunities and challenges of nanotechnology in agriculture. *Critical Reviews in Environmental Science and Technology*. 51(22): 2595–2636.
Mali P, Yang L, Esvelt KM, Aach J, Guell M, DiCarlo JE, … Church GM. 2013. RNA-guided human genome engineering via Cas9. *Science*. 339: 823–826.
Mammadov JA, Chen W, Ren R, Pai R, Marchione W, Yalçin F, … Kumpatla SP. 2010. Development of highly polymorphic SNP markers from the complexity reduced portion of maize [*Zea mays* L.] genome for use in marker-assisted breeding. *Theoretical and Applied Genetics*, 121, 577–588.
Marx H, Minogue C, Jayaraman D. 2016. A proteomic atlas of the legume *Medicago truncatula* and its nitrogen-fixing endosymbiont *Sinorhizobium meliloti*. *Nature Biotechnology*. 34: 1198–1205
Mashaki MK, Garg V, NasrollahnezhadGhomi AA, Kudapa H, Chitikineni A, ZaynaliNezhad K, … Thudi M. 2018. RNA-Seq analysis revealed genes associated with drought stress response in kabuli chickpea (*Cicer arietinum* L.). *PLoS One*. 13(6): e0199774
Matres JM, Hilscher J, Datta A, Armario-Nájera V, Baysal C, He W, ... Slamet-Loedin IH. 2021. Genome editing in cereal crops: an overview. *Transgenic Research*. 30: 461–498.
McGettigan, P. A. 2013. Transcriptomics in the RNA-seq era. *Current Opinion in Chemical Biology*. 17(1): 4–11.
Meena MR, Appunu C, Arun Kumar R, Manimekalai R, Vasantha S, Krishnappa G,...Hemaprabha G. 2022. Recent advances in sugarcane genomics, physiology, and phenomics for superior agronomic traits. *Frontiers in Genetics*. 13: 854936.
Mehta S, James D, Reddy MK. 2019. Omics Technologies for Abiotic Stress Tolerance in Plants: Current Status and Prospects. In: WaniS. (eds) Recent Approaches in Omics for Plant Resilience to Climate Change. Springer, Cham. https://doi.org/10.1007/978-3-030-21687-0_1
Mergner J, Kuster B. 2022. Plant proteome dynamics. *Annual Review of Plant Biology*. 73:67–92.
Mikheenko A, Prjibelski AD, Joglekar A, Tilgner HU. 2022. Sequencing of individual barcoded cDNAs using Pacific Biosciences and Oxford Nanopore Technologies reveals platform-specific error patterns. *Genome Research*. 32(4): 726–737.
Misra BB. 2018. Updates on resources, software tools, and databases for plant proteomics in 2016-2017. *Electrophoresis*. 39(13): 1543–1557. doi:10.1002/elps.201700401
Mohanta TK, Bashir T, Hashem A, Abd Allah EF, Bae H. 2017. Genome editing tools in plants. *Genes*. 8(12): 399. doi: 10.3390/genes8120399.
Müller Paul H, Istanto DD, Heldenbrand J, Hudson ME. 2022. CROPSR: an automated platform for complex genome-wide CRISPR gRNA design and validation. *BMC Bioinformatics*. 23(1): 1–19.
Nawaz G, Usman B, Peng H, Zhao N, Yuan R, Liu Y, Li R. 2020. Knockout of Pi21 by CRISPR/Cas9 and iTRAQ-based proteomic analysis of mutants revealed new insights into *M. oryzae* resistance in elite rice line. *Genes*. 11(7): 735.
Okou DT, Steinberg KM, Middle C, Cutler DJ, Albert TJ, Zwick ME. 2007. Microarray-based genomic selection for high-throughput resequencing. *Nature Methods*. 4(11): 907–909.
Pal G, Bakade R, Deshpande S, Sureshkumar V, Patil SS, Dawane A, … Vemanna RS. 2022. Transcriptomic responses under combined bacterial blight and drought stress in rice reveal potential genes to improve multi-stress tolerance. *BMC Plant Biology*. 22(1): 1–20.
Papikian A, Liu W, Gallego-Bartolomé J, Jacobsen SE. 2019. Site-specific manipulation of Arabidopsis loci using CRISPR-Cas9 SunTag systems. *Nature Communications*. 10(1): 729.
Pazhamala LT, Purohit S, Saxena RK, Garg V, Krishnamurthy L, Verdier J, Varshney RK. 2017. Gene expression atlas of pigeonpea and its application to gain insights into genes associated with pollen fertility implicated in seed formation. *Journal of Experimental Botany*. 68(8): 2037–2054.
Perez-Riverol Y, Bai J, Bandla C, García-Seisdedos D, Hewapathirana S, Kamatchinathan S, … Vizcaíno JA. 2022. The PRIDE database resources in 2022: A hub for mass spectrometry-based proteomics evidences. *Nucleic Acids Research*, 50: 543–552.
Porreca GJ, Zhang K, Li JB, Xie B, Austin D, Vassallo SL, … Shendure J. 2007. Multiplex amplification of large sets of human exons. *Nature Methods*. 4(11): 931–936.

Qi Q, Hu B, Jiang W, Wang Y, Yan J, Ma F, ... Xu J. 2023. Advances in plant epigenome editing research and its application in plants. *International Journal of Molecular Sciences*. 24(4): 3442.

Ran FA, Hsu PD, Lin CY, Gootenberg JS, Konermann S, Trevino AE, ... Zhang F. 2013. Double nicking by RNA-guided CRISPR Cas9 for enhanced genome editing specificity. *Cell*. 154: 1380–1389. doi:10.1016/j.cell.2013.08.021

Ray S, Satya P. 2014. Next generation sequencing technologies for next generation plant breeding. *Frontiers in Plant Science*. 5: 367.

Rich-Griffin C, Stechemesser A, Finch J, Lucas E, Ott S, Schäfer P. 2020. Single-cell transcriptomics: A high-resolution avenue for plant functional genomics. *Trends in Plant Science*. 25(2): 186–197.

Rose JC, Popp NA, Richardson CD, Stephany JJ, Mathieu J, Wei CT, ... Fowler DM. 2020. Suppression of unwanted CRISPR-Cas9 editing by co-administration of catalytically inactivating truncated guide RNAs. *Nature Communications* 11: 2697. doi:10.1038/s41467-020-16542-9

Safari F, Zare K, Negahdaripour M, Barekati-Mowahed M, Ghasemi Y. 2019. CRISPR Cpf1 proteins: Structure, function and implications for genome editing. *Cell & Bioscience* 9: 36. doi:10.1186/s13578-019-0298-7

Sahu BB, Sumit R, Srivastava SK, Bhattacharyya MK. 2012. Sequence based polymorphic (SBP) marker technology for targeted genomic regions: Its application in generating a molecular map of the Arabidopsis thaliana genome. *BMC Genomics*. 13(1): 1–10.

Saifaldeen M, Al-Ansari DE, Ramotar D, Aouida M. 2020. CRISPR FokI Dead Cas9 system: Principles and applications in genome engineering. *Cells*. 9(11): 2518. doi:10.3390/cells9112518.

Senkler J, Senkler M, Eubel H, Hildebrandt T, Lengwenus C, Schertl P, ... Braun HP. 2017. The mitochondrial complexome of *Arabidopsis thaliana*. *The Plant Journal*. 89(6): 1079–1092.

Shi J, Tian Z, Lai J, Huang X. 2023. Plant pan-genomics and its applications. *Molecular Plant*. 16: 168–186

Silva G, Poirot L, Galetto R, Smith J, Montoya G, Duchateau P, Pâques F. 2011. Meganucleases and other tools for targeted genome engineering: Perspectives and challenges for gene therapy. *Current Gene Therapy*. 11; 11–27.

Sinha P, Bajaj P, Pazhamala LT, Nayak SN, Pandey MK, Chitikineni A, ... Zhuang W. 2020. *Arachis hypogaea* gene expression atlas for *fastigiata* subspecies of cultivated groundnut to accelerate functional and translational genomics applications. *Plant Biotechnology Journal*. 18(11): 2187–2200.

Stapley J, Reger J, Feulner PG, Smadja C, Galindo J, Ekblom R, ... Slate J. 2010. Adaptation genomics: The next generation. *Trends in Ecology & Evolution*. 25(12): 705–712.

Strasser R, Seifert GJ, Doblin MS, Johnson KL, Ruprecht C, Pfrengle F, Bacic A, Estevez JM. 2021. Cracking the "Sugar Code": A snapshot of N- and O-glycosylation pathways and functions in plants cells. *Frontiers in Plant Science*. 12: 640919.

Su B, Qian Z, Li T, Zhou Y, Wong A. 2019. PlantMP: A database for moonlighting plant proteins. *Database*, baz050.

Tang D, Jia Y, Zhang J, Li H, Cheng L, Wang P, ... Huang S. 2022. Genome evolution and diversity of wild and cultivated potatoes. *Nature*. 606(7914): 535–541.

Tang F, Barbacioru C, Wang Y, Nordman E, Lee C, Xu N, ... Lao K. 2009. mRNA-Seq whole-transcriptome analysis of a single cell. *Nature Methods*. 6(5): 377–382.

Teer JK, Bonnycastle LL, Chines PS, Hansen NF, Aoyama N, Swift AJ, ... NISC Comparative Sequencing Program. 2010. Systematic comparison of three genomic enrichment methods for massively parallel DNA sequencing. *Genome Research*. 20(10): 1420–1431.

Teer JK, Mullikin JC. 2010. Exome sequencing: The sweet spot before whole genomes. *Human Molecular Genetics*. 19(R2): R145–R151.

Tian J, Wang C, Xia J, Wu L, Xu G, Wu W, ... Tian F. 2019. Teosinte ligule allele narrows plant architecture and enhances high-density maize yields. *Science*. 365(6454): 658–664.

Tian L, Chen F, Macosko EZ. 2023. The expanding vistas of spatial transcriptomics. *Nature Biotechnology*. 41(6): 773–782.

Usman B, Zhao N, Nawaz G, Qin B, Liu F, Liu Y, Li R. 2021. CRISPR/Cas9 guided mutagenesis of grain size 3 confers increased rice (*Oryza sativa* L.) grain length by regulating cysteine proteinase inhibitor and ubiquitin-related proteins. *International Journal of Molecular Sciences*. 22(6): 3225.

Van Bel M, Bucchini F, Vandepoele K. 2019. Gene space completeness in complex plant genomes. *Current Opinion in Plant Biology*. 48: 9–17.

Villate A, San Nicolas M, Gallastegi M, Aulas PA, Olivares M, Usobiaga A, Etxebarria N, Aizpurua-Olaizola O. 2021. Metabolomics as a prediction tool for plants performance under environmental stress. *Plant Science*. 303: 110789.

Vojta A, Dobrinić P, Tadić V, Bočkor L, Korać P, Julg B, ... Zoldoš V. 2016. Repurposing the CRISPR-Cas9 system for targeted DNA methylation. *Nucleic Acids Research*. 44(12): 5615–5628.

Voytik E, Skowronek P, Zeng WF, Tanzer MC, Brunner AD, Thielert M, … Mann M. 2022. AlphaViz: Visualization and validation of critical proteomics data directly at the raw data level. *bioRxiv*, 2022–07.

Wang B, Kumar V, Olson A, Ware D. 2019. Reviving the transcriptome studies: An insight into the emergence of single-molecule transcriptome sequencing. *Frontiers in Genetics*. 10: 384.

Xu L, Sun B, Liu S, Gao X, Zhou H, Li F, Li Y. 2023. The evaluation of active transcriptional repressor domain for CRISPRi in plants. *Gene*. 851: 146967.

Xue H, Zhang Q, Wang P, Cao B, Jia C, Cheng B, … Cheng H. 2022. qPTMplants: An integrative database of quantitative post-translational modifications in plants. *Nucleic Acids Research*. 50(D1): D1491–D1499.

Yan S, Bhawal R, Yin Z, Thannhauser TW, Zhang S. 2022. Recent advances in proteomics and metabolomics in plants. *Molecular Horticulture*. 2(1): 17.

Yang Y, Saand MA, Huang L, Abdelaal WB, Zhang J, Wu Y, Li J, Sirohi MH, Wang F. 2021. Applications of multi-omics technologies for crop improvement. *Frontiers in Plant Science*. 12: 563953.

Yoshida H, Hirano K, Yano K, Wang F, Mori M, Kawamura M, … Yamamoto E. 2022. Genome-wide association study identifies a gene responsible for temperature-dependent rice germination. *Nature Communications*. 13(1): 1–13.

Zanardo LG, de Souza GB, Alves MS. 2019. Transcriptomics of plant-virus interactions: A review. *Theoretical and Experimental Plant Physiology*. 31: 103–125.

Zetsche B, Gootenberg JS, Abudayyeh OO, Slaymaker IM, Makarova KS, Essletzbichler P, … Zhang F. 2015. Cpf1 is a single RNA-guided endonuclease of a class 2 CRISPR-Cas system. *Cell*. 163(3): 759–771. doi:10.1016/j.cell.2015.09.038

Zhang F, Cong L, Lodato S, Kosuri S, Church GM, Arlotta P. 2011. Efficient construction of sequence-specific TAL effectors for modulating mammalian transcription. *Nature Biotechnology* 29: 149–153.

Zhao N, Wang W, Grover CE, Jiang K, Pan Z, Guo B, … Hua, J. 2022. Genomic and GWAS analyses demonstrate phylogenomic relationships of *Gossypium barbadense* in China and selection for fibre length, lint percentage and *Fusarium* wilt resistance. *Plant Biotechnology Journal*. 20(4): 691–710.

Zhou H, Xu L, Li F, Li Y. 2022. Transcriptional regulation by CRISPR/dCas9 in common wheat. *Gene*. 807: 145919.

Zhu FY, Song YC, Zhang KL, Chen X, Chen MX. 2020. Quantifying plant dynamic proteomes by SWATH-based mass spectrometry. *Trends in Plant Science*. 25(11): 1171–1172.

3 Recent Tools of Genome Editing and Functional Genomics in Plants

Present and Future Applications on Crop Improvement

Johni Debbarma, Mamta Bhattacharjee, Trishna Konwar, Channakeshavaiah Chikkaputtaiah, and Dhanawantari L. Singha

3.1 CRISPR/Cas SYSTEM FOR GENOME ALTERATION

The CRISPR/Cas gene-editing technology is a groundbreaking tool that enables researchers to accurately modify the DNA of an organism (El-Mounadi et al. 2020). The acronym "CRISPR" refers to "Clustered Regularly Interspaced Short Palindromic Repeats", while "CRISPR-associated protein Cas" denotes the protein component associated with this technology. The mechanism of CRISPR/Cas activity is a product of a naturally occurring defense system present in bacteria that aids in their protection against viral infections (Montecillo et al. 2020). The CRISPR/Cas9 methodology comprises a pair of fundamental constituents, namely the Cas9 protein and the guide RNA (gRNA) for the target gene (El-Mounadi et al. 2020). The Cas9 protein functions as a dimeric endonuclease, exhibiting the ability to cleave DNA double helices at precise genomic coordinates. The guide RNA (gRNA) can be defined as a small ribonucleic acid segment engineered to bind to a particular target sequence within the DNA selectively. The guide RNA (gRNA) functions as a molecular guide, facilitating the precise targeting of the Cas9 endonuclease to the intended genomic locus. Upon introduction into a cell, the Cas9 protein along with the gRNA combines to generate a complex that surveys the DNA for a corresponding target sequence. Upon identification of a suitable target, the Cas9 protein initiates a cleavage event at the corresponding site on the DNA strands. The incision elicits the cellular innate DNA repair mechanisms, which can be utilized to deliberately induce targeted modifications to the DNA (Shan et al. 2013). Scientists are exploring the technique and introducing targeted alterations to the genome by designing specific gRNAs using the CRISPR/Cas9 editing system. This method has transformed the field of genetic transformation, allowing researchers to investigate gene function, provide new treatments for genetic illnesses, and perhaps improve different elements of biotechnology and agriculture (Feng et al. 2013; Li et al. 2013). With time, researchers have also explored other Cas enzymes, such as Cas12a and Cas13, which offer unique advantages for different applications (Yao et al. 2018; Kavuri et al. 2022). Also, scientists have developed base editors that enable precise changes to individual nucleotides within the genome without creating double-strand breaks (Azameti and Dauda 2021). Beyond editing DNA sequences, researchers have also harnessed CRISPR/Cas systems to regulate gene expression. Furthermore, advancements like CRISPR interference (CRISPRi) and CRISPR activation (CRISPRa) further

DOI: 10.1201/9781003387060-3

revolutionized the field of gene editing. CRISPRi involves using a deactivated Cas9 (dCas9) protein fused to repressor domains, while CRISPRa employs dCas9 fused to activator domains (Khatodia et al. 2016; Ding et al. 2022). Further, CRISPR-based tools have been adapted for modifying the epigenome, which regulates gene expression without altering the underlying DNA sequence (Pulecio et al. 2017). Introducing techniques like prime editing, a newer development further expands the capabilities of CRISPR/Cas9. It combines a catalytically impaired Cas9 (dCas9) with a reverse transcriptase enzyme to introduce precise changes to the genome, including insertions, deletions, and point mutations, without requiring double-strand breaks (Li et al. 2022). The emergence of novel improvisations in various facets of conventional CRISPR/Cas9 mediated gene-editing has enabled researchers to precisely and efficiently manipulate crop-specific characteristics, rendering it the most versatile technology in the domains of crop breeding and functional genomics.

3.2 DRAWBACKS OF TRADITIONAL CRISPR/Cas TOOL

The application of the CRISPR/Cas9 mechanism as a highly accurate technique for manipulating genomes has resulted in a significant revolution in the realm of plant biology investigation. (El-Mounadi et al. 2020). Through quick and reliable gene mutagenesis for reverse genetic investigations, the technology also significantly contributes to our understanding of how biological processes work. Although the system's adaptability and effectiveness in plant genome editing have been depicted in many works of literature (Montecillo et al. 2020; Zhu et al. 2020), some issues and constraints must be addressed to be used effectively. The effectiveness of the CRISR/Cas9 system relies mainly on the construction of the plasmid-based sgRNAs/Cas9 cassette since it dictates the appropriate expression of the Cas9 nuclease and the specificity of the targeting sgRNAs. To ensure the efficient utilization of the genome editing system in plants and mitigate any potential off-target effects, several factors must be taken into account. These include the careful selection of the target sequence, compatibility with the protospacer adjacent motif (PAM), assessment of the activity of the promoter, and the overall design of the expression cassettes (Montecillo et al., 2020).

The precise nature of the gRNAs produced from nucleotides that match a segment of the target DNA loci is essential in guiding the activity of the Cas9 protein toward a specific DNA target. Thus, accurate and effective targeting of the CRISPR/Cas9 system depends on proper target site identification and optimal gRNA design. In general, the construction of a single guide RNA (sgRNA) requires a sequence of 20 nucleotides that corresponds to the target loci of DNA along with a recognizable PAM. The complementarity between the target DNA bases and the sgRNA determines the affinity and effectiveness of the activity of the Cas9 nuclease. When Cas9 interacts with unintended genomic locations, it causes cleavages that could have unfavorable implications. According to Wang et al. (2016), it has been observed that Cas9 can withstand mismatches of up to three nucleotides between the single guide RNA (sgRNA) and DNA, which implies that the off-target sites are often dependent on sgRNA. To identify potential off-target sites throughout the genome and assess the probability of off-target editing, computational methods and *in silico* technologies can be utilized (Naeem et al., 2020). Several studies have accumulated evidence demonstrating the presence of off-target effects that are independent of single guide RNA (sgRNA), thereby highlighting the need for unbiased experimental detection and validation, as noted by Richter et al. (2020). The CRISPR/Cas9 system's capacity to target DNA sequences is also limited by the availability of PAM sequence placement within a gene, which determines the choice of gRNA sequence. The most widely utilized Cas9 i.e., SpCas9, recognizes NGG PAMs, which limits its target sites to a locus that contains this motif (Jinek et al. 2012). In the context of gene editing through Homology-Directed Repair, the PAM requirement of SpCas9 holds considerable importance due to its involvement in the precise placement of DSBs within the proximity (10–20 bp) to the target region (Yao et al. 2018). Additionally, a Cas9 has limited DNA editing options and is primarily used for introducing double-stranded breaks (DSBs) in DNA, which undergo repair mechanisms. While this is a powerful tool for gene disruption, it may not be the ideal method for all types of genetic modifications. For example, precise single-base changes require additional modifications or alternative gene-editing tools (Eid et al. 2018).

3.3 USE OF CRISPR/Cas TECHNOLOGY TO STUDY FUNCTIONAL GENOMICS

A persistent obstacle in the field of biology pertains to the comprehensive delineation of genotype-phenotype associations. Functional genomics aims to enhance the understanding of genetic and protein functions and interactions by utilizing data from various biological levels such as genome, transcriptome, epigenome, proteome, and metabolome, among others (Ford et al. 2019). The escalating volume and intricacy of genomic data have underscored the need for expeditious screening methodologies. By employing high-throughput assays on a large scale, scientists can expedite the process of delineating the role of numerous genes and proteins simultaneously (Liu et al. 2019). To achieve this objective, CRISPR/Cas technology has been used to facilitate the investigation and implementation of functional outcomes through targeted editing approaches. These systems exhibit high efficacy in selectively targeting particular regions of the genome. The discussed technology has resulted in a noteworthy alteration in the respective domain, owing to its capacity to expedite genetic disruptions and permit functional output evaluation in a multiplexed approach Abdelrahman et al. (2021). This phenomenon can be attributed to its ease of use, rapidity of execution, and versatile targeting capabilities. The technology has facilitated the execution of numerous high-throughput functional genomic screenings, leading to the identification of pivotal genes implicated in the process (Ford et al. 2019).

To study functional genomics, a proper CRISPR tool set has to be chosen by the researcher depending on the question to be addressed during the investigation. The CRISPR/Cas systems are categorized into distinct groups and types (Makarova and Koonin 2015). Among all the classes, the most commonly used Cas protein used in functional studies belongs to the Class II family. One of the widely used Cas proteins, i.e., SpCas9, is a type II single protein effector originating from *Streptococcus pyogenes* (Ford et al. 2019). The efficient cleavage of the target gene by SpCas9 is facilitated through the utilization of guide RNA. The aforementioned procedure has the potential to yield double-strand breaks (Sternberg et al. 2014), which may subsequently prompt the occurrence of indels or knockout mutations upon repair. The Non-Homologous End Joining mechanism of repair frequently results in minor deletions or insertions, whereas Homology-Directed Repair can be employed to introduce novel customized regions or rectify particular genetic mutations (Montalbano et al. 2017).

The adaptability of CRISPR/Cas systems has progressively expanded over time. Effector fusions of Cas9 have been devised to bring about targeted modifications of histones, DNA base pairs, gene transcription (CRISPRa/CRISPRi), and DNA methylation/demethylation at particular loci. The utilization of catalytically inactive Cas9 (dCas9) proteins that are fused to effector moieties has been observed to function as DNA targeting platforms for a varied array of functionalities (Dominguez et al. 2016). Nevertheless, the optimization of Cas9 fusions in terms of their efficacy and off-target effects can present difficulties, while their considerable size can impose restrictions on their delivery. Notwithstanding these obstacles, dCas9 fusions offer resilient instruments for exploring genome functionality. Cas9 variants are selected according to the particular functionality under investigation (Ford et al. 2019). The utilization of wild-type Cas9 is optimal for conducting knockout investigations aimed at ascertaining gene indispensability. Conversely, CRISPRa/i systems are capable of regulating gene transcription without inducing permanent modifications to the genome (Khatodia et al. 2016; Ding et al. 2022). Cas9 fusion constructs that modify DNA and histones can alter gene accessibility and expression (Dominguez et al. 2016). On the other hand, CRISPR base editors can precisely modify individual nucleotide base pairs (Pulecio et al. 2017). The incorporation of protein-binding RNA segments into engineered sgRNAs facilitates the recruitment of proteins that possess a wide range of functionalities (Li et al. 2022). The aforementioned tools provide diverse alternatives for executing functional genomic screenings. Notably, every perturbation technique has its own set of limitations and factors to be considered when designing experiments and carrying out subsequent validation. The selection of the perturbation approach is contingent upon the particular research inquiry and intended results (Ford et al. 2019).

Briefly, the steps of CRISPR/Cas technology to study functional genomics include the identification of specific target genomic loci or genes that need to be edited using the CRISPR/Cas technology. This is followed by designing guide RNA specifically targeting the identified genomic loci. Various *in silico* tools can be used to design gRNAs efficiently and ensure they have high specificity and efficacy. Next, the CRISPR/Cas system, which consists of the Cas endonuclease and the designed gRNAs is generated by cloning the gRNA sequences into an appropriate vector having the Cas enzyme. The CRISPR/Cas system is incorporated into plant cells through transformation methodologies, including *Agrobacterium*-mediated transformation or biolistic particle delivery. This will result in the integration of the CRISPR/Cas9 system into the plant genome. This is followed by the screening and selection of transformed plants using appropriate selection markers and techniques. The presence of the CRISPR/Cas system in the plant is confirmed using molecular analysis, such as polymerase chain reaction. To confirm the presence of desired mutations or gene knockouts, molecular techniques like Sanger sequencing are taken into consideration. To evaluate the functional consequences of the altered genes on the growth, development, or particular traits of interest in plants, phenotypic analysis or metabolic investigations are performed on the edited plants and subsequently compared to the wild-type/control plants. Analysis and interpretation of the collected data can help the researchers conclude the function of the perturbed genes. Further, a comparison of the phenotypes and molecular characteristics of the edited plants' can help to gain insights into gene function and plant biology. Annotation of the function of the identified genes using additional techniques or follow-up experiments comes next in line. Further, consideration of complementation studies, transcriptomic analysis, or other functional assays can be further helpful to gain a better knowledge of the roles of different genes in plant biology.

3.4 MULTIPLEX GENOME EDITING (MGE) TOOLS

MGE tools are advanced genetic engineering technologies that allow precise modification of two or more genes in a target genome. MGE facilitates rapid changes in gene sequence by using engineered nucleases and introducing two or more DSBs. MGE can routinely be used to facilitate effective changes in the desirable crop, viz. the production of seedless plant species and disease-resistant crops (Naim et al. 2018; Tripathi et al. 2019) and to skip extensive segregation.

3.4.1 CRISPR/Cas9 System-Based MGE

CRISPR/Cas9 has become a promising technique for modifying DNA sequence, expression, and function whereby target DNA is recognized by specific gRNA introducing DSB, resulting in precise editing. CRISPR/Cas9 system enhances the efficient multiple gene editing by delivering multiple sgRNAs simultaneously and expressing one Cas protein in various plants such as rice, wheat, maize, and *Arabidopsis* (Ma et al. 2015, Rothan et al. 2019). Improved efficiency of saccharification was observed in sugarcane by targeting mutagenesis of more than 100 alleles without affecting biomass yield (Kannan et al. 2018). Golden gate assembly is a widely used method for molecular cloning. This cloning method simultaneously and directionally assembles multiple DNA fragments using specific restriction enzymes and DNA ligase-producing sticky ends in the DNA sequence with sgRNA. Multiple guide RNAs are assembled into a single binary vector and transcribed by adopting various strategies including transcription either by polymerase promoter such as U6/U3 (Ordon et al. 2017; Tian et al. 2017), or through polycistronic mRNA facilitating post-transcriptional excision of individual gRNA by Cys4 (Tsai et al. 2014), tRNAs (Xie et al. 2014) and hammerhead ribozymes mediated self-cleavage or Cpf1-based crRNA (Tang et al. 2018).

3.4.1.1 Expression of Multiple gRNAs Driven by Polymerase III Promoter

Expression of multiple gRNAs in a binary vector can be transcribed with a single promoter and terminator cassettes or with a single promoter and terminator. RNA polymerase III promoters transcribe the multiple sgRNA expression in a single binary vector (Figure 3.1a). The CRISPR/Cas9 construct with multiple sgRNA driven by U3/U6 promoter was designed and cloned in a cloning vector by the Golden Gate Cloning method using restriction enzyme (Type IIS), BsaI to knock out multiple genes in Arabidopsis and Rice. However, its limitation is that this method results in the assembly of two to eight gRNA in one expression cassette and multiple polymerase III promoters may often direct variation in the expression of mRNA and induce silencing of transgene resulting in plasmid instability in *E. coli* and *Agrobacterium* (Ma et al. 2015).

3.4.1.2 Polycistronic Cys4-Based Excision

Expression of multiple gRNA from a single transcript was adopted to overcome the limitation of the above-mentioned method of transcribing multiple gRNA by polymerase III promoter which allows multiplexing of up to 24 gRNAs (Stuttmann et al. 2021) transcribed by polymerase II promoter which produces longer transcript (Arimbasseri et al. 2013) compared to polymerase III promoter. RNA-cleaving enzymes process post-transcriptional modification of polycistronic mRNAs into individual gRNAs. (Figure 3.1b). Cys 4 crRNA endonucleases from *Pseudomonas aeruginosa.* Each gRNA is assembled and flanked by a cleavage site for cys4 RNAse. In this method, RNA polymerase II is used to transcribe the gRNA. The gRNA has ends with 3′ and Csy4 sequences that facilitate cleavage of the transcript, thereby releasing long sequences of the gRNAs (Tsai et al. 2014).

3.4.1.3 Polycistronic tRNA-gRNA

An artificial polycistronic tRNA-gRNA (PTG) gene method is used for multiple gRNAs expression driven by RNA polymerase in an individual transcript. The gRNAs are linked together by tRNA sequences (Figure 3.1c) which act as an enhancer and facilitate transcription of the promoter. The endogenous RNase enzymes like RNaseP and RNaseZ recognize the tRNAs and cleave the single gRNAs. These gRNAs direct Cas9 to their targets for gene modification by editing (Xie et al. 2014).

3.4.1.4 Ribozyme Mediated Self-Cleavage

In an alternative approach, a CRISPR/Cas system was designed to facilitate the expression of both Cas9 and sgRNAs by a single pol II promoter in which a hammerhead (HH) ribozyme is successfully introduced and enhanced to produce multiple gRNAs (Tang et al. 2020). A ribozyme cleavage sequence such as hammerhead type ribozyme (HH) (Pley et al. 1994) or hepatitis delta virus ribozyme (HDV) (Ferré-D'Amaré et al. 1998) is inserted in between Cas9 and sgRNAs which facilitates cleavage and transcription of a ribozyme-sgRNA-ribozyme gene enhancing site-specific gene editing (Figure 3.1d).

3.4.1.5 CRISPR/Cas12a (Cpf1) Mediated crRNA

CRISPR/Cas9 system has limitations in multiplexing experiments when the gene sequence to be edited is large as Cas9 requires GC- rich PAM sequences. Expression of multiple gRNAs in a single polycistronic method requires additional sequences for excising gRNAs and their transcription.

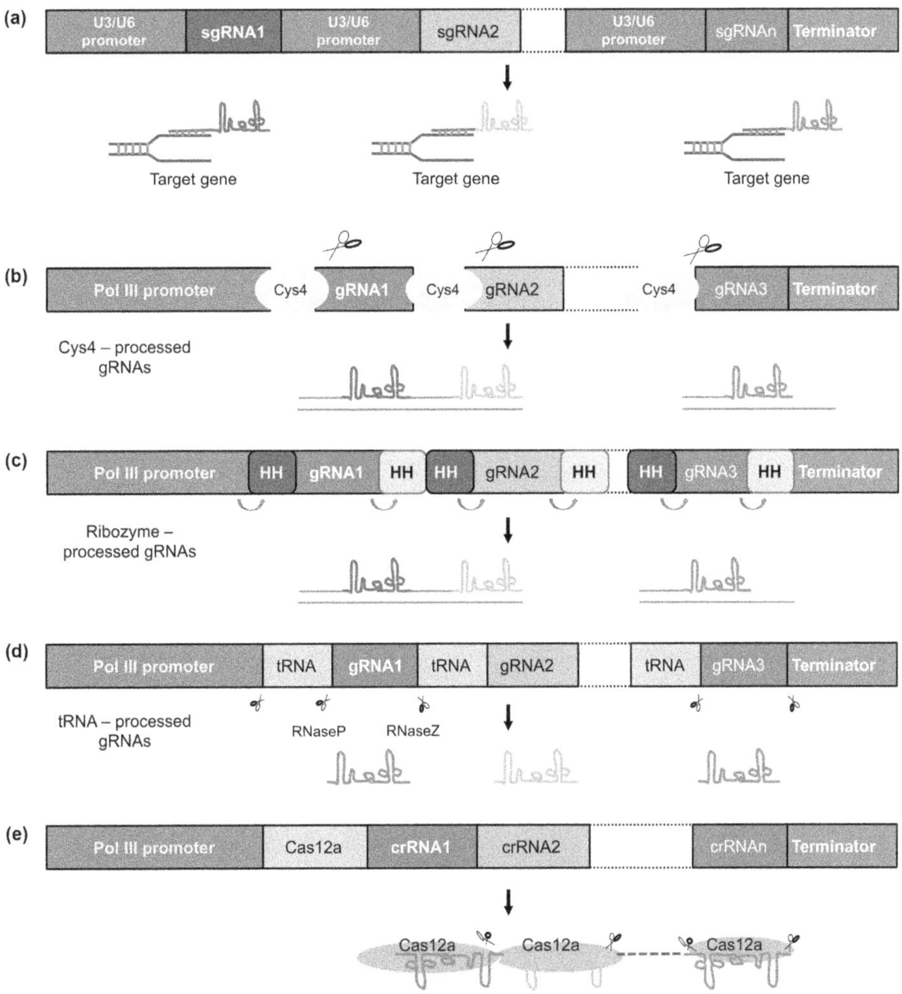

FIGURE 3.1 (a) Individual gRNAs are generated by cloning small gRNAs with U3/U6 promoters driven by RNA polymerase III(b–d). Small gRNAs are transcribed into a single transcript with Csy4, tRNAs, and hammerhead ribozymes for post-transcriptional processing of gRNA separations (e). Casl2a system where polymerase III promoter drives mature crRNAs.

These limitations may be controlled by using the Cas protein, Casl2a/Cpf1 which has a relatively small size and requires short crRNA (42 nucleotides) and PAM sequence that targets T-rich regions which enhances multiplexing (Figure 3.1e). One of its major advantages is that the addition of polymerase promoters or the addition of specific sequences is not required while using Casl2a, this system requires a single polymerase III promoter to drive and process crRNAs. Casl2a results in double-strand breaks at cohesive ends away from the PAM site facilitating subsequent DNA cleavage (Zetsche et al. 2017).

3.5 BASE EDITING

Base editing (BE) is the process of introducing single nucleotide variants (SNVs) through gene editing in the desired locations of the DNA or RNA of an organism. Unlike conventional GE (CGE), BE works without introducing a break in the double-strand DNA in the target genome and donor repair template. BE is a recently developed alternative approach to HDR-mediated CGE. Further, BE received quick acceptance because of its precise targeting, simplicity compared to CGE, and ability to multiplex. Recently available base editors are mentioned here.

3.5.1 Transition BE

Cytidine base editors (CBEs) and Adenine base editors (ABEs) are the BE techniques in plants. The CBE brings C.G to T.A and ABE brings A.T to G.C mutations. They can create precise single-base transitions (pyrimidine to pyrimidine or purine to purine). Therefore, four types of transitions are possible through CBEs and ABEs.

3.5.1.1 Cytidine Base Editors (CBEs)

The CBE tool consists of two components: a catalytically inactive Cas nuclease protein (dCas9), which is fused to the second component cytidine deaminase (Figure 3.2a). CBE allows precise C.G to T.A transitions by deaminating deoxycytidine to deoxyuridine (Komor et al. 2016). During DNA replication the deoxyuridine is read as deoxythymidine. The application of CBE for crop improvement is represented in Table 3.1a. Several modifications were made to the existing base editors to achieve a more efficient system (mentioned in Table 3.2a).

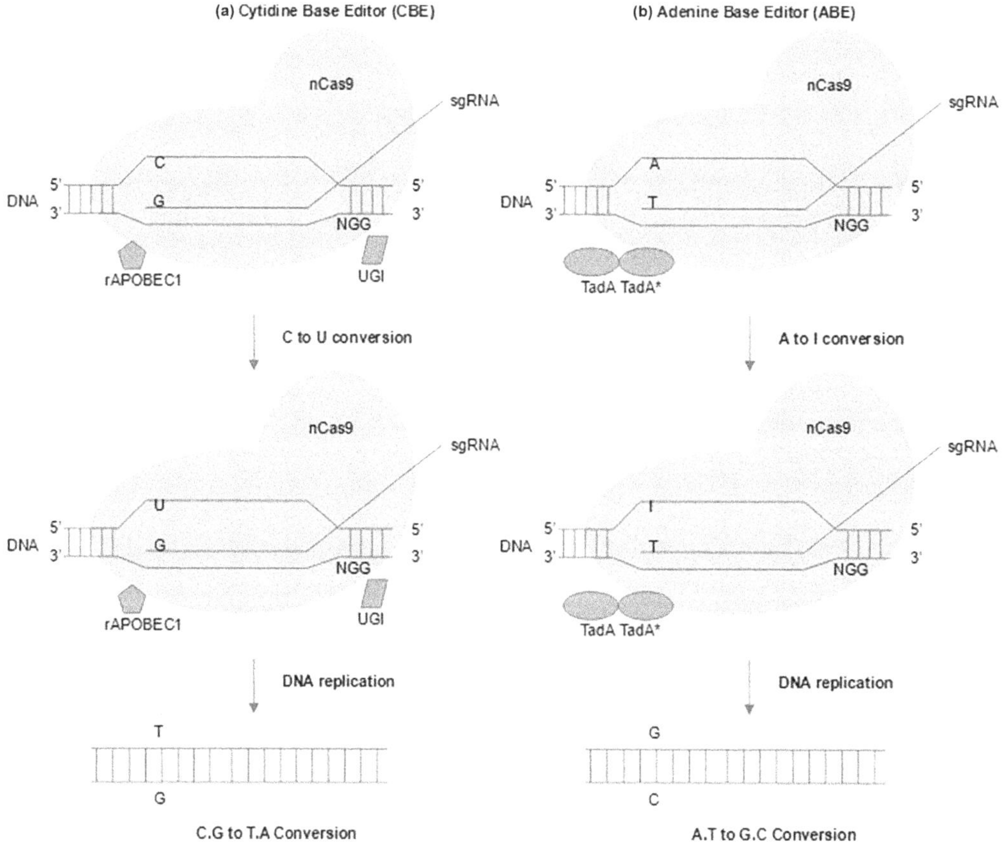

FIGURE 3.2 The mechanism of base editing: (a) the components of cytidine base editing are i. cytidine deaminase rAPOBEC1, ii. uracil glycosylase inhibitor (UGI) and iii. nickase Cas9 (nCas9). The CBE introduces C.G to T.A; (b) the components of adenine base editing are i. adenine deaminase TadA and ii. nCas9 and facilitate the change of A.T to G.C.

3.5.1.2 Adenine Base Editors (ABEs)

In ABE, the adenosine deaminase is fused to the dCas9. The adenosine deaminase catalyzes the deamination of deoxyadenosine to deoxyinosine (deoxyinosine mimics deoxyguanosine) (Figure 3.2b). During DNA replication this deoxyinosine is read as deoxyguanosine. Thus, the deamination causes the transition of A.T to G.C (Nishida et al. 2016). Further advancement in adenine base editor leads to the introduction of the PhieABE toolbox, an adenine base editor that does not depend on PAM in plants (Tan et al. 2022). The application of ABEs and their advanced tools for crop improvement is listed in Tables 3.1a and 3.2a.

3.5.1.3 Transition Base Editing of RNA

Precise editing at the RNA level is also possible. In RNA, programmable editing is possible through transition deamination of A to I or C to U. The two important types of RNA base editors are i. The REPAIR (RNA editing for programmable A to I replacement) and ii. RESCUE (RNA editing for specific C-to-U exchange).

3.5.1.4 REPAIR RNA Base Editing

REPAIR system consists of two important components: i. RNA-targeting CRISPR effector Cas13 and ii. the catalytic domain of ADAR2 (deaminase domain of human ADAR), allowing A-to-I base conversion in RNA transcripts (Cox et al. 2018).

3.5.1.5 The RESCUE RNA Base Editing

The RESCUE RNA base editing system also utilizes the dead Cas dCas13 fused to an evolved adenine deaminase domain with E488Q mutations (ADAR2DD). RESCUE editing system enables the conversion of C to U in the RNA (Abudayyeh et al. 2019).

3.5.2 Transversion Base Editing

The widely used CBE and ABE have major limitations as they can only cause transition conversions. Therefore, only four possible types of base transitions are available. However, transversion base editing is also possible. An example of a transversion base editor is the C to G base editor (CGBE). CGBE comprises Cas9 nickase, uracil DNA N-glycosylase (UNG), and a cytidine deaminase (Kurt et al. 2021).

3.5.2.1 Dual Base Editors

The capabilities of single-base editors (CBE and ABE) are extended by using both the C to T and A to G transitions together as dual base editors by adding both the cytidine deaminase and adenine deaminase to the Cas9. Examples of such dual base editors for plants are the "saturated targeted endogenous mutagenesis editors" (STEMEs) (Li et al. 2020) and base-editing-mediated gene evolution (BEMGE) (Kuang et al. 2020).

3.5.2.2 PAMless Base Editors

In general, NGG is the PAM recognition sequence of SpCas9. However, the modification in the specificity of PAM can increase the scope of using the CRISPR/Cas system. XCas9-3.7 is one such example where the NGG PAM specificity of SpCas9 is shifted to a single guanine nucleotide (Hu et al. 2018). SpCas9-NG is another such example (Nishimasu et al. 2018).

The change in bases has the possibility of developing new crop varieties, thereby enhancing crop improvement processes. The important applications of base editors for crop improvements are listed in Table 3.1.

TABLE 3.1
Examples of Base Editing for Crop Improvement

Category of the Editor	Plant	Gene Target	Specifications	References
		a. Base Editing		
CBE	Rice	*OsNRT1.1B, OsSLR1*	High nitrogen use efficiency; Dwarf	Lu and Zhu (2017)
CBE	Potato	*StGBSSI*	Edited the amylose biosynthesis gene in the tetraploid potato using gene and base-editing strategies making the plant impaired with amylose biosynthesis	Veillet et al. (2019a)
CBE & ABE	Rice protoplast	*OsACC*	In rice protoplasts, saturated targeted endogenous mutagenesis editors (STEME-1) edited cytosine and adenine at the same target site and obtained herbicide resistance	Li et al. (2020)
CBE	strawberry	*FvebZIPs1.1*	Seven novel alleles are generated by using base editor A3A-PBE and the edited lines show higher sugar content than the wild types	Xing et al. (2020)
CBE	Tomato & Potato	*ALS*	Efficiently edited the cytidine bases in the ALS gene resulting in chlorsulfuron-resistant plants	Veillet et al. (2019b)
CBE & ABE	Rice	*OsALS1*	*OsALS1* was artificially evolved in rice cells using the base-editing-mediated gene evolution (BEMGE) method	Kuang et al. (2020)
		b. Prime Editing		
Prime editor	Rice	*OsALS, PvuII, OsTB1*	Herbicide resistance traits in rice were achieved via nucleotide substitutions through prime editing	Butt et al. (2020)
Prime editor	Maize	*ZmALS1, ZmALS2*	Efficiently generated double mutations in maize in the two ALS genes	Jiang et al. (2020)
Prime editor	Tomato	*NanoLucM*	The genome editing efficiency was improved by codon and promoter optimization in tomato	Lu et al. (2021)
Prime editor	Rice, Peanut, chickpea, and cowpea protoplasts	GFP	Using the dual pegRNA system, the editing efficiency in rice was 16 times higher than in peanut, chickpea, and cowpea	Biswas et al. (2022)
Prime editor	Rice	*OsCDC48, OsNRT1.1B* and *OsALS*	Prime editing efficiency was enhanced by using modified prime binding sites and designed Plant Peg Designer, a web application for the design of pegRNAs	Lin et al. (2021)

3.6 PRIME EDITING

Prime editing is another evolved system of the traditional CRISPR/Cas technique. Prime editors can create insertions or deletions in the targeted genome without creating DSBs and donor DNA (Figure 3.3). Prime editing is a fusion product of three components, nickase Cas9 (nCas9), engineered reverse transcriptase enzyme from Moloney– murine leukemia virus (M-MLV RT), and

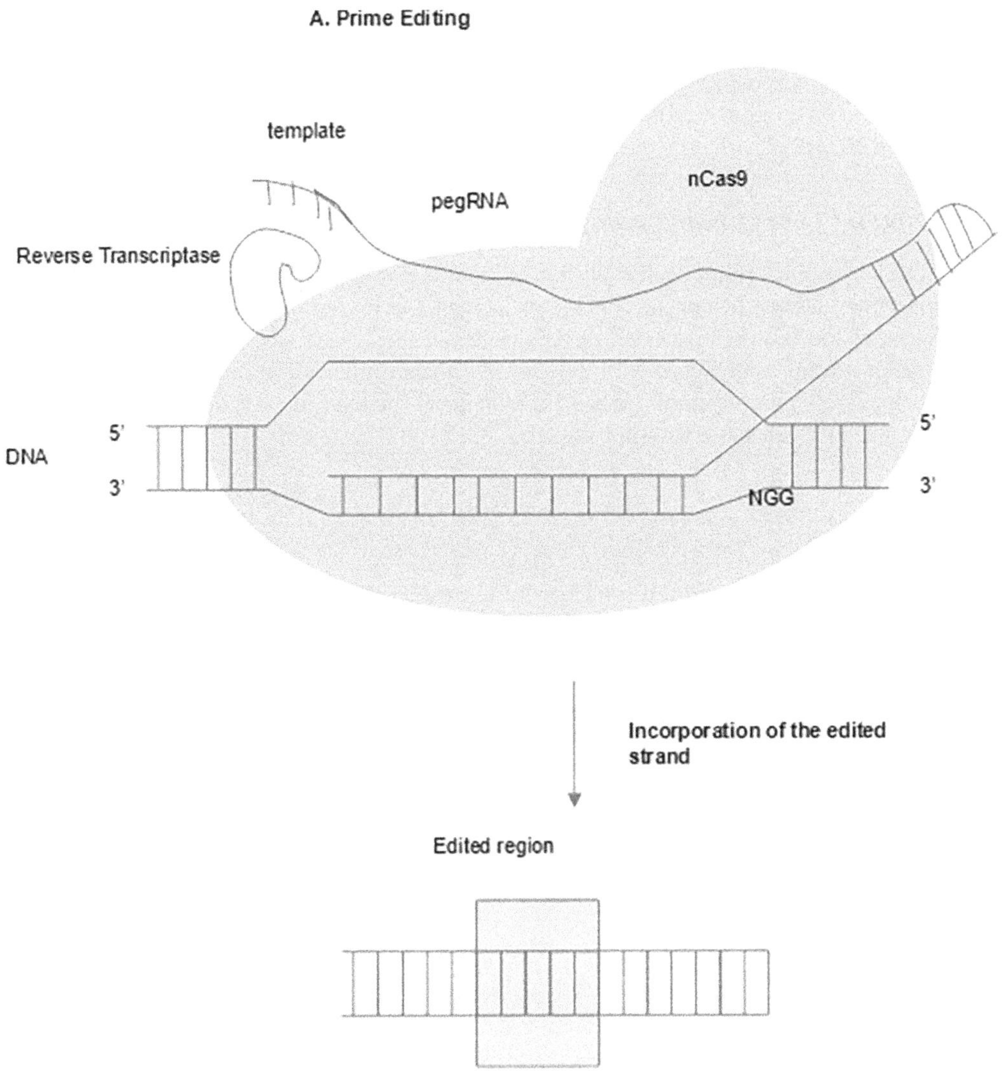

FIGURE 3.3 The mechanism of prime editing in plants.

prime-editing gRNA (pegRNA). The three components have individual unique functions such as nCas9 causes nicks in the DNA strands without creating DSBs, M-MLV RT component is responsible for the incorporation of nucleotides in the edited extension of pegRNA using the nicked DNA as a primer (Anzalone et al. 2019; Lin et al. 2020; Ahmad et al. 2023). The 3′end of the pegRNA has a tail reason consisting of a primer binding site. Prime editors can perform all 12 types of possible transitions and transversions.

3.7 PASTE 'DRAG-AND-DROP' EDITING FOR LARGE INSERTIONS

Programmable Addition via Site-specific Targeting Elements (PASTE) is an advanced genome editing tool developed by Jonathan Gootenberg and Omar Abudayyeh. The components of PASTE are CRISPR/Cas9 nickase and two enzymes, namely a reverse transcriptase and a serine integrase. The enzymes catalyze the excision and integration of foreign DNA. PASTE enables any site in the genome to be targeted for insertion of integrase landing sites using the appropriate guide RNA

(Yarnall et al. 2023). Integrases then insert their DNA payloads at these sites, eliminating the need for double-stand breaks. The advantage of the system is that it allows large DNA sequence insertion of about 36kb without generating DSBs. The editing efficiency of PASTE is better than that of HDR and NHEJ-based integration. In simple terms, the PASTE works via drag-and-drop gene integration (Yarnall et al. 2023).

3.8 APPROACH FOR IMPROVEMENT IN PRIME EDITING EFFICIENCY

Prime editing is an emerging genome editing tool that is powerful and target-specific. However, it suffers from low editing efficiency in plants. Enhancing the prime editing efficiency can be achieved by manipulating the existing prime editing approaches. Various ways were developed to modify the existing prime editing approaches with the aim of enhancing its editing efficiency. Several such important approaches are listed in Table 3.2b. Increasing the dual peg RNA molecule, and codon optimization of PPE genes two such approaches.

TABLE 3.2
Advanced Techniques Developed for Improving Base Editing and Prime Editing in Plants

Category	Target Plant	Target Gene	Specifications	References
		a. Base Editing		
ABE	Rice	*Wx*	A heterodimer consisting of *Escherichia coli* TadA (*ec*TadA) mutant and wild-type TadA was formed into a modified Cas9 enzyme N-terminal *Sp*Cas9 (D10A) nickase caused efficient A·T to G·C conversion in the rice genome	Hao et al. (2019)
ABE	Cotton	*CLA1*	Different ABE vectors were developed for allotetraploid cotton by engineering adenosine deaminase (TadA) proteins which are fused to Cas9 variants	Wang et al. (2022)
CBE	Rice	*OsPAL5, OsGSK4, OsCERK1, OsETR2, OsRLCK185*	SpRY-fused hAID*Δ for cytidine deaminase and SpRY-fused TadA8e for adenosine deaminase is constructed that efficiently induces C-to-T and A-to-G conversions in the target genes	Xu et al. (2020a)
CBE & ABE	Rice	*OsSUS6, OsSPS1, Os*07g0134700	Base transversions through C to G base editors (CGBEs) were achieved in rice	Zeng et al. (2022)
CBE	Rice protoplast	*OsCDC48-SaT1, OsDEP1-SaT2, OsNRT1.1B-SaT1,* and *OsDEP1-SaT1*	Established the high-throughput nSaCas9-mediated assay for analyzing the sgRNA-independent off-target effects caused by CBE in rice protoplasts.	Jin et al. (2020)
ABE	Rice	*OsSPL14* and *SLR1*	The base editor ABE-P1S with ecTadA*7.10-nSpCas9 (D10A) fusion display highly targeted gene editing in callus and protoplast	Hua et al. (2020)

(*Continued*)

TABLE 3.2 (*Continued*)
Advanced Techniques Developed for Improving Base Editing and Prime Editing in Plants

Category	Target Plant	Target Gene	Specifications	References
		b. Prime Editing		
Prime editor	Rice	*ALS*	Prime editing efficiency in rice was increased by the addition of T5 exonuclease that increases the herbicide-resistant through substitutions at two target sites of the *ALS* gene	Liang et al. (2023)
Prime editor	Rice & Wheat	*OsCDC48, ALS*	Prime editors were used in rice & wheat protoplast through codon, promoter, and editing-condition optimization resulting in targeted point mutations, insertions, and deletions	Lin et al. (2020)
Prime editor	Rice	*hptll, OsEPSPS*	An efficient prime editing system was established for precision editing in the rice genome	Li et al. (2022)
Prime editor	Rice	*OsACC-D2176G, OsACC-I1879V, and OsALS-S627I.a.*	The MS2 RNA aptamer enhanced the prime editing efficiency in edited rice	Chai et al. (2021)
Prime editor	Rice	*OsCDC48, OsIPA1, OsALS, and OsPDS*	Developed enhanced plant prime editor 2 system with higher editing efficiency in rice	Li et al. (2022)
Prime editor	Rice	*OsALS, OsKO2, OsDEP1, OsPDS*	Single nucleotide substitution, and insertion enabling precise prime editing	Tang et al. (2020)
Prime editor	Rice	*OsALS, OsACC, OsDEP1*	Single and multiple nucleotide edits were obtained using the improved prime editing system	Xu et al. (2020b)
Prime editor	Rice	*OsPDS, OsACC1, OsWx*	Induce versatile and flexible programmable editing at different genome sites	Xu et al. (2020a)

3.9 RECENT ADVANCES IN PLANT FUNCTIONAL GENOMICS

Functional genomics has played an important role in basic research and plant breeding (Maghuly et al. 2022). The post-genomic era has developed by fully utilizing the enormous amount of genome sequence data (Fahad et al. 2017). Functional genomics is the key strategy for converting quantity into quality in this field. By putting new functions on undiscovered genes, functional genomics is a generic way to understand how an organism's genes interact (Fahad et al. 2017) (Figure 3.4).

3.9.1 Functional Genomics Analysis

ATAC-seq methodology (assay for transposase-accessible chromatin utilizing sequencing), transposases are applied to create DNA fragments that are sequenced to detect open genomic areas in a particular cell type (Buenrostro et al. 2013). Immunoprecipitate methylated chromatin and sequencing are methods for identifying methylated DNA status (ChIP-seq) (Laird 2010).

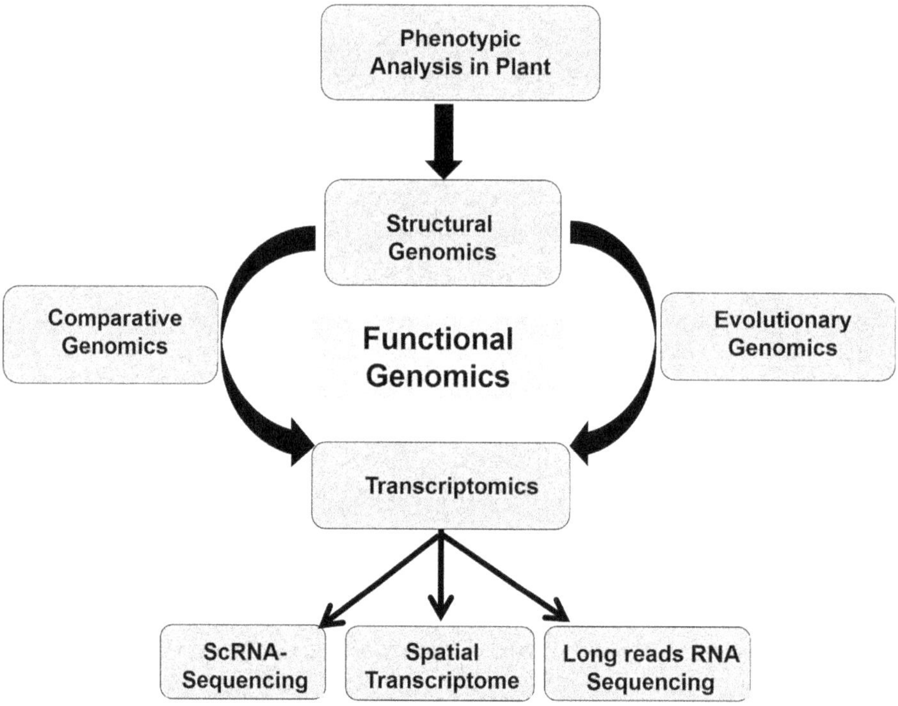

FIGURE 3.4 A flow chart of genomics application in plants.

However, ChIP-seq-based techniques are impacted by chromosomal changes in CpG density (Guigo and de Hoon 2018; Garg and Jain* 2013). With the development of throughput, cost-cutting, and long-read sequencing technologies like PacBio single-molecule real-time (SMRT) sequencing, Oxford nanopore sequencing, 10X chromium genome sequencing, and Bionano optical mapping, a highly contiguous genome assembly can now be completed in a single laboratory setup (Rai et al. 2019).

3.9.2 Transcriptome Analysis

Without knowing the sequence beforehand, it is feasible to learn more about how genes are expressed using RNA-seq (Garg and Jain* 2013; Yan et al. 2020). In the medicinal plant *Artemisia argyi*, transcriptomics was applied to identify novel genes under abiotic stress conditions (Rastogi et al. 2019). We can better comprehend the intricacy of the transcriptome in NGS technology (Guo et al. 2021). NGS was used to collect various plant species' complete gene sequences (Tyagi et al. 2022). An in-depth analysis of the transcriptome of *Tetrastigma hemsleyanum* leaves has shown gene expression patterns in anthocyanin biosynthesis (Yan et al. 2020). Recently, transcriptomics data has shown that both the transcription factors (TFs) of *OsWRKY62* and *OsWRKY45* might negatively regulate resistance to brown spot diseases in rice plants (Marwein et al. 2022).

3.9.2.1 Single-Cell RNA seq Technologies (scRNA-seq)

Technology for high-throughput single-cell RNA sequencing (scRNA-seq) has advanced significantly (Giacomello 2021; Xie et al. 2022). Individual cell transcriptional heterogeneity can be analyzed using scRNA-seq (Giacomello 2021). Single-cell resolution measurements of DNA, mRNA, and protein profiles are now possible via recent developments in single-cell isolation and barcoding technologies (Buenrostro et al. 2013). Technologies for single-cell research have improved our comprehension of the intricate and dynamic regulatory networks that underlie cellular phenotypes (Klemm et al. 2019). A recent single-cell transcriptomic study of potential marker genes involving

4,000 root xylem cells has shown several novel links to their functional activity (Shulse et al. 2019). Several studies demonstrated the applicability of scRNA-seq in root tips of *Arabidopsis thaliana* (Zhang et al. 2019; Shulse et al. 2019).

3.9.2.2 Spatial Transcriptomes

With single-cell transcriptomics, spatial transcriptomics techniques have advanced to study transcriptomes of plant tissues at the cellular level (Yu et al. 2023). To completely comprehend metabolites' functions in controlling specific biological processes in individual plant tissues or cells, spatial analysis is necessary (Peng et al. 2020). A molecular imaging technique called mass spectrometry imaging (MSI) can be used to map the spatial distribution of several metabolites in particular tissues (Peng et al. 2020; Yu et al. 2023). A spatial transcriptome study of peanut tissues revealed the peg's heterogeneous cell composition (Peng et al. 2020).

3.9.2.3 Long Read Sequencing

Recently technologies like long-read RNA sequencing (lrRNA-seq) have emerged as powerful tools for functional genomics (Logsdon et al. 2020). HiFi sequencing, the most sophisticated long-read sequencing technology, allows for reliable and reproducible variants (Lee et al. 2023). In particular, HiFi assembly-based variant identification is successful for detecting lengthy insertions, which is technically tough with HiFi reads. Furthermore, high-quality indel detection was possible with a 10 HiFi assembly to species *Arabidopsis thaliana* L. (Lee et al. 2023). Moreover, PacBio HiFi and Nanopore long-read technologies were able to unravel the complete genome sequences in *Solanum lycopersicum* L (van Rengs et al. 2022) (Table 3.3).

TABLE 3.3
Recent Genomics Tools Applied in Plants

Sl.No.	Tools	Plant Species	Applications	References
1.	Next-generation sequencing (NGS)	*T. hemsleyanum*	Anthocyanin biosynthesis	Yan et al. (2020)
2.	Single-cell transcriptome	*Populus trichocarpa*	Identify markers genes in cell types of wood tissue	Lin et al. (2014); Xie et al. (2022)
3.	Spatially enhanced resolution omics-sequencing (Stereo-seq)	Peanuts	Identify the cells heterogeneity of pegs	Liu et al. (2022)
4.	Spatial transcriptome	*Arabidopsis, P. tremula, p.abies*	Expression profiling of dormant leaf buds	Giacomello et al. (2017)
5.	scRNA Sequencing	*Arabidopsis*	Root cells differentiation	Zhang et al. (2019)
6.	10 Hifi Sequencing Assembly	*Arabidopsis*	Analysis of genome sequences	Boris et al. (2000); de Souza et al. (2023)
7.	Pac Bio Hifi and ONT nanopore sequencing	*Solanum lycopersicum* L.	Identification of TMV-resistant genes	van Rengs et al. (2022)
8.	Pac Bio long read sequencing	*S. pimpinellifolium* L. and *S. lycopersicum* var. *cerasiformae*	Analysis of inter-variation and intra-specific variations of genes	Takei et al. (2021)
9.	RNA-Sequencing	Wheat	Assessment of susceptible genes upon fungal infection by *Blumeria graminis* f. sp. *tritici* and *Puccinia triticina*	Poretti et al. (2021)
10.	Transcriptome analysis	*Oryza sativa* L.	Early response to the pathogen in suppressing the disease response to *Bipolaris oryzae* infection	Marwein et al. (2022)

3.10 CONCLUSION

In recent years, several plant functional genomics tools and techniques have been developed for crop improvement. The important advanced functional genomics tools are single-cell RNA-seq technologies, spatial transcriptomes, long-read sequencing, and ten Hifi sequencing assembly. The different approaches to crop improvement are further complemented by novel genome editing tools with improved efficiency and precision for targeting the desired gene of interest. Advanced GE tools such as multiplex GE, base editing, prime editing, and PASTE have further complemented the existing CRISPR/Cas technology. Apart from boosting crop improvement, these advanced techniques also offer biologists in elucidating the role of different genes involved in governing desirable traits in plants. In context, this chapter offers a comprehensive representation of the different improvements brought about in the classical CRISPR/Cas technology to improve genome editing efficiency and study functional genomics in plants. Even though immense progress has been observed, more work needs to be done to overcome the currently existing limitations of CRISPR/Cas technology in investigating functional genomics and genome editing.

REFERENCES

Arimbasseri AG, Rijal K, Maraia RJ. Transcription termination by the eukaryotic RNA polymerase III. Biochim Biophys Acta. 2013, 1829(3-4):318-30. doi: 10.1016/j.bbagrm.2012.10.006. Epub 2012 Oct 23. PMID: 23099421; PMCID: PMC3568203.

Abdelrahman M, Wei Z, Rohila JS, Zhao K (2021) Multiplex genome-editing technologies for revolutionizing plant biology and crop improvement. *Front Plant Science* 12:721203. doi: 10.3389/fpls.2021.721203.

Abudayyeh OO, Gootenberg JS, Franklin B, et al (2019) A cytosine deaminase for programmable single-base RNA editing. *Science* 365:382–386. doi: 10.1126/science.aax7063

Ahmad N, Awan MJA, Mansoor S (2023) Improving editing efficiency of prime editor in plants. *Trends in Plant Science* 28:1–3. doi: 10.1016/j.tplants.2022.09.001

Anzalone AV., Randolph PB, Davis JR, et al (2019) Search-and-replace genome editing without double-strand breaks or donor DNA. *Nature* 576:149–157.

Azameti MK, Dauda WP (2021) Base editing in plants: Applications, challenges, and future prospects. *Frontiers in Plant Science* 12:664997. doi: 10.3389/fpls.2021.664997

Biswas S, Bridgeland A, Irum S, et al (2022) Optimization of prime editing in rice, peanut, chickpea, and cowpea protoplasts by restoration of GFP activity. *International Journal of Molecular Sciences* 23:9809. doi: 10.3390/ijms23179809

Boris KV, Ryzhova NN, Kochieva EZ (2000) Analysis of the genome sequence of the flowering plant *Arabidopsis thaliana*. *Nature* 408:796–815. doi: 10.1038/35048692

Buenrostro JD, Giresi PG, Zaba LC, et al (2013) Transposition of native chromatin for fast and sensitive epigenomic profiling of open chromatin, DNA-binding proteins and nucleosome position. *Nature Methods* 10:1213–1218. doi: 10.1038/nmeth.2688

Butt H, Rao GS, Sedeek K, et al (2020) Engineering herbicide resistance via prime editing in rice. *Plant Biotechnology Journal* 18:2370–2372. doi: 10.1111/pbi.13399

Chai Y, Jiang Y, Wang J, et al. (2021) MS2 RNA aptamer greatly enhances prime editing in rice. *Research Square*. DOI: 10.21203/rs.3.rs-900088/v1.

Cox DBT, Gootenberg JS, Abudayyeh OO, et al (2018) RNA editing with CRISPR-Cas13. *Science*. 358:1019–1027. doi: 10.1126/science.aaq0180

de Souza VBC, Jordan BT, Tseng E, et al (2023) Transformation of alignment files improves performance of variant callers for long-read RNA sequencing data. *Genome Biology* 24:91. doi: 10.1186/s13059-023-02923-y

Ding X, Yu L, Chen L, et al (2022) Recent progress and future prospect of CRISPR/Cas-derived transcription activation (CRISPRa) system in plants. *Cells* 11:3045. doi: 10.3390/cells11193045

Dominguez AA, Lim WA, Qi LS (2016) Beyond editing: Repurposing CRISPR-Cas9 for precision genome regulation and interrogation. *Nature Reviews Molecular Cell Biology* 17:5–15. doi: 10.1038/nrm.2015.2

Eid A, Alshareef S, Mahfouz MM (2018) CRISPR base editors: Genome editing without double-stranded breaks. *Biochemical Journal* 475:1955–1964. doi: 10.1042/BCJ20170793

El-Mounadi K, Morales-Floriano ML, Garcia-Ruiz H (2020) Principles, applications, and biosafety of plant genome editing using CRISPR-Cas9. *Frontiers in Plant Science* 11:1–16. doi: 10.3389/fpls.2020.00056

Fahad S, Bajwa AA, Nazir U, et al (2017) Crop production under drought and heat stress: Plant responses and management options. *Frontiers in Plant Science* 8:1–16. doi: 10.3389/fpls.2017.01147

Feng Z, Zhang B, Ding W, et al (2013) Efficient genome editing in plants using a CRISPR/Cas system. *Cell Research* 23:1229–1232. doi: 10.1038/cr.2013.114

Ferré-D'Amaré AR, Zhou K, Doudna JA (1998) Crystal structure of a hepatitis delta virus ribozyme. *Nature* 395:567–574. doi: 10.1038/26912

Ford K, McDonald D, Mali P (2019) Functional genomics via CRISPR-Cas. *Journal of Molecular Biology* 431:48–65. doi: 10.1016/j.jmb.2018.06.034

Garg R, Jain* M (2013) Transcriptome analyses in legumes: A resource for functional genomics. *The Plant Genome* 6:1–9. doi: 10.3835/plantgenome2013.04.0011

Giacomello S (2021) A new era for plant science: Spatial single-cell transcriptomics. *Current Opinion in Plant Biology* 60:102041. doi: 10.1016/j.pbi.2021.102041

Giacomello S, Salmén F, Terebieniec BK, et al (2017) Spatially resolved transcriptome profiling in model plant species. *Nature Plants* 3:1–11. doi: 10.1038/nplants.2017.61

Guigo R, de Hoon M (2018) Recent advances in functional genome analysis. *F1000Research* 7:1968. doi: 10.12688/f1000research.15274.1

Guo J, Huang Z, Sun J, et al (2021) Research progress and future development trends in medicinal plant transcriptomics. *Frontiers in Plant Science* 12:1–10. doi: 10.3389/fpls.2021.691838

Hao L, Ruiying Q, Xiaoshuang L, et al (2019) CRISPR/Cas9-mediated adenine base editing in rice genome. *Rice Science* 26:125–128. doi: 10.1016/j.rsci.2018.07.002

Hu JH, Miller SM, Geurts MH, et al (2018) Evolved Cas9 variants with broad PAM compatibility and high DNA specificity. *Nature* 556:57–63. doi: 10.1038/nature26155

Hua K, Tao X, Liang W, et al (2020) Simplified adenine base editors improve adenine base editing efficiency in rice. *Plant Biotechnology Journal* 18:770–778. doi: 10.1111/pbi.13244

Jiang YY, Chai YP, Lu MH, et al (2020) Prime editing efficiently generates W542L and S621I double mutations in two ALS genes in maize. *Genome Biology* 21:1–10. doi: 10.1186/s13059-020-02170-5

Jin S, Fei H, Zhu Z, et al (2020) Rationally designed APOBEC3B cytosine base editors with improved specificity. *Molecular Cell* 79:728–740.e6. doi: 10.1016/j.molcel.2020.07.005

Jinek M, Chylinski K, Fonfara I, Hauer M, Doudna JA, Charpentier E (2012) A programmable dual-RNA-guided DNA endonuclease in adaptive bacterial immunity. *Science* 337(6096):816–21. doi: 10.1126/science.1225829.

Kannan, B, Jung, JH, Moxley, GW, et al. (2018). TALEN-mediated targeted mutagenesis of more than 100 COMT copies/alleles in highly polyploid sugarcane improves saccharification efficiency without compromising biomass yield. *Plant Biotechnology Journal* 16(4): 856–866.

Kavuri NR, Ramasamy M, Qi Y, Mandadi K (2022) Applications of CRISPR/Cas13-based RNA editing in plants. *Cells* 11:2665. doi: 10.3390/cells11172665

Khatodia S, Bhatotia K, Passricha N, et al (2016) The CRISPR/Cas genome-editing tool: Application in improvement of crops. *Frontiers in Plant Science* 7:1–13. doi: 10.3389/fpls.2016.00506

Klemm SL, Shipony Z, Greenleaf WJ (2019) Chromatin accessibility and the regulatory epigenome. *Nature Reviews Genetics* 20:207–220. doi: 10.1038/s41576-018-0089-8

Komor AC, Kim YB, Packer MS, et al (2016) Programmable editing of a target base in genomic DNA without double-stranded DNA cleavage. *Nature* 533:420–424. doi: 10.1038/nature17946

Kuang Y, Li S, Ren B, et al (2020) Base-editing-mediated artificial evolution of OsALS1 in planta to develop novel herbicide-tolerant rice germplasms. *Molecular Plant* 13:565–572. doi: 10.1016/j.molp.2020.01.010

Kurt IC, Zhou R, Iyer S, et al (2021) CRISPR C-to-G base editors for inducing targeted DNA transversions in human cells. *Nature Biotechnology* 39:41–46. doi: 10.1038/s41587-020-0609-x

Laird PW (2010) Principles and challenges of genome-wide DNA methylation analysis. *Nature Reviews Genetics* 11:191–203. doi: 10.1038/nrg2732

Lee H, Kim J, Lee J (2023) Benchmarking datasets for assembly-based variant calling using high-fidelity long reads. *BMC Genomics* 24:148. doi: 10.1186/s12864-023-09255-y

Li C, Zhang R, Meng X, et al (2020) Targeted, random mutagenesis of plant genes with dual cytosine and adenine base editors. *Nature Biotechnology* 38:875–882. doi: 10.1038/s41587-019-0393-7

Li J, Chen L, Liang J, et al (2022) Development of a highly efficient prime editor 2 system in plants. *Genome Biology* 23:161. doi: 10.1186/s13059-022-02730-x

Li W, Teng F, Li T, Zhou Q (2013) Simultaneous generation and germline transmission of multiple gene mutations in rat using CRISPR-Cas systems. *Nature Biotechnology* 31:684–686. doi: 10.1038/nbt.2652

Liang Z, Wu Y, Guo Y, Wei S (2023) Addition of the T5 exonuclease increases the prime editing efficiency in plants. *Journal of Genetics and Genomics* 50:582–588. doi: 10.1016/j.jgg.2023.03.008

Lin Q, Jin S, Zong Y, et al (2021) High-efficiency prime editing with optimized, paired pegRNAs in plants. *Nature Biotechnology* 39:923–927. doi: 10.1038/s41587-021-00868-w

Lin Q, Zong Y, Xue C, et al (2020) Prime genome editing in rice and wheat. *Nature Biotechnology* 38:582–585. doi: 10.1038/s41587-020-0455-x

Lin YC, Li W, Chen H, et al (2014) A simple improved-throughput xylem protoplast system for studying wood formation. *Nature Protocols* 9:2194–2205. doi: 10.1038/nprot.2014.147

Liu D, Chen M, Mendoza B, et al (2019) CRISPR/Cas9-mediated targeted mutagenesis for functional genomics research of crassulacean acid metabolism plants. *Journal of Experimental Botany* 70:6621–6629. doi: 10.1093/jxb/erz415

Liu Y, Li C, Han Y, et al (2022) Spatial transcriptome analysis on peanut tissues shed light on cell heterogeneity of the peg. *Plant Biotechnology Journal* 20:1648–1650. doi: 10.1111/pbi.13884

Logsdon GA, Vollger MR, Eichler EE (2020) Long-read human genome sequencing and its applications. *Nature Reviews Genetics* 21:597–614. doi: 10.1038/s41576-020-0236-x

Lu Y, Tian Y, Shen R, et al (2021) Precise genome modification in tomato using an improved prime editing system. *Plant Biotechnology Journal* 19:415–417. doi: 10.1111/pbi.13497

Lu Y, Zhu J-K (2017) Precise editing of a target base in the rice genome using a modified CRISPR/Cas9 system. *Molecular Plant* 10:523–525. doi: 10.1016/j.molp.2016.11.013

Ma X, Zhang Q, Zhu Q, et al (2015) A robust CRISPR/Cas9 system for convenient, high-efficiency multiplex genome editing in monocot and dicot plants. *Molecular Plant* 8:1274–1284. doi: 10.1016/j.molp.2015.04.007

Maghuly F, Molin EM, Saxena R, Konkin DJ (2022) Editorial: Functional genomics in plant breeding 2.0. *International Journal of Molecular Sciences* 23:5–8. doi: 10.3390/ijms23136959

Makarova KS, Koonin EV (2015) Annotation and classification of CRISPR-Cas systems. *CRISPR: Methods and Protocols* 1311:47–75. doi: 10.1007/978-1-4939-2687-9_4.

Marwein R, Singh S, Maharana J, et al (2022) Transcriptome-wide analysis of North-East Indian rice cultivars in response to *Bipolaris oryzae* infection revealed the importance of early response to the pathogen in suppressing the disease progression. *Gene* 809:146049. doi: 10.1016/j.gene.2021.146049

Montalbano A, Canver MC, Sanjana NE (2017) High-throughput approaches to pinpoint function within the noncoding genome. *Molecular Cell* 68:44–59. doi: 10.1016/j.molcel.2017.09.017

Montecillo JA V., Chu LL, Bae H (2020) CRISPR-Cas9 system for plant genome editing: Current approaches and emerging developments. *Agronomy* 10. doi: 10.3390/agronomy10071033

Naeem M, Majeed S, Hoque MZ, Ahmad I (2020) Latest developed strategies to minimize the off-target effects in CRISPR-Cas-mediated genome editing. *Cells* 9:1608. doi: 10.3390/cells9071608

Naim F, Dugdale B, Kleidon J, et al (2018) Gene editing the phytoene desaturase alleles of Cavendish banana using CRISPR/Cas9. *Transgenic Research* 27:451–460. doi: 10.1007/s11248-018-0083-0

Nishida K, Arazoe T, Yachie N, et al. (2016) Targeted nucleotide editing using hybrid prokaryotic and vertebrate adaptive immune systems. *Science* 353(6305):aaf8729. doi: 10.1126/science.aaf8729.

Nishimasu H, Shi X, Ishiguro S, et al (2018) Engineered CRISPR-Cas9 nuclease with expanded targeting space. *Science* 361:1259–1262. doi: 10.1126/science.aas9129

Ordon J, Gantner J, Kemna J, et al (2017) Generation of chromosomal deletions in dicotyledonous plants employing a user-friendly genome editing toolkit. *The Plant Journal* 89:155–168. doi: 10.1111/tpj.13319

Peng G, Cui G, Ke J, Jing N (2020) Using single-cell and spatial transcriptomes to understand stem cell lineage specification during early embryo development. *Annual Review of Genomics and Human Genetics* 21:163–181. doi: 10.1146/annurev-genom-120219-083220

Pley HW, Flaherty KM, McKay DB (1994) Three-dimensional structure of a hammerhead ribozyme. *Nature* 372:68–74. doi: 10.1038/372068a0

Poretti M, Sotiropoulos AG, Graf J, et al (2021) Comparative transcriptome analysis of wheat lines in the field reveals multiple essential biochemical pathways suppressed by obligate pathogens. *Frontiers in Plant Science* 12. doi: 10.3389/fpls.2021.720462

Pulecio J, Verma N, Mejía-Ramírez E, et al (2017) CRISPR/Cas9-based engineering of the epigenome. *Cell Stem Cell* 21:431–447. doi: 10.1016/j.stem.2017.09.006

Rai A, Yamazaki M, Saito K (2019) A new era in plant functional genomics. *Current Opinion in Systems Biology* 15:58–67. doi: 10.1016/j.coisb.2019.03.005

Rastogi S, Shah S, Kumar R, et al (2019) Ocimum metabolomics in response to abiotic stresses: Cold, flood, drought and salinity. *PLoS ONE* 14:1–26. doi: 10.1371/journal.pone.0210903

Richter MF, Zhao KT, Eton E, et al (2020) Phage-assisted evolution of an adenine base editor with improved Cas domain compatibility and activity. *Nature Biotechnology* 38:883–891. doi: 10.1038/s41587-020-0453-z

Rothan C, Diouf I, Causse M (2019) Trait discovery and editing in tomato. *The Plant Journal* 97:73–90. doi: 10.1111/tpj.14152

Shan Q, Wang Y, Li J, et al (2013) Targeted genome modification of crop plants using a CRISPR-Cas system. *Nature Biotechnology* 31:686–688. doi: 10.1038/nbt.2650

Shulse CN, Cole BJ, Ciobanu D, et al (2019) High-throughput single-cell transcriptome profiling of plant cell types. *Cell Reports* 27:2241–2247.e4. doi: 10.1016/j.celrep.2019.04.054

Sternberg SH, Redding S, Jinek M, et al (2014) DNA interrogation by the CRISPR RNA-guided endonuclease Cas9. *Nature* 507:62–67. doi: 10.1038/nature13011

Stuttmann J, Barthel K, Martin P, et al. (2021). Highly efficient multiplex editing: one-shot generation of 8× Nicotiana benthamiana and 12× Arabidopsis mutants. *The Plant Journal* 106 (1):8–22.

Takei H, Shirasawa K, Kuwabara K, et al (2021) De novo genome assembly of two tomato ancestors, *Solanum pimpinellifolium* and *Solanum lycopersicum* var. *cerasiforme*, by long-read sequencing. *DNA Research* 28:dsaa029. doi: 10.1093/dnares/dsaa029

Tan J, Zeng D, Zhao Y, et al (2022) PhieABEs: A PAM-less/free high-efficiency adenine base editor toolbox with wide target scope in plants. *Plant Biotechnology Journal* 20:934–943. doi: 10.1101/2020.07.21.213827

Tang X, Sretenovic S, Ren Q, et al (2020) Plant prime editors enable precise gene editing in rice cells. *Molecular Plant* 13:667–670. doi: 10.1016/j.molp.2020.03.010

Tang Y, Fu Y (2018) Class 2 CRISPR/Cas: an expanding biotechnology toolbox for and beyond genome editing. *Cell Bioscience* 8:59. doi: 10.1186/s13578-018-0255-x.

Tian S, Jiang L, Gao Q, et al (2017) Efficient CRISPR/Cas9-based gene knockout in watermelon. *Plant Cell Reports* 36:399–406. doi: 10.1007/s00299-016-2089-5

Tripathi L, Dhugga KS, Ntui VO, Runo S, Syombua ED, Muiruri S, Tripathi JN (2022). Genome editing for sustainable agriculture in Africa. *Frontiers in Genome Editing* 4:876697.

Tsai SQ, Wyvekens N, Khayter C, et al (2014) Dimeric CRISPR RNA-guided FokI nucleases for highly specific genome editing. *Nature Biotechnology* 32:569–576. doi: 10.1038/nbt.2908

Tyagi P, Singh D, Mathur S, et al (2022) Upcoming progress of transcriptomics studies on plants: An overview. *Frontiers in Plant Science* 13:1–15. doi: 10.3389/fpls.2022.1030890

van Rengs WMJ, Schmidt MHW, Effgen S, et al (2022) A chromosome scale tomato genome built from complementary PacBio and Nanopore sequences alone reveals extensive linkage drag during breeding. *Plant Journal* 110:572–588. doi: 10.1111/tpj.15690

Veillet F, Chauvin L, Paule M, et al (2019a) The *Solanum tuberosum GBSSI gene*: A target for assessing gene and base editing in tetraploid potato. *Plant Cell Reports*. doi: 10.1007/s00299-019-02426-w

Veillet F, Perrot L, Chauvin L, et al (2019b) Transgene-free genome editing in tomato and potato plants using *Agrobacterium*-mediated delivery of a CRISPR/Cas9 cytidine base editor. *International Journal of Molecular Sciences* 20. doi: 10.3390/ijms20020402

Wang G, Xu Z, Wang F, et al (2022) Development of an efficient and precise adenine base editor (ABE) with expanded target range in allotetraploid cotton (*Gossypium hirsutum*). *BMC Biology* 20:45. doi: 10.1186/s12915-022-01232-3

Wang H, La Russa M, Qi LS (2016) CRISPR/Cas9 in genome editing and beyond. *Annual Review of Biochemistry* 85:227–264. doi: 10.1146/annurev-biochem-060815-014607

Xie J, Li M, Zeng J, et al (2022) Single-cell RNA sequencing profiles of stem-differentiating xylem in poplar. *Plant Biotechnology Journal* 20:417–419. doi: 10.1111/pbi.13763

Xie K, Zhang J, Yang Y (2014) Genome-wide prediction of highly specific guide RNA spacers for CRISPR-Cas9-mediated genome editing in model plants and major crops. *Molecular Plant* 7:923–926. doi: 10.1093/mp/ssu009

Xing S, Chen K, Zhu H, et al (2020) Fine-tuning sugar content in strawberry. *Genome Biology* 21:230. doi: 10.1186/s13059-020-02146-5

Xu R, Li J, Liu X, et al (2020a) Development of plant prime-editing systems for precise genome editing. *Plant Communications* 1:100043. doi: 10.1016/j.xplc.2020.100043

Xu W, Zhang C, Yang Y, et al (2020b) Versatile nucleotides substitution in plant using an improved prime editing system. *Molecular Plant* 13:675–678. doi: 10.1016/j.molp.2020.03.012

Yan J, Qian L, Zhu W, et al (2020) Integrated analysis of the transcriptome and metabolome of purple and green leaves of *Tetrastigma hemsleyanum* reveals gene expression patterns involved in anthocyanin biosynthesis. *PLoS ONE* 15:e0230154. doi: 10.1371/journal.pone.0230154

Yan L, Wei S, Wu Y, et al (2015) High-efficiency genome editing in arabidopsis using YAO promoter-driven CRISPR/Cas9 system. *Molecular Plant* 8:1820–1823. doi: 10.1016/j.molp.2015.10.004

Yao R, Liu D, Jia X, et al (2018) CRISPR-Cas9/Cas12a biotechnology and application in bacteria. *Synthetic and Systems Biotechnology* 3:135–149. doi: 10.1016/j.synbio.2018.09.004

Yarnall MTN, Ioannidi EI, Schmitt-Ulms C, et al (2023) Drag-and-drop genome insertion of large sequences without double-strand DNA cleavage using CRISPR-directed integrases. *Nature Biotechnology* 41:500–512. doi: 10.1038/s41587-022-01527-4

Yu X, Liu Z, Sun X (2023) Single-cell and spatial multi-omics in the plant sciences: Technical advances, applications, and perspectives. *Plant Communications* 100508. doi: 10.1016/j.xplc.2022.100508

Zeng D, Zheng Z, Liu Y, et al (2022) Exploring C-to-G and A-to-Y base editing in rice by using new vector tools. *International Journal of Molecular Sciences* 23:7990. doi: 10.3390/ijms23147990

Zetsche B, Heidenreich M, Mohanraju P, et al (2017) Multiplex gene editing by CRISPR-Cpf1 using a single crRNA array. *Nature Biotechnology* 35:31–34. doi: 10.1038/nbt.3737

Zhang TQ, Xu ZG, Shang GD, Wang JW (2019) A single-cell RNA sequencing profiles the developmental landscape of *Arabidopsis* root. *Molecular Plant* 12:648–660. doi: 10.1016/j.molp.2019.04.004

Zhu H, Li C, Gao C (2020) Applications of CRISPR-Cas in agriculture and plant biotechnology. *Nature Reviews Molecular Cell Biology* 21:661–677. doi: 10.1038/s41580-020-00288-9

4 Strategies and Applications of Genomic Editing in Plants with CRISPR/Cas9

Ishfaq Majid Hurrah, Tabasum Mohiuddin, Sayanti Mandal, Mimosa Ghorai, Sayan Bhattacharya, Potshangbam Nongdam, Abdel Rahman Al-Tawaha, Ercan Bursal, Mahipal S Shekhawat, Devendra Kumar Pandey, and Abhijit Dey

4.1 INTRODUCTION

Genome editing in plants refers to the use of precise techniques to make targeted modifications in the DNA sequences of plants. One of the most widely used and powerful genome editing technologies is CRISPR (Clustered Regularly Interspaced Short Palindromic Repeats) (Liu et al., 2017a; El-Mounadi et al., 2020). CRISPR-Cas9, the most commonly used CRISPR system in genome editing, consists of two main components: the Cas9 protein and a guide RNA (gRNA). CRISPR/Cas9 is a revolutionary gene-editing technology that allows scientists to make precise changes to the DNA of living organisms (Zhang and Showalter, 2020; Wada et al., 2020). It has gained immense popularity and has transformed the field of genetic engineering since its development in 2012. CRISPR is a naturally occurring system found in bacteria and archaea that helps them defend against viral infections (Tahir et al., 2020; Deb et al., 2022). It consists of short, repeated DNA sequences interspersed with unique sequences called spacers derived from viral DNA. Cas9 is an enzyme that acts as a molecular pair of scissors in the CRISPR system (Bao et al., 2019; Zhang and Showalter, 2020). It is an RNA-guided endonuclease, which means it can cut DNA at specific locations based on the instructions provided by a guide RNA molecule (Sun et al., 2019; Cram et al., 2019). The guide RNA (gRNA) is a synthetic molecule designed to be complementary to a specific target DNA sequence. It consists of two components: CRISPR RNA (crRNA) is a short RNA sequence derived from the repeated DNA sequences in the CRISPR region (Demirci et al., 2018; Bao et al., 2019). The Trans-activating CRISPR RNA (tracrRNA) is an engineered RNA molecule that helps the crRNA and Cas9 form a functional complex.

When the Cas9 protein is combined with the guide RNA and is introduced into a cell, it searches for a DNA sequence that matches the guide RNA (Schindele et al., 2018; Bao et al., 2019). Once a match is found, Cas9 binds to the target DNA and creates a double-strand break (DSB) at that location. The cell's natural repair mechanisms then come into play to repair the DSB (Feng et al., 2018; Erpen-Dalla et al., 2019). The DNA repair mechanism, known as non-homologous end joining (NHEJ), often results in small insertions or deletions (indels) at the site of the break. These indels can disrupt the function of the target gene (Basso et al., 2021). In the presence of a DNA repair template, such as a synthetic DNA molecule, the homology-directed repair (HDR) pathway can be utilized to introduce precise changes to the DNA sequence. This allows researchers to insert or

DOI: 10.1201/9781003387060-4

replace specific DNA sequences with high accuracy (Vu et al., 2020; Tang et al., 2023). CRISPR/Cas9 has opened up new possibilities in various fields, including biomedical research, agriculture, and environmental conservation. It enables scientists to study the functions of genes, develop disease models, engineer crops with desired traits, and potentially treat genetic diseases in the future (Bessoltane et al., 2022; Impens et al., 2022).

It's important to note that while CRISPR/Cas9 holds tremendous promise, its applications and ethical considerations are still being actively researched and discussed. Regulatory frameworks are being developed to ensure the responsible and safe use of this technology (Naz et al., 2022; Deb et al., 2022). CRISPR/Cas9 is a powerful gene-editing tool that has revolutionized the field of genetic engineering, including its application in plants. It allows scientists to make precise changes to the DNA of plants by targeting specific genes and altering their sequence or expression (Li et al., 2022a; Rasheed et al., 2022). Researchers first identify the gene they intend to modify in a plant. This gene could be associated with traits such as disease resistance, improved yield, or enhanced nutritional content. A small RNA molecule called a guide RNA (gRNA) is designed to match the target gene's sequence (Zhao et al., 2022; Haq et al., 2022). The gRNA acts as a guide to direct the Cas9 protein to the desired location in the plant's genome. The Cas9 protein and the gRNA are introduced into plant cells. This can be achieved through various methods such as *Agrobacterium*-mediated transformation, particle bombardment, or direct delivery using nanoparticles. Once inside the plant cells, the Cas9 protein creates a DSB at the target site of the DNA (Razzaq et al., 2022; Boubakri, 2023).

Researchers can provide a DNA template along with the Cas9 and gRNA components, which the cell can use as a template for precise gene replacement or insertion. Plant cells that have undergone successful gene editing are identified and selected using various methods, such as molecular markers or fluorescent reporters linked to the target gene (Hillary and Ceasar, 2023; Akram et al., 2023). The edited cells are further cultured and developed into whole plants using tissue culture techniques. These plants are then grown under regulated conditions to produce the desired traits. CRISPR/Cas9 offers several advantages for plant breeding compared to traditional methods like mutagenesis or transgenic approaches. It enables targeted and precise modifications, reduces off-target effects, and accelerates the breeding process (Angon and Habiba, 2023; Alamillo et al., 2023). It has been used successfully to introduce disease resistance, improve crop yield, enhance nutritional content, and develop stress-tolerant varieties in various plant species. It's important to note that the use of CRISPR/Cas9 in plants is a rapidly evolving field, and research and regulations surrounding its applications are subject to change (Mandal et al., 2022; Hemalatha et al., 2023; Kumar et al., 2023a).

4.2 APPLICATIONS OF CRISPR/Cas9 IN PLANTS

CRISPR (Clustered Regularly Interspaced Short Palindromic Repeats) is a revolutionary gene-editing technology that has the potential to make precise modifications in the DNA of various organisms, including plants. By using CRISPR, scientists can edit specific genes, add new genetic material, or disable existing genes within the plant's genome. The applications of CRISPR in plants are extensive and have immense potential in agriculture and crop improvement (Figure 4.1).

Some potential uses of CRISPR in plants include:

4.2.1 Enhancing Crop Yield and Quality

CRISPR can be used to modify genes involved in plant growth, development, and stress response, leading to increased crop productivity and improved traits such as disease resistance, nutritional value, and shelf life. Enhancing crop yield and quality is one of the key goals of agricultural research, and

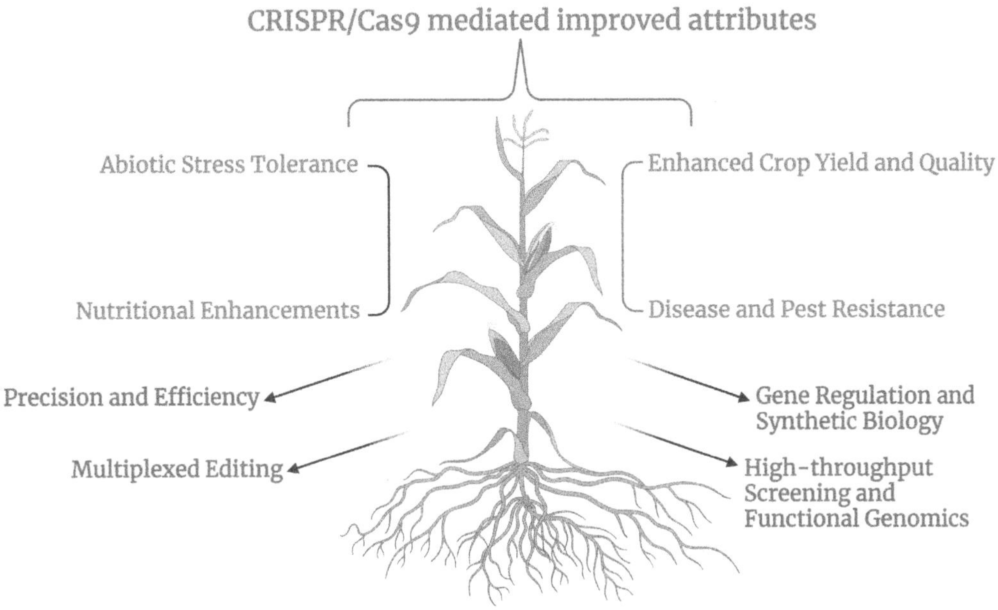

FIGURE 4.1 Enhanced plant attributes via CRISPR/Cas9 modification tool.

CRISPR/Cas9 technology offers a powerful tool for achieving these objectives (Liu et al., 2021a; Rasheed et al., 2021). Several applications of CRISPR/Cas9 to enhance crop yield and quality are discussed as follows:

4.2.1.1 Disease Resistance

Crop plants are susceptible to various diseases caused by pathogens such as viruses, bacteria, and fungi. By using CRISPR/Cas9, researchers can modify specific genes involved in plant defense mechanisms, thereby enhancing resistance to diseases. This can reduce yield losses, and the overall quality of the harvested crops can be substantially improved (Ahmad et al., 2020; Boubakri, 2023).

4.2.1.2 Pest Resistance

Pests like insects and nematodes can cause significant damage to crops. CRISPR/Cas9 can be utilized to introduce genetic modifications that make plants more resistant to pests. By targeting genes involved in pest recognition, defense responses, or toxin production, researchers can develop crops that are less susceptible to pest damage, leading to higher yields and improved quality (Douris et al., 2020; Zhang et al., 2022a).

4.2.1.3 Abiotic Stress Tolerance

Environmental stresses such as drought, heat, salinity, and cold can severely impact crop productivity and quality. CRISPR/Cas9 can be employed to modify genes associated with stress responses and tolerance, enabling plants to withstand and thrive under adverse conditions. By enhancing abiotic stress tolerance, crop yields can be improved, and the quality of harvested produce can be maintained (Mandal et al., 2022; Li et al., 2022b; Razzaq et al., 2022).

4.2.1.4 Nutritional Enhancements

CRISPR/Cas9 can be used to improve the nutritional composition of crops. For example, researchers can target genes involved in the biosynthesis of essential nutrients such as vitamins, minerals, or amino acids and enhance their production in crops. This can result in crops with improved nutritional value, addressing deficiencies in human diets and promoting better health outcomes (Fiaz et al., 2019; Priti et al., 2023).

4.2.1.5 Yield-Related Traits

CRISPR/Cas9 can be employed to modify genes involved in plant growth, development, and yield determination. By targeting genes associated with flowering time, fruit size, seed development, or other yield-related traits, researchers can optimize these characteristics to increase crop yields without compromising quality (Zeng et al., 2020; Liu et al., 2021b).

It's worth noting that the application of CRISPR/Cas9 in crop improvement is a rapidly advancing field of research. However, it's essential to consider the regulatory frameworks and public acceptance surrounding genetically modified crops in different regions. Responsible and ethical use of CRISPR/Cas9 technology, along with appropriate safety assessments, is crucial for ensuring sustainable and safe production.

4.2.2 Disease Resistance

Disease resistance in plants is a crucial aspect of crop protection and can be enhanced using CRISPR/Cas9 technology. CRISPR/Cas9 offers a precise and efficient method to modify specific genes involved in plant defense mechanisms, thereby improving their resistance to various pathogens (Li et al., 2020a). CRISPR can help create plants with enhanced resistance to diseases caused by viruses, bacteria, fungi, or pests (Ahmad et al., 2020; Boubakri, 2023). Applications of CRISPR/Cas9 to develop disease-resistant plants are discussed as follows:

4.2.2.1 Identification of Target Genes

The first step in developing disease-resistant plants through CRISPR/Cas9 is to identify the specific genes that play key roles in plant–pathogen interactions. These genes can be involved in pathogen recognition, defense signal transduction, or the production of antimicrobial compounds (Zhou et al., 2021; Jiang et al., 2021).

4.2.2.2 Gene Knockout

Once the target genes are identified, CRISPR/Cas9 can be used to disrupt or knock out those genes. By introducing DSBs in the target gene sequences using CRISPR/Cas9, the plant's natural DNA repair mechanisms will attempt to repair the breaks. This repair process often introduces errors, resulting in gene mutations or disruptions. Knocking out these genes can impair the pathogen's ability to infect the plant or weaken its virulence, thus enhancing the plant's resistance to the disease (Huang et al., 2018; Bellinvia et al., 2022).

4.2.2.3 Gene Modification

Alternatively, CRISPR/Cas9 can be used to introduce specific modifications in the target genes to enhance plant defense responses. This approach can involve altering regulatory elements, modifying protein-coding regions, or introducing mutations that lead to enhanced defense-related traits (Li et al., 2022a; Sami et al., 2021).

4.2.2.4 Multiplex Editing

CRISPR/Cas9 allows for the simultaneous targeting of multiple genes or gene regions. This capability enables the development of plants with enhanced resistance against multiple pathogens or strains simultaneously, offering broader and long-term disease resistance (Liu et al., 2017b; Uranga et al., 2021).

4.2.2.5 Non-Host Resistance

CRISPR/Cas9 can be used to introduce genes from non-host organisms into plants. Non-host resistance involves transferring genes from species that are not susceptible to a particular pathogen into the target plant. CRISPR/Cas9 facilitates the precise integration of these non-host resistance genes into the plant genome, enabling the development of disease-resistant plants that can defend against pathogens they would not naturally encounter (Rai et al., 2022; Tek et al., 2022).

By employing CRISPR/Cas9 technology to modify genes associated with plant–pathogen interactions, researchers can develop crop varieties with enhanced resistance to viral, bacterial, and fungal diseases. This approach provides a promising avenue for reducing yield losses, minimizing the need for chemical pesticides, and promoting sustainable and environmentally friendly agriculture. However, it's important to note that the release and cultivation of genetically modified crops, including those developed using CRISPR/Cas9, may be subject to specific regulations and considerations depending on the geographical location.

4.2.3 Improved Nutritional Content

CRISPR/Cas9 technology offers a promising approach to improving the nutritional content of plants by precisely modifying specific genes involved in nutrient synthesis, accumulation, or metabolism. CRISPR can be utilized to enhance the nutritional profile of plants by increasing the production of certain vitamins, minerals, or other beneficial compounds (Achary and Reddy, 2021; Chaudhary et al., 2022). This has the potential to address nutritional deficiencies and improve human health. The applications of CRISPR/Cas9 to enhance the nutritional content of plants are:

4.2.3.1 Vitamin and Mineral Enhancement

CRISPR/Cas9 can be employed to increase the production or accumulation of essential vitamins and minerals in plants. By targeting genes involved in the biosynthesis pathways of specific vitamins (e.g., vitamin A, vitamin C) or minerals (e.g., iron, zinc), researchers can introduce precise modifications to enhance their production. This can lead to crops with improved nutritional quality, addressing deficiencies in human diets (Kaur et al., 2020; Wan et al., 2021).

4.2.3.2 Amino Acid and Protein Enrichment

CRISPR/Cas9 can be used to modify genes involved in amino acid biosynthesis or protein metabolism pathways. By enhancing the expression or activity of key genes in these pathways, researchers can increase the levels of essential amino acids or improve protein quality in crops. This can result in plants with improved protein content or improved amino acid profiles, providing better nutrition to humans and livestock (Wang et al., 2021; Zhang et al., 2022b).

4.2.3.3 Fatty Acid Modification

CRISPR/Cas9 can be applied to alter the fatty acid composition of crops. By targeting genes involved in fatty acid biosynthesis or metabolism, researchers can modify the production of specific fatty acids, such as omega-3 fatty acids or monounsaturated fats. This can lead to crops with healthier lipid profiles and improved nutritional benefits (Huang et al., 2020a; Ma et al., 2021).

4.2.3.4 Antioxidant Enhancement

Plants produce various antioxidants that contribute to their nutritional value and health benefits. CRISPR/Cas9 can be utilized to increase the expression of genes involved in antioxidant synthesis pathways, thereby enhancing the production of antioxidants in crops. This can result in plants with improved antioxidant capacity, which is beneficial for human health (Bouzroud et al., 2020; Riu et al., 2023).

4.2.3.5 Anti-Nutrient Reduction

Some plants naturally contain anti-nutrients, such as phytic acid or tannins, which can reduce the bioavailability of certain nutrients. CRISPR/Cas9 can be used to target and modify genes involved in the synthesis or accumulation of these anti-nutrients, leading to crops with reduced levels of anti-nutritional compounds and improved nutrient absorption (Kumar et al., 2023b; Sahu et al., 2023).

It's important to note that the application of CRISPR/Cas9 to improve the nutritional content of crops has been one of the most significant research domains in recent times. Regulatory considerations and public acceptance of genetically modified crops may vary depending on the region. Responsible and ethical use of CRISPR/Cas9 technology, along with appropriate safety assessments, is crucial to ensure that crops with improved nutritional content are developed sustainably and safely.

4.2.4 Environmental Adaptation

Climate change poses significant challenges to agriculture. CRISPR/Cas9 technology offers a powerful tool for enhancing the environmental adaptation of plants. By modifying specific genes involved in stress responses and tolerance mechanisms, CRISPR/Cas9 can help develop plants that are better suited to withstand adverse environmental conditions (Sarma et al., 2021; Bhattacharyya et al., 2023). Several applications of CRISPR/Cas9 for developing environmental adaptation in plants are discussed as follows:

4.2.4.1 Drought Tolerance

Drought is a significant environmental stress that negatively impacts crop productivity. CRISPR/Cas9 can be employed to modify genes involved in water-use efficiency, stomatal regulation, or osmotic stress responses. By enhancing the expression or function of these genes, plants can exhibit improved drought tolerance, maintaining better water status and productivity under water-deficient conditions (Shinwari et al., 2020; Abdallah et al., 2022).

4.2.4.2 Heat and Cold Tolerance

Extreme temperatures, both high and low, can severely affect plant growth and development. CRISPR/Cas9 can be used to modify genes associated with temperature stress responses, heat shock proteins, or cold acclimation pathways. By optimizing the expression or activity of these genes, plants can exhibit enhanced tolerance to temperature extremes, ensuring better growth and survival under challenging thermal conditions (Li et al., 2022b; Kumar et al., 2023a).

4.2.4.3 Salinity Tolerance

Salinity stress, caused by high levels of salt in the soil, is a major constraint to agricultural productivity. CRISPR/Cas9 can be utilized to modify genes involved in ion transport, osmotic regulation, or ion homeostasis. By optimizing the expression or function of these genes, plants can develop improved salinity tolerance, allowing them to thrive in saline environments and maintain productivity (Chauhan et al., 2022; Peng et al., 2022). This can help cope with adverse situations like sea level rise and seawater intrusion in coastal areas.

4.2.4.5 Disease and Pest Resistance in Changing Climates

Climate change can lead to shifts in disease and pest dynamics, challenging plant health. CRISPR/Cas9 can be applied to modify genes involved in plant defense responses, disease resistance pathways, or insect resistance mechanisms. By enhancing the expression or effectiveness of these genes, plants can exhibit improved resistance against evolving pathogens and pests, minimizing yield losses and maintaining crop quality (Zhang and Batley, 2020; Mubarik et al., 2020).

4.2.4.6 Adaptation to Nutrient-Poor Soils

Some regions have soils with limited nutrient availability, thereby affecting plant growth and nutrient uptake. CRISPR/Cas9 can be used to modify genes involved in nutrient acquisition, utilization, or symbiotic interactions. By optimizing these genes, plants can exhibit enhanced adaptation to nutrient-poor soils, leading to improved growth and yield in challenging soil conditions (Wang et al., 2020a; Panchal et al., 2023).

4.2.5 Weed Control

Weed control is a critical aspect of agriculture, and while CRISPR/Cas9 technology has tremendous potential in various areas of plant research, its direct application for weed control is still in the early stages of development (Han and Kim 2019; Lyu et al., 2022). Applications of CRISPR/Cas9 for weed management are discussed as follows:

4.2.5.1 Herbicide Resistance

Herbicide-resistant weeds pose a significant challenge to agriculture. CRISPR/Cas9 can potentially be used to confer herbicide resistance in crops. By modifying specific genes involved in herbicide susceptibility or detoxification pathways, researchers can develop crops that are resistant to specific herbicides. This would be effective in controlling weeds without damaging the cultivated plants (Wu et al., 2020a; Dong et al., 2021).

4.2.5.2 Weed-Specific Gene Modification

CRISPR/Cas9 can be used to target and modify genes that are unique to weeds. By identifying weed-specific genes involved in growth, development, or survival, researchers can design CRISPR/Cas9 strategies to disrupt or modify those genes, potentially impairing weed growth and reducing their competitive advantage (Gressel, 2018; Fernández-Aparicio et al., 2020).

4.2.5.3 Targeted Weed Control Approaches

CRISPR/Cas9 can potentially be applied to develop targeted weed control approaches. For example, researchers can target genes involved in weed seed germination or weed-specific traits that make them invasive or competitive. By modifying these genes, it may be possible to reduce weed populations or hinder their growth without affecting desirable crops (Naz et al., 2022; Kumar et al., 2023b).

It's important to note that the application of CRISPR/Cas9 for weed control is a complex and evolving field of research. Several challenges need to be addressed, such as identifying suitable target genes, ensuring the specificity of gene modifications to weed species, and considering potential off-target effects. Additionally, regulatory considerations and public acceptance of genetically modified organisms, including weed control strategies, play a significant role in determining the feasibility and implementation of CRISPR/Cas9-based approaches for weed management. Further research and development are necessary to explore the potential of CRISPR/Cas9 for effective and sustainable weed control strategies.

4.3 STRATEGIES UTILIZED VIA CRISPR/Cas9 FOR PLANT DEVELOPMENT

CRISPR/Cas9 strategies in plants involve the use of the CRISPR/Cas9 gene-editing technology to introduce targeted genetic modifications. Genome editing techniques like CRISPR offer precise and efficient tools for plant breeders and researchers to accelerate the development of new plant varieties with desired traits, contributing to more sustainable and resilient agricultural practices. The applications of genome editing in plants are broad and diverse. They include crop improvement for enhanced yield, nutritional quality, and stress tolerance, as well as the development of plants with resistance to diseases, pests, and environmental challenges (Mandal and Verma, 2021).

4.3.1 Designing Guide RNAs (gRNAs)

The first step is to design gRNAs that specifically target the gene of interest. The gRNA consists of two components: a targeting sequence and a scaffold sequence. The targeting sequence is a short sequence of about 20 nucleotides that is complementary to the DNA sequence at the target site. The scaffold sequence is a constant sequence derived from the CRISPR system, which helps in the stability and functioning of the gRNA. The targeting sequence should be designed to maximize specificity and efficiency. Regions with high GC content or secondary structures should be avoided, as these can affect gRNA stability and efficiency. A target site should be chosen where the first few nucleotides are purine (A or G), as it has been shown to improve Cas9 cleavage efficiency. Target sites near the start codon or critical functional domains of the gene should be avoided, as disruptions in these regions can have more significant effects on gene function (Sun et al., 2019; Uniyal et al., 2019).

4.3.2 Constructing the CRISPR/Cas9 Vector

The gRNA sequence is typically cloned into a vector along with the Cas9 gene or Cas9 protein-coding sequence. Choosing a suitable plasmid or vector backbone would be useful; that will serve as the platform for constructing the CRISPR/Cas9 vector. The backbone should have features such as the origin of replication, selectable markers (e.g., antibiotic resistance genes), and appropriate promoter and terminator sequences. A cassette for expressing the Cas9 protein can be designed and incorporated. This cassette typically includes a promoter to drive Cas9 expression (e.g., U6, Ubi, or other strong promoters), the Cas9 coding sequence, and a terminator sequence. The Cas9 gene can be obtained from a bacterial source (e.g., *Streptococcus pyogenes* Cas9) or other Cas9 orthologs with different properties (e.g., Cas9 from *Staphylococcus aureus* or Cpf1 from *Francisella novicida*). A cassette can be designed and incorporated for expressing the guide RNA (gRNA) that will guide the Cas9 enzyme to the target site in the genome. This cassette typically includes a promoter (e.g., U6 or other RNA polymerase III promoters) to drive gRNA expression, the gRNA sequence, and a terminator sequence. The Cas9 expression cassette and the gRNA expression cassette can be cloned into the vector backbone using standard molecular biology techniques such as restriction enzyme digestion, ligation, or Gibson assembly (Bai et al., 2020; Wu et al., 2020b).

4.3.3 Delivery of CRISPR/Cas9 into Plant Cells

The CRISPR/Cas9 construct is delivered into plant cells, typically through tissue culture methods. *Agrobacterium tumefaciens* is a natural plant pathogen that can be used as a delivery system for introducing foreign DNA into plant cells. The CRISPR/Cas9 components, including the Cas9 gene and the gRNA expression cassette, are introduced into Agrobacterium cells through bacterial transformation (Schmitz et al., 2020). Particle bombardment involves gold or tungsten particles coated with the CRISPR/Cas9 components that are shot into plant tissues using a gene gun or biolistic device. The high-velocity particles penetrate the cell walls and deliver the CRISPR/Cas9 system into the plant cells (Liang et al., 2019). Protoplasts are plant cells that have had their cell walls removed enzymatically. The CRISPR/Cas9 components are directly introduced into protoplasts by methods such as polyethylene glycol (PEG) or electroporation. The protoplasts can then regenerate into whole plants under appropriate conditions (Sant'Ana et al., 2020). Plant viruses can be engineered to carry the CRISPR/Cas9 components and infect plant cells. The viral vector delivers the CRISPR/Cas9 system, allowing gene editing to occur. The choice of viral vector depends on the target plant species and the viral host range (Liu and Zhang, 2020).

4.3.4 Cas9-Induced DSB

Once inside the plant cells, the Cas9 protein and gRNA complex recognize and bind to the target gene's DNA sequence. Cas9 then introduces DSBs at the targeted site, generating DNA breaks (Figure 4.2).

The Cas9 protein, when guided by the gRNA, scans the genome for a specific DNA sequence known as the protospacer adjacent motif (PAM) followed by a region complementary to the gRNA's targeting sequence. Once the Cas9-gRNA complex recognizes the target site, it induces a DSB by introducing precise cuts in the DNA strands. The repair outcome at the DSB site determines the gene editing result in knockout mutations. If the DSB is repaired through error-prone NHEJ, it can result in the disruption of the target gene's function, leading to gene knockout. If an appropriate repair template is provided and utilized through HDR, precise edits can be introduced at the target site, such as point mutations, gene insertions, or gene replacements (Ben Shlush et al., 2020; Ben-Tov et al., 2023).

4.3.5 DNA Repair Mechanisms

The plant's natural DNA repair mechanisms come into play to repair the DSBs induced by Cas9. The two primary DNA repair pathways are NHEJ and HDR (Přibylová et al., 2022). The NHEJ pathway repairs the DSBs by joining the DNA ends, often resulting in the introduction of insertions or deletions (indels) at the target site (Tan et al., 2020). These indels can cause frameshift mutations, premature stop codons, or other disruptions in the target gene, effectively knocking it out. The Cas9 protein and gRNA complex recognize and bind to the target DNA sequence, creating DSBs. In this case, the goal is to exploit the HDR pathway, which can use the provided DNA template as a guide to repair the DSBs (Singh et al., 2023).

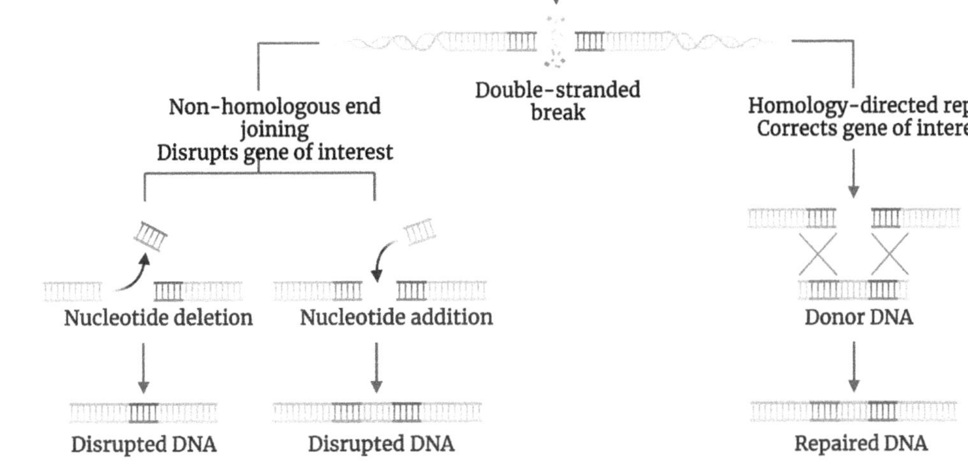

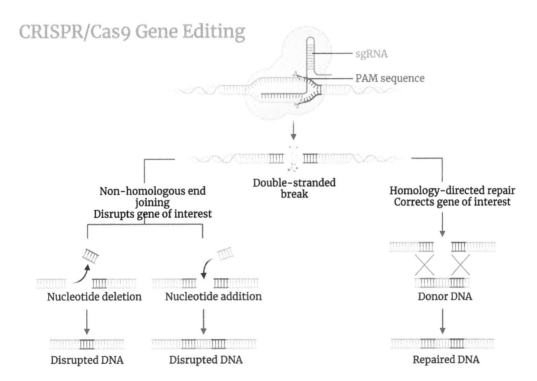

FIGURE 4.2 General work module of CRISPR/Cas9.

4.3.6 HDR with Gene Insertion

To insert a gene or DNA fragment at the target site, a repair template is designed. The repair template typically consists of the desired gene sequence flanked by homology arms that are complementary to regions adjacent to the Cas9-induced DSB site. The homology arms are essential for guiding the HDR process. The Cas9 enzyme, along with the gRNA targeting the desired site, is introduced into the cells or organisms of interest (Ali et al., 2020; Luo et al., 2023). Simultaneously, the repair template containing the gene of interest flanked by homology arms is delivered into the cells. The repair template aligns with the genomic DNA near the DSB site using the homology arms as a guide. Enzymes involved in HDR, such as exonucleases and DNA polymerases, facilitate the copying of the repair template to fill the gap created by the DSB. As the repair template is copied into the genomic DNA, the gene of interest is inserted precisely at the target site. The HDR process may also involve additional steps to finalize the repair, such as DNA ligation and DNA replication (Wei et al., 2021; Chen et al., 2022).

4.3.7 Selection and Identification of Knockout Plants

After gene editing, the transformed plant cells are typically subjected to selection markers, such as antibiotic or herbicide resistance genes, to identify cells that have successfully integrated the CRISPR/Cas9 construct. These cells can then be regenerated into whole plants. Plants can be regenerated from the transformed cells through tissue culture techniques. The transformed plants can be selected thereafter by incorporating selectable markers (e.g., antibiotic resistance genes) into the CRISPR/Cas9 construct. Selective pressure can be applied by adding the appropriate antibiotics to the growth media to eliminate non-transformed cells (Ren et al., 2020; Uetz et al., 2022).

Certain molecular analyses, such as PCR or genomic DNA sequencing, can be performed to detect and verify the presence of mutations at the target site. PCR amplification followed by sequencing can reveal insertions, deletions, or other mutations induced by CRISPR/Cas9 at the specific target site (Ren et al., 2020). Functional analyses can be conducted to assess the impact of the mutations on the target gene's expression or protein function. This can involve techniques like qRT-PCR (quantitative reverse transcription PCR) to measure gene expression levels or Western blotting to analyze protein expression (Kim et al., 2022). The phenotypes of the knockout plants can be observed and compared with those of wild-type or control plants to identify any observable changes or disruptions in the target gene's function or associated traits (Huang et al., 2018; Wang et al., 2020b).

4.3.8 Point Mutations or Indels

The repair process using NHEJ can result in small insertions or deletions (indels) at the target site, leading to modifications in the gene sequence. Alternatively, if a DNA template is provided during the repair process, HDR can be exploited to introduce precise point mutations or insertions at the target site. Accurate and specific gRNA design is crucial for minimizing off-target effects, including point mutations (Xu et al., 2023). Bioinformatics tools can be utilized to select gRNA sequences with minimal off-target potential, ensuring high specificity for the intended target site. Before proceeding with experiments, it's essential to experimentally validate the specificity of the selected gRNAs. This can be done using techniques like deep sequencing or surveyor assays to identify potential off-target sites and quantify their occurrence (Tanihara et al., 2023). Truncated gRNAs with shorter lengths can help reduce the likelihood of gRNA off-target binding and subsequent point mutations. Shorter gRNAs have a decreased chance of imperfect target complementarity, which can contribute to off-target effects (Lee et al., 2022).

4.3.9 Characterization of Transformed Plants

The identified transformed plants are further analyzed to confirm the presence and stability of the desired gene insertion. This can involve molecular techniques such as DNA sequencing, genotyping, and expression analysis to assess the integration and expression of the inserted gene or genetic material (Hyde et al., 2023). Molecular analyses can be done in order to confirm the presence of the desired genetic modifications. This can be done through PCR amplification followed by sequencing or other genotyping techniques. Specifically, amplify and sequence the target genomic region to identify the presence of insertions, deletions, or point mutations introduced by CRISPR/Cas9 (Brant et al., 2021; Juma et al., 2022). The editing efficiency can be quantified by analyzing the ratio of edited alleles to wild-type alleles in the transformed plants. This can be achieved through methods like TIDE analysis (Tracking of Indels by DEcomposition) or deep sequencing. These analyses provide insights into the frequency and type of genetic modifications present in the plant population (Campbell et al., 2019; Huang et al., 2021).

4.3.10 Epigenome Editing

CRISPR/Cas9 can be employed to target and modify epigenetic marks, such as DNA methylation or histone modifications, in the genome. This strategy allows for the modulation of gene expression patterns by altering the epigenetic landscape of specific genomic regions (Goell and Hilton 2021; Nakamura et al., 2021).

4.3.10.1 Recruitment of Epigenetic Modifiers

To achieve the desired epigenetic modifications, Cas9 can be fused or co-expressed with enzymes or proteins that are capable of modifying epigenetic marks. For example, DNA methyltransferases or histone-modifying enzymes can be recruited to specific sites to introduce DNA methylation or histone modifications, respectively (Ding et al., 2019; Yu et al., 2020).

4.3.10.2 Evaluation of Epigenetic Changes

The edited plants are further analyzed to confirm the presence and stability of the desired epigenetic modifications. This can involve techniques such as bisulfite sequencing to assess DNA methylation changes or chromatin immunoprecipitation followed by sequencing (ChIP-seq) to evaluate histone modifications (González-Benito et al., 2020; Pucci et al., 2021).

Epigenome editing in plants holds great potential for studying gene regulation and manipulating plant traits without altering the DNA sequence. However, it's important to note that epigenome editing is a complex process, and the impact of specific epigenetic modifications on plant traits and development can be challenging to predict. Additionally, ethical considerations and potential off-target effects should be carefully evaluated when applying epigenome editing techniques in plants (Liu et al., 2022; Nakamura et al., 2021).

4.3.11 Promoter Editing

CRISPR/Cas9 can be used to modify the regulatory regions of genes, including promoters, enhancers, or repressors, to alter gene expression patterns. By targeting and modifying these regions, researchers can modulate gene expression levels or confer tissue-specific or inducible gene expression (Huang et al., 2020b; Tang and Zhang, 2023). Promoter editing using CRISPR/Cas9 enables precise control over gene expression, allowing researchers to modulate gene activity and study gene functions in specific tissues or under particular conditions. It offers potential applications in trait improvement, metabolic engineering, and crop biotechnology by fine-tuning gene expression profiles in plants (Ren et al., 2021; Zhou et al., 2023).

4.3.12 Genome-Wide Screening

Genome-wide screening using CRISPR/Cas9 is a powerful approach that allows researchers to systematically interrogate gene function across the entire genome (Jason and Yusa, 2019; Nguyen et al., 2023). This strategy, often referred to as CRISPR-based functional genomics, enables the identification and characterization of genes involved in specific biological processes or traits. CRISPR/Cas9 can be employed in large-scale screens to systematically interrogate gene function across the genome. By creating libraries of gRNAs targeting different genes, researchers can simultaneously edit multiple genes and evaluate their phenotypic consequences (Selma et al., 2019; Wu et al., 2022). This approach, known as CRISPR-based functional genomics, enables the identification and characterization of genes involved in specific biological processes or traits. It provides valuable insights into the roles of specific genes in biological processes, disease mechanisms, and cellular pathways. This approach has the potential to accelerate the discovery of novel gene functions and therapeutic targets (Grant et al., 2020; Sturme et al., 2022).

4.3.13 Gene Stacking

Gene stacking, also known as trait stacking, refers to the process of introducing multiple desired traits or genes into a single organism. This technique is commonly used in agricultural biotechnology to create plants or animals with improved characteristics, such as increased resistance to pests, diseases, or environmental stresses (Gao et al., 2020; Kumar et al., 2021). Genome editing technologies enable the stacking of multiple genetic modifications in a single plant. This allows for the simultaneous introduction of multiple traits or modifications, such as disease resistance, improved nutritional content, and environmental adaptation, into a single plant variety (Zaidi et al., 2020; Singh et al., 2020). CRISPR/Cas9 offers precise and efficient gene editing capabilities, allowing for the targeted integration of transgenes during gene stacking in plants. It enables the simultaneous introduction of multiple transgenes into the plant genome, offering opportunities for crop improvement, trait enhancement, and the creation of novel plant varieties (Li et al., 2020b; Lacchini et al., 2020).

4.4 FUTURE PROSPECTS

The prospects of the CRISPR/Cas9 tool in plant genome editing are incredibly promising. The technology has already revolutionized the field, and current research advancements suggest even greater potential.

4.4.1 Increased Precision and Efficiency

Efforts are underway to enhance the precision and efficiency of CRISPR/Cas9 editing in plants. Improving the accuracy of gRNA design, minimizing off-target effects, and optimizing delivery methods will further enhance the precision of genome editing. Additionally, advancements in Cas9 variants with improved specificity and reduced off-target effects will contribute to more precise editing (Clement et al., 2020; Javaid and Choi, 2021).

4.4.2 Multiplexed Editing

The ability to simultaneously edit multiple target sites in the plant genome (multiplex editing) is a highly desirable feature. The continued development of CRISPR/Cas9 tools and methodologies will allow researchers to efficiently edit multiple genes, regulatory elements, or genomic regions in a single experiment. Multiplex editing will facilitate the stacking of multiple traits, the study of gene interactions, and the engineering of complex traits in plants (McCarty et al., 2020; Yang et al., 2020).

4.4.3 Base Editing and Epigenome Editing

The application of base editing techniques will enable precise nucleotide substitutions without inducing DSBs. Base editors can introduce single-base changes, expanding the range of possible edits beyond indels. Furthermore, advancements in epigenome editing using CRISPR/Cas9 will allow researchers to precisely modify DNA methylation patterns and histone modifications, offering new opportunities for studying gene regulation and manipulating gene expression patterns (Musunuru, 2022; Bloomer et al., 2022).

4.4.4 Non-Coding RNA Manipulation

CRISPR/Cas9 can be utilized to target and manipulate non-coding RNAs (ncRNAs), including microRNAs and long non-coding RNAs. Manipulating ncRNAs can have significant impacts on gene expressions and regulatory networks, providing avenues for fine-tuning complex traits and developmental processes in plants (Ghorbanzadeh et al., 2022; Hazan and Bester, 2021).

4.4.5 Gene Regulation and Synthetic Biology

CRISPR/Cas9 tools can be employed to precisely control gene expression levels or dynamics. Techniques such as CRISPR interference (CRISPRi) and CRISPR activation (CRISPRa) enable targeted repression or activation of gene expression, respectively. These tools will allow researchers to modulate gene networks, understand regulatory mechanisms, and engineer synthetic gene circuits for improved plant traits (Vigouroux and Bikard, 2020; Gao et al., 2022).

4.4.6 High-Throughput Screening and Functional Genomics

The combination of CRISPR/Cas9 with high-throughput screening technologies will enable large-scale functional genomics studies in plants. Genome-wide screens using gRNA libraries will provide insights into the functions of previously uncharacterized genes, gene interactions, and the genetic basis of complex traits (Blay et al., 2020; Huang et al., 2022).

4.4.7 Disease Resistance and Stress Tolerance

CRISPR/Cas9 offers tremendous potential for enhancing disease resistance and stress tolerance in crops. Targeted editing of susceptibility genes or the introduction of genes conferring resistance to pests, diseases, or abiotic stresses can contribute to the development of more resilient and sustainable crop varieties (Zaidi et al., 2020; Dogar et al., 2022). These are significant and relevant approaches in the era of global warming, climate change, and environmental pollution.

These prospects highlight the ongoing advancements and expanding capabilities of CRISPR/Cas9 in plant genome editing. As research and technology continue to progress, CRISPR/Cas9 will play a pivotal role in shaping the future of plant biotechnology, crop improvement, and sustainable agriculture.

4.5 CONCLUSION

CRISPR/Cas9 has revolutionized the field of genome editing in plants, offering unprecedented precision, efficiency, and versatility. Its ability to target specific DNA sequences and introduce precise modifications has opened up numerous possibilities for crop improvement, functional genomics, and basic plant research (Demirer et al., 2021; Wada et al., 2022). CRISPR/Cas9 allows researchers to precisely target specific genomic loci by designing complementary gRNA sequences. This specificity enables accurate editing of desired genes while minimizing off-target effects, providing more control over the modifications introduced in the plant genome. CRISPR/Cas9 can be used for various types of genome modifications in plants, including gene knockout, gene editing, promoter editing, and the introduction of specific point mutations or insertions (Pan et al., 2022; Son and Park, 2022). It allows researchers to create precise genetic changes, enabling the study of gene function, trait improvement, and metabolic engineering. CRISPR/Cas9 enables highly efficient genome editing in plants, often resulting in a high frequency of edited cells or transformed plants. This efficiency reduces the time and resources required for generating edited plant lines compared to traditional breeding or mutagenesis methods (Karmakar et al., 2022; Deng et al., 2022).

CRISPR/Cas9 facilitates the simultaneous editing or introduction of multiple genes, allowing for gene stacking and trait stacking in plants. This capability enables the integration of multiple desirable traits into a single plant line, promoting trait combinations and accelerating the development of improved crop varieties (Robertson et al., 2022; Kavuri et al., 2022). CRISPR/Cas9 enables researchers to study gene functions and dissect complex biological processes through reverse genetic approaches. By targeting and disrupting specific genes, researchers can assess the impact of gene loss-of-function on plant phenotypes, thus unraveling gene functions and regulatory networks. CRISPR/Cas9 has immense potential for crop improvement, enabling the development of plants with enhanced agronomic traits, disease resistance, abiotic stress tolerance, and nutritional quality (Chauhan et al., 2022; Ceasar et al., 2022). By precisely modifying key genes, researchers can address challenges in agriculture and contribute to sustainable food production. CRISPR/Cas9 facilitates genome-wide screening approaches, allowing for the systematic analyses of gene functions across the entire plant genome (Hassan et al., 2021; Pan et al., 2021). Large-scale screens using gRNA libraries enable the identification of genes involved in specific biological processes or traits, accelerating the discovery of novel gene functions. CRISPR/Cas9 can be employed to modify epigenetic marks, such as DNA methylation or histone modifications, providing tools for studying epigenetic regulation and modifying gene expression patterns in plants (Gaillochet et al., 2021; Pan et al., 2023).

CRISPR/Cas9 has significantly advanced our ability to precisely and efficiently edit plant genomes. Its impact extends beyond basic research, offering enormous potential for crop improvement, functional genomics, and addressing global agricultural challenges. As the technology continues to evolve, CRISPR/Cas9 is poised to play a transformative role in shaping the future of plant biotechnology and agriculture.

ACKNOWLEDGMENT

Image created with BioRender.com

REFERENCES

Abdallah, N.A., Elsharawy, H., Abulela, H.A., Thilmony, R., Abdelhadi, A.A. and Elarabi, N.I., 2022. Multiplex CRISPR/Cas9-mediated genome editing to address drought tolerance in wheat. *GM Crops & Food*, pp. 1–17. doi: 10.1080/21645698.2022.2120313. Epub ahead of print. PMID: 36200515.

Achary, V.M.M. and Reddy, M.K., 2021. CRISPR-Cas9 mediated mutation in *GRAIN WIDTH* and *WEIGHT2 (GW2)* locus improves aleurone layer and grain nutritional quality in rice. *Scientific Reports*, 11(1), p. 21941.

Ahmad, S., Wei, X., Sheng, Z., Hu, P. and Tang, S., 2020. CRISPR/Cas9 for development of disease resistance in plants: Recent progress, limitations and future prospects. *Briefings in Functional Genomics*, 19(1), pp. 26–39.

Akram, F., Sahreen, S., Aamir, F., Haq, I.U., Malik, K., Imtiaz, M., Naseem, W., Nasir, N. and Waheed, H.M., 2023. An insight into modern targeted genome-editing technologies with a special focus on CRISPR/Cas9 and its applications. *Molecular Biotechnology*, 65(2), pp. 227–242.

Alamillo, J.M., López, C.M., Rivas, F.J.M., Torralbo, F., Bulut, M. and Alseekh, S., 2023. Clustered regularly interspaced short palindromic repeats/CRISPR-associated protein and hairy roots: A perfect match for gene functional analysis and crop improvement. *Current Opinion in Biotechnology*, 79, p. 102876.

Ali, Z., Shami, A., Sedeek, K., Kamel, R., Alhabsi, A., Tehseen, M., Hassan, N., Butt, H., Kababji, A., Hamdan, S.M. and Mahfouz, M.M., 2020. Fusion of the Cas9 endonuclease and the VirD2 relaxase facilitates homology-directed repair for precise genome engineering in rice. *Communications Biology*, 3(1), p. 44.

Angon, P.B. and Habiba, U., 2023. Application of the CRISPR/Cas9 gene-editing system and its participation in plant and medical science. *Current Applied Science and Technology*, 23(3), p. 1–13.

Bai, M., Yuan, J., Kuang, H., Gong, P., Li, S., Zhang, Z., Liu, B., Sun, J., Yang, M., Yang, L. and Wang, D., 2020. Generation of a multiplex mutagenesis population via pooled CRISPR-Cas9 in soya bean. *Plant Biotechnology Journal*, 18(3), pp. 721–731.

Bao, A., Burritt, D.J., Chen, H., Zhou, X., Cao, D. and Tran, L.S.P., 2019. The CRISPR/Cas9 system and its applications in crop genome editing. *Critical Reviews in Biotechnology*, 39(3), pp. 321–336.

Basso, M.F., Duarte, K.E., Santiago, T.R., de Souza, W.R., de Oliveira Garcia, B., da Cunha, B.D.B., Kobayashi, A.K. and Molinari, H.B.C., 2021. Efficient genome editing and gene knockout in *Setaria viridis* with CRISPR/Cas9 directed gene editing by the non-homologous end-joining pathway. *Plant Biotechnology*, 38(2), pp. 227–238.

Bellinvia, E., García-González, J., Cifrová, P., Martinek, J., Sikorová, L., Havelková, L. and Schwarzerová, K., 2022. CRISPR-Cas9 Arabidopsis mutants of genes for ARPC1 and ARPC3 subunits of ARP2/3 complex reveal differential roles of complex subunits. *Scientific Reports*, 12(1), p. 18205.

Ben Shlush, I., Samach, A., Melamed-Bessudo, C., Ben-Tov, D., Dahan-Meir, T., Filler-Hayut, S. and Levy, A.A., 2020. CRISPR/Cas9 induced somatic recombination at the CRTISO locus in tomato. *Genes*, 12(1), p. 59.

Ben-Tov, D., Mafessoni, F., Cucuy, A., Honig, A., Bessudo, C. and Levy, A.A., 2023. Uncovering the dynamics of precise repair at CRISPR/Cas9-induced double-strand breaks. *bioRxiv*, p. 2023–01.

Bessoltane, N., Charlot, F., Guyon-Debast, A., Charif, D., Mara, K., Collonnier, C., Perroud, P.F., Tepfer, M. and Nogué, F., 2022. Genome-wide specificity of plant genome editing by both CRISPR-Cas9 and TALEN. *Scientific Reports*, 12(1), p. 9330.

Bhattacharyya, N., Anand, U., Kumar, R., Ghorai, M., Aftab, T., Jha, N.K., Rajapaksha, A.U., Bundschuh, J., Bontempi, E. and Dey, A., 2023. Phytoremediation and sequestration of soil metals using the CRISPR/Cas9 technology to modify plants: A review. *Environmental Chemistry Letters*, 21(1), pp. 429–445.

Blay, V., Tolani, B., Ho, S.P. and Arkin, M.R., 2020. High-throughput screening: Today's biochemical and cell-based approaches. *Drug Discovery Today*, 25(10), pp. 1807–1821.

Bloomer, H., Khirallah, J., Li, Y. and Xu, Q., 2022. CRISPR/Cas9 ribonucleoprotein-mediated genome and epigenome editing in mammalian cells. *Advanced Drug Delivery Reviews*, 181, p. 114087.

Boubakri, H., 2023. Recent progress in CRISPR/Cas9-based genome editing for enhancing plant disease resistance. *Gene*, 866, p. 147334.

Bouzroud, S., Gasparini, K., Hu, G., Barbosa, M.A.M., Rosa, B.L., Fahr, M., Bendaou, N., Bouzayen, M., Zsögön, A., Smouni, A. and Zouine, M., 2020. Down regulation and loss of auxin response factor 4 function using CRISPR/Cas9 alters plant growth, stomatal function and improves tomato tolerance to salinity and osmotic stress. *Genes*, 11(3), p. 272.

Brant, E.J., Baloglu, M.C., Parikh, A. and Altpeter, F., 2021. CRISPR/Cas9 mediated targeted mutagenesis of LIGULELESS-1 in sorghum provides a rapidly scorable phenotype by altering leaf inclination angle. *Biotechnology Journal*, 16(11), p. 2100237.

Campbell, B.W., Hoyle, J.W., Bucciarelli, B., Stec, A.O., Samac, D.A., Parrott, W.A. and Stupar, R.M., 2019. Functional analysis and development of a CRISPR/Cas9 allelic series for a CPR5 ortholog necessary for proper growth of soybean trichomes. *Scientific Reports*, 9(1), p. 14757.

Ceasar, S.A., Maharajan, T., Hillary, V.E. and Krishna, T.A., 2022. Insights to improve the plant nutrient transport by CRISPR/Cas system. *Biotechnology Advances*, 59, p. 107963.

Chaudhary, M., Mukherjee, T.K., Singh, R., Gupta, M., Goyal, S., Singhal, P., Kumar, R., Bhusal, N. and Sharma, P., 2022. CRISPR/Cas technology for improving nutritional values in the agricultural sector: An update. *Molecular Biology Reports*, 49(7), pp. 7101–7110.

Chauhan, P.K., Upadhyay, S.K., Tripathi, M., Singh, R., Krishna, D., Singh, S.K. and Dwivedi, P., 2022. Understanding the salinity stress on plant and developing sustainable management strategies mediated salt-tolerant plant growth-promoting rhizobacteria and CRISPR/Cas9. *Biotechnology and Genetic Engineering Reviews*, pp. 1–37. doi: 10.1080/02648725.2022.2131958. Epub ahead of print. PMID: 36254096.

Chen, J., Li, S., He, Y., Li, J. and Xia, L., 2022. An update on precision genome editing by homology-directed repair in plants. *Plant Physiology*, 188(4), pp. 1780–1794.

Clement, K., Hsu, J.Y., Canver, M.C., Joung, J.K. and Pinello, L., 2020. Technologies and computational analysis strategies for CRISPR applications. *Molecular Cell*, 79(1), pp. 11–29.

Cram, D., Kulkarni, M., Buchwaldt, M., Rajagopalan, N., Bhowmik, P., Rozwadowski, K., Parkin, I.A., Sharpe, A.G. and Kagale, S., 2019. WheatCRISPR: A web-based guide RNA design tool for CRISPR/Cas9-mediated genome editing in wheat. *BMC Plant Biology*, 19, pp. 1–8.

Deb, S., Choudhury, A., Kharbyngar, B. and Satyawada, R.R., 2022. Applications of CRISPR/Cas9 technology for modification of the plant genome. *Genetica*, 150(1), pp. 1–12. doi: 10.1007/s10709-021-00146-2. Epub 2022 Jan 12. PMID: 35018532.

Demirci, Y., Zhang, B. and Unver, T., 2018. CRISPR/Cas9: An RNA-guided highly precise synthetic tool for plant genome editing. *Journal of Cellular Physiology*, 233(3), pp. 1844–1859.

Demirer, G.S., Silva, T.N., Jackson, C.T., Thomas, J.B., Ehrhardt, D.W, Rhee, S.Y., Mortimer, J.C. and Landry, M.P., 2021. Nanotechnology to advance CRISPR-Cas genetic engineering of plants. *Nature Nanotechnology*, 16(3), pp. 243–250.

Deng, F., Zeng, F., Shen, Q., Abbas, A., Cheng, J., Jiang, W., Chen, G., Shah, A.N., Holford, P., Tanveer, M. and Zhang, D., 2022. Molecular evolution and functional modification of plant miRNAs with CRISPR. *Trends in Plant Science*, 27(9), pp. 890–907.

Ding, N., Maiuri, A.R. and O'Hagan, H.M., 2019. The emerging role of epigenetic modifiers in repair of DNA damage associated with chronic inflammatory diseases. *Mutation Research/Reviews in Mutation Research*, 780, pp. 69–81.

Dogar, A., Ali, M., Riaz, Z., Ali, Q., Ahmad, S. and Javed, M., 2022. Role of CRISPR to improve stress tolerance in plants. *Biological and Clinical Sciences Research Journal*, 2022(1). https://doi.org/10.54112/bcsrj.v2022i1.97

Dong, H., Huang, Y. and Wang, K., 2021. The development of herbicide resistance crop plants using CRISPR/Cas9-mediated gene editing. *Genes*, 12(6), p. 912.

Douris, V., Denecke, S., Van Leeuwen, T., Bass, C., Nauen, R. and Vontas, J., 2020. Using CRISPR/Cas9 genome modification to understand the genetic basis of insecticide resistance: Drosophila and beyond. *Pesticide Biochemistry and Physiology*, 167, p. 104595.

El-Mounadi, K., Morales-Floriano, M.L. and Garcia-Ruiz, H., 2020. Principles, applications, and biosafety of plant genome editing using CRISPR-Cas9. *Frontiers in Plant Science*, 11, p. 56.

Erpen-Dalla Corte, L., Mahmoud, L.M, Moraes, T.S., Mou, Z., Grosser, J.W. and Dutt, M., 2019. Development of improved fruit, vegetable, and ornamental crops using the CRISPR/Cas9 genome editing technique. *Plants*, 8(12), p. 601.

Feng, S., Song, W., Fu, R., Zhang, H., Xu, A. and Li, J., 2018. Application of the CRISPR/Cas9 system in *Dioscorea zingiberensis*. *Plant Cell, Tissue and Organ Culture*, 135, pp. 133–141.

Fernández-Aparicio, M., Delavault, P. and Timko, M.P., 2020. Management of infection by parasitic weeds: A review. *Plants*, 9(9), p. 1184.

Fiaz, S., Ahmad, S., Noor, M.A., Wang, X., Younas, A., Riaz, A., Riaz, A. and Ali, F., 2019. Applications of the CRISPR/Cas9 system for rice grain quality improvement: Perspectives and opportunities. *International Journal of Molecular Sciences*, 20(4), p. 888.

Gaillochet, C., Develtere, W. and Jacobs, T.B., 2021. CRISPR screens in plants: Approaches, guidelines, and future prospects. *The Plant Cell*, 33(4), pp. 794–813.

Gao, H., Mutti, J., Young, J.K., Yang, M., Schroder, M., Lenderts, B., Wang, L., Peterson, D., St. Clair, G., Jones, S. and Feigenbutz, L., 2020. Complex trait loci in maize enabled by CRISPR-Cas9 mediated gene insertion. *Frontiers in Plant Science*, 11, p. 535.

Gao, J., Xu, J., Zuo, Y., Ye, C., Jiang, L., Feng, L., Huang, L., Xu, Z. and Lian, J., 2022. Synthetic biology toolkit for marker-less integration of multigene pathways into *Pichia pastoris* via CRISPR/Cas9. *ACS Synthetic Biology*, 11(2), pp. 623–633.

Ghorbanzadeh, Z., Hamid, R., Jacob, F., Asadi, S., Salekdeh, G.H. and Ghaffari, M.R., 2022. Non-coding RNA: Chief architects of drought-resilient roots. *Rhizosphere*, 23, p. 100572.

Goell, J.H. and Hilton, I.B., 2021. CRISPR/Cas-based epigenome editing: Advances, applications, and clinical utility. *Trends in Biotechnology*, 39(7), pp. 678–691.

González-Benito, M.E., Ibáñez, M.Á., Pirredda, M., Mira, S. and Martín, C., 2020. Application of the MSAP technique to evaluate epigenetic changes in plant conservation. *International Journal of Molecular Sciences*, 21(20), p. 7459.

Grant, C.V., Cai, S., Risinger, A.L., Liang, H., O'Keefe, B.R., Doench, J.G., Cichewicz, R.H. and Mooberry, S.L., 2020. CRISPR-Cas9 genome-wide knockout screen identifies mechanism of selective activity of dehydrofalcarinol in mesenchymal stem-like triple-negative breast cancer cells. *Journal of Natural Products*, 83(10), pp. 3080–3092.

Gressel, J., 2018. Intractable weed problems need innovative solutions using all available technologies. *Indian Journal of Weed Science*, 50(3), pp. 201–208.

Han, Y.J. and Kim, J.I., 2019. Application of CRISPR/Cas9-mediated gene editing for the development of herbicide-resistant plants. *Plant Biotechnology Reports*, 13, pp. 447–457.

Haq, S.I.U., Zheng, D., Feng, N., Jiang, X., Qiao, F., He, J.S. and Qiu, Q.S., 2022. Progresses of CRISPR/Cas9 genome editing in forage crops. *Journal of Plant Physiology*, 279, p. 153860.

Hassan, M.M., Zhang, Y., Yuan, G., De, K., Chen, J.G., Muchero, W., Tuskan, G.A., Qi, Y. and Yang, X., 2021. Construct design for CRISPR/Cas-based genome editing in plants. *Trends in Plant Science*, 26(11), pp. 1133–1152.

Hazan, J. and Bester, A.C., 2021. CRISPR-based approaches for the high-throughput characterization of long non-coding RNAs. *Non-Coding RNA*, 7(4), p. 79.

Hemalatha, P., Abda, E.M., Shah, S., Prabhu, S.V., Jayakumar, M., Karmegam, N., Kim, W. and Govarthanan, M., 2023. Multi-faceted CRISPR-Cas9 strategy to reduce plant based food loss and waste for sustainable bio-economy-A review. *Journal of Environmental Management*, 332, p. 117382.

Hillary, V.E. and Ceasar, S.A., 2023. A review on the mechanism and applications of CRISPR/Cas9/Cas12/Cas13/Cas14 proteins utilized for genome engineering. *Molecular Biotechnology*, 65(3), pp. 311–325.

Huang, H., Cui, T., Zhang, L., Yang, Q., Yang, Y., Xie, K., Fan, C. and Zhou, Y., 2020a. Modifications of fatty acid profile through targeted mutation at *BnaFAD2* gene with CRISPR/Cas9-mediated gene editing in *Brassica napus*. *Theoretical and Applied Genetics*, 133, pp. 2401–2411.

Huang, J., Li, J., Zhou, J., Wang, L., Yang, S., Hurst, L.D., Li, W.H. and Tian, D., 2018. Identifying a large number of high-yield genes in rice by pedigree analysis, whole-genome sequencing, and CRISPR-Cas9 gene knockout. *Proceedings of the National Academy of Sciences*, 115(32), pp. E7559–E7567.

Huang, L., Li, Q., Zhang, C., Chu, R., Gu, Z., Tan, H., Zhao, D., Fan, X. and Liu, Q., 2020b. Creating novel *Wx* alleles with fine-tuned amylose levels and improved grain quality in rice by promoter editing using CRISPR/Cas9 system. *Plant Biotechnology Journal*, 18(11), p. 2164.

Huang, T.K., Armstrong, B., Schindele, P. and Puchta, H., 2021. Efficient gene targeting in *Nicotiana tabacum* using CRISPR/SaCas9 and temperature tolerant LbCas12a. *Plant Biotechnology Journal*, 19(7), pp. 1314–1324.

Huang, Y., Shang, M., Liu, T. and Wang, K., 2022. High-throughput methods for genome editing: The more the better. *Plant Physiology*, 188(4), pp. 1731–1745.

Hyde, L., Osman, K., Winfield, M., Sanchez-Moran, E., Higgins, J.D., Henderson, I.R., Sparks, C., Franklin, F.C.H. and Edwards, K.J., 2023. Identification, characterization, and rescue of CRISPR/Cas9 generated wheat SPO11-1 mutants. *Plant Biotechnology Journal*, 21(2), pp. 405–418.

Impens, L., Jacobs, T.B., Nelissen, H., Inzé, D. and Pauwels, L., 2022. Mini-review: Transgenerational CRISPR/Cas9 gene editing in plants. *Frontiers in Genome Editing*, 4, p. 825042.

Jason, S.L. and Yusa, K., 2019. Genome-wide CRISPR-Cas9 screening in mammalian cells. *Methods*, 164, pp. 29–35.

Javaid, N. and Choi, S., 2021. CRISPR/Cas system and factors affecting its precision and efficiency. *Frontiers in Cell and Developmental Biology*, 9, p. 761709.

Jiang, Y., An, X., Li, Z., Yan, T., Zhu, T., Xie, K., Liu, S., Hou, Q., Zhao, L., Wu, S. and Liu, X., 2021. CRISPR/Cas9-based discovery of maize transcription factors regulating male sterility and their functional conservation in plants. *Plant Biotechnology Journal*, 19(9), pp. 1769–1784.

Juma, B.S., Mukami, A., Mweu, C., Ngugi, M.P. and Mbinda, W., 2022. Targeted mutagenesis of the CYP79D1 gene via CRISPR/Cas9-mediated genome editing results in lower levels of cyanide in cassava. *Frontiers in Plant Science*, 13, p. 1009860.

Karmakar, S., Das, P., Panda, D., Xie, K., Baig, M.J. and Molla, K.A., 2022. A detailed landscape of CRISPR-Cas-mediated plant disease and pest management. *Plant Science*, 323, p. 111376.

Kaur, N., Awasthi, P. and Tiwari, S., 2020. Fruit crops improvement using CRISPR/Cas9 system. In Vijai Singh and Pawan K. Dhar (eds) *Genome Engineering via CRISPR-Cas9 System* (pp. 131–145). Cambridge, MA: Academic Press.

Kavuri, N.R., Ramasamy, M., Qi, Y. and Mandadi, K., 2022. Applications of CRISPR/Cas13-based RNA editing in plants. *Cells*, 11(17), p. 2665.

Kim, J.Y., Kim, J.H., Jang, Y.H., Yu, J., Bae, S., Kim, M.S., Cho, Y.G., Jung, Y.J. and Kang, K.K., 2022. Transcriptome and metabolite profiling of tomato SGR-knockout null lines using the CRISPR/Cas9 system. *International Journal of Molecular Sciences*, 24(1), p. 109.

Kumar, A., Dash, G.K., Sahoo, S.K., Lal, M.K., Sahoo, U., Sah, R.P., Ngangkham, U., Kumar, S., Baig, M.J., Sharma, S. and Lenka, S.K., 2023a. Phytic acid: A reservoir of phosphorus in seeds plays a dynamic role in plant and animal metabolism. *Phytochemistry Reviews*, 22, pp. 1281–1304.

Kumar, M., Prusty, M.R., Pandey, M.K., Singh, P.K., Bohra, A., Guo, B. and Varshney, R.K., 2023b. Application of CRISPR/Cas9-mediated gene editing for abiotic stress management in crop plants. *Frontiers in Plant Science*, 14, p. 1157678.

Kumar, S., Rymarquis, L.A., Ezura, H. and Nekrasov, V., 2021. CRISPR-Cas in agriculture: Opportunities and challenges. *Frontiers in Plant Science*, 12, p. 672329.

Lacchini, E., Kiegle, E., Castellani, M., Adam, H., Jouannic, S., Gregis, V. and Kater, M.M., 2020. CRISPR-mediated accelerated domestication of African rice landraces. *PLoS One*, 15(3), p. e0229782.

Lee, H.J., Kim, H.J., Park, Y.J. and Lee, S.J., 2022. Efficient single-nucleotide microbial genome editing achieved using CRISPR/Cpf1 with maximally 3′-end-truncated crRNAs. *ACS Synthetic Biology*, 11(6), pp. 2134–2143.

Li, B., Fu, C., Zhou, J., Hui, F., Wang, Q., Wang, F., Wang, G., Xu, Z., Che, L., Yuan, D. and Wang, Y., 2022a. Highly efficient genome editing using Geminivirus-based CRISPR/Cas9 system in cotton plant. *Cells*, 11(18), p. 2902.

Li, C., Li, W., Zhou, Z., Chen, H., Xie, C. and Lin, Y., 2020a. A new rice breeding method: CRISPR/Cas9 system editing of the Xa13 promoter to cultivate transgene-free bacterial blight-resistant rice. *Plant Biotechnology Journal*, 18(2), p. 313.

Li, C., Zong, Y., Jin, S., Zhu, H., Lin, D., Li, S., Qiu, J.L., Wang, Y. and Gao, C., 2020b. SWISS: Multiplexed orthogonal genome editing in plants with a Cas9 nickase and engineered CRISPR RNA scaffolds. *Genome Biology*, 21, pp. 1–15.

Li, X., Xu, S., Fuhrmann-Aoyagi, M.B., Yuan, S., Iwama, T., Kobayashi, M. and Miura, K., 2022b. CRISPR/Cas9 technique for temperature, drought, and salinity stress responses. *Current Issues in Molecular Biology*, 44(6), pp. 2664–2682.

Liang, Z., Chen, K. and Gao, C., 2019. Biolistic delivery of CRISPR/Cas9 with ribonucleoprotein complex in wheat. In Qi, Y. (ed) *Plant Genome Editing with CRISPR Systems: Methods and Protocols*, vol. 1917, pp. 327–335. New York, NY: Humana.

Liu, H. and Zhang, B., 2020. Virus-based CRISPR/Cas9 genome editing in plants. *Trends in Genetics*, 36(11), pp. 810–813.

Liu, L., Gallagher, J., Arevalo, E.D., Chen, R., Skopelitis, T., Wu, Q., Bartlett, M. and Jackson, D., 2021a. Enhancing grain-yield-related traits by CRISPR-Cas9 promoter editing of maize CLE genes. *Nature Plants*, 7(3), pp. 287–294.

Liu, Q., Yang, F., Zhang, J., Liu, H., Rahman, S., Islam, S., Ma, W. and She, M., 2021b. Application of CRISPR/Cas9 in crop quality improvement. *International Journal of Molecular Sciences*, 22(8), p. 4206.

Liu, S., Sretenovic, S., Fan, T., Cheng, Y., Li, G., Qi, A., Tang, X., Xu, Y., Guo, W., Zhong, Z. and He, Y., 2022. Hypercompact CRISPR-Cas12j2 (CasΦ) enables genome editing, gene activation, and epigenome editing in plants. *Plant Communications*, 3(6): 100453. doi: 10.1016/j.xplc.2022.100453.

Liu, X., Wu, S., Xu, J., Sui, C. and Wei, J., 2017a. Application of CRISPR/Cas9 in plant biology. *Acta Pharmaceutica Sinica B*, 7(3), pp. 292–302.

Liu, X., Xie, C., Si, H. and Yang, J., 2017b. CRISPR/Cas9-mediated genome editing in plants. *Methods*, 121, pp. 94–102.

Luo, W., Suzuki, R. and Imai, R., 2023. Precise in planta genome editing via homology-directed repair in wheat. *Plant Biotechnology Journal*, 21(4), p. 668.

Lyu, Y.S., Cao, L.M., Huang, W.Q., Liu, J.X. and Lu, H.P., 2022. Disruption of three polyamine uptake transporter genes in rice by CRISPR/Cas9 gene editing confers tolerance to herbicide paraquat. *Abiotech*, 3(2), pp. 140–145.

Ma, J., Sun, S., Whelan, J. and Shou, H., 2021. CRISPR/Cas9-mediated knockout of GmFATB1 significantly reduced the amount of saturated fatty acids in soybean seeds. *International Journal of Molecular Sciences*, 22(8), p. 3877.

Mandal, S. and Verma, A.K. 2021. Wheat breeding, fertilizers, and pesticides: Do they contribute to the increasing immunogenic properties of modern wheat? *Gastrointestinal Disorders*, 3, 247–264. https://doi.org/10.3390/gidisord3040023

Mandal, S., Ghorai, M., Anand, U., Roy, D., Kant, N., Mishra, T., Mane, A.B., Jha, N.K., Lal, M.K., Tiwari, R.K., Kumar, M., Radha and Ghosh, A., Bhattacharjee, R., Proćków, J. and Dey, A. 2022. Cytokinins: A genetic target for increasing yield potential in the CRISPR era. *Frontiers in Genetics*, 26(13), p. 883930. https://doi.org/10.3389/fgene.2022.883930.

McCarty, N.S., Graham, A.E., Studená, L. and Ledesma-Amaro, R., 2020. Multiplexed CRISPR technologies for gene editing and transcriptional regulation. *Nature Communications*, 11(1), p. 1281.

Mubarik, M.S., Ma, C., Majeed, S., Du, X. and Azhar, M.T., 2020. Revamping of cotton breeding programs for efficient use of genetic resources under changing climate. *Agronomy*, 10(8), p. 1190.

Musunuru, K., 2022. Moving toward genome-editing therapies for cardiovascular diseases. *The Journal of Clinical Investigation*, 132(1): e148555. doi: 10.1172/JCI148555.

Nakamura, M., Gao, Y., Dominguez, A.A. and Qi, L.S., 2021. CRISPR technologies for precise epigenome editing. *Nature Cell Biology*, 23(1), pp. 11–22.

Naz, M., Benavides-Mendoza, A., Tariq, M., Zhou, J., Wang, J., Qi, S., Dai, Z. and Du, D., 2022. CRISPR/Cas9 technology as an innovative approach to enhancing the phytoremediation: Concepts and implications. *Journal of Environmental Management*, 323, p. 116296.

Nguyen, N.H., Rafiee, R., Tagmount, A., Sobh, A., Loguinov, A., de Jesus Sosa, A.K., Elsayed, A.H., Gbadamosi, M., Seligson, N., Cogle, C.R. and Rubnitz, J., 2023. Genome-wide CRISPR/Cas9 screen identifies etoposide response modulators associated with clinical outcomes in pediatric AML. *Blood Advances*, 7(9), pp. 1769–1783.

Pan, C., Li, G., Bandyopadhyay, A. and Qi, Y., 2023. Guide RNA library-based CRISPR screens in plants: Opportunities and challenges. *Current Opinion in Biotechnology*, 79, p. 102883.

Pan, C., Li, G., Malzahn, A.A., Cheng, Y., Leyson, B., Sretenovic, S., Gurel, F., Coleman, G.D. and Qi, Y., 2022. Boosting plant genome editing with a versatile CRISPR-Combo system. *Nature Plants*, 8(5), pp. 513–525.

Pan, C., Wu, X., Markel, K., Malzahn, A.A., Kundagrami, N., Sretenovic, S., Zhang, Y., Cheng, Y., Shih, P.M. and Qi, Y., 2021. CRISPR–Act3.0 for highly efficient multiplexed gene activation in plants. *Nature Plants*, 7(7), pp. 942–953.

Panchal, A., Singh, R.K. and Prasad, M., 2023. Recent advancements and future perspectives of foxtail millet genomics. *Plant Growth Regulation*, 99(1), pp. 11–23.

Peng, Y., Chen, L., Zhu, L., Cui, L., Yang, L., Wu, H. and Bie, Z., 2022. CsAKT1 is a key gene for the CeO2 nanoparticle's improved cucumber salt tolerance: A validation from CRISPR-Cas9 lines. *Environmental Science: Nano*, 9(12), pp. 4367–4381.

Přibylová, A., Fischer, L., Pyott, D.E., Bassett, A. and Molnar, A., 2022. DNA methylation can alter CRISPR/Cas9 editing frequency and DNA repair outcome in a target-specific manner. *The New Phytologist*, 235(6), p. 2285.

Priti, D.K., Chaudhary, V., Baliyan, N., Rani, R. and Jangra, S., 2023. Nutritional enhancement in horticultural crops by CRISPR/Cas9: Status and future prospects. In Prakash, C.S., Fiaz, S., Nadeem, M.A., Baloch, F.S., Qayyum, A. (eds) *Sustainable Agriculture in the Era of the OMICs Revolution*, pp. 399–430. Cham: Springer.

Pucci, G., Forte, G.I. and Cavalieri, V., 2021. Evaluation of epigenetic and radiomodifying effects during radiotherapy treatments in zebrafish. *International Journal of Molecular Sciences*, 22(16), p. 9053.

Rai, A., Sivalingam, P.N. and Senthil-Kumar, M., 2022. A spotlight on non-host resistance to plant viruses. *PeerJ*, 10, p. e12996.

Rasheed, A., Barqawi, A.A., Mahmood, A., Nawaz, M., Shah, A.N., Bay, D.H., Alahdal, M.A., Hassan, M.U. and Qari, S.H., 2022. CRISPR/Cas9 is a powerful tool for precise genome editing of legume crops: A review. *Molecular Biology Reports*, 49(6), pp. 5595–5609.

Rasheed, A., Gill, R.A., Hassan, M.U., Mahmood, A., Qari, S., Zaman, Q.U., Ilyas, M., Aamer, M., Batool, M., Li, H. and Wu, Z., 2021. A critical review: Recent advancements in the use of CRISPR/Cas9 technology to enhance crops and alleviate global food crises. *Current Issues in Molecular Biology*, 43(3), pp. 1950–1976.

Razzaq, M.K., Akhter, M., Ahmad, R.M., Cheema, K.L., Hina, A., Karikari, B., Raza, G., Xing, G., Gai, J. and Khurshid, M., 2022. CRISPR-Cas9 based stress tolerance: New hope for abiotic stress tolerance in chickpea (*Cicer arietinum*). *Molecular Biology Reports*, 49(9), pp. 8977–8985.

Ren, C., Guo, Y., Kong, J., Lecourieux, F., Dai, Z., LI, S. and Liang, Z., 2020. Knockout of VvCCD8 gene in grapevine affects shoot branching. *BMC Plant Biology*, 20(1), pp. 1–8.

Ren, C., Liu, Y., Guo, Y., Duan, W., Fan, P., Li, S. and Liang, Z., 2021. Optimizing the CRISPR/Cas9 system for genome editing in grape by using grape promoters. *Horticulture Research*, 8, p. 52.

Riu, Y.S., Kim, G.H., Chung, K.W. and Kong, S.G., 2023. Enhancement of the CRISPR/Cas9-based genome editing system in lettuce (*Lactuca sativa* L.) using the endogenous U6 promoter. *Plants*, 12(4), p. 878.

Robertson, G., Burger, J. and Campa, M., 2022. CRISPR/Cas-based tools for the targeted control of plant viruses. *Molecular Plant Pathology*, 23(11), pp. 1701–1718.

Sahu, A., Verma, R., Gupta, U., Kashyap, S. and Sanyal, I., 2023. An overview of targeted genome editing strategies for reducing the biosynthesis of phytic acid: An anti-nutrient in crop plants. *Molecular Biotechnology*, pp. 1–15. https://doi.org/10.1007/s12033-023-00722-1

Sami, A., Xue, Z., Tazein, S., Arshad, A., He Zhu, Z., Ping Chen, Y., Hong, Y., Tian Zhu, X. and Jin Zhou, K., 2021. CRISPR-Cas9-based genetic engineering for crop improvement under drought stress. *Bioengineered*, 12(1), pp. 5814–5829.

Sant'Ana, R.R.A., Caprestano, C.A., Nodari, R.O. and Agapito-Tenfen, S.Z., 2020. PEG-delivered CRISPR-Cas9 ribonucleoproteins system for gene-editing screening of maize protoplasts. *Genes*, 11(9), p. 1029.

Sarma, H., Islam, N.F., Prasad, R., Prasad, M.N.V., Ma, L.Q. and Rinklebe, J., 2021. Enhancing phytoremediation of hazardous metal (loid) s using genome engineering CRISPR-Cas9 technology. *Journal of Hazardous Materials*, 414, p. 125493.

Schindele, P., Wolter, F. and Puchta, H., 2018. Transforming plant biology and breeding with CRISPR/Cas9, Cas12 and Cas13. *FEBS Letters*, 592(12), pp. 1954–1967.

Schmitz, D.J., Ali, Z., Wang, C., Aljedaani, F., Hooykaas, P.J., Mahfouz, M. and de Pater, S., 2020. CRISPR/Cas9 mutagenesis by translocation of Cas9 protein into plant cells via the *Agrobacterium* type IV secretion system. *Frontiers in Genome Editing*, 2, p. 6.

Selma, S., Bernabé-Orts, J.M., Vazquez-Vilar, M., Diego-Martin, B., Ajenjo, M., Garcia-Carpintero, V., Granell, A. and Orzaez, D., 2019. Strong gene activation in plants with genome-wide specificity using a new orthogonal CRISPR/Cas9-based programmable transcriptional activator. *Plant Biotechnology Journal*, 17(9), p. 1703.

Shinwari, Z.K., Jan, S.A., Nakashima, K. and Yamaguchi-Shinozaki, K., 2020. Genetic engineering approaches to understanding drought tolerance in plants. *Plant Biotechnology Reports*, 14, pp. 151–162.

Singh, S., Chaudhary, R., Deshmukh, R. and Tiwari, S., 2023. Opportunities and challenges with CRISPR-Cas mediated homologous recombination based precise editing in plants and animals. *Plant Molecular Biology*, 111(1–2), pp. 1–20.

Singh, S., Singh, A., Kumar, S., Mittal, P. and Singh, I.K., 2020. Protease inhibitors: Recent advancement in its usage as a potential biocontrol agent for insect pest management. *Insect Science*, 27(2), pp. 186–201.

Son, S. and Park, S.R., 2022. Challenges facing CRISPR/Cas9-based genome editing in plants. *Frontiers in Plant Science*, 13, p. 902413.

Sturme, M.H., van der Berg, J.P., Bouwman, L.M., De Schrijver, A., de Maagd, R.A., Kleter, G.A. and Battaglia-de Wilde, E., 2022. Occurrence and nature of off-target modifications by CRISPR-Cas genome editing in plants. *ACS Agricultural Science & Technology*, 2(2), pp. 192–201.

Sun, J., Liu, H., Liu, J., Cheng, S., Peng, Y., Zhang, Q., Yan, J., Liu, H.J. and Chen, L.L., 2019. CRISPR-local: A local single-guide RNA (sgRNA) design tool for non-reference plant genomes. *Bioinformatics*, 35(14), pp. 2501–2503.

Tahir, T., Ali, Q., Rashid, M.S. and Malik, A., 2020. The journey of CRISPR-Cas9 from bacterial defense mechanism to a gene editing tool in both animals and plants. *Biological and Clinical Sciences Research Journal*, 2020(1).

Tan, J., Zhao, Y., Wang, B., Hao, Y., Wang, Y., Li, Y., Luo, W., Zong, W., Li, G., Chen, S. and Ma, K., 2020. Efficient CRISPR/Cas9-based plant genomic fragment deletions by microhomology-mediated end joining. *Plant Biotechnology Journal*, 18(11), p. 2161.

Tang, X. and Zhang, Y., 2023. Beyond knockouts: Fine-tuning regulation of gene expression in plants with CRISPR-Cas-based promoter editing. *New Phytologist*, 239, pp. 868–874.

Tang, Y., Zhang, Z., Yang, Z. and Wu, J., 2023. CRISPR/Cas9 and *Agrobacterium tumefaciens* virulence proteins synergistically increase efficiency of precise genome editing via homology directed repair in plants. *Journal of Experimental Botany*, 74, pp. 3518–3530.

Tanihara, F., Hirata, M., Namula, Z., Do, L.T.K., Yoshimura, N., Lin, Q., Takebayashi, K., Sakuma, T., Yamamoto, T. and Otoi, T., 2023. Pigs with an INS point mutation derived from zygotes electroporated with CRISPR/Cas9 and ssODN. *Frontiers in Cell and Developmental Biology*, 11, p. 884340.

Tek, M.I., Calis, O., Fidan, H., Shah, M.D., Celik, S. and Wani, S.H., 2022. CRISPR/Cas9 based mlo-mediated resistance against *Podosphaera xanthii* in cucumber (*Cucumis sativus* L.). *Frontiers in Plant Science*, 13, p. 1081506.

Uetz, P., Melnik, S., Grünwald-Gruber, C., Strasser, R. and Stoger, E., 2022. CRISPR/Cas9-mediated knockout of a prolyl-4-hydroxylase subfamily in *Nicotiana benthamiana* using DsRed2 for plant selection. *Biotechnology Journal*, 17(7), p. 2100698.

Uniyal, A.P., Mansotra, K., Yadav, S.K. and Kumar, V., 2019. An overview of designing and selection of sgRNAs for precise genome editing by the CRISPR-Cas9 system in plants. *3 Biotech*, 9(6), p. 223.

Uranga, M., Aragonés, V., Selma, S., Vázquez-Vilar, M., Orzáez, D. and Daròs, J.A., 2021. Efficient Cas9 multiplex editing using unspaced sgRNA arrays engineering in a *Potato virus X* vector. *The Plant Journal*, 106(2), pp. 555–565.

Vigouroux, A. and Bikard, D., 2020. CRISPR tools to control gene expression in bacteria. *Microbiology and Molecular Biology Reviews*, 84(2), pp. 10–1128.

Vu, T.V., Sivankalyani, V., Kim, E.J., Doan, D.T.H., Tran, M.T., Kim, J., Sung, Y.W., Park, M., Kang, Y.J. and Kim, J.Y., 2020. Highly efficient homology-directed repair using CRISPR/Cpf1-geminiviral replicon in tomato. *Plant Biotechnology Journal*, 18(10), pp. 2133–2143.

Wada, N., Osakabe, K. and Osakabe, Y., 2022. Expanding the plant genome editing toolbox with recently developed CRISPR-Cas systems. *Plant Physiology*, 188(4), pp. 1825–1837.

Wada, N., Ueta, R., Osakabe, Y. and Osakabe, K., 2020. Precision genome editing in plants: State-of-the-art in CRISPR/Cas9-based genome engineering. *BMC Plant Biology*, 20, pp. 1–12.

Wan, L., Wang, Z., Tang, M., Hong, D., Sun, Y., Ren, J., Zhang, N. and Zeng, H., 2021. CRISPR-Cas9 gene editing for fruit and vegetable crops: Strategies and prospects. *Horticulturae*, 7(7), p. 193.

Wang, B., Lin, Z., Li, X., Zhao, Y., Zhao, B., Wu, G., Ma, X., Wang, H., Xie, Y., Li, Q. and Song, G., 2020a. Genome-wide selection and genetic improvement during modern maize breeding. *Nature Genetics*, 52(6), pp. 565–571.

Wang, S., Chen, A., Xie, K., Yang, X., Luo, Z., Chen, J., Zeng, D., Ren, Y., Yang, C., Wang, L. and Feng, H., 2020b. Functional analysis of the OsNPF4. 5 nitrate transporter reveals a conserved mycorrhizal pathway of nitrogen acquisition in plants. *Proceedings of the National Academy of Sciences*, 117(28), pp. 16649–16659.

Wang, T., Zhang, C., Zhang, H. and Zhu, H., 2021. CRISPR/Cas9-mediated gene editing revolutionizes the improvement of horticulture food crops. *Journal of Agricultural and Food Chemistry*, 69(45), pp. 13260–13269.

Wei, Z., Abdelrahman, M., Gao, Y., Ji, Z., Mishra, R., Sun, H., Sui, Y., Wu, C., Wang, C. and Zhao, K., 2021. Engineering broad-spectrum resistance to bacterial blight by CRISPR-Cas9-mediated precise homology directed repair in rice. *Molecular Plant*, 14(8), pp. 1215–1218.

Wu, J., Chen, C., Xian, G., Liu, D., Lin, L., Yin, S., Sun, Q., Fang, Y., Zhang, H. and Wang, Y., 2020a. Engineering herbicide-resistant oilseed rape by CRISPR/Cas9-mediated cytosine base-editing. *Plant Biotechnology Journal*, 18(9), p. 1857.

Wu, N., Lu, Q., Wang, P., Zhang, Q., Zhang, J., Qu, J. and Wang, N., 2020b. Construction and analysis of GmFAD2-1A and GmFAD2-2A soybean fatty acid desaturase mutants based on CRISPR/Cas9 technology. *International Journal of Molecular Sciences*, 21(3), p. 1104.

Wu, Y., Ren, Q., Zhong, Z., Liu, G., Han, Y., Bao, Y., Liu, L., Xiang, S., Liu, S., Tang, X. and Zhou, J., 2022. Genome-wide analyses of PAM-relaxed Cas9 genome editors reveal substantial off-target effects by ABE8e in rice. *Plant Biotechnology Journal*, 20(9), pp. 1670–1682.

Xu, X., Zhang, X., Peng, X., Liu, C., Li, W. and Liu, M., 2023. Introduction of the FecGF mutation in GDF9 gene via CRISPR/Cas9 system with single-stranded oligodeoxynucleotide. *Theriogenology*, 197, pp. 177–185.

Yang, Z., Edwards, H. and Xu, P., 2020. CRISPR-Cas12a/Cpf1-assisted precise, efficient and multiplexed genome-editing in *Yarrowia lipolytica. Metabolic Engineering Communications*, 10, p. e00112.

Yu, Z., Ai, M., Samanta, S.K., Hashiya, F., Taniguchi, J., Asamitsu, S., Ikeda, S., Hashiya, K., Bando, T., Pandian, G.N. and Isaacs, L., 2020. A synthetic transcription factor pair mimic for precise recruitment of an epigenetic modifier to the targeted DNA locus. *Chemical Communications*, 56(15), pp. 2296–2299.

Zaidi, S.S.E.A., Mahas, A., Vanderschuren, H. and Mahfouz, M.M., 2020. Engineering crops of the future: CRISPR approaches to develop climate-resilient and disease-resistant plants. *Genome Biology*, 21(1), pp. 1–19.

Zeng, Y., Wen, J., Zhao, W., Wang, Q. and Huang, W., 2020. Rational improvement of rice yield and cold tolerance by editing the three genes OsPIN5b, GS3, and OsMYB30 with the CRISPR-Cas9 system. *Frontiers in Plant Science*, 10, p. 1663.

Zhang, F. and Batley, J., 2020. Exploring the application of wild species for crop improvement in a changing climate. *Current Opinion in Plant Biology*, 56, pp. 218–222.

Zhang, J., Li, J., Saeed, S., Batchelor, W.D., Alariqi, M., Meng, Q., Zhu, F., Zou, J., Xu, Z., Si, H. and Wang, Q., 2022a. Identification and functional analysis of lncRNA by CRISPR/Cas9 during the cotton response to sap-sucking insect infestation. *Frontiers in Plant Science*, 13, p. 784511.
Zhang, Y. and Showalter, A.M., 2020. CRISPR/Cas9 genome editing technology: A valuable tool for understanding plant cell wall biosynthesis and function. *Frontiers in Plant Science*, 11, p. 589517.
Zhang, Z., Wang, W., Ali, S., Luo, X. and Xie, L., 2022b. CRISPR/Cas9-mediated multiple knockouts in abscisic acid receptor genes reduced the sensitivity to ABA during soybean seed germination. *International Journal of Molecular Sciences*, 23(24), p. 16173.
Zhao, F., Lyu, X., Ji, R., Liu, J., Zhao, T., Li, H., Liu, B. and Pei, Y., 2022. CRISPR/Cas9-engineered mutation to identify the roles of phytochromes in regulating photomorphogenesis and flowering time in soybean. *The Crop Journal*, 10(6), pp. 1654–1664.
Zhou, J., Liu, G., Zhao, Y., Zhang, R., Tang, X., Li, L., Jia, X., Guo, Y., Wu, Y., Han, Y. and Bao, Y., 2023. An efficient CRISPR-Cas12a promoter editing system for crop improvement. *Nature Plants*, 9(4), pp. 588–604.
Zhou, J., Yuan, M., Zhao, Y., Quan, Q., Yu, D., Yang, H., Tang, X., Xin, X., Cai, G., Qian, Q. and Qi, Y., 2021. Efficient deletion of multiple circle RNA loci by CRISPR-Cas9 reveals Os06circ02797 as a putative sponge for OsMIR408 in rice. *Plant Biotechnology Journal*, 19(6), pp. 1240–1252.

5 Virus-Induced Genome Editing
Methods and Applications in Plant Breeding

Mireia Uranga

5.1 INTRODUCTION

Crops are essential to humans for providing food, feed, fuel, and other consumable resources. Current agricultural production can satisfy food demands for most of the global population, thanks to technical and scientific advances derived from the 1960s Green Revolution. However, it is estimated that 9.6 billion people will inhabit planet Earth by 2050, increasing the demand for staple crops by up to 60% (FAO, 2017). It is also expected that factors such as climate change, limited availability of arable land, and shortage of water resources will cause dramatic reductions in crop yields. Therefore, there is an urgent need for scientific innovations to develop a more sustainable, productive, and resilient agriculture in the future.

From the beginnings of agriculture, traditional cross-breeding has been used to obtain novel cultivars by crossing edible crops with wild relatives harboring traits of agronomic interest (Scheben et al., 2017) (Figure 5.1a). As a result of a prolonged directed evolution, crops consumed nowadays contain large parts of fixed genomic regions, so alternative approaches are needed to artificially increase genetic variability in elite germplasms. Mutation breeding uses physical or chemical agents to induce random genetic mutations (Holme et al., 2019) (Figure 5.1b). Its main drawback is that identifying individuals with desired modifications requires screening large populations and time-consuming breeding programs, which cannot satisfy the demands for increased crop production. Alternatively, genes from other organisms can be introduced into crop genomes to obtain the desired traits (Prado et al., 2014) (Figure 5.1c). Although transgenesis considerably enhances genetic variation over conventional approaches, it generates social apprehension about the biosafety of genetically modified organisms (GMOs) because the integration of the exogenous genes at random genomic locations can provoke unintended effects in the host plant. Therefore, the commercialization of GMOs is currently hindered by restrictive regulatory frameworks in many countries worldwide (Turnbull et al., 2021).

5.1.1 CRISPR-Cas Technology: Toward a New Era in Plant Genome Editing

In the past few decades, rapid progress in sequencing technologies has enabled to obtain genomic information for an increasing number of plant species, thus offering new opportunities for the precise and predictable modification of plant genomes. This approach relies on a site-specific nuclease (SSN) that specifically binds to a user-selected genomic region and induces targeted modifications (Voytas & Gao, 2014). The main SSNs developed so far are meganucleases, zinc-finger nucleases, and transcription activator-like effectors, each with different editing efficiencies. A decade ago, appearance of the Clustered Regularly Interspaced Short Palindromic Repeats (CRISPR)-CRISPR associated protein (Cas) systems brought about a revolution in the field of genome editing (Cong et al., 2013; Gasiunas et al., 2012; Jinek et al., 2012; Mali et al., 2013) (Figure 5.1d). While previously

DOI: 10.1201/9781003387060-5

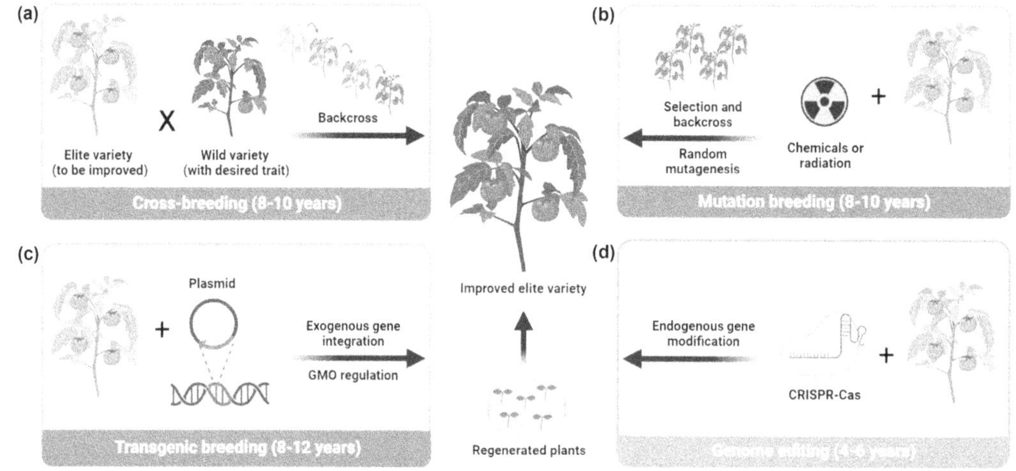

FIGURE 5.1 Plant breeding techniques that commonly used for crop improvement. (A) Cross-breeding exploits natural genetic variability by crossing an elite variety (recipient) with a wild variety (donor) harboring the desired trait and selecting the outstanding progeny. Then, this progeny is backcrossed with the elite variety for several generations to introduce the desired trait into the recipient and eliminate unexpectedly linked traits. (B) Mutation breeding uses physical or chemical agents to induce random mutations genome-wide and artificially expand genetic variability. Its stochastic nature requires screening large mutant populations to identify individuals with the desired traits. (C) Transgenic breeding is based on the random integration of exogenous genes into the plant genome, raising public concern about the biosecurity of transgenic crops and restricting their commercialization. (D) Genome editing allows precise and predictable genomic modifications in the elite variety following a transgene-free approach. (Figure created with BioRender©.)

developed SSNs rely on protein–DNA interactions, sequence specificity in the CRISPR-Cas systems is achieved by a customizable guide RNA (gRNA) that can target any DNA sequence. An additional advantage is that a single Cas nuclease can be combined with several gRNAs to target multiple genomic sites (i.e., multiplexing). Owing to its specificity, design simplicity, and efficiency, CRISPR-Cas systems have been adopted to modify numerous plant species, offering the opportunity to perform plant breeding at an unprecedented pace and minimal cost (Chen et al., 2019; Liu et al., 2022a; Zhu et al., 2020).

The most popular CRISPR-Cas systems for genome editing purposes are the class-2 type-II Cas9 from *Streptococcus pyogenes* (SpCas9, hereinafter referred as Cas9) (Jinek et al., 2012) and the class-2 type-V Cas12a from *Lachnospiraceae* bacterium ND2006 (LbCas12a) (Zetsche et al., 2015), both being single effector proteins with intrinsic nuclease activity (Figure 5.2a). The 20- to 23-nucleotide (nt) region located in the 5′ end of the gRNA, commonly known as protospacer, directs the nuclease to a complementary target DNA site next to the protospacer adjacent motif (PAM; 5′-NGG-3′ for Cas9 and 5′-TTTV-3′ for Cas12a). Upon DNA binding, the Cas nuclease produces a double-strand break (DSB) that activates the host DNA repair mechanisms (Que et al., 2019) (Figure 5.2a, bottom panel). The favored pathway in plants is non-homologous end joining (NHEJ), which generates small insertions or deletions (indels) at the junction point as a consequence of minimal end processing. The indels can vary in length and sequence, potentially leading to gene disruption due to frameshift mutations (Chang et al., 2017; Seol et al., 2018). Despite its high efficiency, NHEJ cannot be used for sophisticated engineering purposes owing to its inaccuracy. Alternatively, homology-directed repair (HDR) is favored when an exogenous DNA template with high similarity to the DSB is provided, being useful to produce point mutations and sequence replacements or insertions in the target gene (Ceccaldi et al., 2016). However, the application of HDR-based approaches in plants is limited by their little efficiency and the challenging delivery of donor DNA (Atkins & Voytas, 2020; Huang & Puchta, 2021).

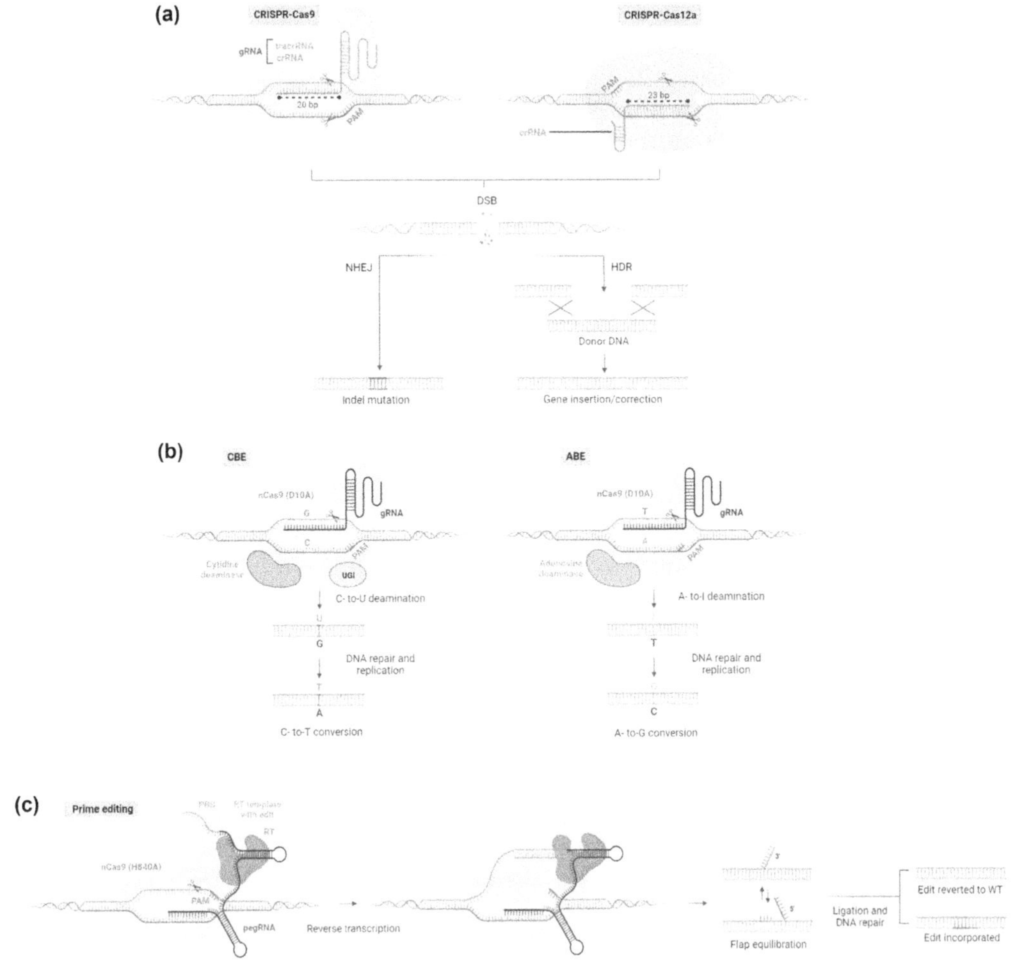

FIGURE 5.2 Overview of genetic modifications produced by CRISPR-Cas systems. (A) Mechanism of action of class-2 CRISPR-Cas systems. Cas9 or Cas12a produce double-strand breaks (DSBs) in the target genomic region, which have different outcomes depending on the DNA repair pathways: non-homologous end joining (NHEJ) is predominant and generates small insertions or deletions (indels) leading to gene disruption (left); in the presence of a DNA donor template, precise gene insertion or correction occurs via homology-directed repair (HDR) (right). (B) Base editing. Fusion of a Cas9 nickase (nCas9 D10A) to a cytidine (left) or adenosine (right) deaminase generates a cytosine base editor (CBE) or adenosine base editor (ABE), respectively. The catalytic activity of CBE produces C-to-T conversions with the aid of the uracil glycosylase inhibitor (UGI), while the ABE generates A-to-G transitions. (C) Prime editing. The prime editor is composed of an impaired Cas9 nickase (nCas9 H840) fused to a reverse transcriptase (RT). Upon target site binding, the nicked PAM strand hybridizes with the primer-binding site (PBS) located at the 3′ end of the prime-editing gRNA (pegRNA). The PBS acts as an RT template, transferring the desired sequence edits to the non-target strand. The editing complex uncouples from the target site, leaving a 3′ flap (the modified PAM strand) and a 5′ flap (the original PAM strand). After endonucleases degrade the 5′ flap, the desired edits get incorporated into the genome through DNA repair mechanisms. crRNA, CRISPR RNA; tracrRNA; trans-activating crRNA; PAM, protospacer adjacent motif; gRNA, guide RNA. (Figure created with BioRender©.)

In addition to canonical DSB-mediated genome editing, base editors are useful tools for generating precise single-base changes (Komor et al., 2016; Nishida et al., 2016). These systems consist of a catalytically impaired Cas9 'nickase' (nCas9, bearing the D10A mutation) fused to a single-stranded DNA deaminase (Figure 5.2b). Two main types of base editors exist: cytosine base editors (CBEs) introduce C:G>T:A transitions (Figure 5.2b, left panel), whereas adenine base editors (ABEs) mediate A:T> G:C conversions (Figure 5.2b, right panel). In the case of CBEs, a uracil DNA glycosylase inhibitor (UGI) is added to the complex for manipulating the DNA repair mechanisms and thus increase the editing efficiency. Both CBEs and ABEs have been successfully used in plants (Shimatani et al., 2017; Tang et al., 2018; Zong et al., 2018). Furthermore, dual base editors have been developed to simultaneously induce C-to-T and A-to-G changes at the target site using a unique gRNA (Li et al., 2020). Nevertheless, the main disadvantage of current base editing systems is the impossibility of producing additional modifications, such as base transversions and specific sequence insertions or deletions. Prime editing, a CRISPR-Cas technological breakthrough that enables the generation of up to 12 different types of base substitutions and small insertions or deletions, recently overcame this limitation (Anzalone et al., 2019). Prime editors consist of a nCas9 (H840A) fused to a reverse transcriptase (RT) (Figure 5.2c). The prime-editing gRNA (pegRNA) has a primer-binding site (PBS) at the 3′ end and an RT template including the desired modifications. Upon target site recognition, the nickase generates a single-stranded DNA that hybridizes with the PBS and acts as a primer for the RT. After transferring the sequence edits from the pegRNA to the non-target strand by reverse transcription, these changes get finally incorporated into the genome by DNA repair mechanisms. Prime editors have been rapidly adapted for use in rice and wheat (Lin et al., 2020; Xu et al., 2020), albeit with considerably lower efficiency than base editors, meaning that further optimization of these systems is needed.

5.1.2 Novel Approaches for DNA-Free Editing in Plants

Unique plant characteristics have hindered the development of a robust and universal delivery method of CRISPR-Cas components, which is essential for a broad application of genome editing. The rigid cell wall hinders the entry of foreign molecules into plant cells; moreover, polyploidy and other genomic rearrangements have been shown to affect the expression of heterologous genes (Varanda et al., 2021). Conventional approaches for delivering plasmids encoding the editing reagents into plants rely on genetic transformation via *Agrobacterium tumefaciens* or particle bombardment (Laforest & Nadakuduti, 2022) (Figure 5.3). Then, the CRISPR-Cas DNA cassette integrates into the plant genome and induces targeted modification with exceptional efficiency. Nevertheless, the main disadvantage of transformation technologies is that random integration of the transgene can disrupt essential endogenous genes and increase off-target events (Sun et al., 2016; Van Kregten et al., 2016). Moreover, crops carrying the CRISPR-Cas transgene are considered GMOs in many countries; thus, they require breeding procedures like selfing or crossing to remove foreign DNA before they are approved as marketable (Peng et al., 2017; Pyott et al., 2016). Alternatively, obtaining modified plants without any trace of foreign DNA is possible by transient expression of CRISPR-Cas reagents and a selection-free regeneration of edited tissue (Chen et al., 2018). However, this strategy only partially eradicates transgene integration, because the host genome may still accommodate degraded DNA fragments.

Lately, ribonucleoproteins (RNPs) have gained substantial attention as a DNA-free editing strategy in plants (Banakar et al., 2020; Liang et al., 2017; Woo et al., 2015; Zhang et al., 2016a) (Figure 5.3). Preassembled Cas-gRNA complexes are directly delivered into protoplasts through polyethylene glycol, particle bombardment, or electroporation. RNPs are functional upon delivery, and they get promptly degraded inside the host cell, which considerably reduces off-target frequency compared with stable expression of editing reagents. However, plant regeneration from protoplasts is only possible in a few species, and this process frequently leads to genome instability (Yue et al., 2021).

Nanoparticles are a novel alternative approach for the delivery of DNA, RNA, and proteins for plant genetic engineering (Figure 5.3). They show several advantages like low toxicity in host cells, availability of multiple nanomaterials, and lack of dependency on plant species or genotypes

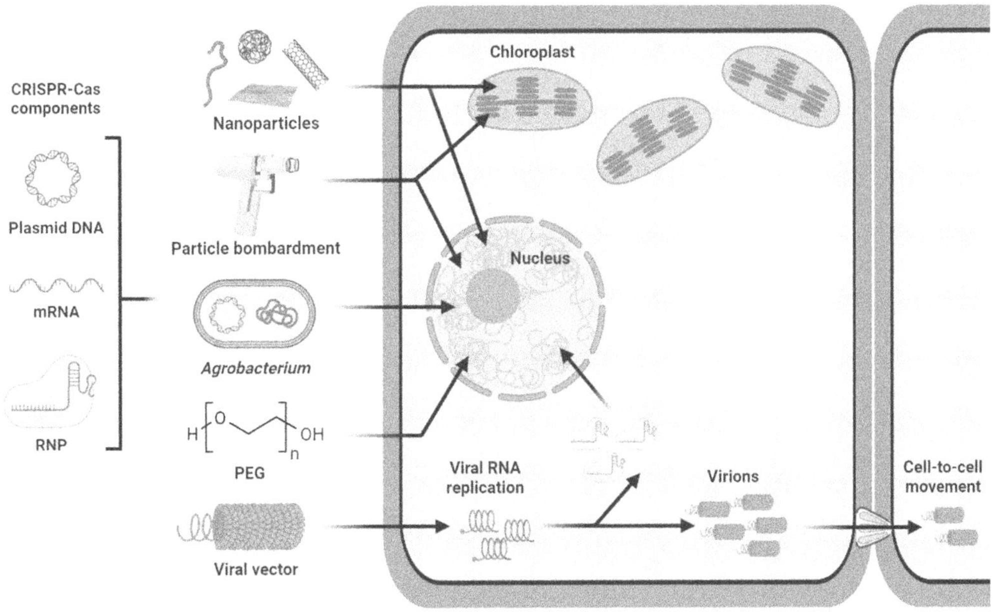

FIGURE 5.3 CRISPR-Cas delivery strategies in plants. CRISPR-Cas components (Cas nuclease and gRNAs) can be delivered into plant cells as plasmid DNA, messenger RNA (mRNA), or ribonucleoproteins (RNP) using nanoparticles, particle bombardment, or Agrobacterium, which can all go across the cell wall. Polyethylene glycol (PEG)-mediated delivery is only used for protoplast transformation. Alternatively, DNA-encoding CRISPR-Cas components can be introduced into the genomes of plant viruses; recombinant viral vectors can infect host cells, replicate in them and produce virions that spread to neighboring cells through plasmodesmata, thus leading to systemic expression of the editing reagents. (Figure created with BioRender©.)

(Demirer et al., 2019, 2020). Interestingly, researchers have used magnetic nanoparticles to deliver DNA plasmids expressing CRISPR-Cas components into the pollen grains of dicot species, enabling the production of genetically modified seeds (Zhao et al., 2017). Nanoparticle-mediated expression of CRISPR-Cas RNPs has also been accomplished in mammalian cells (Lee et al., 2017), thus opening attractive research possibilities for transgene-free editing in plants.

5.2 PLANT VIRUSES FOR THE DELIVERY OF CRISPR-Cas COMPONENTS

Plant viruses have contributed to functional genetics from the beginning of genetic engineering. During the last few decades, the emergence of molecular biology and high-throughput sequencing technologies have facilitated the manipulation of viral genomes as expression vectors of heterologous proteins and RNAs in plants (Wang et al., 2020). More recently, the use of plant viruses in biotechnology has reached the next level, thanks to several studies demonstrating their potential as transient delivery vehicles of CRISPR-Cas components (Figure 5.3). Commonly named virus-induced genome editing (VIGE), this approach shows several advantages over long-established delivery strategies (Cody & Scholthof, 2019). First, prolific viral replication promotes an increased accumulation of CRISPR-Cas components within the host cell, resulting in a higher editing efficiency, which enables to evaluate the efficacy and specificity of gRNA in a time-effective way. Second, essential genes can be targeted because inoculations are performed in adult plants, and the gene is disrupted before any severe physiological effect appears. Finally, using RNA viruses prevents foreign genetic material from getting integrated into the host genome, avoiding potential regulatory and ethical concerns.

A critical step for engineering viral vectors is assembling their genome into appropriate expression plasmids that will trigger plant infection. Both DNA and RNA plant viruses present a diversity of genomic organizations that can be single- (ss) or double-stranded (ds), or linear or circular (Rampersad & Tennant, 2018). Constructing infectious clones of DNA viruses is relatively straightforward. Linear DNA can be simply cloned in a suitable plasmid that mediates *in planta* expression. In contrast, the usual procedure for circular DNA is to clone tandem repeats of the viral genome under the control of two different replication origins, thus releasing the complete viral genome into the host cell (Stenger et al., 1991).

Cloning RNA viruses requires complementary DNA (cDNA) synthesis and the addition of regulatory sequences for the correct expression of viral parts. A common strategy consists of cloning the corresponding cDNA of the viral genome in a plasmid under the control of a bacterial promoter, which following *in vitro* transcription, can be mechanically inoculated into the host plant (Duff-Farrier et al., 2019). It should be noted that RNA transcripts must be carefully handled to prevent their degradation, and this strategy cannot be applied with phloem-limited or negative-strand RNA viruses. Moreover, it has been observed that viral proteins can be toxic or unstable in *E. coli*, which can be prevented either by fragmenting the toxic gene or by inserting plant introns to interrupt the coding sequence (Tran et al., 2019). As an alternative to *E. coli*, yeast has also been shown to enable binary clone assembly of RNA viruses containing toxic components (Youssef et al., 2011). Fragments of the viral genome are amplified by PCR containing 20 to 30nt overlaps, and once inside yeast cells, they get assembled through homologous recombination into the linearized destination plasmid.

Its susceptibility to a wide range of viruses makes *Nicotiana benthamiana* an ideal species for virus-based genetic studies, and different methodologies can be used to deliver infectious viral clones into the plant cells (Table 5.1). Agroinoculation is a preferred method owing to its cost-effectiveness, homogenous infection of plant tissues, and modularity for large-scale applications (Gleba et al., 2014). It consists of inoculating leaves with *A. tumefaciens* cultures previously transformed with recombinant plasmids. *A. tumefaciens* is generally introduced in the plant through syringe infiltration of the leaf abaxial surface, but other procedures can also be used, such as vacuum infiltration,

TABLE 5.1
Main Methodologies for the Delivery of Infectious Viral Clones into Plants

Methodology	Advantages	Disadvantages	Reference
Agroinoculation	• Straightforward and cost-effective	• Multiple delivery is time-consuming	Gleba et al. (2014) Chen et al. (2013)
i. Syringe infiltration	• Applicable to many plant species	• Requires specialized equipment	Chen et al. (2013)
ii. Vacuum infiltration	• Allows simultaneous delivery of multiple viral vectors	• Optimized protocols for different plant species	Torti et al. (2021)
iii. Agrospray			Ryu et al. (2004)
iv. Agrodench	• Robust and efficient	• Risk of virus escape (in the field)	
	• Minimal plant damage	• Risk of virus escape (in the field)	
	• High scalability		
	• Fast and efficient		
	• Large scalability		
Rub inoculation	• Applicable to many plant species • Allows to use diverse sources of inoculum (viral transcripts, plasmids, or infected plant tissue)	• Plant damage and wound-related gene expression are frequently observed • A large inoculum is needed to initiate viral infection	Kang et al. (2016)

spraying, or drenching. If *N. benthamiana* is a susceptible host of the virus, leaves showing infection symptoms can be harvested, homogenized and used to mechanically inoculate those plant species for which infiltration is not possible (Bedoya et al., 2010; Kang et al., 2016). Abrasive agents generate wounds in the leaf that allow the virus to enter the plant cells, thus resembling the natural process in which viral infections are initiated. Once inside the host cell, the viral replication cycle produces virions that systematically spread throughout the plant.

Over the last few years, research efforts have focused on engineering viral vectors to efficiently deliver CRISPR-Cas components into plant cells. Nevertheless, the molecular biology properties and the specific host range of each viral vector frequently hinder the use of VIGE. The following section describes the most relevant contributions for expanding the VIGE toolbox to diverse plant species and applications (Table 5.2).

5.2.1 Geminiviruses as Pioneers for VIGE

Geminiviridae is one of the largest families of plant viruses conformed by single-stranded, circular DNA with monopartite or bipartite genomes (Aguilar et al., 2020). The use of geminiviruses in plant biotechnology attracts much attention owing to following remarkable properties: (1) a wide host range ranging from staple to fiber crops; (2) a DNA genome highly reduced in size (2.7–5.5 kb) that encodes five to eight functional proteins on both forward and reverse strands; (3) the capability for replicating in the host cell using a few proteins (only replication-associated protein, Rep, is needed in the Mastrevirus genera); (4) regulation of viral DNA expression by an intergenic endogenous promoter; and (5) production of large amounts of amplicons inside the host cell. A heterologous sequence of up to 800–1,000 bp can partially replace the coat protein (CP) of some bipartite geminiviruses while maintaining the capacity for viral movement and replication (Lozano-Durán, 2016). This manipulation has enabled the use of geminiviruses for heterologous protein expression (Rybicki & Martin, 2014) as well as gene silencing (Carrillo-Tripp et al., 2006), and was also proposed to be useful for genome editing.

Although the partial replacement of the CP allows the production of gRNA, it is not enough for Cas9 nucleases owing to their large coding sequences. The geminivirus cargo capacity can be further increased by removing both the CP- and movement protein (MP)-encoding genes. However, because the ssDNA can no longer be packed into virions, cell-to-cell movement and transmission to neighboring plants are eliminated, leading to non-infectious replicons. Baltes et al. (2014) were the first to demonstrate virus-mediated expression of the Cas9 nuclease and DNA repair templates using *Bean yellow dwarf virus* (BeYDV). As a consequence of viral replication, repair templates accumulate to high levels in the host cell nucleus and allow HDR to outcompete NHEJ, thereby notably increasing the editing efficiency compared to *A. tumefaciens*-mediated delivery. These promising results motivated the use of BeYDV for targeted modifications in solanaceous crops like tomato (Čermák et al., 2015; Dahan-Meir et al., 2018) and potato (Butler et al., 2016). Shortly after, *Wheat dwarf virus* (WDV) was engineered for editing complex cereal genomes, including wheat (Gil-Humanes et al., 2017) and rice (Wang et al., 2017). Additional works with *Cabbage leaf curly virus* (CaLCuV) (Yin et al., 2015) and *Cotton leaf crumple* virus (CLCrV) (Lei et al., 2021) provide further evidence that geminiviral replicons promote gene targeting.

5.2.2 gRNA Delivery with RNA Viruses

Because replication and transcription of DNA viruses require the host's machinery, there is a risk of unintended integration of foreign genetic material into the plant genome. On the contrary, the infectious cycle of RNA viruses is carried out exclusively in the cytoplasm and using only viral RNA-synthesizing enzymes, allowing the recovery of plants without any trace of viral genetic material. Such an advantage has favored the manipulation of diverse RNA viruses to deliver gRNAs into plant cells (Table 5.2).

TABLE 5.2
Plant Viruses Engineered as Delivery Vectors of CRISPR-Cas Components

Viral Vector	Virus Family/Genus	Nuclease	gRNA Type	Plant Species/Transgene	Systemic Expression	Heritability of Gene Editing	Ref.
				+ Single-stranded DNA			
BeYDV	*Geminiviridae/ Mastrevirus*	ZFN, TALEN, Cas9	AtU6-gRNA	Tobacco (*N. tabacum*)	No	Not determined	Baltes et al. (2014)
		TALEN, Cas9	AtU6-gRNA	Tomato (*Solanum lucopersicum*)	No	Not determined	Čermák et al. (2015)
				Potato (*S. tuberosum*)	No	Not determined	Butler et al. (2016)
		Cas9	AtU6-gRNA	Potato (*S. tuberosum*)	No	Not determined	Butler et al. (2015)
				Tomato (*S. lucopersicum*)	No	Not determined	Dahan-Meir et al. (2018)
WDV	*Geminiviridae/ Mastrevirus*	Cas9	OsU6-gRNA	Rice (*Oryza sativa*)	No	Not determined	Wang et al. (2017)
			TaU6-gRNA	Wheat (*Triticum aestivum*)	No	Not determined	Gil-Humanes et al. (2017)
CaLCuV	*Geminiviridae/ Begomovirus*	-	AtU6-gRNA	*N. benthamiana*/Cas9	Yes	Not determined	Yin et al. (2015)
CLCrV	*Geminiviridae/ Begomovirus*	-	AtU6-gRNA (+/- *FT*)	*A. thaliana*/Cas9	Yes	Low frequency (4.4–8.8%; progeny of infected plants)	Lei et al. (2021)
				+ Single-stranded RNA			
TRV	*Virgaviridae/Tobravirus*	Mega-nuclease	-	*N. alata*	Yes	Low frequency (% not specified)	Honig et al. (2015)
		ZFN, TALEN	-	Tobacco (*N. tabacum*), petunia (*P. hybrida*)	Yes	Not determined	Marton et al. (2010)
		-	PEBV-gRNA	*N. benthamiana*/Cas9	Yes	Low frequency (% not specified; progeny of infected plants)	Ali et al. (2015)
		-	PEBV-gRNA	*A. thaliana, N. benthamiana*/Cas9	Yes	Not determined	Ali et al. (2018)
		-	PEBV-gRNA (+/- *FT* and $tRNA^{Met/Gly/Ile}$)	*N. benthamiana*/ Cas9	Yes	High frequency (65–100%; progeny of infected plants)	Ellison et al. (2020)
		-	PEBV-gRNA (+ *FT*)	*N. benthamiana*/ Cas9	Yes	High frequency (57.5%; progeny of infected plants)	Aragonés et al. (2021)

(Continued)

TABLE 5.2 (*Continued*)
Plant Viruses Engineered as Delivery Vectors of CRISPR-Cas Components

Viral Vector	Virus Family/Genus	Nuclease	gRNA Type	Plant Species/Transgene	Systemic Expression	Heritability of Gene Editing	Ref.
		-	PEBV-gRNA (+ $tRNA^{Ile}$)	*A. thaliana*/Cas9	Yes	High frequency (55%–100%; progeny of infected plants)	Nagalakshmi et al. (2022)
		-	PEBV-gRNA (+ $tRNA^{Ile}$)	*A. thaliana*/CBE	Yes	Low frequency (0.8%–7.6%; progeny of infected plants)	Liu et al. (2022b)
		-	PEBV-gRNA (+ $tRNA^{Ile}$)	*A. thaliana*/SunTag-VP64	Yes	Low frequency (6.6%–8%; progeny of infected plants)	Ghoshal et al. (2020)
TMV	*Virgaviridae*/ *Tobamovirus*	-	TMV-gRNA-ribozyme	*N. benthamiana*	No	Not determined	Cody et al. (2017)
		Cas9	TMV-gRNA-ribozyme	*N. benthamiana*	No	Not determined	Chiong et al. (2021)
PEBV	*Virgaviridae/Tobravirus*	-	PEBV-gRNA	*A. thaliana, N. benthamiana*/Cas9	Yes	Not determined	Ali et al. (2018)
PVX	*Alphaflexiviridae*/ *Potexvirus*	-	BaMV-gRNA (+/- *FT*)	*N. benthamiana*/Cas9	Yes	High frequency (100%, via tissue regeneration; 22%, progeny of infected plants)	Uranga, et al. (2021a)
		-	BaMV-gRNA	*N. benthamiana*/Cas9	Yes	Not determined	Selma et al. (2022)
		Cas9, CBE	AtU6-gRNA	*N. benthamiana*	Yes	High frequency (62%; via tissue regeneration)	Ariga et al. (2020)
BSMV	*Virgaviridae/Hordeivirus*	-	BSMV-gRNA	*N. benthamiana*, wheat (*Triticum aestivum*), maize (*Zea mays*)/ Cas9	Yes	*N. benthamiana*: low frequency (% not specified; via tissue regeneration); not determined for wheat and maize	Hu et al. (2019)
			BSMV-gRNA (+/– *FT*, $tRNA^{Met}$)	Wheat (*T. aestivum*)/Cas9	Yes	Yes (46–69%; progeny of infected plants)	T. Li et al. (2021)
		-	BSMV-gRNA (+/– *FT*, $tRNA^{Met/Ile}$)	Wheat (*T. aestivum*)/Cas9	Yes	Yes (0–27%; progeny of infected plants)	Wang et al. (2022)

(*Continued*)

TABLE 5.2 (*Continued*)
Plant Viruses Engineered as Delivery Vectors of CRISPR-Cas Components

Viral Vector	Virus Family/Genus	Nuclease	gRNA Type	Plant Species/Transgene	Systemic Expression	Heritability of Gene Editing	Ref.
BNYVV	*Benyviridae/Benyvirus*	-	p31-gRNA	*N. benthamiana*/ Cas9	Yes	Not determined	Jiang et al. (2019)
FoMV	*Alphaflexiviridae/ Potexvirus*	-	FoMV-gRNA	*N. benthamiana*, maize (*Z. mays*), foxtail (*Setaria viridis*)/Cas9	Yes	Not determined	Mei et al. (2019)
		Cas9	AtU6-gRNA	*N. benthamiana*	Yes	Not determined	X. Zhang et al. (2020)
ALSV	*Secoviridae/CHeravirus*	-	ALSV-gRNA	*N. benthamiana*, soybean (*Glycine max*)	Yes	Not determined	Luo et al. (2021)
TEV	*Potyviridae /Potyvirus*	LbCas12a	BaMV-gRNA (in PVX vector)	*N. benthamiana*	Yes	Not determined	Uranga et al. (2021b)
		AcrIIA4	BaMV-gRNA-*FT* (in PVX vector)	*N. benthamiana*/Cas9	Yes	Not determined	Calvache et al. (2021)
				– Single-stranded RNA			
BYSMV	*Rhabdoviridae/ Cytorhabdovirus*	Cas9	BYSMV-gRNA	*N. benthamiana*	No	Not determined	Gao et al. (2019)
SYNV	*Rhabdoviridae/ Betanucleo-rhabdovirus*	Cas9	SYNV-gRNA-$tRNA^{Gly}$	*N. benthamiana*	Yes	Yes (90%–100%; via tissue regeneration)	Ma et al. (2020)
TSWV	*Tospoviridae/ Orthotospovirus*	Cas9, LbCas12a, ABE, CBE	TSWV-gRNA	*N. benthamiana*	Yes	Yes (55.6%–95%; via tissue regeneration)	Liu et al. (2023)

Gly, glycine; Ile, isoleucine; Met, methionine.

Tobacco rattle virus (TRV; genus *Tobravirus*) is a bipartite positive-strand RNA virus consisting of TRV1 and TRV2 subgenomes. Genes essential for viral replication and movement are encoded in TRV1, whereas TRV2 contains the CP and other non-structural proteins involved in nematode transmission (Macfarlane, 2010). By introducing a multiple cloning site in the non-essential region of TRV2, researchers have demonstrated the expression of heterologous proteins and fragments of host-plant genes for gene silencing (Senthil-Kumar & Mysore, 2014). The main advantages of TRV over geminiviruses are its wide host range (>400 plant species) and the capacity to enter developing plant tissues.

In preliminary studies related to plant genome engineering, TRV-based delivery of SSNs into tobacco and petunia produced modifications transmitted to the next generation (Marton et al., 2010). Later, Ali et al. (2015) developed a TRV-gRNA delivery system for targeted mutagenesis in *A. thaliana* and *N. benthamiana*. These modifications were partially transmitted to the progeny of infected plants, reinforcing the evidence that TRV can infect germline cells. TRV-based gRNA delivery is a versatile system with various applications, like transcriptional activation and epigenomic editing (Ghoshal et al., 2020). Recently, TRV-gRNA vectors have been used to inoculate a transgenic *A. thaliana* line expressing a cytidine deaminase that resulted in heritable base editing (Liu et al., 2022b). This seems a good system for gene function studies based on loss- and gain-of-function mutants. Another tobravirus engineered for VIGE is *Pea early browning virus* (PEBV). Research findings have demonstrated that PEBV-mediated gRNA delivery is more efficient than TRV and can infect meristematic tissues, allowing the inheritance of the desired modifications (Ali et al., 2018).

Tobacco mosaic virus (TMV; genus *Tobamovirus*) is a monopartite positive-strand RNA virus in which a strong viral subgenomic promoter mediates the expression of CP. Most biotechnological approaches focus on partially replacing CP with heterologous genes to obtain large amounts of their products (Gleba et al., 2007). This same approach was applied to develop a TMV-based vector for gRNA delivery (Cody et al., 2017). Although the virus could only infect locally owing to CP deletion, gRNA accumulation in infected tissues resulted in effective editing of *N. benthamiana* plants previously infiltrated with a Cas9-expressing plasmid.

Potato virus X (PVX; genus *Potexvirus*) is a monopartite positive-strand RNA virus with more than 60 host plant species from different families, including important solanaceous crops such as potato, tomato, pepper, and tobacco (Loebenstein & Gaba, 2012). PVX was engineered for multiplex gRNAs delivery resulting in highly efficient genome editing in *N. benthamiana* (Uranga et al., 2021a). Remarkably, tandemly arrayed gRNAs without processable spacers could induce modifications in multiple genes. In addition, plants regenerated from infected tissue produced virus-free progeny with biallelic modifications. Later, the PVX-gRNA delivery system was tested for transcriptional regulation using a *N. benthamiana* transgenic line expressing the remaining components for CRISPR-mediated gene activation (Selma et al., 2022). Spray inoculation of the viral vector induced a robust activation of the target genes both locally and systemically, leading to program-specific metabolic fingerprints. All these findings postulate PVX as a practical tool for diverse applications in functional genomics in a broad range of solanaceous crops.

The feasibility of a particular viral vector in genome editing can be easily tested using transgenic plants constitutively expressing Cas9. Following this approach, a *Beet necrotic yellow vein virus* (BNYVV; genus *Benyvirus*) vector formerly engineered for heterologous protein expression was also shown to allow gRNA delivery for targeted mutagenesis in *N. benthamiana* (Jiang et al., 2019). *Barley stripe mosaic virus* (BSMV; genus *Hordeivirus*) is a positive-strand RNA with a tripartite genome designated α, β, and γ that infects staple monocot crops. In a first report, BSMV replicons were engineered for gRNA delivery in wheat and maize (Hu et al., 2019), and further improvement of this system evidenced that inoculation with a pool of BSMV vectors harboring individual gRNAs could induce multiplex, heritable editing (Li et al., 2021). Moreover, pollen from Cas9-expressing wheat was crossed with wild-type plants, resulting in Cas9-free mutant individuals in the second generation. Therefore, BSMV opens new possibilities to simplify genome editing in cereal crops in the upcoming years. Additional viral vectors with promising applications in precision crop breeding include *Foxtail mosaic virus* (FoMV; genus *Potexvirus*) in maize and green foxtail (Mei et al., 2019), and *Apple latent spherical virus* (ALSV, genus *Cheravirus*) in soybean (Luo et al., 2021).

5.2.3 Virus-Mediated Delivery of Cas Proteins

All the VIGE studies mentioned earlier rely on virus-mediated gRNA expression in Cas9-expressing transgenic plants, so backcrossing with the wild-type is essential to obtain transgene-free, edited individuals. It is known that the expression of large heterologous sequences impairs the stability and cell-to-cell movement in positive-strand RNA viruses (Avesani et al., 2007). For this reason, researchers were unsuccessful in their initial attempts to engineer a virus vector that could deliver all CRISPR-Cas components into the plant while maintaining systemic movement. However, two independent reports have recently overcome this limitation by using negative-strand RNA viruses with large cargo capacities for *in planta* expression of all editing reagents (Table 5.2 and Figure 5.4a).

Rhabdoviridae is a family of negative-strand RNA viruses able to stably express heterologous genes up to 6 kb in vertebrates (Jackson & Li, 2016), an attribute that creates the opportunity to deliver large Cas9 proteins into plants. In a pioneering report, *Barley yellow striate mosaic virus* (BYSMV) successfully delivered Cas9 and gRNAs for targeted mutagenesis in *N. benthamiana* (Gao et al., 2019). Next, Ma et al. (2020) further demonstrated that a vector based on *Sonchus yellow net virus* (SYNV) could deliver all CRISPR-Cas components to modify single or multiple genes at the whole-plant level. However, tissue culture is still required to obtain edited progeny from infected plants because it is infrequent that rhabdoviruses reach the germline. Moreover, their restricted host range and difficulties in genetic manipulation highlight the need to develop other viral vectors with broad applicability in plant genome engineering (Zhou et al., 2019).

Concurrently, two potexviruses with positive-strand RNA genomes were engineered to deliver Cas9 protein with the assistance of RNA interference suppressor p19, derived from *Tomato bushy stunt virus*. Unlike most virus families, the gene insert size is not physically limited in potexviruses

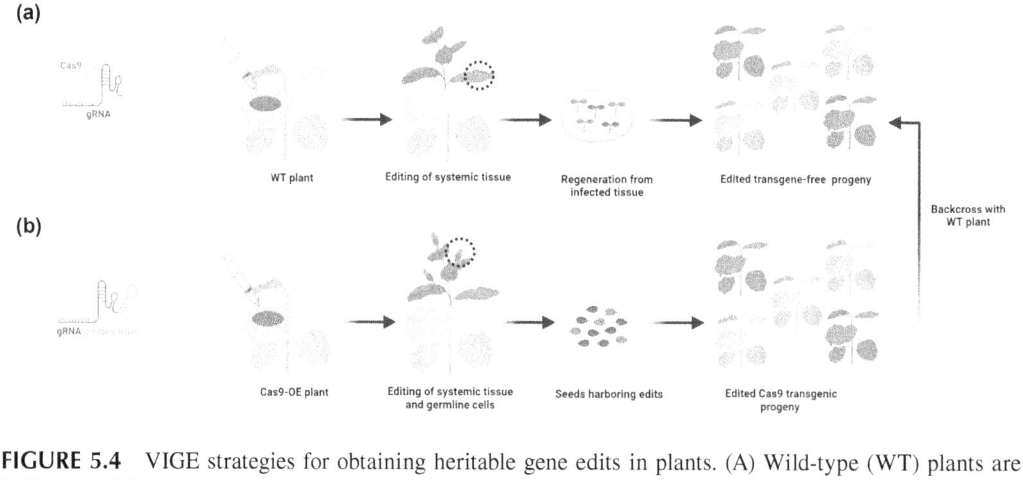

FIGURE 5.4 VIGE strategies for obtaining heritable gene edits in plants. (A) Wild-type (WT) plants are infiltrated with *Agrobacterium* carrying viral vectors expressing Cas9 and gRNAs. Somatic tissues are edited as the viral infection spreads systemically through the plant. Following tissue culture, infected leaves can be used to regenerate transgene-free edited progeny. (B) *Agrobacterium* carrying viral vectors expressing gRNAs fused to RNA mobility signals (*FT* or transfer RNAs) are infiltrated into Cas9-overexpressing (Cas9-OE) plants. Mobile gRNAs spread systemically through the plant and can also enter the shoot apical meristem to induce targeted modifications in the germline. Seeds are collected from infected plants to obtain edited Cas9 progeny bypassing tissue culture procedures. To segregate the Cas9 transgene, the progeny must be backcrossed with wild-type plants. Edited plants or plant tissues are highlighted. (Figure created with BioRender©.)

owing to their flexible filamentous structure (Kendall et al., 2008). In a first study, systemic editing was observed when FoMV viral replicons harboring either Cas9 or the gRNA were delivered simultaneously in *N. benthamiana* (Zhang et al., 2020). Nonetheless, the inheritance of the mutations could not be studied because plants exhibited severe symptoms of viral infection. Alternatively, the CRISPR-Cas cassette was inserted into PVX and delivered into *N. benthamiana* through either agroinfiltration or mechanical inoculation (Ariga et al., 2020). PVX infected most cells and mediated a strong Cas9 expression, while almost no integration of viral DNA into the host genome was observed. Shoots were regenerated from infected tissue bypassing antibiotic selection, which produced virus-free progeny that inherited the modifications. In addition, the authors used this same PVX vector to express a CBE and efficiently induce C-to-T substitutions.

Potyvirus is the most extensive genus among positive-strand RNA viruses. Many members establish synergistic interactions with unrelated viruses, favoring the accumulation of the non-potyviral partner (Wylie et al., 2017). Based on this peculiarity, a two-virus vector system was recently developed to simultaneously express the Cas nuclease and gRNA (Uranga et al., 2021b). First, the RNA-dependent RNA polymerase *NIb* gene in *Tobacco etch virus* (TEV) was replaced with a heterologous sequence encoding LbCas12a; then this vector was combined with PVX that mediated gRNA delivery. A *N. benthamiana* transgenic line constitutively expressing the potyviral *NIb* was used in the co-inoculation experiments to enable replication of the defective TEV. The viral-mediated delivery of all CRISPR-Cas12a components resulted in efficient single and multiplex editing at the whole plant. However, the supplementation of TEV *NIb* from a transgene limits the applicability of the two-virus system. Alternatively, a PVX variant was engineered to simultaneously express gRNA and TEV *NIb*, allowing transgene-free mutagenesis in wild-type plants. This innovative two-virus vector system has also been validated for additional applications beyond conventional gene knockouts. For instance, phage-derived inhibitors of CRISPR-Cas immunity, known as anti-CRISPR proteins, allow to control gene editing temporarily (Borges et al., 2017; Marino et al., 2020; Trasanidou et al., 2019). Work by Calvache et al. (2021) showed that anti-CRISPR proteins delivered via TEV entirely abolished any edits produced by the PVX-gRNA system.

Most recently, a new milestone was reached by engineering *Tomato spotted wilt virus* (TSWV; genus *Orthotospovirus*) for non-transgenic delivery of CRISPR-Cas components. Three RNA segments form the TSWV genome with a complex coding strategy. The large (L) segment is minus-sense and contains the RNA-dependent RNA polymerase gene, while the medium (M) and small (S) segments are ambisense, and they encode two proteins each (Pappu, 2008). This virus has one of the broadest host ranges, including over 1,000 monocotyledonous and dicotyledonous species, and is naturally transmitted by polyphagous trips (Oliver & Whitfield, 2016). In their work, Liu et al. (2023) eliminated the insect transmission genes within the TSWV genome to accommodate large genetic cargoes without compromising the systemic movement of the virus. The TSWV glycoprotein precursor (GP) in segment M was removed to express Cas9, LbCas12a, or base editors. In contrast, the gRNA replaced the non-structural protein S (NSs) gene in the S segment. This non-insect transmissible TSWV vector effectively induced targeted mutagenesis in multiple crop species (i.e. tobacco, tomato, ground cherry, and diverse pepper and peanut varieties), and biallelic edits were transmitted to the next generation by tissue culture of infected leaves. Notably, adding ribavirin to the regeneration medium inhibited TSWV replication and allowed the production of virus-free edited progeny.

5.2.4 RNA Mobility Signals for Heritable Genome Editing

Recovery of edited progeny from infected plants is rare in VIGE approaches because most plant viruses are excluded from entering the shoot apical meristem (SAM), a strongly protected region of pluripotent cells. Therefore, additional strategies are needed to promote access of the gRNA to the germline and generate heritable edits circumventing tissue culture procedures. Plants present many endogenous mobile RNA sequences that can be useful for this task. For example, *Flowering locus*

T (*FT*) is transcribed in leaves and then migrates to the SAM to promote flowering (Li et al., 2009; Lu et al., 2012). A 102-nt fragment located in the 5′ end of the *A. thaliana FT* mRNA has been described to enable cell-to-cell movement of other coding sequences and viral RNAs. Several plant species contain FT orthologs, thus suggesting a conserved function among them (Lifschitz et al., 2006; Meng et al., 2011; Tamaki et al., 2007). In addition, plant phloem exudates contain many transfer RNA (tRNA)-like structures that migrate between leaves and roots (Zhang et al., 2016b).

In an outstanding work, Ellison et al. (2020) were the first to propose that fusing mobility signals such as FT or tRNAs to the 3′ end of virus-produced gRNAs would promote the trafficking of the heterologous RNAs into the SAM to induce germline editing (Figure 5.4b). They infected Cas9-expressing *N. benthamiana* plants with TRV vectors containing mobile gRNAs, observing that up to 65% of individuals within the progeny inherited the targeted modifications. Moreover, no evidence of TRV transmission to the progeny was observed, probably due to differences in the migration rates between gRNAs and the large viral genome. The TRV-based delivery of mobile gRNAs was later tested in *A. thaliana* to induce multiplex, biallelic heritable editing (Nagalakshmi et al., 2022), and base editing (Liu et al., 2022a). The *FT* strategy has been successfully expanded to other viral vectors derived from PVX (Uranga et al., 2021a) and the DNA virus *Cotton leaf crumple virus* (CLCrV) (Lei et al., 2021), achieving heritable editing in seedlings of infected *N. benthamiana* and *A. thaliana* plants (22% and 8%, respectively). However, the biological properties of each virus might partly determine the outcome of mobile RNA sequences on heritable. A study carried out with FoMV in maize found that fusion of the gRNA to *FT* or tRNAs sequences improved the efficiency of somatic editing, but they could not enable germline modifications (Mei et al., 2019). In addition, it was shown that BSMV-gRNA replicons could produce efficient heritable editing in wheat without any modification, whereas the fusion of either *FT* or tRNA sequences reduced the presence of edits in the progeny (T. Li et al., 2021). This is probably a consequence of the exceptional ability of BSMV to reach apical meristems and its high-rate seed transmissibility. In the same study, the Cas9 transgene was quickly removed from the progeny by pollinating wild-type wheat with pollen from BMSV-infected, Cas9-overexpressing plants.

5.3 APPLICATIONS OF GENOME EDITING IN PRECISION PLANT BREEDING

CRISPR-Cas technologies open a novel toolkit to perform plant breeding at an unprecedented pace and cost-effective way, bypassing extensive genetic crossing and large-scale genotyping (Lassoued et al., 2019). Genome editing has mainly contributed to plant breeding by disrupting 'deleterious' genes to improve major agronomic traits such as yield, nutritional quality, and disease resistance (Ahmad et al., 2020; Basu et al., 2023; Garg et al., 2018). The molecular mechanisms controlling complex traits are often conserved among plant species; basic research provides with genetic and biological information that can be directly applied to generate ideal germplasms in elite varieties. *MILDEW RESISTANCE LOCUS* (*MLO*) is an excellent example of cross-species trait sharing. Researchers have proven that editing of *MLO* generates lasting, broad-spectrum resistance to powdery mildew in diverse crops like wheat (Wang et al., 2014), tomato (Nekrasov et al., 2017), and grapevine (Wan et al., 2020).

Separating tightly linked loci by cross-breeding is particularly challenging when one of them is beneficial and the other deleterious. Undoing this genetic linkage involves a time-consuming backcrossing process that might be impracticable in certain plant species (Lee & Wang, 2020). In this case, the deleterious locus can be removed through editing following two different approaches: (1) by directly knocking out the unwanted allele (Roldan et al., 2017); or (2) by inducing chromosomal rearrangements (i.e. translocations or inversions) that favor genetic recombination events (Beying et al., 2020; Zhang et al., 2013). Recently, a novel RecQ helicase was found in wheat that directs genome-wide conversions during DSB repair (Gardiner et al., 2019). This mechanism could provide a novel approach to breaking genetic linkages in plants via genome editing.

5.3.1 Multiplex Gene Editing for Trait Stacking

CRISPR-Cas systems have the potential to simultaneously express various gRNAs in the same cell for targeting multiple genomic sites. This unique feature can be applied to accelerate gene stacking for valuable agronomic traits. It is estimated that around 25% of vascular plants are polyploids, many of which are crops (Abe et al., 2019). In polyploid species, most genes contain multiple copies (i.e. homeologs) that play similar biological functions to regulate particular traits. Following multiplexing approaches, it is possible to simultaneously mutate homeologs to produce a phenotypic effect. Another common attribute of plant genomes is the abundance of paralogous gene families, whose members usually conserve similar structures and overlapping functions (Hyams et al., 2018). Although such redundancy provides genetic robustness, it is usually requisite to mutate more than one paralog to observe a change in the phenotype. For example, its underlying genetic complexity makes it challenging to produce gluten-free wheat through conventional breeding. At least 29 α-gliadins, 18 γ-gliadins, and 10 ω-gliadins, together with 16 low-molecular weight and 6 high-molecular-weight glutenins, are included in the gluten gene family in wheat (Jouanin et al., 2020). Sánchez-León et al. (2018) successfully used CRISPR-Cas9 to disrupt various α- and γ-gliadins in bread wheat. Remarkably, they obtained a single wheat line carrying mutations in 35 genes that showed an 85% reduction in immunoreactivity.

Quantitative trait loci (QTLs) are defined as genetic regions that control the phenotypic variation of quantitative agronomic traits through interactions with additional genes and the environment (Powder, 2020). Owing to their polygenic nature, QTLs are not inherited following Mendelian rules, making them difficult to study or manipulate. According to deep sequencing-based genome-wide association studies and pan-genomics studies, there is a correlation between QTLs and single-nucleotide polymorphisms or structural variants, many of which are found in non-coding or regulatory genetic regions (Huang & Han, 2014). Multiplex editing allows modifying combinations of candidate QTLs in a specific region to cause measurable phenotypic changes while bypassing crossing procedures. Based on this approach, a systematic analysis was carried out for the correlation between *cis*-regulatory regions and phenotypic variations in tomato (Rodríguez-Leal et al., 2017), and candidate QTLs were also identified in maize (Liu et al., 2020a).

5.3.2 *De novo* Domestication

Major staple crops consumed nowadays originate from wild ancestors domesticated at the beginning of agriculture. Domestication aims to enrich traits that enhance crop productivity by crossing elite cultivars with wild relatives harboring favorable properties (Scheben et al., 2017). However, directed evolution leads to the fixation of large portions in plant genomes and a loss of genetic diversity. Moreover, this strategy is only viable for monogenic traits, while many beneficial properties are polygenic and fixing them by segregation is laborious. As a potential alternative for crop improvement, genome editing offers the opportunity to *de novo* domesticate wild species (López-Marqués et al., 2020). As a proof of concept, multiple yield-related genes were simultaneously edited to partially domesticate wild tomato (*Solanum pimpinellifolium*) (Li et al., 2018; Zsögön et al., 2018). The wild rice species *Oryza alta* is also an attractive candidate for domestication because of its large biomass and stress resistance. Recently, CRISPR-Cas was employed to create mutant *O. alta* lines with desirable agronomic traits (Yu et al., 2021). These works have set the principles for accelerating the domestication of wild relatives of crops.

Orphan crops are grown by local farmers in marginal areas and usually attract little attention because they are not traded internationally. They have a rich nutritional profile, are suitable for medicinal use, and are well adapted to growing in suboptimal conditions (Talabi et al., 2022). Although orphan crops show a considerably lower yield than cultivated varieties, they can play a fundamental role in regional food security. Lemmon et al. (2018) demonstrated that targeted editing of beneficial traits could be used to domesticate ground cherry, a remote ancestor of tomato.

Therefore, *de novo* domestication is a promising approach to develop new super crops that might contribute to ensure the global food supply.

5.3.3 Genome-Wide Screenings for Trait Discovery

A deep knowledge of how beneficial agronomic traits are regulated is essential to exploit the potential of genome editing in plant breeding. CRISPR knock-out screens allow for rapidly generating an ordered, known mutant collection for large-scale characterizations of plant genomes (Gaillochet et al., 2021). Typically, thousands of unique gRNAs are delivered into plants via transformation technologies, resulting in a mutant progeny that can be easily tracked by sequencing the integrated gRNAs and modifications. In addition, complex phenotypes can be quickly evaluated to identify their underlying genetic causes. CRISPR screens have been developed in several crops, such as rice (Lu et al., 2017; Meng et al., 2017), tomato (Jacobs et al., 2017), soybean (Bai et al., 2020), and maize (Liu et al., 2020b). As a drawback, this method is currently restricted to a few genotypes because many plants are recalcitrant to transformation and/or regeneration. Virus-based gRNA delivery systems have not yet been applied for CRISPR screens and therefore are highly attractive. Generating genome-wide mutant populations in a broader range of plant genotypes without requiring tissue culture procedures will take CRISPR screens to the next level.

Another relevant large-scale application is saturation genome editing, which uses CRISPR-Cas to introduce all possible single nucleotide polymorphisms into a target gene, the primary source of natural genetic variation (Findlay et al., 2014). This approach efficiently establishes relationships between the mutations and their corresponding phenotypes. However, the usefulness of CRISPR-Cas9 in directed protein evolution is limited because the predominant mutations are indel-associated frameshifts (Capdeville et al., 2020). Base editors are ideal for conquering this challenge because they generate single base transitions to produce all possible amino acid substitutions. For instance, CBEs or ABEs were used to deliver a gRNA library promoting herbicide resistance in rice (Liu et al., 2020a). Methods for iterative mutagenesis and single cell- or protoplast-based screenings could expand the utility of CRISPR-directed evolution studies in the future.

5.3.4 Plant Synthetic Biology

Synthetic biology applies engineering principles to extract, validate and reassemble biological components from living organisms to create systems with novel abilities (Liu & Stewart, 2015). These parts are interchangeable between different species and have a natural origin (e.g. DNA sequences, proteins, etc.); however, their assembly is artificial. Plant design and synthetic biology can be significantly improved by using CRISPR-Cas technologies. For instance, it is possible to redirect or create novel metabolic networks in plants by targeted modifications in endogenous genes or by introducing exogenous genes from a specific biosynthetic pathway. This strategy has allowed obtaining plants supplemented with natural or artificial compounds (Zhu et al., 2021). Another attractive purpose is to engineer photosynthesis. This process is the primary propulsor of plant growth and biomass production but is far from performing optimally. The C3-type metabolism in which RuBisCO fixes CO_2 is found in around 85% of plants, including monocot and dicot staple crops (Leister, 2019). The main issue is that RuBisCO has a low affinity for CO_2 and also accepts O_2 as a substrate in a process known as photorespiration, resulting in huge losses of carbon, nitrogen, and energy. Concordantly, C3 plants maintain RuBisCO levels up to 50% of leaf total protein to maximize CO_2 capture, leading to high photorespiration rates and water loss. Several reports have shown that editing RuBisCO for higher activity, specificity, and activation considerably increases photosynthesis efficiency (Batista-Silva et al., 2020; Gunn et al., 2020). Moreover, carbon-concentration mechanisms from cyanobacteria and algae can be introduced into C3 plants to limit photorespiration and thus promote carbon fixation (South et al., 2019; Weber & Bar-Even, 2019). The successful application of synthetic biology approaches would increment crop yield and production necessary to meet future food demands.

5.4 CHALLENGES AND FUTURE DIRECTIONS FOR VIGE

CRISPR-Cas is a powerful technology for manipulating plant genomes and is opening new opportunities to define the next generation of precision crop breeding. As an alternative to conventional delivery methods, recent advances in virus biotechnology have evidenced that plant viruses can be repurposed for the expression of editing reagents in compatible host species. Since the first reports less than a decade ago, many plant viruses have been engineered for efficient genome editing in diverse plant species (see Table 5.2). Moreover, various studies have proven that the progeny of infected plants is virus-free unless vegetatively propagated (Ali et al., 2015; Ellison et al., 2020; Lei et al., 2021; Li et al., 2021; Uranga et al., 2021a).

The main constraint of VIGE relies on the packaging limit of viral vectors that precludes the expression of Cas9 nucleases (~1,000–1,400 amino acids (AA)). Only a few viruses with exceptional genome stability, mainly negative-stranded RNA viruses (Gao et al., 2019; Ma et al., 2020; Liu et al., 2023) and certain potexviruses (Ariga et al., 2020; Zhang et al., 2020), have been engineered for the delivery of Cas9 and the gRNA in wild-type plants. Consequently, research efforts are now focusing on developing small-sized CRISPR-Cas systems that would raise new opportunities for viral-based, transgene-free editing in plants. One of the first hypercompact candidates was CRISPR-CasΦ from huge bacteriophages. This system comprises a single Cas protein considerably smaller than Cas9 or Cas12a (~700–800 AA) and can generate edits when delivered via RNPs into *A. thaliana* protoplasts, but at extremely low efficiency (0.85%) (Pausch et al., 2020). Further optimization via protein engineering led to the production of two nuclease variants (vCasΦ and nCasΦ) with an increased editing efficiency (~10%–20%) and able to generate heritable modifications in *A. thaliana* (Li et al., 2023). Latest works on miniature-size Cas nucleases have focused on the type V-F Cas12f system from uncultivated archaea, which contains a very small nuclease (422–603 AA) and is also reported to cleave dsDNA (Karvelis et al., 2020). After subjecting this system to RNA and protein engineering, Xu et al. (2021) found that the fusion of Cas12f to a transcriptional activator could switch on genes as efficiently as Cas12a in mammalian cells. Just recently, the miniature Cas12f derived from *Syntrophomonas palmitatica* (SpCas12f) has been applied for gene editing in maize and rice (Bigelyte et al., 2021; Sukegawa et al., 2023), making it a promising candidate for viral delivery in plants. Last, the obligate mobile element-guided activity (OMEGA) proteins, a widespread class of transposon-encoded RNA-guided nucleases with reduced size (~400 AA), are also being tested for genome editing purposes (Altae-Tran et al., 2021).

Particular aspects must considered for the effective bench-to-field transfer of editing technologies. First, plant model species have been used as a proof of concept in many VIGE studies, so adapting and optimizing these systems across agronomic crops is a priority. Conveniently, the increasing diversity of viral vectors allows them to reach a great number of species and genotypes. Second, biosafety concerns regarding the unwanted spread of viral vectors are currently restricting their field application (Eckerstorfer et al., 2019). The most common precautionary measure relies on mutating or deleting essential genes for virus transmission. However, recombining these genes from wild-type viruses might create synthetic viral strains with restored transmissibility (Bedoya et al., 2010; Tepfer et al., 2015). Risks associated with viral mutation and recombination must be evaluated case by case, so initial experiments are limited to small-scale greenhouses with strict biocontainment facilities. An alternative preventive strategy is to eliminate the replication machinery from the viral vector and relocate it into the host plant genome so that defective viruses cannot infect wild-type plants. Engineering synergistic viral communities where each virus performs a specific function are an alternative to prevent the use of transgenic plants. In this condition, unwanted virus release is unlikely to happen because all community members would have to escape or evolve simultaneously. A last approach proposes to use enzymes that cleave RNA backbones in the presence of a particular trigger molecule, known as aptazymes (Liu et al., 2019; Zhong et al., 2016). If integrated into the viral genome, aptazymes can eliminate the virus from the host plant upon induction by specific agrochemicals.

If viral vectors want to gain popularity for plant genome editing, it is crucial to establish a reliable method for generating heritable modifications without tissue culture. However, most plant viruses are excluded from entering apical meristems. RNA mobility signals have attracted much attention, because they can promote gRNA access to the germline and mutation transmission to the next generation (see Section 5.2.4). However, this strategy still relies on Cas9-expressing plants and further research is needed to study whether the viral-delivered nuclease can reach the germline. It has been proposed that other host and viral factors can cooperate with mobile RNAs to facilitate germline mutations. Recent studies have found that either viral suppressor of RNA silencing from seed-transmissible viruses (Bradamante et al., 2021) or silencing of the WUSCHEL transcription factor (Wu et al., 2020) can favor the access of viral vectors to the SAM. Therefore, it is evident that obtaining heritable modifications using viral delivery systems requires much more than just adding mobility signals to the gRNA. A profound knowledge of viral meristem exclusion and entry mechanisms will allow the development of VIGE systems applicable to wide-ranging plant species.

In summary, viral vectors hold an enormous potential for genome editing in plants owing to their simpleness, versatility, and rapidity. This research field has lately evolved at tremendous speed, and we foretell that advances in the field will lead to the development of a highly efficient, DNA-free, and heritable system that will allow to implement VIGE as the method of choice for plant genome engineering.

ACKNOWLEDGMENT

This work is supported by the Marie Skłodowska Curie Actions (HORIZON-MSCA-2022-PF-01-101110621) from the European Commision.

REFERENCES

Abe, F., Haque, E., Hisano, H., Tanaka, T., Kamiya, Y., Mikami, M., Kawaura, K., Endo, M., Onishi, K., Hayashi, T., & Sato, K. (2019). Genome-edited triple-recessive mutation alters seed dormancy in wheat. *Cell Reports*, 28(5), 1362–1369. https://doi.org/10.1016/j.celrep.2019.06.090

Aguilar, E., Garnelo Gomez, B., & Lozano-Duran, R. (2020). Recent advances on the plant manipulation by geminiviruses. *Current Opinion in Plant Biology*, *56*, 56–64. https://doi.org/10.1016/j.pbi.2020.03.009

Ahmad, S., Wei, X., Sheng, Z., Hu, P., & Tang, S. (2020). CRISPR/Cas9 for development of disease resistance in plants: Recent progress, limitations and future prospects. *Briefings in Functional Genomics*, *19*(1), 26–39. https://doi.org/10.1093/bfgp/elz041

Ali, Z., Abulfaraj, A., Idris, A., Ali, S., Tashkandi, M., & Mahfouz, M. M. (2015). CRISPR/Cas9-mediated viral interference in plants. *Genome Biology*, *16*, 238. https://doi.org/10.1186/s13059-015-0799-6

Ali, Z., Eid, A., Ali, S., & Mahfouz, M. M. (2018). *Pea early-browning virus*-mediated genome editing via the CRISPR/Cas9 system in *Nicotiana benthamiana* and Arabidopsis. *Virus Research*, *244*, 333–337. https://doi.org/10.1016/j.virusres.2017.10.009

Altae-Tran, H., Kannan, S., Demircioglu, F. E., Oshiro, R., Nety, S. P., McKay, L. J., Dlakić, M., Inskeep, W. P., Makarova, K. S., Macrae, R. K., Koonin, E. V., & Zhang, F. (2021). The widespread IS200/IS605 transposon family encodes diverse programmable RNA-guided endonucleases. *Science*, *374*(6563), 57–65. https://doi.org/10.1126/science.abj6856

Anzalone, A. V., Randolph, P. B., Davis, J. R., Sousa, A. A., Koblan, L. W., Levy, J. M., Chen, P. J., Wilson, C., Newby, G. A., Raguram, A., & Liu, D. R. (2019). Search-and-replace genome editing without double-strand breaks or donor DNA. *Nature*, *576*(7785), 149–157. https://doi.org/10.1038/s41586-019-1711-4

Aragonés, V., Aliaga, F., Pasin, F., & Daròs, J. A. (2021). Simplifying plant gene silencing and genome editing logistics by a one-*Agrobacterium* system for simultaneous delivery of multipartite virus vectors. *Biotechnology Journal*, 2100504. https://doi.org/10.1002/biot.202100504

Ariga, H., Toki, S., & Ishibashi, K. (2020). Potato virus X vector-mediated DNA-free genome editing in plants. *Plant and Cell Physiology*, *61*(11), 1946–1953. https://doi.org/10.1093/pcp/pcaa123

Atkins, P. A., & Voytas, D. F. (2020). Overcoming bottlenecks in plant gene editing. *Current Opinion in Plant Biology*, *54*, 79–84. https://doi.org/10.1016/j.pbi.2020.01.002

Avesani, L., Marconi, G., Morandini, F., Albertini, E., Bruschetta, M., Bortesi, L., Pezzotti, M., & Porceddu, A. (2007). Stability of *Potato virus X* expression vectors is related to insert size: Implications for replication models and risk assessment. *Transgenic Research*, *16*(5), 587–597. https://doi.org/10.1007/s11248-006-9051-1

Bai, M., Yuan, J., Kuang, H., Gong, P., Li, S., Zhang, Z., Liu, B., Sun, J., Yang, M., Yang, L., Wang, D., Song, S., & Guan, Y. (2020). Generation of a multiplex mutagenesis population via pooled CRISPR-Cas9 in soya bean. *Plant Biotechnology Journal*, *18*(3), 721–731. https://doi.org/10.1111/pbi.13239

Baltes, N. J., Gil-Humanes, J., Cermak, T., Atkins, P. A., & Voytas, D. F. (2014). DNA replicons for plant genome engineering. *Plant Cell*, *26*(1), 151–163. https://doi.org/10.1105/tpc.113.119792

Banakar, R., Schubert, M., Collingwood, M., Vakulskas, C., Eggenberger, A. L., & Wang, K. (2020). Comparison of CRISPR-Cas9/Cas12a ribonucleoprotein complexes for genome editing efficiency in the rice phytoene desaturase (OsPDS) gene. *Rice*, *13*(1), 1–7. https://doi.org/10.1186/S12284-019-0365-z

Basu, U., Riaz Ahmed, S., Bhat, B. A., Anwar, Z., Ali, A., Ijaz, A., Gulzar, A., Bibi, A., Tyagi, A., Nebapure, S. M., Goud, C. A., Ahanger, S. A., Ali, S., & Mushtaq, M. (2023). A CRISPR way for accelerating cereal crop improvement: Progress and challenges. *Frontiers in Genetics*, *13*, 866976. https://doi.org/10.3389/fgene.2022.866976

Batista-Silva, W., da Fonseca-Pereira, P., Martins, A. O., Zsögön, A., Nunes-Nesi, A., & Araújo, W. L. (2020). Engineering improved photosynthesis in the era of synthetic biology. *Plant Communications*, *1*(2), 100032. https://doi.org/10.1016/j.xplc.2020.100032

Bedoya, L., Martínez, F., Rubio, L., & Daròs, J.-A. (2010). Simultaneous equimolar expression of multiple proteins in plants from a disarmed potyvirus vector. *Journal of Biotechnology*, *150*(2), 268–275. https://doi.org/10.1016/j.jbiotec.2010.08.006

Beying, N., Schmidt, C., Pacher, M., Houben, A., & Puchta, H. (2020). CRISPR-Cas9-mediated induction of heritable chromosomal translocations in Arabidopsis. *Nature Plants*, *6*(6), 638–645. https://doi.org/10.1038/s41477-020-0663-x

Bigelyte, G., Young, J. K., Karvelis, T., Budre, K., Zedaveinyte, R., Djukanovic, V., Van Ginkel, E., Paulraj, S., Gasior, S., Jones, S., Feigenbutz, L., Clair, G. S., Barone, P., Bohn, J., Acharya, A., Zastrow-Hayes, G., Henkel-Heinecke, S., Silanskas, A., Seidel, R., & Siksnys, V. (2021). Miniature type V-F CRISPR-Cas nucleases enable targeted DNA modification in cells. *Nature Communications*, *12*(1), 1–8. https://doi.org/10.1038/s41467-021-26469-4

Borges, A. L., Davidson, A. R., & Bondy-Denomy, J. (2017). The discovery, mechanisms, and evolutionary impact of anti-CRISPRs. *Annual Review of Virology*, *4*, 37–59. https://doi.org/10.1146/annurev-virology-101416-041616

Bradamante, G., Scheid, O. M., & Incarbone, M. (2021). Under siege: Virus control in plant meristems and progeny. *The Plant Cell*, *33*(8), 2523–2537. https://doi.org/10.1093/plcell/koab140

Butler, N. M., Atkins, P. A., Voytas, D. F., & Douches, D. S. (2015). Generation and inheritance of targeted mutations in potato (*Solanum tuberosum* L.) using the CRISPR/Cas system. *PLoS ONE*, *10*(12), e0144591. https://doi.org/10.1371/journal.pone.0144591

Butler, N. M., Baltes, N. J., Voytas, D. F., & Douches, D. S. (2016). Geminivirus-mediated genome editing in potato (*Solanum tuberosum* L.) using sequence-specific nucleases. *Frontiers in Plant Science*, 7, 1045. https://doi.org/10.3389/fpls.2016.01045

Calvache, C., Vazquez-Vilar, M., Selma, S., Uranga, M., Fernández-del-Carmen, A., Daròs, J., & Orzáez, D. (2021). Strong and tunable anti-CRISPR/Cas activities in plants. *Plant Biotechnology Journal*, *20*(2), 399–408. https://doi.org/10.1111/pbi.13723

Capdeville, N., Schindele, P., & Puchta, H. (2020). Application of CRISPR/Cas-mediated base editing for directed protein evolution in plants. *Science China Life Sciences*, *63*(4), 613–616. https://doi.org/10.1007/s11427-020-1655-9/metrics

Carrillo-Tripp, J., Shimada-Beltrán, H., & Rivera-Bustamante, R. (2006). Use of geminiviral vectors for functional genomics. *Current Opinion in Plant Biology*, *9*(2), 209–215. https://doi.org/10.1016/j.pbi.2006.01.012

Ceccaldi, R., Rondinelli, B., & D'Andrea, A. D. (2016). Repair pathway choices and consequences at the double-strand break. *Trends in Cell Biology*, *26*(1), 52–64. https://doi.org/10.1016/j.tcb.2015.07.009

Čermák, T., Baltes, N. J., Čegan, R., Zhang, Y., & Voytas, D. F. (2015). High-frequency, precise modification of the tomato genome. *Genome Biology*, *16*(1), 232. https://doi.org/10.1186/s13059-015-0796-9

Chang, H. H. Y., Pannunzio, N. R., Adachi, N., & Lieber, M. R. (2017). Non-homologous DNA end joining and alternative pathways to double-strand break repair. *Nature Reviews Molecular Cell Biology,* 18(8), 495–506. https://doi.org/10.1038/nrm.2017.48

Chen, K., Wang, Y., Zhang, R., Zhang, H., & Gao, C. (2019). CRISPR/Cas genome editing and precision plant breeding in agriculture. *Annual Review of Plant Biology*, *70*(1), 667–697. https://doi.org/10.1146/annurev-arplant-050718-100049

Chen, L., Li, W., Katin-Grazzini, L., Ding, J., Gu, X., Li, Y., Gu, T., Wang, R., Lin, X., Deng, Z., McAvoy, R. J., Gmitter, F. G., Deng, Z., Zhao, Y., & Li, Y. (2018). A method for the production and expedient screening of CRISPR/Cas9-mediated non-transgenic mutant plants. *Horticulture Research*, 5(1). https://doi.org/10.1038/S41438-018-0023-4/6486748

Chen, Q., Lai, H., Hurtado, J., Stahnke, J., Leuzinger, K., & Dent, M. (2013). Agroinfiltration as an effective and scalable strategy of gene delivery for production of pharmaceutical proteins. *Advanced Techniques in Biology & Medicine*, *1*(1), 103. https://doi.org/10.4172/atbm.1000103

Chiong, K. T., Cody, W. B., & Scholthof, H. B. (2021). RNA silencing suppressor-influenced performance of a virus vector delivering both guide RNA and Cas9 for CRISPR gene editing. *Scientific Reports*, *11*(1), 6769. https://doi.org/10.1038/S41598-021-85366-4

Cody, W. B., & Scholthof, H. B. (2019). Plant virus vectors 3.0: Transitioning into synthetic genomics. *Annual Review of Phytopathology*, *57*, 211–230. https://doi.org/10.1146/annurev-phyto-082718-100301

Cody, W. B., Scholthof, H. B., & Mirkov, T. E. (2017). Multiplexed gene editing and protein overexpression using a tobacco mosaic virus viral vector. *Plant Physiology*, *175*(1), 23–35. https://doi.org/10.1104/pp.17.00411

Cong, L., Ran, F. A., Cox, D., Lin, S., Barretto, R., Habib, N., Hsu, P. D., Wu, X., Jiang, W., Marraffini, L. A., & Zhang, F. (2013). Multiplex genome engineering using CRISPR/Cas systems. *Science*, *339*(6121), 819–823. https://doi.org/10.1126/science.1231143

Dahan-Meir, T., Filler-Hayut, S., Melamed-Bessudo, C., Bocobza, S., Czosnek, H., Aharoni, A., & Levy, A. A. (2018). Efficient in planta gene targeting in tomato using geminiviral replicons and the CRISPR/Cas9 system. *Plant Journal*, *95*(1), 5–16. https://doi.org/10.1111/tpj.13932

Demirer, G. S., Zhang, H., Goh, N. S., Pinals, R. L., Chang, R., & Landry, M. P. (2020). Carbon nanocarriers deliver siRNA to intact plant cells for efficient gene knockdown. *Science Advances*, *6*(26), eaaz0495. https://doi.org/10.1126/sciadv.aaz0495

Demirer, G. S., Zhang, H., Matos, J. L., Goh, N. S., Cunningham, F. J., Sung, Y., Chang, R., Aditham, A. J., Chio, L., Cho, M. J., Staskawicz, B., & Landry, M. P. (2019). High aspect ratio nanomaterials enable delivery of functional genetic material without DNA integration in mature plants. *Nature Nanotechnology*, *14*(5), 456–464. https://doi.org/10.1038/s41565-019-0382-5

Duff-Farrier, C. R. A., Mbanzibwa, D. R., Nanyiti, S., Bunawan, H., Pablo-Rodriguez, J. L., Tomlinson, K. R., James, A. M., Alicai, T., Seal, S. E., Bailey, A. M., & Foster, G. D. (2019). Strategies for the construction of cassava brown streak disease viral infectious clones. *Molecular Biotechnology, 61*(2). https://doi.org/10.1007/s12033-018-0139-7

Eckerstorfer, M. F., Heissenberger, A., Reichenbecher, W., Steinbrecher, R. A., & Waßmann, F. (2019). An EU perspective on biosafety considerations for plants developed by genome editing and other new genetic modification techniques (nGMs). *Frontiers in Bioengineering and Biotechnology*, *7*, 31. https://doi.org/10.3389/fbioe.2019.00031

Ellison, E. E., Nagalakshmi, U., Gamo, M. E., Huang, P. J., Dinesh-Kumar, S., & Voytas, D. F. (2020). Multiplexed heritable gene editing using RNA viruses and mobile single guide RNAs. *Nature Plants*, *6*(6), 620–624. https://doi.org/10.1038/s41477-020-0670-y

FAO (2017). *The Future of Food and Agriculture: Trends and Challenges*. Rome.

Findlay, G. M., Boyle, E. A., Hause, R. J., Klein, J. C., & Shendure, J. (2014). Saturation editing of genomic regions by multiplex homology-directed repair. *Nature*, *513*(7516), 120–123. https://doi.org/10.1038/nature13695

Gaillochet, C., Develtere, W., & Jacobs, T. B. (2021). CRISPR screens in plants: Approaches, guidelines, and future prospects. *The Plant Cell*, *33*(4), 794–813. https://doi.org/10.1093/plcell/koab099

Gao, Q., Xu, W. Y., Yan, T., Fang, X. D., Cao, Q., Zhang, Z. J., Ding, Z. H., Wang, Y., & Wang, X. B. (2019). Rescue of a plant cytorhabdovirus as versatile expression platforms for planthopper and cereal genomic studies. *New Phytologist*, *223*(4), 2120–2133. https://doi.org/10.1111/nph.15889

Gardiner, L. J., Wingen, L. U., Bailey, P., Joynson, R., Brabbs, T., Wright, J., Higgins, J. D., Hall, N., Griffiths, S., Clavijo, B. J., & Hall, A. (2019). Analysis of the recombination landscape of hexaploid bread wheat reveals genes controlling recombination and gene conversion frequency. *Genome Biology*, 20(1). https://doi.org/10.1186/s13059-019-1675-6

Garg, M., Sharma, N., Sharma, S., Kapoor, P., Kumar, A., Chunduri, V., & Arora, P. (2018). Biofortified crops generated by breeding, agronomy, and transgenic approaches are improving lives of millions of people around the world. *Frontiers in Nutrition*, *5*, 12. https://doi.org/10.3389/fnut.2018.00012

Gasiunas, G., Barrangou, R., Horvath, P., & Siksnys, V. (2012). Cas9-crRNA ribonucleoprotein complex mediates specific DNA cleavage for adaptive immunity in bacteria. *Proceedings of the National Academy of Sciences of the United States of America*, *109*(39), E2579–E2586. https://doi.org/10.1073/pnas.1208507109

Ghoshal, B., Vong, B., Picard, C. L., Feng, S., Tam, J. M., & Jacobsen, S. E. (2020). A viral guide RNA delivery system for CRISPR-based transcriptional activation and heritable targeted DNA demethylation in *Arabidopsis thaliana*. *PLoS Genetics*, *16*(12), e1008983. https://doi.org/10.1371/journal.pgen.1008983

Gil-Humanes, J., Wang, Y., Liang, Z., Shan, Q., Ozuna, C. V., Sánchez-León, S., Baltes, N. J., Starker, C., Barro, F., Gao, C., & Voytas, D. F. (2017). High-efficiency gene targeting in hexaploid wheat using DNA replicons and CRISPR/Cas9. *Plant Journal*, *89*(6), 1251–1262. https://doi.org/10.1111/tpj.13446

Gleba, Y. Y., Tusé, D., & Giritch, A. (2014). Plant viral vectors for delivery by *Agrobacterium*. *Current Topics in Microbiology and Immunology*, *375*, 155–192. https://doi.org/10.1007/82_2013_352

Gleba, Y., Klimyuk, V., & Marillonnet, S. (2007). Viral vectors for the expression of proteins in plants. *Current Opinion in Biotechnology*, *18*(2), 134–141. https://doi.org/10.1016/j.copbio.2007.03.002

Gunn, L. H., Avila, E. M., Birch, R., & Whitney, S. M. (2020). The dependency of red Rubisco on its cognate activase for enhancing plant photosynthesis and growth. *Proceedings of the National Academy of Sciences of the United States of America*, *117*(41), 25890–25896. https://doi.org/10.1073/pnas.2011641117

Holme, I. B., Gregersen, P. L., & Brinch-Pedersen, H. (2019). Induced genetic variation in crop plants by random or targeted mutagenesis: Convergence and differences. *Frontiers in Plant Science*, *10*, 1468. https://doi.org/10.3389/fpls.2019.01468

Honig, A., Marton, I., Rosenthal, M., Smith, J. J., Nicholson, M. G., Jantz, D., Zuker, A., & Vainstein, A. (2015). Transient expression of virally delivered meganuclease in planta generates inherited genomic deletions. *Molecular Plant*, *8*(8), 1292–1294. https://doi.org/10.1016/j.molp.2015.04.001

Hu, J., Li, S., Li, Z., Li, H., Song, W., Zhao, H., Lai, J., Xia, L., Li, D., & Zhang, Y. (2019). A barley stripe mosaic virus-based guide RNA delivery system for targeted mutagenesis in wheat and maize. *Molecular Plant Pathology*, *20*(10), 1463–1474. https://doi.org/10.1111/mpp.12849

Huang, T. K., & Puchta, H. (2021). Novel CRISPR/Cas applications in plants: From prime editing to chromosome engineering. *Transgenic Research*, *30*(4), 529–549. https://doi.org/10.1007/s11248-021-00238-x

Huang, X., & Han, B. (2014). Natural variations and genome-wide association studies in crop plants. *Annual Review of Plant Biology*, *65*, 531–551. https://doi.org/10.1146/annurev-arplant-050213-035715

Hyams, G., Abadi, S., Lahav, S., Avni, A., Halperin, E., Shani, E., & Mayrose, I. (2018). CRISPys: Optimal sgRNA design for editing multiple members of a gene family using the CRISPR system. *Journal of Molecular Biology*, *430*(15), 2184–2195. https://doi.org/10.1016/j.jmb.2018.03.019

Jackson, A. O., & Li, Z. (2016). Developments in plant negative-strand RNA virus reverse genetics. *Annual Review of Phytopathology*, *54*, 469–498. https://doi.org/10.1146/annurev-phyto-080615-095909

Jacobs, T. B., Zhang, N., Patel, D., & Martin, G. B. (2017). Generation of a collection of mutant tomato lines using pooled CRISPR libraries. *Plant Physiology,* 174(4), 2023–2037. https://doi.org/10.1104/pp.17.00489

Jiang, N., Zhang, C., Liu, J. Y., Guo, Z. H., Zhang, Z. Y., Han, C. G., & Wang, Y. (2019). Development of *Beet necrotic yellow vein virus*-based vectors for multiple-gene expression and guide RNA delivery in plant genome editing. *Plant Biotechnology Journal*, *17*(7), 1302–1315. https://doi.org/10.1111/pbi.13055

Jinek, M., Chylinski, K., Fonfara, I., Hauer, M., Doudna, J. A., & Charpentier, E. (2012). A programmable dual-RNA-guided DNA endonuclease in adaptive bacterial immunity. *Science*, *337*(6096), 816–821. https://doi.org/10.1126/science.1225829

Jouanin, A., Gilissen, L. J. W. J., Schaart, J. G., Leigh, F. J., Cockram, J., Wallington, E. J., Boyd, L. A., van den Broeck, H. C., van der Meer, I. M., America, A. H. P., Visser, R. G. F., & Smulders, M. J. M. (2020). CRISPR/Cas9 gene editing of gluten in wheat to reduce gluten content and exposure-reviewing methods to screen for coeliac safety. *Frontiers in Nutrition*, *7*, 51. https://doi.org/10.3389/fnut.2020.00051

Kang, M., Seo, J. K., Choi, H., Choi, H. S., & Kim, K. H. (2016). Establishment of a simple and rapid gene delivery system for cucurbits by using engineered of *Zucchini yellow mosaic virus*. *The Plant Pathology Journal*, *32*(1), 70–76. https://doi.org/10.5423/ppj.nt.08.2015.0173

Karvelis, T., Bigelyte, G., Young, J. K., Hou, Z., Zedaveinyte, R., Budre, K., Paulraj, S., Djukanovic, V., Gasior, S., Silanskas, A., Venclovas, C., & Siksnys, V. (2020). PAM recognition by miniature CRISPR-Cas12f nucleases triggers programmable double-stranded DNA target cleavage. *Nucleic Acids Research*, *48*(9), 5016–5023. https://doi.org/10.1093/nar/gkaa208

Kendall, A., McDonald, M., Bian, W., Bowles, T., Baumgarten, S. C., Shi, J., Stewart, P. L., Bullitt, E., Gore, D., Irving, T. C., Havens, W. M., Ghabrial, S. A., Wall, J. S., & Stubbs, G. (2008). Structure of flexible filamentous plant viruses. *Journal of Virology*, *82*(19), 9546–9554. https://doi.org/10.1128/jvi.00895-08

Komor, A. C., Kim, Y. B., Packer, M. S., Zuris, J. A., & Liu, D. R. (2016). Programmable editing of a target base in genomic DNA without double-stranded DNA cleavage. *Nature*, *533*(7603), 420–424. https://doi.org/10.1038/nature17946

Laforest, L. C., & Nadakuduti, S. S. (2022). Advances in delivery mechanisms of CRISPR gene-editing reagents in plants. *Frontiers in Genome Editing*, *4*, 830178. https://doi.org/10.3389/fgeed.2022.830178

Lassoued, R., Macall, D. M., Hesseln, H., Phillips, P. W. B., & Smyth, S. J. (2019). Benefits of genome-edited crops: Expert opinion. *Transgenic Research*, *28*(2), 247–256. https://doi.org/10.1007/s11248-019-00118-5

Lee, K., & Wang, K. (2020). Level up to chromosome restructuring. *Nature Plants*, *6*(6), 600–601. https://doi.org/10.1038/s41477-020-0669-4

Lee, K., Conboy, M., Park, H. M., Jiang, F., Kim, H. J., Dewitt, M. A., Mackley, V. A., Chang, K., Rao, A., Skinner, C., Shobha, T., Mehdipour, M., Liu, H., Huang, W. C., Lan, F., Bray, N. L., Li, S., Corn, J. E., Kataoka, K., … Murthy, N. (2017). Nanoparticle delivery of Cas9 ribonucleoprotein and donor DNA in vivo induces homology-directed DNA repair. *Nature Biomedical Engineering*, *1*(11), 889–901. https://doi.org/10.1038/s41551-017-0137-2

Lei, J., Dai, P., Li, Y., Zhang, W., Zhou, G., Liu, C., & Liu, X. (2021). Heritable gene editing using FT mobile guide RNAs and DNA viruses. *Plant Methods*, *17*(1), 20. https://doi.org/10.1186/s13007-021-00719-4

Leister, D. (2019). Genetic engineering, synthetic biology and the light reactions of photosynthesis. *Plant Physiology*, *179*(3), 778–793. https://doi.org/10.1104/pp.18.00360

Lemmon, Z. H., Reem, N. T., Dalrymple, J., Soyk, S., Swartwood, K. E., Rodriguez-Leal, D., Van Eck, J., & Lippman, Z. B. (2018). Rapid improvement of domestication traits in an orphan crop by genome editing. *Nature Plants*, *4*(10), 766–770. https://doi.org/10.1038/s41477-018-0259-x

Li, C., Zhang, K., Zeng, X., Jackson, S., Zhou, Y., & Hong, Y. (2009). A *cis* element within *flowering locus T* mRNA determines its mobility and facilitates trafficking of heterologous viral RNA. *Journal of Virology*, *83*(8), 3540–3548. https://doi.org/10.1128/jvi.02346-08

Li, C., Zhang, R., Meng, X., Chen, S., Zong, Y., Lu, C., Qiu, J. L., Chen, Y. H., Li, J., & Gao, C. (2020). Targeted, random mutagenesis of plant genes with dual cytosine and adenine base editors. *Nature Biotechnology*, *38*(7), 875–882. https://doi.org/10.1038/s41587-019-0393-7

Li, T., Hu, J., Sun, Y., Li, B., Zhang, D., Li, W., Liu, J., Li, D., Gao, C., Zhang, Y., & Wang, Y. (2021). Highly efficient heritable genome editing in wheat using an RNA virus and bypassing tissue culture. *Molecular Plant*, *14*(11), 1787–1798. https://doi.org/10.1016/j.molp.2021.07.010

Li, T., Yang, X., Yu, Y., Si, X., Zhai, X., Zhang, H., Dong, W., Gao, C., & Xu, C. (2018). Domestication of wild tomato is accelerated by genome editing. *Nature Biotechnology*, *36*(12), 1160–1163. https://doi.org/10.1038/nbt.4273

Li, Z., Zhong, Z., Wu, Z., Pausch, P., Al-Shayeb, B., Amerasekera, J., Doudna, J. A., & Jacobsen, S. E. (2023). Genome editing in plants using the compact editor CasΦ. *Proceedings of the National Academy of Sciences of the United States of America*, *120*(4), e2216822120. https://doi.org/10.1073/pnas.2216822120

Liang, Z., Chen, K., Li, T., Zhang, Y., Wang, Y., Zhao, Q., Liu, J., Zhang, H., Liu, C., Ran, Y., & Gao, C. (2017). Efficient DNA-free genome editing of bread wheat using CRISPR/Cas9 ribonucleoprotein complexes. *Nature Communications*, 8. https://doi.org/10.1038/ncomms14261

Lifschitz, E., Eviatar, T., Rozman, A., Shalit, A., Goldshmidt, A., Amsellem, Z., Alvarez, J. P., & Eshed, Y. (2006). The tomato FT ortholog triggers systemic signals that regulate growth and flowering and substitute for diverse environmental stimuli. *Proceedings of the National Academy of Sciences of the United States of America*, *103*(16), 6398–6403. https://doi.org/10.1073/pnas.0601620103

Lin, Q., Zong, Y., Xue, C., Wang, S., Jin, S., Zhu, Z., Wang, Y., Anzalone, A. V., Raguram, A., Doman, J. L., Liu, D. R., & Gao, C. (2020). Prime genome editing in rice and wheat. *Nature Biotechnology*, 38(5). https://doi.org/10.1038/s41587-020-0455-x

Liu, D., Xuan, S., Prichard, L. E., Donahue, L. I., Pan, C., Nagalakshmi, U., Ellison, E. E., Starker, C. G., Dinesh-Kumar, S. P., Qi, Y., & Voytas, D. F. (2022a). Heritable base-editing in Arabidopsis using RNA viral vectors. *Plant Physiology*, *189*(4), 1920–1924. https://doi.org/10.1093/plphys/kiac206

Liu, G., Lin, Q., Jin, S., & Gao, C. (2022b). The CRISPR-Cas toolbox and gene editing technologies. *Molecular Cell*, *82*(2), 333–347. https://doi.org/10.1016/j.molcel.2021.12.002

Liu, H. J., Jian, L., Xu, J., Zhang, Q., Zhang, M., Jin, M., Peng, Y., Yan, J., Han, B., Liu, J., Gao, F., Liu, X., Huang, L., Wei, W., Ding, Y., Yang, X., Li, Z., Zhang, M., Sun, J., … Yan, J. (2020a). High-throughput CRISPR/Cas9 mutagenesis streamlines trait gene identification in maize. *The Plant Cell*, *32*(5), 1397–1413. https://doi.org/10.1105/tpc.19.00934

Liu, M., Khan, A., Wang, Z., Liu, Y., Yang, G., Deng, Y., & He, N. (2019). Aptasensors for pesticide detection. *Biosensors and Bioelectronics*, *130*, 174–184. https://doi.org/10.1016/j.bios.2019.01.006

Liu, Q., Zhao, C., Sun, K., Deng, Y., & Li, Z. (2023). Engineered biocontainable RNA virus vectors for non-transgenic genome editing across crop species and genotypes. *Molecular Plant*, *16*(3), 616–631. https://doi.org/10.1016/j.molp.2023.02.003

Liu, W., & Stewart, C. N. (2015). Plant synthetic biology. *Trends in Plant Science*, *20*(5), 309–317. https://doi.org/10.1016/j.tplants.2015.02.004

Liu, X., Qin, R., Li, J., Liao, S., Shan, T., Xu, R., Wu, D., & Wei, P. (2020b). A CRISPR-Cas9-mediated domain-specific base-editing screen enables functional assessment of ACCase variants in rice. *Plant Biotechnology Journal*, *18*(9), 1845–1847. https://doi.org/10.1111/pbi.13348

Loebenstein, G., & Gaba, V. (2012). Viruses of potato. In *Advances in Virus Research* (Vol. 84, pp. 209–246). Academic Press Inc. https://doi.org/10.1016/B978-0-12-394314-9.00006-3

López-Marqués, R. L., Nørrevang, A. F., Ache, P., Moog, M., Visintainer, D., Wendt, T., Østerberg, J. T., Dockter, C., Jørgensen, M. E., Salvador, A. T., Hedrich, R., Gao, C., Jacobsen, S. E., Shabala, S., & Palmgren, M. (2020). Prospects for the accelerated improvement of the resilient crop quinoa. *Journal of Experimental Botany*, *71*(18), 5333–5347. https://doi.org/10.1093/jxb/eraa285

Lozano-Durán, R. (2016). Geminiviruses for biotechnology: The art of parasite taming. *New Phytologist*, *210*(1), 58–64. https://doi.org/10.1111/nph.13564

Lu, K. J., Huang, N. C., Liu, Y. S., Lu, C. A., & Yu, T. S. (2012). Long-distance movement of Arabidopsis *FLOWERING LOCUS T* RNA participates in systemic floral regulation. *RNA Biology*, *9*(5), 653–662. https://doi.org/10.4161/rna.19965

Lu, Y., Ye, X., Guo, R., Huang, J., Wang, W., Tang, J., Tan, L., Zhu, J. kang, Chu, C., & Qian, Y. (2017). Genome-wide targeted mutagenesis in rice using the CRISPR/Cas9 system. *Molecular Plant*, *10*(9), 1242–1245. https://doi.org/10.1016/j.molp.2017.06.007

Luo, Y., Na, R., Nowak, J. S., Qiu, Y., Lu, Q. S., Yang, C., Marsolais, F., & Tian, L. (2021). Development of a Csy4-processed guide RNA delivery system with soybean-infecting virus ALSV for genome editing. *BMC Plant Biology*, *21*(1), 419. https://doi.org/10.1186/s12870-021-03138-8

Ma, X., Zhang, X., Liu, H., & Li, Z. (2020). Highly efficient DNA-free plant genome editing using virally delivered CRISPR-Cas9. *Nature Plants*, *6*(7), 773–779. https://doi.org/10.1038/s41477-020-0704-5

Macfarlane, S. A. (2010). Tobraviruses--plant pathogens and tools for biotechnology. *Molecular Plant Pathology*, *11*(4), 577–583. https://doi.org/10.1111/j.1364-3703.2010.00617.X

Mali, P., Yang, L., Esvelt, K. M., Aach, J., Guell, M., DiCarlo, J. E., Norville, J. E., & Church, G. M. (2013). RNA-guided human genome engineering via Cas9. *Science*, *339*(6121), 823–826. https://doi.org/10.1126/science.1232033

Marino, N. D., Pinilla-Redondo, R., Csörgő, B., & Bondy-Denomy, J. (2020). Anti-CRISPR protein applications: Natural brakes for CRISPR-Cas technologies. *Nature Methods*, *17*(5), 471–479. https://doi.org/10.1038/S41592-020-0771-6

Marton, I., Zuker, A., Shklarman, E., Zeevi, V., Tovkach, A., Roffe, S., Ovadis, M., Tzfira, T., & Vainstein, A. (2010). Nontransgenic genome modification in plant cells. *Plant Physiology*, *154*(3), 1079–1087. https://doi.org/10.1104/pp.110.164806

Mei, Y., Beernink, B. M., Ellison, E. E., Konečná, E., Neelakandan, A. K., Voytas, D. F., & Whitham, S. A. (2019). Protein expression and gene editing in monocots using foxtail mosaic virus vectors. *Plant Direct*, *3*(11), 1–16. https://doi.org/10.1002/pld3.181

Meng, X., Muszynski, M. G., & Danilevskaya, O. N. (2011). The FT-like ZCN8 gene functions as a floral activator and is involved in photoperiod sensitivity in maize. *Plant Cell*, *23*(3), 942–960. https://doi.org/10.1105/tpc.110.081406

Meng, X., Yu, H., Zhang, Y., Zhuang, F., Song, X., Gao, S., Gao, C., & Li, J. (2017). Construction of a genome-wide mutant library in rice using CRISPR/Cas9. *Molecular Plant*, *10*(9), 1238–1241. https://doi.org/10.1016/J.molp.2017.06.006

Nagalakshmi, U., Meier, N., Liu, J. Y., Voytas, D. F., & Dinesh-Kumar, S. P. (2022). High-efficiency multiplex biallelic heritable editing in Arabidopsis using an RNA virus. *Plant Physiology*, *189*(3), 1241–1245. https://doi.org/10.1093/plphys/kiac159

Nekrasov, V., Wang, C., Win, J., Lanz, C., Weigel, D., & Kamoun, S. (2017). Rapid generation of a transgene-free powdery mildew resistant tomato by genome deletion. *Scientific Reports*, *7*(1), 482. https://doi.org/10.1038/s41598-017-00578-x

Nishida, K., Arazoe, T., Yachie, N., Banno, S., Kakimoto, M., Tabata, M., Mochizuki, M., Miyabe, A., Araki, M., Hara, K. Y., Shimatani, Z., & Kondo, A. (2016). Targeted nucleotide editing using hybrid prokaryotic and vertebrate adaptive immune systems. *Science*, *353*(6305), aaf8729. https://doi.org/10.1126/science.aaf8729

Oliver, J. E., & Whitfield, A. E. (2016). The genus tospovirus: Emerging bunyaviruses that threaten food security. *Annual Review in Biology*, *3*(1), 101–124. https://doi.org/10.1146/annurev-virology-100114-055036

Pappu, H. R. (2008). Tomato spotted wilt virus. In *Encyclopedia of Virology* (pp. 133–138). https://doi.org/10.1016/b978-012374410-4.00648-8

Pausch, P., Al-Shayeb, B., Bisom-Rapp, E., Tsuchida, C. A., Li, Z., Cress, B. F., Knott, G. J., Jacobsen, S. E., Banfield, J. F., & Doudna, J. A. (2020). Crispr-casΦ from huge phages is a hypercompact genome editor. *Science*, *369*(6501), 333–337. https://doi.org/10.1126/science.abb1400

Peng, A., Chen, S., Lei, T., Xu, L., He, Y., Wu, L., Yao, L., & Zou, X. (2017). Engineering canker-resistant plants through CRISPR/Cas9-targeted editing of the susceptibility gene *CsLOB1* promoter in citrus. *Plant Biotechnology Journal*, *15*(12), 1509–1519. https://doi.org/10.1111/pbi.12733

Powder, K. E. (2020). Quantitative trait loci (QTL) mapping. *Methods in Molecular Biology*, *2082*, 211–229. https://doi.org/10.1007/978-1-0716-0026-9_15

Prado, J. R., Segers, G., Voelker, T., Carson, D., Dobert, R., Phillips, J., Cook, K., Cornejo, C., Monken, J., Grapes, L., Reynolds, T., & Martino-Catt, S. (2014). Genetically engineered crops: From idea to product. *Annual Review of Plant Biology*, *65*, 769–790. https://doi.org/10.1146/annurev-arplant-050213-040039

Pyott, D. E., Sheehan, E., & Molnar, A. (2016). Engineering of CRISPR/Cas9-mediated potyvirus resistance in transgene-free *Arabidopsis* plants. *Molecular Plant Pathology*, *17*(8), 1276–1288. https://doi.org/10.1111/mpp.12417

Que, Q., Chen, Z., Kelliher, T., Skibbe, D., Dong, S., & Chilton, M. D. (2019). Plant DNA repair pathways and their applications in genome engineering. *Methods in Molecular Biology*, 1917, 3–24. https://doi.org/10.1007/978-1-4939-8991-1_1

Rampersad, S., & Tennant, P. (2018). Replication and expression strategies of viruses. *Viruses: Molecular Biology, Host Interactions, and Applications to Biotechnology*, 55–82. https://doi.org/10.1016/b978-0-12-811257-1.00003-6

Rodríguez-Leal, D., Lemmon, Z. H., Man, J., Bartlett, M. E., & Lippman, Z. B. (2017). Engineering quantitative trait variation for crop improvement by genome editing. *Cell*, *171*(2), 470–480. https://doi.org/10.1016/j.cell.2017.08.030

Roldan, M. V. G., Périlleux, C., Morin, H., Huerga-Fernandez, S., Latrasse, D., Benhamed, M., & Bendahmane, A. (2017). Natural and induced loss of function mutations in SlMBP21 MADS-box gene led to jointless-2 phenotype in tomato. *Scientific Reports*, *7*(1), 4402. https://doi.org/10.1038/s41598-017-04556-1

Rybicki, E. P., & Martin, D. P. (2014). Virus-derived ssDNA vectors for the expression of foreign proteins in plants. *Current Topics in Microbiology and Immunology*, *375*, 19–45. https://doi.org/10.1007/82-2011-185

Ryu, C. M., Anand, A., Kang, L., & Mysore, K. S. (2004). Agrodrench: A novel and effective agroinoculation method for virus-induced gene silencing in roots and diverse Solanaceous species. *The Plant Journal*, *40*(2), 322–331. https://doi.org/10.1111/j.1365-313x.2004.02211.x

Sánchez-León, S., Gil-Humanes, J., Ozuna, C. V., Giménez, M. J., Sousa, C., Voytas, D. F., & Barro, F. (2018). Low-gluten, nontransgenic wheat engineered with CRISPR/Cas9. *Plant Biotechnology Journal*, *16*(4), 902–910. https://doi.org/10.1111/pbj.12837

Scheben, A., Wolter, F., Batley, J., Puchta, H., & Edwards, D. (2017). Towards CRISPR/Cas crops - Bringing together genomics and genome editing. *New Phytologist*, *216*(3), 682–698. https://doi.org/10.1111/nph.14702

Selma, S., Gianoglio, S., Uranga, M., Vázquez-Vilar, M., Espinosa-Ruiz, A., Drapal, M., Fraser, P. D., Daròs, J. A., & Orzáez, D. (2022). *Potato virus X*-delivered CRISPR activation programs lead to strong endogenous gene induction and transient metabolic reprogramming in *Nicotiana benthamiana*. *The Plant Journal*, *111*(6), 1550–1564. https://doi.org/10.1111/tpj.15906

Senthil-Kumar, M., & Mysore, K. S. (2014). Tobacco rattle virus-based virus-induced gene silencing in *Nicotiana benthamiana*. *Nature Protocols*, *9*(7), 1549–1562. https://doi.org/10.1038/nprot.2014.092

Seol, J. H., Shim, E. Y., & Lee, S. E. (2018). Microhomology-mediated end joining: Good, bad and ugly. *Mutation Research - Fundamental and Molecular Mechanisms of Mutagenesis*, *809*, 81–87. https://doi.org/10.1016/j.mrfmmm.2017.07.002

Shimatani, Z., Kashojiya, S., Takayama, M., Terada, R., Arazoe, T., Ishii, H., Teramura, H., Yamamoto, T., Komatsu, H., Miura, K., Ezura, H., Nishida, K., Ariizumi, T., & Kondo, A. (2017). Targeted base editing in rice and tomato using a CRISPR-Cas9 cytidine deaminase fusion. *Nature Biotechnology*, *35*(5), 441–443. https://doi.org/10.1038/nbt.3833

South, P. F., Cavanagh, A. P., Liu, H. W., & Ort, D. R. (2019). Synthetic glycolate metabolism pathways stimulate crop growth and productivity in the field. *Science*, *363*(6422), eaat9077. https://doi.org/10.1126/science.aat9077

Stenger, D. C., Revington, G. N., Stevenson, M. C., & Bisaro, D. M. (1991). Replicational release of geminivirus genomes from tandemly repeated copies: Evidence for rolling-circle replication of a plant viral DNA. *Proceedings of the National Academy of Sciences of the United States of America*, *88*(18), 8029–8033. https://doi.org/10.1073/pnas.88.18.8029

Sukegawa, S., Nureki, O., Toki, S., & Saika, H. (2023). Genome editing in rice mediated by miniature size Cas nuclease SpCas12f. *Frontiers in Genome Editing*, *5*, 1138843. https://doi.org/10.3389/FGEED.2023.1138843

Sun, Y., Li, J., & Xia, L. (2016). Precise genome modification via sequence-specific nucleases-mediated gene targeting for crop improvement. *Frontiers in Plant Science*, *7*, 1928. https://doi.org/10.3389/fpls.2016.01928

Talabi, A. O., Vikram, P., Thushar, S., Rahman, H., Ahmadzai, H., Nhamo, N., Shahid, M., & Singh, R. K. (2022). Orphan crops: A best fit for dietary enrichment and diversification in highly deteriorated marginal environments. *Frontiers in Plant Science*, 13, 839704. https://doi.org/10.3389/fpls.2022.839704

Tamaki, S., Matsuo, S., Hann, L. W., Yokoi, S., & Shimamoto, K. (2007). Hd3a protein is a mobile flowering signal in rice. *Science*, *316*(5827), 1033–1036. https://doi.org/10.1126/science.1141753

Tang, X., Liu, G., Zhou, J., Ren, Q., You, Q., Tian, L., Xin, X., Zhong, Z., Liu, B., Zheng, X., Zhang, D., Malzahn, A., Gong, Z., Qi, Y., Zhang, T., & Zhang, Y. (2018). A large-scale whole-genome sequencing analysis reveals highly specific genome editing by both Cas9 and Cpf1 (Cas12a) nucleases in rice. *Genome Biology*, *19*(1), 84. https://doi.org/10.1186/S13059-018-1458-5

Tepfer, M., Jacquemond, M., & García-Arenal, F. (2015). A critical evaluation of whether recombination in virus-resistant transgenic plants will lead to the emergence of novel viral diseases. *New Phytologist*, *207*(3), 536–541. https://doi.org/10.1111/nph.13358

Torti, S., Schlesier, R., Thümmler, A., Bartels, D., Römer, P., Koch, B., Werner, S., Panwar, V., Kanyuka, K., Wirén, N. von, Jones, J. D. G., Hause, G., Giritch, A., & Gleba, Y. (2021). Transient reprogramming of crop plants for agronomic performance. *Nature Plants*, *7*(2), 159–171. https://doi.org/10.1038/s41477-021-00851-y

Tran, P. T., Fang, M., Widyasari, K., & Kim, K. H. (2019). A plant intron enhances the performance of an infectious clone in planta. *Journal of Virological Methods*, *265*, 26–34. https://doi.org/10.1016/j.jviromet.2018.12.012

Trasanidou, D., Gerós, A. S., Mohanraju, P., Nieuwenweg, A. C., Nobrega, F. L., & Staals, R. H. J. (2019). Keeping crispr in check: Diverse mechanisms of phage-encoded anti-crisprs. *FEMS Microbiology Letters*, *366*(9), fnz098. https://doi.org/10.1093/femsle/fnz098

Turnbull, C., Lillemo, M., & Hvoslef-Eide, T. A. K. (2021). Global regulation of genetically modified crops amid the gene edited crop boom - A review. *Frontiers in Plant Science*, *12*, 630396. https://doi.org/10.3389/fpls.2021.630396

Uranga, M., Aragonés, V., Selma, S., Vázquez-Vilar, M., Orzáez, D., & Daròs, J. (2021a). Efficient Cas9 multiplex editing using unspaced sgRNA arrays engineering in a *Potato virus X* vector. *The Plant Journal*, *106*(2), 555–565. https://doi.org/10.1111/tpj.15164

Uranga, M., Vazquez-Vilar, M., Orzáez, Di., & Daròs, J. A. (2021b). CRISPR-Cas12a genome editing at the whole-plant level using two compatible RNA virus vectors. *The CRISPR Journal*, *4*(5), 761–769. https://doi.org/10.1089/crispr.2021.0049

Van Kregten, M., De Pater, S., Romeijn, R., Van Schendel, R., Hooykaas, P. J. J., & Tijsterman, M. (2016). T-DNA integration in plants results from polymerase-θ-mediated DNA repair. *Nature Plants*, *2*(11), 1–6. https://doi.org/10.1038/nplants.2016.164

Varanda, C. M., Félix, M. do R., Campos, M. D., Patanita, M., & Materatski, P. (2021). Plant viruses: From targets to tools for CRISPR. *Viruses*, *13*(1), 141. https://doi.org/10.3390/v13010141

Voytas, D. F., & Gao, C. (2014). Precision genome engineering and agriculture: Opportunities and regulatory challenges. *PLoS Biology*, *12*(6), 1–6. https://doi.org/10.1371/journal.pbio.1001877

Wan, D. Y., Guo, Y., Cheng, Y., Hu, Y., Xiao, S., Wang, Y., & Wen, Y. Q. (2020). CRISPR/Cas9-mediated mutagenesis of *VvMLO3* results in enhanced resistance to powdery mildew in grapevine (*Vitis vinifera*). *Horticulture Research*, *7*(1), 116. https://doi.org/10.1038/s41438-020-0339-8

Wang, M., Gao, S., Zeng, W., Yang, Y., Ma, J., & Wang, Y. (2020). Plant virology delivers diverse toolsets for biotechnology. *Viruses*, *12*(11), 1338. https://doi.org/10.3390/v12111338

Wang, M., Lu, Y., Botella, J. R., Mao, Y., Hua, K., & Zhu, J. kang. (2017). Gene targeting by homology-directed repair in rice using a geminivirus-based CRISPR/Cas9 system. *Molecular Plant*, *10*(7), 1007–1010. https://doi.org/10.1016/j.molp.2017.03.002

Wang, W., Yu, Z., He, F., Bai, G., Trick, H. N., Akhunova, A., & Akhunov, E. (2022). Multiplexed promoter and gene editing in wheat using a virus-based guide RNA delivery system. *Plant Biotechnology Journal*, *20*(12), 2332–2341. https://doi.org/10.1111/PBI.13910

Wang, Y., Cheng, X., Shan, Q., Zhang, Y., Liu, J., Gao, C., & Qiu, J. L. (2014). Simultaneous editing of three homoeoalleles in hexaploid bread wheat confers heritable resistance to powdery mildew. *Nature Biotechnology*, *32*(9), 947–951. https://doi.org/10.1038/nbt.2969

Weber, A. P. M., & Bar-Even, A. (2019). Update: Improving the efficiency of photosynthetic carbon reactions. *Plant Physiology*, *179*(3). https://doi.org/10.1104/pp.18.01521

Woo, J. W., Kim, J., Kwon, S. Il, Corvalán, C., Cho, S. W., Kim, H., Kim, S. G., Kim, S. T., Choe, S., & Kim, J. S. (2015). DNA-free genome editing in plants with preassembled CRISPR-Cas9 ribonucleoproteins. *Nature Biotechnology*, *33*(11), 1162–1164. https://doi.org/10.1038/nbt.3389

Wu, H., Qu, X., Dong, Z., Luo, L., Shao, C., Forner, J., Lohmann, J. U., Su, M., Xu, M., Liu, X., Zhu, L., Zeng, J., Liu, S., Tian, Z., & Zhao, Z. (2020). WUSCHEL triggers innate antiviral immunity in plant stem cells. *Science*, 370(6513). https://doi.org/10.1126/science.abb7360

Wylie, S. J., Adams, M., Chalam, C., Kreuze, J., López-Moya, J. J., Ohshima, K., Praveen, S., Rabenstein, F., Stenger, D., Wang, A., & Zerbini, F. M. (2017). ICTV virus taxonomy profile: Potyviridae. *Journal of General Virology*, *98*(3), 352–354. https://doi.org/10.1099/jgv.0.000740

Xu, W., Zhang, C., Yang, Y., Zhao, S., Kang, G., He, X., Song, J., & Yang, J. (2020). Versatile nucleotides substitution in plant using an improved prime editing system. *Molecular Plant*, *13*(5), 675–678. https://doi.org/10.1016/j.molp.2020.03.012

Xu, X., Chemparathy, A., Zeng, L., Kempton, H. R., Shang, S., Nakamura, M., & Qi, L. S. (2021). Engineered miniature CRISPR-Cas system for mammalian genome regulation and editing. *Molecular Cell, 81*, 4333–4345. https://doi.org/10.1016/j.molcel.2021.08.008

Yin, K., Han, T., Liu, G., Chen, T., Wang, Y., Yu, A. Y. L., & Liu, Y. (2015). A geminivirus-based guide RNA delivery system for CRISPR/Cas9 mediated plant genome editing. *Scientific Reports*, *5*, 14926. https://doi.org/10.1038/srep14926

Youssef, F., Marais, A., Faure, C., Gentit, P., & Candresse, T. (2011). Strategies to facilitate the development of uncloned or cloned infectious full-length viral cDNAs: Apple chlorotic leaf spot virus as a case study. *Virology Journal*, *8*, 488. https://doi.org/10.1186/1743-422x-8-488

Yu, H., Lin, T., Meng, X., Du, H., Zhang, J., Liu, G., Chen, M., Jing, Y., Kou, L., Li, X., Gao, Q., Liang, Y., Liu, X., Fan, Z., Liang, Y., Cheng, Z., Chen, M., Tian, Z., Wang, Y., … Li, J. (2021). A route to de novo domestication of wild allotetraploid rice. *Cell*, *184*(5), 1156–1170. https://doi.org/10.1016/j.cell.2021.01.013

Yue, J.-J., Yuan, J.-L., Wu, F.-H., Yuan, Y.-H., Cheng, Q.-W., Hsu, C.-T., & Lin, C.-S. (2021). Protoplasts: From isolation to CRISPR/Cas genome editing application. *Frontiers in Genome Editing*, *3*, 717017. https://doi.org/10.3389/fgeed.2021.717017

Zetsche, B., Gootenberg, J. S., Abudayyeh, O. O., Slaymaker, I. M., Makarova, K. S., Essletzbichler, P., Volz, S. E., Joung, J., van der Oost, J., Regev, A., Koonin, E. V, & Zhang, F. (2015). Cpf1 is a single RNA-guided endonuclease of a class 2 CRISPR-Cas system. *Cell*, *163*(3), 759–771. https://doi.org/10.1016/j.cell.2015.09.038

Zhang, W., Thieme, C. J., Kollwig, G., Apelt, F., Yang, L., Winter, N., Andresen, N., Walther, D., & Kragler, F. (2016a). TRNA-related sequences trigger systemic mRNA transport in plants. *Plant Cell*, *28*(6). https://doi.org/10.1105/tpc.15.01056

Zhang, X., Kang, L., Zhang, Q., Meng, Q., Pan, Y., Yu, Z., Shi, N., Jackson, S., Zhang, X., Wang, H., Tor, M., & Hong, Y. (2020). An RNAi suppressor activates in planta virus-mediated gene editing. *Functional and Integrative Genomics*, *20*(4), 471–477. https://doi.org/10.1007/s10142-019-00730-y

Zhang, Y., Liang, Z., Zong, Y., Wang, Y., Liu, J., Chen, K., Qiu, J. L., & Gao, C. (2016b). Efficient and transgene-free genome editing in wheat through transient expression of CRISPR/Cas9 DNA or RNA. *Nature Communications*, *7*, 12617. https://doi.org/10.1038/ncomms12617

Zhang, Y., Shan, Q., Wang, Y., Chen, K., Liang, Z., Li, J., Zhang, Y., Zhang, K., Liu, J., Voytas, D. F., Zheng, X., & Gao, C. (2013). Rapid and efficient gene modification in rice and brachypodium using TALENs. *Molecular Plant*, *6*(4), 1365–1368. https://doi.org/10.1093/mp/sss162

Zhao, X., Meng, Z., Wang, Y., Chen, W., Sun, C., Cui, B., Cui, J., Yu, M., Zeng, Z., Guo, S., Luo, D., Cheng, J. Q., Zhang, R., & Cui, H. (2017). Pollen magnetofection for genetic modification with magnetic nanoparticles as gene carriers. *Nature Plants*, *3*(12), 956–964. https://doi.org/10.1038/s41477-017-0063-z

Zhong, G., Wang, H., Bailey, C. C., Gao, G., & Farzan, M. (2016). Rational design of aptazyme riboswitches for efficient control of gene expression in mammalian cells. *ELife*, *5*, e18858. https://doi.org/10.7554/eLife.18858

Zhou, X., Sun, K., Zhou, X., Jackson, A. O., & Li, Z. (2019). The matrix protein of a plant rhabdovirus mediates superinfection exclusion by inhibiting viral transcription. *Journal of Virology*, *93*(20). https://doi.org/10.1128/jvi.00680-19

Zhu, H., Li, C., & Gao, C. (2020). Applications of CRISPR-Cas in agriculture and plant biotechnology. *Nature Reviews Molecular Cell Biology*, *21*(11), 661–677). https://doi.org/10.1038/s41580-020-00288-9

Zhu, X., Liu, X., Liu, T., Wang, Y., Ahmed, N., Li, Z., & Jiang, H. (2021). Synthetic biology of plant natural products: From pathway elucidation to engineered biosynthesis in plant cells. *Plant Communications*, *2*(5), 100229. https://doi.org/10.1016/j.xplc.2021.100229

Zong, Y., Song, Q., Li, C., Jin, S., Zhang, D., Wang, Y., Qiu, J. L., & Gao, C. (2018). Efficient C-to-T base editing in plants using a fusion of nCas9 and human APOBEC3A. *Nature Biotechnology*, *36*(10), 950–953. https://doi.org/10.1038/nbt.4261

Zsögön, A., Čermák, T., Naves, E. R., Notini, M. M., Edel, K. H., Weinl, S., Freschi, L., Voytas, D. F., Kudla, J., & Peres, L. E. P. (2018). De novo domestication of wild tomato using genome editing. *Nature Biotechnology*, *36*(12), 1211–1216. https://doi.org/10.1038/nbt.4272

6 CRISPR/Cas9

The Revolutionary Genome Editing Tools for Crop Genetic Tweak to Render It 'Climate Smart'

Abira Chaudhuri, Asis Datta, Koushik Halder, and Malik Z. Abdin

6.1 INTRODUCTION

The primary staple food crops of the world, like rice, wheat, potato, millet, etc., provide nutrition and calories to more than half the world's population. In the present times, we barely need to mention that the world's population is increasing by leaps and bounds, and it's a fully challenging task to cater to the constantly growing demand for food grains for the flock, which is inferred to reach 10 billion by 2050. Abiotic and biotic stresses hinder farmers from achieving the required high-yield target and superior quality of agricultural produce. Temperature and optimum rainfall are the two major environmental factors that affect the growth, development, and output of food crops.[1] Asia is a continent renowned for natural calamities like drought. According to data, the average global surface temperature has risen by 0.85°C over the years, and this hike poses to be a potential menace to agriculture worldwide.[2] Without the application of fertilizers adequately and the generation of genetically altered varieties with improved traits, it can be justified accurately that worldwide crop production will decrease drastically with the gradual increase in the mean global temperature. As the optimum temperature for crop growth and development peaks, other natural calamities such as drought, flood, hurricane, etc. follow and ruin agriculture wherever they strike. The miserable effects of these natural disasters, so-called 'acts of God,' will compromise agriculture drastically, subsequently leading to an unfathomable disbalance in equilibrium between the demand/supply of food and cash crops. The overall impact of such catastrophies will double the miseries of the population inhabiting the underprivileged countries (Figure 6.1). After years of research and analysis by agricultural scientists, economists, and strategy makers, worldwide, it was concluded that designing and developing climate-smart crops is the best way to diminish climate change and its damaging consequences.

We are incredibly privileged to have access to the revolutionary genome editing tool CRISPR/Cas, which has proved versatile and expanded the horizons of modifying genomes. Though its discovery dates back less than a decade, this genome editing tool has majorly revolutionized crop yield. Now, crop researchers can tweak the gene functions, enabling the plants to adapt to extreme temperatures and increase the annual harvest, imparting better grain quality.

The CRISPR/Cas tool was discovered in primitive bacteria whose primary function was to impart immunity against bacteriophages. Tiny DNA segments (Spacers) are incorporated into the bacterial genome and are transcribed, generating CRISPR RNA (crRNA). The latter brace up with the trans-activating crRNAs (tracrRNAs) generating guide RNA (gRNA). These two can be fused and expressed as solely one guide RNA (sgRNA).[3] This gRNA orchestrates the Cas9 endonuclease enzyme (CRISPR-associated protein 9) to the target site. For target recognition, the Protospacer

DOI: 10.1201/9781003387060-6

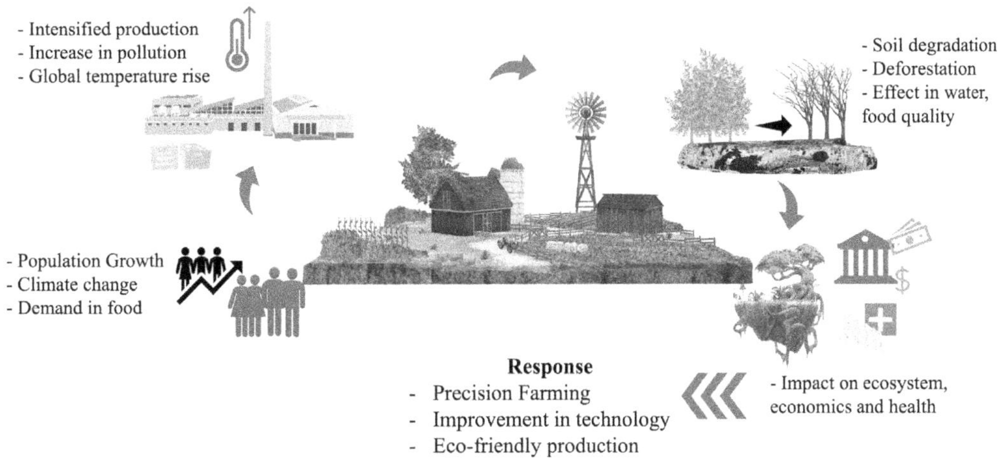

FIGURE 6.1 Impact of climate change on agriculture.

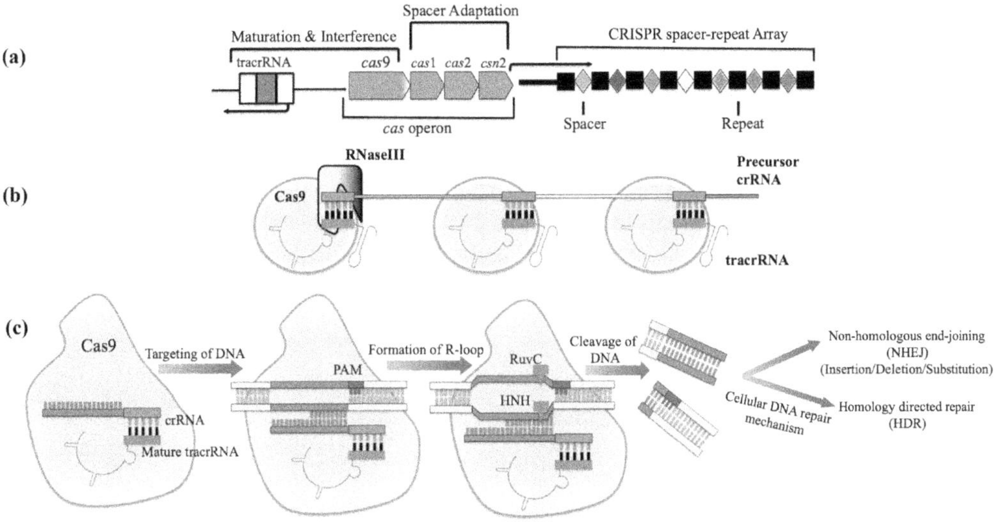

FIGURE 6.2 Type IIA CRISPR/Cas system in *Streptococcus pyogenes*. (a) CRISPR locus in the genome: CRISPR locus consists of CRISPR repeats and spacer array along with tracrRNA and cas operon. New spacers from the invading organism are incorporated into this CRISPR array (adaptation) (b) crRNA:tracrRNA co-maturation and complex formation with Cas9: pre-crRNA becomes matured crRNA with the help of tracrRNA along with Csn1 and Rnase III. Matured crRNA with Cas9 gets incorporated into ribonucleoprotein complexes which start scanning for nucleic acids complementary to the sequence coded by crRNA (maturation) (c) RNA-guided Cas9-mediated cleavage of targeted DNA: Upon recognition, complementary foreign sequences are cleaved by this Cas protein complex (Interference).

Adjacent Motif (PAM) must be placed adjacent to the spacer sequence. Cas9, also known as the RNA-guided endonuclease (RGEN), cuts the DNA approximately 3 bp upstream of PAM, forming a sequence-specific double-strand break. This activity boosts up the natural repair machinery of the cell, resulting in re-ligating the DNA cut ends by the process known as Non-Homologous End-Joining (NHEJ) (Figure 6.2).[4] After thorough research, scientists have concluded that NHEJ is not a fool-proof process, since it is prone to insert tiny insertions-deletions (INDELs) and even can replace one base with another at the cleavage site. This entire function can harm the cell since it can form random stop codons, suddenly, causing the polypeptide chain to terminate. Therefore, by altering the 20 bp spacer sequence of the gRNA, Cas9 can be utilized to target random genomic sequences in plants.[5] The DSBs can also be repaired by the Homology-Directed-Repair (HDR) process. The thumb rule of this process is that a separate donor DNA template is required, having matching sequences around the corresponding break sites of the DNA. The added specialty of HDR is that: any mutated gene or defective allele can be repaired or replaced with a similar copy of the same gene, thus bringing in a new or improved trait.[2]

Here, we focus on the impacts of climate change on agriculture, the need and development of climate-smart agriculture as the savior for an impending disaster of famine and hunger, and how the revolutionary technology CRISPR/Cas9 (along with its pros and cons) has been proving itself as a potential game changer by aiding in the generation of climate-smart crops, worldwide.

6.2 DETRIMENTAL EFFECTS OF CLIMATE CHANGE ON AGRICULTURE AND RELATED FACTORS

Climate change has been gradually unfurling its devilish grip over agriculture and food security. It has been predicted, emphatically, by climate experts at the Intergovernmental Panel on Climate Change (IPCC) that the global crop yield will indeed worsen over the century if appropriate measures are not adopted. At present, tropical crops are the worst hit, and the temperate climate crops have shown an increase in yield, contrarily.[6] Gradually, this trend will lead to soaring food prices due to a stark disbalance in demand and supply. The arid regions of the globe will lead progressively from partial to complete desertification, and the equatorial regions will witness a severe soar in temperature, especially the continents like Asia and Africa, being most affected since they have considerable population pressure. Global warming, increased desertification, land degradation, rising sea level, and the recurrence of hurricanes and tropical storms have already been attributed to climate change. All these factors have started to reduce agricultural production and biodiversity and have increased the risk of salinization.[7]

Greenhouse gases, mainly carbon dioxide (CO_2), are benefiting plants because the gradual increase in CO_2 also surges the photosynthetic rate and carbon assimilation, called CO_2 fertilization.[8] However, the food quality decreases with the gradual rise in CO_2. This increase in agricultural yield may stop due to other climatic factors like shifting nutrient concentrations and poor availability of water.[8] The climatic concerns discussed above, like the threat of high temperature; and irregular precipitation added with innumerable biotic stress factors[9,10] will lead to detrimental effects on crop yield, which has already started to impart its menacing shadow on the global output of crops such as soybean, maize, wheat, etc.[6] Beyond climate change and its harmful effects on agriculture, another domain which has been significantly affected is biodiversity. Nunez et al. (2019)[11] reported in their research that even a moderate spike in global temperature will lead to severe impacts on biodiversity. Increased agricultural demand of the worldwide population and the adverse effects of climate on biodiversity have imposed added pressure on acquiring land for agriculture and other human activities at the cost of natural landscapes. Scientists can preempt that the effect of climate as an individual factor and its impact on other overlapping biological factors have become quite complex. On a broad scale, the adverse effects of climate change are proving to be highly challenging to plants and animals in agriculture and food production. The burning issue threatening environmentalists, agricultural researchers, farmers, economists, and policymakers is how to mitigate the harmful effects of climate change on agriculture and biodiversity. In this regard, the most promising tool of genome editing CRISPR/Cas has proved as a boon. Agricultural products have been customized to changing climatic conditions through gene editing.

FIGURE 6.3 Climate-smart agriculture (CSA) by sustainably increasing productivity and income, reducing agricultural contribution to climate change, and strengthening resilience to climate change and variability.

Thus, climate-innovative crops and climate-smart agriculture are the most sought terms in this age of global climatic catastrophe.

Agricultural scientists have deployed gene editing to modify essential traits to address climate change adversities and a booming population. But with the increasing threat now, they have embarked on more ambitious and challenging adventures in genomics. They are using new technologies to generate tailormade crops having superior and evolved traits, in short, generate climate-smart crops. Climate Smart Agriculture (CSA) is a novel approach developed where recent technology (CRISPR/Cas), strategy, and investment guidelines amalgamate to reach a tangible solution for growing climate-smart crops for food security.[12] The natural calamities that arise due to climate change demand earnest endeavors by scientists to ensure an extensive, global, and integrated summary of the cause-and-effect cycle and how to mitigate the damage through agricultural planning and investment programs (Figure 6.3). CSA is the need of the hour since it is reshaping and modernizing the present sustainable farming systems that will support food security under the prevailing menace of climate change. CSA is being implemented in our agricultural systems to peak productivity, ensure food security, and lessen the emission of greenhouse gases.[13]

6.3 CRISPR/CAS TECHNOLOGY—THE POTENTIAL WAY OUT FOR GENERATION OF CLIMATE RESILIENT CROPS

6.3.1 CRISPR/Cas and Abiotic Stress

Global research regarding the application of the CRISPR/Cas system for the betterment of agriculture is increasing daily compared to many other new plant breeding technologies (NPBTs). This novel tool has introduced vital traits related to abiotic stress tolerance, biotic stress resistance, and many more in food and cash crops such as wheat, rice, maize, potato, tomato, tobacco, cotton, soybean, and mustard.[14] Several research groups have accomplished these alterations in the genome of economically essential crops using a transgene-free system.

Climate change is a potential reason that leads to other stressful conditions for the plants affecting agriculture drastically. Some plant genes have harmful effects of abiotic stresses and are called 'sensitive' (*Se* genes).[15] CRISPR/Cas technology has been vastly used in various plant species (grain, vegetables, and fruit crops) to enhance stress tolerance capacity by jeopardizing the fatal *Se* genes (Table 6.1). A notable example is the research with the *β-Amylase* genes, which are associated with cold tolerance in rice. They protect the plants against cold stress by regulating starch degradation and aggregation of maltose. BMY protein function is jeopardized when an OsMYB30 (R2R3-type MYB transcription factor, which is cold-responsive) along with *OsJAZ9* binds cold-responsive regions of the *BMY* genes. Here, CRISPR-Cas disrupted *OsMYB30* to generate a rice variety possessing chilling tolerance.[16] The *DST* gene (*Drought and Salt Tolerance* gene) was mutated in the indica variety (*indica rice cv.* MTU1010) with CRISPR/Cas. All the mutant lines were tailormade by a 366bp deletion in the coding sequence. These lines were robust and resistant to drought, osmotic, and saline stress since the above mutation caused a reduction in stomatal density, and broad leaf area resulting in better light-use efficiency, and pronounced water retention capacity.[17] Maize was the researchers' choice when it came to drought tolerance. The *Auxin-Regulated Gene Involved in Organ Size 8* (*ARGOS8*) was disrupted using the CRISPR/Cas9 tool to enhance drought tolerance.[18] Tomato, a model plant in plant genetics, was mutated using CRISPR-based editing to get surged sensitivity to heat and saline stress. The editing of orthologues *Mitogen-Activated Protein Kinase 3* (*SlMAPK3*) along with *Agamous-like 6* (*SlAGL6*) increases sensitivity to heat stress. Another gene mutation of the *ADP-ribosylation factor 4* (*SlARF4*) generated mutated tomatoes with sensitivity to saline stress. All these mutated tomato plants developed through CRISPR/Cas had climate-smart traits like increased resistance to abiotic stresses with refined agronomic characteristics[19]: climate change and alteration in precipitation lead to heavy metal stress in soil. Polluted soils pose a potential hazard to crops where heavy metal contamination (cadmium,[20] arsenic, etc.) moves up the food chain and accumulates in the food grains, thus making it risky for consumption. CRISPR/Cas was used to disrupt heavy metal transporter genes so that plants could grow safely in contaminated soil. The cadmium transporter gene *Natural Resistance-Associated Macrophage Protein 5* (*OsNRAMPS*) was edited in rice to decrease cadmium accumulation and promote the growth of food and cash crops in cadmium-contaminated soils.[21] Farmers are now much more aware of the drastic effects of climate change and are trained accordingly. Global climate change has also led to unwanted plants and weeds in cultivable lands, harming crops by competing with them for light, nutrients, and water. Instead of targeting the main crops directly, farmers are now using herbicides to abolish weeds and lessen competition in the farmlands. So, plants are edited through CRISPR/Cas and made herbicide-tolerant, and these crops have proved highly advantageous in modern agriculture. CRISPR/Cas base editing and prime editing systems successfully edited *Acetolactate Synthase 1* and *2* (*ZmALS1*, *ZmALS2*) to generate herbicide-tolerant maize.[22] Scientists have fused Cas9 endonuclease with the *Vir2* gene in rice to facilitate *OsALS* gene disruption with simultaneous repair (HDR-mediated) of this gene-generated herbicide-resistant variety of rice.[23] *ALS* gene was edited using a CRISPR-based gene editing system in oilseeds (*Brassica napus*) and tomatoes to increase their herbicide tolerance.

Natural calamities such as floods or excessive rains restrict crop yield on a yearly basis.[47] Though rice grows in flooded fields mainly during the rainy season, overflooding affects rice cultivation in many areas. The literature survey doesn't provide us with cases where CRISPR/Cas was applied to generate flood-tolerant lines of rice or other crops. Some other techniques were tried on rice, such as the admittance of the *Submergence 1 A-1* (*SUB1A-1*) gene in superior rice varieties using the marker-assisted breeding technique, which consequently gave rise to submergence-tolerant rice varieties.[48] Therefore, it can be inferred that the entrée of genes such as *SUB1A-1*, *Trehalose-6-Phosphate Phosphatase-7* (*TPP7*), *Shikimate Kinase 1* (*SK1*), *SK2*, *Semi-Dwarfing 1* (*SD1*), using the CRISPR/Cas-based editing to submergence susceptible varieties of crops can be made submergence-tolerant.

TABLE 6.1
Application of CRISPR/Cas in the Development of Abiotic Stress-Resistant Crops

Abiotic Stress	Crop Species	Technology Used	Targeted Gene	Gene Function	Result	Transgene-Free Plant	References
Heat	Tomato	CRISPR/Cas9	*SlMAPK3*	Sensitivity	Enhanced heat tolerance	No	24
		CRISPR/Cas9	*SlAGL6*	Sensitivity	Enhanced heat tolerance	No	25
Drought	Maize	CRISPR/Cas9	*ARGOS8*	Sensitivity	Enhanced drought tolerance	No	18
	Chickpea	CRISPR/Cas9	*Coumarate ligase (4CL), Reveille 7 (RVE7)*	Sensitivity	Enhanced drought tolerance	Yes	26
	Rice	CRISPR/Cas9	*OsSRL1*, *OsSRL2*	Sensitivity	Enhanced drought tolerance	Yes	27
		CRISPR/Cas9	*OsDST*	Sensitivity	Moderate drought tolerance	No	17
		CRISPR/Cas9	*OsSAPK2*	Sensitivity	Enhanced drought tolerance	No	28
		CRISPR/Cas9	*OsEBP89*	Sensitivity	Direct seeding on wet-land Enhanced drought tolerance	No	29
		CRISPR/Cas9	*OsMYB30*	Sensitivity	Improved cold tolerance	No	16
		CRISPR/Cas9	*TIFY1a, TIFY1b*	Sensitivity	Cold adaptive plants	No	30
Cold	Tomato	CRISPR/Cas9	*SlCBF1*	Sensitivity	Improved cold acclimation	No	31
		CRISPR/Cas9	*SlARF4*	Sensitivity	Improved salinity tolerance	No	19
Salinity	Rice	CRISPR/Cas9	*OsDST*	Control tolerance	Improved salinity tolerance	No	17
		CRISPR/Cas9	*qSOR1*	Control root growth angle	Improved root structure for saline soil	No	32
		CRISPR/Cas9	*OsRR22*	Sensitivity	Improved salinity tolerance	Yes	33

(*Continued*)

TABLE 6.1 (*Continued*)
Application of CRISPR/Cas in the Development of Abiotic Stress-Resistant Crops

Abiotic Stress	Crop Species	Technology Used	Targeted Gene	Gene Function	Result	Transgene-Free Plant	References
Herbicide	Maize	CRISPR/Cas9	*ZmALS2*	Sensitivity	Herbicide-resistant	No	34
		CRISPR/Cas- mediated base editing	*ZmALS1*, *ZmALS2*	Sensitivity	Sulfonylurea herbicide-resistant plant	Yes	35
		CRISPR- nCas9-RT	*ZmALS1*, *ZmALS2*	Sensitivity	Herbicide-resistant	No	36
	Rice	Base-editing mediated gene evolution	*OsALS1*	Sensitivity	Herbicide-resistant	No	37
		CRISPR- Cpf1	*OsALS*	Sensitivity	Herbicide-resistant	Yes	38
		CRISPR/Cas9	*OsALS*	Sensitivity	Herbicide-resistant	No	39
		CRISPR/Cas9	*OsALS* (Novel allele *G628W*)	Sensitivity	Herbicide-resistant	Yes	40
		CRISPR- nCas9-RT	*OsALS*	Sensitivity	Herbicide-resistant	Yes	41
		CRISPR- Cas9-based cytosine base editing	*OsALS*	Sensitivity	Herbicide-resistant	Yes	42
	Tomato	CRISPR/Cas9	*SlALS*	Sensitivity	Herbicide-resistant	Yes	43
	Oilseed rape	CRISPR/Cas-mediated base editing	*BnALS1*	Sensitivity	Herbicide-resistant	Yes	44
Heavy metal	Rice	CRISPR/Cas9	*OsNramp5*	Sensitivity	Decreased Cd accumulation	No	45
		CRISPR/Cas9	*OsLCT1*, *OsNramp5*	Sensitivity	Decreased Cd accumulation	Yes	46
		CRISPR/Cas9	*OsNramp5*	Sensitivity	Decreased Cd accumulation	Yes	21

6.3.2 CRISPR/Cas and Biotic Stress

Biotic stress factors, i.e., bacterial, fungal, and viral attacks, are the causal reasons for numerous plant diseases consequently leading to significant yield losses. Moreover, other threats like herbivorous insects such as whitefly act as vectors for many pathogens and cause direct harm to crops eventually slashing their yield. Climate change is the primary reason why the tolerance level of plants to various biotic stresses has been adversely affected. Reports suggest that biotic stress is responsible for a direct 20%–40% loss in crop production.[15] Many research groups worldwide have discussed the potential of the CRISPR/Cas system in developing insect-resistant and disease-resistant crops[49] (Table 6.2). The safe and successful deployment of the CRISPR/Cas system to produce climate-smart crops, especially food crops, is gradually paving the path toward achieving Sustainable Development Goal 2 (SDG2) of the United Nations (UN) as the zero hunger goal.

Among other primary approaches, one practical approach is admitting dominant resistance genes (*R* genes) through traditional breeding techniques or recombinant DNA technology. But the drawback of this process is that these R genes are usually target specific and are redundant avirulent genes inside the pathogen. So, scientists inferred that the R gene concept would gradually impart resistance traits to the pathogen during evolution and would fail to impart long-term and broad-spectrum resistance. So, CRISPR/Cas was the tool that targeted the susceptibility (*Su*) genes of the plant genome, whose primary function is to encode factors that aid in pathogen infection. Knocking out these *Su* genes through CRISPR/Cas-aided targeted mutagenesis reduced disease susceptibility of the host plant significantly. Bacterial blight disease caused by *Xanthomonas oryzae* pv *oryzae* causes immense loss in rice production under optimal conditions of pathogenesis. The pathogen releases Transcription Activator Like Effectors (TALEs), which increase the transcription of sugar transport genes (*OsSWEET11*, *OsSWEET13*, *OsSWEET14*, etc.), and these *SWEET* genes are potential *Su* genes for the proliferation of the bacterial blight diseases. CRISPR/Cas tool was utilized in this case to break the bond between the TALEs and *SWEET* genes, which eventually generated rice lines having broad-spectrum resistance to the deadly blight disease.[74] Here, the scientists used CRISPR/Cas to mutate the TALE binding site of the promoter region of *OsSWEET11* (*Xa13*),[66] *OsSWEET13*,[67] and *OsSWEET14*[68] and got progeny plants with enhanced resistance to the blight-causing pathogen. Another fatal threat to rice production is the deadly blast disease caused by the fungus *Magnaportha oryzae*. CRISPR/Cas has also been used to knock out specific rice *Su* genes (e.g., *OsERF92*[73] and *Pi21*[71]), generating blast-resistant rice lines. The *Mildew Locus O* of powdery mildew (deadly disease of wheat) causing fungus *Podosphaera xanthii* was targeted and disrupted by the CRISPR/Cas tool to generate resistance to powdery mildew.[64]

The *Su* genes have also been modified in plants prone to attack by deadly viral pathogens. Rice tungro disease (RTD) is a burning example of a fatal virus attack on rice caused by the interaction of the *Rice tungro spherical virus* (RTSV) and *Rice tungro bacilliform virus* (RTBV), visibly compromising rice production. The plant's natural resistance against RTD is a recessive trait regulated by a translation initiation factor 4 gamma gene (*eIF4G*). The specific residues ($Y^{1059}V^{1060}V^{1061}$) of *eIF4G* are responsible for the pathogenesis of RTSV. So, scientists mutated *eIF4G* by the CRISPR/Cas9 system, and the lines generated are resistant to RTV[65]. Knockout of *Su* loci in polyploid crops like wheat is a considerably cumbrous process. Thanks to the CRISPR/Cas system, now scientists can target multiple homologs simultaneously. Therefore, the disease resistance capacity of wheat can be enhanced by the knockout of the potential *Su* genes. CRISPR/Cas-mediated knockout of all three homologs of *MLO*[64] and *Enhanced disease resistance I*[63] gene-controlled bread wheat's deadly powdery mildew disease resulted in enhanced disease resistance in the host. One fascinating phenomenon observed in the mutated wheat plants is that they showed improved yield, thus providing an opportunity for agriculturists to achieve food security and zero hunger goals. Besides working with the CRISPR/Cas system only on host plant genes, simultaneously scientists have also tried modifying pathogen genes responsible for infecting hosts. To contain the spread of rice blast disease/false smut disease and side by side to enhance their yield, the following pathogen genes

TABLE 6.2
Application of CRISPR/Cas in the Development of Biotic Stress-Resistant Crops

Biotic Stress	Crop Species	Technology Used	Targeted Gene	Gene Function	Result	Transgene-Free Plant	References
Powdery mildew	Tomato	CRISPR/Cas9	*SlMlo1*/Host S-gene	Facilitate disease spreading	Resistant plant	Yes	49
		CRISPR/Cas9	*PMR4*/Host S-gene	Facilitate disease spreading	Resistant plant	No	50
Botrytis cinerea		CRISPR/Cas9	*SlMYC2*/Host S-gene	Inhibits disease spread	Susceptible plant	Yes	51
Bacterial speck		CRISPR/Cas9	*SlJAZ2*/Host S-gene	Facilitate disease spreading	Resistant plant	Not defined	52
Bacterial speck		CRISPR/Cas9	*SlDMR6-1*/Host S-gene	Facilitate disease spreading	Resistant plant	No	53
Phytophthora blight		CRISPR/Cas9	*SlDMR6-1*/Host S-gene	Facilitate disease spreading	Resistant plant	No	54
yellow leaf curl virus		CRISPR/Cas9	IR and CP/virus genome	Viral infection	Resistant plant	Yes	55
TYLCV		CRISPR/Cas9	CP, Rep/Virus genome	Virus spread	Resistant plant	No	56
PVX, TMV, TMV		CRISPR/Cas9	DCL2/Virus genome	Virus spread	Susceptible plant	No	57,58
Pepper Mottle Virus		CRISPR/Cas9	*eIF4E1*/Host S-gene	Viral infection	Resistant plant	Yes	59
Soybean mosaic virus	Soybean	CRISPR/Cas9	*GmF3H1*, *GmF3H2* and *GmFNSII-1*	Synthesis of isoflavonoid	Resistant plant	Yes	60
DNA viral disease	Barley	CRISPR/Cas9	MP, CP, Rep/Rep, IR/Virus genome	Virus spread	Resistant plant	No	61
RNA viral disease	Cucumber	CRISPR/Cas9	*elf4E*/Host S- gene	Disease facilitation	Resistant plant	Yes	62
Powdery mildew	Wheat	CRISPR/Cas9	*TaEDR1*/Host S-gene	Disease facilitation	Resistant plant	No	63
		CRISPR/Cas9	*TaMLO*/Host S-gene	Disease facilitation	Resistant plant	Yes	64
Rice tungro disease	Rice	CRISPR/Cas9	*eIF4G*	Disease facilitation	Resistant plant	Yes	65
Bacterial blight		CRISPR/Cas9	*Xa13*/Host S- gene	Disease facilitation	Resistant plant	Yes	66
		CRISPR/Cas9	*OsSWEET13*/Host S-gene	Disease facilitation	Resistant plant	No	67
		CRISPR/Cas9	*OsSWEET14*/Host S-gene	Disease facilitation	Resistant plant	Yes	68,69
Rice blast		CRISPR/Cas9	*ALB1*, *RSY1*/Fungal gene	Facilitate growth of pathogens	Improved resistance	Yes	70
		CRISPR/Cas9	*Pi21*/Host S- gene	Disease facilitation	Resistant plant	Yes	71,72
		CRISPR/Cas9	*OsERF922*/Host S-gene	Disease facilitation	Resistant plant	Yes	73

ALB1 (*ALBINA1*), *RSY1*,[70] *UvSLT2*,[75] and *USTA* were knocked out by the CRISPR/Cas system and resistance against the targeted pathogen was obtained. The CRISPR/Cas system directly attacks the viral genome, and various categories of Cas can select and attack different DNA and RNA viruses, to develop viral resistance in plants. *Wheat dwarf virus*, a geminivirus, attacks barley (*Hordeium vulgare*). This virus is disarmed through the CRISPR/Cas system, and resistance against the virus is achieved by jeopardizing the replication of this virus.[61] Similarly, using CRISPR/Cas, disease resistance can be completed in other fruits and vegetables such as cucumber, watermelon, tomato, and other citrus fruits by knockout of the host *Su* genes or mutating the pathogen genomes.

6.4 FUTURE PROSPECTS

Genome engineering has significant applications in the domain of plant molecular breeding to overcome a plethora of challenges and produce climate-smart crops. CRISPR technology's effortless field application is the knockout of a potential gene through NHEJ. Next Generation Sequencing (NGS) and other bioinformatic tools have opened new horizons by predicting novel genes and complex gene networks. However, the practical application of these bioinformatic data to corroborate the gene functions poses a significant challenge to scientists while designing experiments for their *in vivo* application in samples. Here CRISPR/Cas system is used for gene knockout, which helps to support these computational data very efficiently and identify the putative genes responsible for abiotic and biotic stress tolerance (Figure 6.4). Among numerous enticing functions and advantages of CRISPR/Cas technology, one of the most convincing activities is the allelic replacement or allelic substitution, by HDR, of the inferior copies of genes possessing agronomic importance. Point mutations can be done, or sequence segments can be altered to generate efficient knock-in or knockout lines methodically. Even restriction sites can be inserted by this tool inside the point of interest, and the gene/protein can also be tagged. CRISPR/Cas technology can also be utilized for epigenetic modulation and genome imaging. The expression of any target gene can be strategized and planned differently through the CRISPR/Cas system by eliminating the promoter permanently

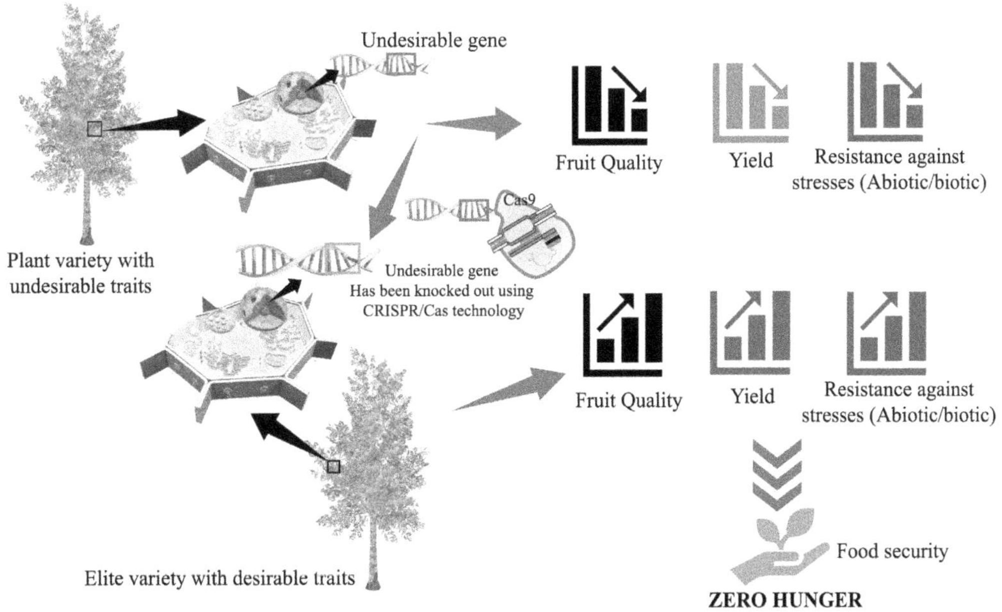

FIGURE 6.4 Application of CRISPR/Cas for developing desirable traits (improved food quality, production and resistance against biotic and abiotic stresses) in plants.

or affixing an enhancer to the dCas9 enzyme so mapped to append in the vicinity of the promoter of the target gene. Scientists have also inserted transgene inside the host genome at the specific site, which results in safe and stable expression of the transgene without jeopardizing the cellular and nuclear machinery.

The CRISPR/Cas T-DNA gradually wanes out of the plant genome by the law of segregation (meiotic purification), thus generating clean lines. Artificial genetic diversity in crop cultivars can be caused by utilizing base editors. With advantages also come disadvantages for any new technique. Similarly, CRISPR/Cas is deployed, covering numerous risks and bottlenecks, and it overcomes the threats looming large on agriculture and food security due to climate change.[2]

CRISPR/Cas is being used to introduce desired traits in economically important crops, including disease resistance, and the generation of disease resistance is not confined only to laboratory tissue cultures and greenhouses. Still, many are in the pipeline waiting for commercialization, like false flax (*C. sativa*), and they are generally observed to be broad-spectrum and resist multiple pathogens.[76] Thus CRISPR/Cas-mediated climate-innovative crops with increased nutritional value will reach the common market shortly. Recently gene editing has witnessed several developments, and CRISPR/Cas technology has been the guiding light on the horizon. Temperature-tolerant CRISPR/LbCas12a has become prevalent to enhance gene targeting and transformation efficiency.

CRISPR technology is a recently developed one, and the generation cum commercialization of food and cash crops is time-consuming after passing through ethical and biosafety guidelines. The plight of the CRISPR/Cas generated yields is pathetic in the European Union (EU) since it has been categorized as a Genetically Modified Crop (GMO) even though it is transgene-free.[77,78] Here, scientists and policymakers need to work together and develop comprehensive plans for the entry of CRISPR-edited climate-smart crops into the market. Apart from improvising policies for the smooth commercialization of CRISPR-edited crops, a few technological bottlenecks must be resolved. One is its low innate HDR efficiency, which impedes numerous potential applications such as massive chromosomal deletions, gene replacement, etc. However, scientists are developing new methods to enhance HDR efficiency in plants and make gene insertions less challenging.[23,79,80] Work is being done worldwide to create novel ways of transporting genome engineering machinery inside germ cells, ultimately resulting in foreign DNA-free edited plant genomes. In the long run, ample field trials are the ultimate requirement to judge the performance of climate-smart crops for sustained resistance to abiotic and biotic stresses.

6.5 CONCLUSION

Under the present climate change scenario, crop production is a challenge, and this is both a boon and a bane for scientists and the common man. Massive loss in the annual yield of crops is the potential disadvantage of fluctuating climatic conditions and challenges to devise new technologies so that plants can adapt to changing climatic conditions and other threats which follow as a natural course and mitigate their effects to a noticeable extent. Climate-smart agriculture and the generation of climate-smart crops have shifted the agricultural system from an unstable state toward a more productive, tolerant, resilient, and sustainable approach with the help of efficient and effective mechanisms. The utilization of newly developed tools and techniques is readily available to the farmers; climate-change adaptation doesn't only deal with the application of modernized tools and methods but mainly with their acceptance by society and immense support from the consumers involved. It's of utmost necessity to sufficiently educate the farmers, making them aware of climate-smart agriculture and crops. They should be provided with all the facilities and infrastructure for agriculture that have been put forth by research institutions and the country's policies. Strong and farsighted agricultural policies should be chalked out by the policymakers, which will help the farming sector generate revenue for the country and income for themselves in developing and underdeveloped countries. Strong commitment from the country's government is in dire need, whose work is to set up an intricate and transparent network among the agricultural policymakers, the scientists/research institutes, agricultural startups, and farmers. Working in close coordination can generate climate-smart crops and mitigate the harmful effects of climate change on agriculture and deal efficiently with further unpredictable climate changes, thus making the whole system productive and sustainable.

Modern tools and genome editing techniques like CRISPR/Cas have revolutionized crop improvement in huge capacity by making economically important crops stress-resistant and increasing their productivity. Researchers from all over the world have published sufficient data that reveals the climate-smart nature of the plants after gene editing, which shows high tolerance and resistance levels upon exposure to stresses making the plants perfectly climate-resilient. Modern-day genomics has opened new avenues by which scientists can identify the molecular markers linked with significant agronomic traits, leading to the generation of a select variety of crops with high productivity, substantial abiotic stress tolerance, and biotic stress resistance. This CRISPR/Cas technology will not only lead to the production of climate-smart crops but will also lead the world toward food security and attaining zero hunger in the coming years.

REFERENCES

[1] Krishnan P, Ramakrishnan B, Reddy KR, Reddy VR. High-temperature effects on rice growth, yield, and grain quality. *Advances in Agronomy.* 2011;111, 87–205. doi:10.1016/B978-0-12-387689-8.00004-7

[2] Biswal AK, Mangrauthia SK, Reddy MR, Yugandhar P. CRISPR mediated genome engineering to develop climate smart rice: Challenges and opportunities. *Semin Cell Dev Biol.* 2019;96:100–106. doi:10.1016/j.semcdb.2019.04.005

[3] Jinek M, Chylinski K, Fonfara I, Hauer M, Doudna JA, Charpentier E. A programmable dual-RNA - guided. *Science* (1979). 2012;337(August):816–822.

[4] Chaudhuri A, Halder K, Datta A. Classification of CRISPR/Cas system and its application in tomato breeding. *Theor Appl Genet.* 2022;135:367–387. doi:10.1007/s00122-021-03984-y

[5] Jiang W, Zhou H, Bi H, Fromm M, Yang B, Weeks DP. Demonstration of CRISPR/Cas9/sgRNA-mediated targeted gene modification in Arabidopsis, tobacco, sorghum and rice. *Nucleic Acids Res.* 2013;41(20):1–12. doi:10.1093/nar/gkt780

[6] Iizumi T, Shiogama H, Imada Y, Hanasaki N, Takikawa H, Nishimori M. Crop production losses associated with anthropogenic climate change for 1981-2010 compared with preindustrial levels. *Int J Climatol.* 2018;38(14):5405–5417. doi:10.1002/joc.5818

[7] Karavolias NG, Horner W, Abugu MN, Evanega SN. Application of gene editing for climate change in agriculture. *Front Sustain Food Syst.* 2021;5. doi:10.3389/fsufs.2021.685801

[8] Wang S, Zhang Y, Ju W, Chen JM, Ciais P, Cescatti A, et al. *Recent Global Decline of CO2 Fertilization Effects on Vegetation Photosynthesis.* https://www.science.org

[9] Fisher MC, Henk DA, Briggs CJ, Brownstein JS, Madoff LC, McCraw SL, et al. Emerging fungal threats to animal, plant and ecosystem health. *Nature.* 2012;484(7393):186–194. doi:10.1038/nature10947

[10] Bett B, Kiunga P, Gachohi J, Sindato C, Mbotha D, Robinson T, et al. Effects of climate change on the occurrence and distribution of livestock diseases. *Prev Vet Med.* 2017;137:119–129. doi:10.1016/j.prevetmed.2016.11.019

[11] Nunez S, Arets E, Alkemade R, Verwer C, Leemans R. Assessing the impacts of climate change on biodiversity: is below 2°C enough? *Clim Change.* 2019;154(3-4):351–365. doi:10.1007/s10584-019-02420-x

[12] Rashid S, Bin Mushtaq M, Farooq I, Khan Z. Climate smart crops for food security. In: *The Nature, Causes, Effects and Mitigation of Climate Change on the Environment.* IntechOpen; 2022. doi:10.5772/intechopen.99164

[13] Campbell BM, Thornton P, Zougmoré R, van Asten P, Lipper L. Sustainable intensification: What is its role in climate smart agriculture? *Curr Opin Environ Sustain.* 2014;8:39–43. doi:10.1016/j.cosust.2014.07.002

[14] Chen K, Wang Y, Zhang R, Zhang H, Gao C. CRISPR/Cas genome editing and precision plant breeding in agriculture. *Annu Rev Plant Biol.* 2019;70:667–697. doi:10.1146/annurev-arplant-050718

[15] Ahmad S, Tang L, Shahzad R, Mawia AM, Rao GS, Jamil S, et al. CRISPR-based crop improvements: A way forward to achieve zero hunger. *J Agric Food Chem.* 2021;69(30):8307–8323. doi:10.1021/acs.jafc.1c02653

[16] Zeng Y, Wen J, Zhao W, Wang Q, Huang W. Rational improvement of rice yield and cold tolerance by editing the three genes OsPIN5b, GS3, and OsMYB30 with the CRISPR-Cas9 system. *Front Plant Sci.* 2020;10. doi:10.3389/fpls.2019.01663
[17] Santosh Kumar VV, Verma RK, Yadav SK, Yadav P, Watts A, Rao MV, et al. CRISPR-Cas9 mediated genome editing of drought and salt tolerance (OsDST) gene in indica mega rice cultivar MTU1010. *Physiol Mol Biol Plants.* 2020;26(6):1099–1110. doi:10.1007/s12298-020-00819-w
[18] Shi J, Gao H, Wang H, Lafitte HR, Archibald RL, Yang M, et al. ARGOS8 variants generated by CRISPR-Cas9 improve maize grain yield under field drought stress conditions. *Plant Biotechnol J.* 2017;15(2):207–216. doi:10.1111/pbi.12603
[19] Bouzroud S, Gasparini K, Hu G, Barbosa MAM, Rosa BL, Fahr M, et al. Down regulation and loss of auxin response factor 4 function using CRISPR/Cas9 alters plant growth, stomatal function and improves tomato tolerance to salinity and osmotic stress. *Genes.* 2020;11(3). doi:10.3390/genes11030272
[20] Khan H, McDonald MC, Williams SJ, Solomon PS. Assessing the efficacy of CRISPR/Cas9 genome editing in the wheat pathogen *Parastagonspora nodorum. Fungal Biol Biotechnol.* 2020;7(1). doi:10.1186/s40694-020-00094-0
[21] Yang CH, Zhang Y, Huang CF. Reduction in cadmium accumulation in japonica rice grains by CRISPR/Cas9-mediated editing of OsNRAMP5. *J Integr Agric.* 2019;18(3):688–697. doi:10.1016/S2095-3119(18)61904-5
[22] Nuccio ML, Claeys H, Heyndrickx KS. CRISPR-Cas technology in corn: A new key to unlock genetic knowledge and create novel products. *Mol Breed.* 2021;41(2). doi:10.1007/s11032-021-01200-9
[23] Ali Z, Shami A, Sedeek K, Kamel R, Alhabsi A, Tehseen M, et al. Fusion of the Cas9 endonuclease and the VirD2 relaxase facilitates homology-directed repair for precise genome engineering in rice. *Commun Biol.* 2020;3(1). doi:10.1038/s42003-020-0768-9
[24] Voesenek LACJ, Bailey-Serres J. Flood adaptive traits and processes: An overview. *New Phytol.* 2015;206(1):57–73. doi:10.1111/nph.13209
[25] Dar MH, Zaidi NW, Waza SA, Verulkar SB, Ahmed T, Singh PK, et al. No yield penalty under favorable conditions paving the way for successful adoption of flood tolerant rice. *Sci Rep.* 2018;8(1). doi:10.1038/s41598-018-27648-y
[26] Ahmad S, Wei X, Sheng Z, Hu P, Tang S. CRISPR/Cas9 for development of disease resistance in plants: Recent progress, limitations and future prospects. *Brief Funct Genomics.* 2018;19(1):26–39. doi:10.1093/bfgp/elz041
[27] Oliva R, Ji C, Atienza-Grande G, Huguet-Tapia JC, Perez-Quintero A, Li T, et al. Broad-spectrum resistance to bacterial blight in rice using genome editing. *Nat Biotechnol.* 2019;37(11):1344–1350. doi:10.1038/s41587-019-0267-z
[28] Li C, Li W, Zhou Z, Chen H, Xie C, Lin Y. A new rice breeding method: CRISPR/Cas9 system editing of the Xa13 promoter to cultivate transgene-free bacterial blight-resistant rice. *Plant Biotechnol J.* 2020;18(2):313–315. doi:10.1111/pbi.13217
[29] Zhou J, Peng Z, Long J, Sosso D, Liu B, Eom JS, et al. Gene targeting by the TAL effector PthXo2 reveals cryptic resistance gene for bacterial blight of rice. *Plant J.* 2015;82(4):632–643. doi:10.1111/tpj.12838
[30] Zeng X, Luo Y, Vu NTQ, Shen S, Xia K, Zhang M. CRISPR/Cas9-mediated mutation of *OsSWEET14* in rice cv. Zhonghua11 confers resistance to *Xanthomonas oryzae pv. oryzae* without yield penalty. *BMC Plant Biol.* 2020;20(1). doi:10.1186/s12870-020-02524-y
[31] Wang F, Wang C, Liu P, Lei C, Hao W, Gao Y, et al. Enhanced rice blast resistance by CRISPR/Cas9-Targeted mutagenesis of the ERF transcription factor gene *OsERF922. PLoS One.* 2016;11(4). doi:10.1371/journal.pone.0154027
[32] Nawaz G, Usman B, Peng H, Zhao N, Yuan R, Liu Y, et al. Knockout of *Pi21* by Crispr/Cas9 and iTRAQ-based proteomic analysis of mutants revealed new insights into *M. oryzae* resistance in elite rice line. *Genes.* 2020;11(7):1–24. doi:10.3390/genes11070735
[33] Wang Y, Cheng X, Shan Q, Zhang Y, Liu J, Gao C, et al. Simultaneous editing of three homoeoalleles in hexaploid bread wheat confers heritable resistance to powdery mildew. *Nat Biotechnol.* 2014;32(9):947–951. doi:10.1038/nbt.2969
[34] Macovei A, Sevilla NR, Cantos C, Jonson GB, Slamet-Loedin I, Čermák T, et al. Novel alleles of rice eIF4G generated by CRISPR/Cas9-targeted mutagenesis confer resistance to Rice tungro spherical virus. *Plant Biotechnol J.* 2018;16(11):1918–1927. doi:10.1111/pbi.12927

[35] Zhang Y, Bai Y, Wu G, Zou S, Chen Y, Gao C, et al. Simultaneous modification of three homoeologs of TaEDR1 by genome editing enhances powdery mildew resistance in wheat. *Plant J.* 2017;91(4):714–724. doi:10.1111/tpj.13599
[36] Foster AJ, Martin-Urdiroz M, Yan X, Wright HS, Soanes DM, Talbot NJ. CRISPR-Cas9 ribonucleoprotein-mediated co-editing and counterselection in the rice blast fungus. *Sci Rep.* 2018;8(1). doi:10.1038/s41598-018-32702-w
[37] Liang Y, Han Y, Wang C, Jiang C, Xu JR. Targeted deletion of the *USTA* and *UvSLT2* genes efficiently in *Ustilaginoidea virens* with the CRISPR-Cas9 system. *Front Plant Sci.* 2018;9. doi:10.3389/fpls.2018.00699
[38] Kis A, Hamar É, Tholt G, Bán R, Havelda Z. Creating highly efficient resistance against wheat dwarf virus in barley by employing CRISPR/Cas9 system. *Plant Biotechnol J.* 2019;17(6):1004–1006. doi:10.1111/pbi.13077
[39] Zaidi SSEA, Mahas A, Vanderschuren H, Mahfouz MM. Engineering crops of the future: CRISPR approaches to develop climate-resilient and disease-resistant plants. *Genome Biol.* 2020;21(1). doi:10.1186/s13059-020-02204-y
[40] Custers R. The regulatory status of gene-edited agricultural products in the EU and beyond. *Emerg Top Life Sci.* 2017;1(2):221–229. doi:10.1042/ETLS20170019
[41] Globus R, Qimron U. A technological and regulatory outlook on CRISPR crop editing. *J Cell Biochem.* 2018;119(2):1291–1298. doi:10.1002/jcb.26303
[42] Van Vu T, Sung YW, Kim J, Doan DTH, Tran MT, Kim JY. Challenges and perspectives in homology-directed gene targeting in monocot plants. *Rice.* 2019;12(1). doi:10.1186/s12284-019-0355-1
[43] Liu M, Rehman S, Tang X, Gu K, Fan Q, Chen D, et al. Methodologies for improving HDR efficiency. *Front Genet.* 2019;10(JAN). doi:10.3389/fgene.2018.00691
[44] Yu W, Wang L, Zhao R, Sheng J, Zhang S, Li R, et al. Knockout of SlMAPK3 enhances tolerance to heat stress involving ROS homeostasis in tomato plants. *BMC Plant Biol.* 2019;19(1). doi:10.1186/s12870-019-1939-z
[45] Klap C, Yeshayahou E, Bolger AM, Arazi T, Gupta SK, Shabtai S, et al. Tomato facultative parthenocarpy results from SlAGAMOUS-LIKE 6 loss of function. *Plant Biotechnol J.* 2017;15(5):634–647. doi:10.1111/pbi.12662
[46] Badhan S, Ball AS, Mantri N. First report of CRISPR/Cas9 mediated DNA-free editing of 4CL and RVE7 genes in chickpea protoplasts. *Int J Mol Sci.* 2021;22(1):1–15. doi:10.3390/ijms22010396
[47] Liao S, Qin X, Luo L, Han Y, Wang X, Usman B, et al. CRISPR/Cas9-induced mutagenesis of semi-rolled Leaf1,2 confers curled leaf phenotype and drought tolerance by influencing protein expression patterns and ROS scavenging in rice (*Oryza sativa* L.). *Agronomy.* 2019;9(11). doi:10.3390/agronomy9110728
[48] Lou D, Wang H, Liang G, Yu D. OsSAPK2 confers abscisic acid sensitivity and tolerance to drought stress in rice. *Front Plant Sci.* 2017;8. doi:10.3389/fpls.2017.00993
[49] Zhang Y, Li J, Chen S, Ma X, Wei H, Chen C, et al. An APETALA2/ethylene responsive factor, OsEBP89 knockout enhances adaptation to direct-seeding on wet land and tolerance to drought stress in rice. *Mol Genet Genom.* 2020;295(4):941–956. doi:10.1007/s00438-020-01669-7
[50] Xiaozhen H, Xiaofang Z, Jianrong L, Degang Z. Construction and analysis of tify1a and tify1b mutants in rice (*Oryza sativa*) based on CRISPR/Cas9 technology. *J Agric Biotechnol.* 2017;25:1003–1012.
[51] Li R, Zhang L, Wang L, Chen L, Zhao R, Sheng J, et al. Reduction of tomato-plant chilling tolerance by CRISPR-Cas9-mediated SlCBF1 mutagenesis. *J Agric Food Chem.* 2018;66(34):9042–9051. doi:10.1021/acs.jafc.8b02177
[52] Kitomi Y, Hanzawa E, Kuya N, Inoue H, Hara N, Kawai S, et al. Root angle modifications by the DRO1 homolog improve rice yields in saline paddy fields. doi:10.1073/pnas.2005911117/-/DCSupplemental
[53] Zhang A, Liu Y, Wang F, Li T, Chen Z, Kong D, et al. Enhanced rice salinity tolerance via CRISPR/Cas9-targeted mutagenesis of the OsRR22 gene. *Mol Breed.* 2019;39(3):47. doi:10.1007/s11032-019-0954-y
[54] Svitashev S, Young JK, Schwartz C, Gao H, Falco SC, Cigan AM. Targeted mutagenesis, precise gene editing, and site-specific gene insertion in maize using Cas9 and guide RNA. *Plant Physiol.* 2015;169(2):931–945. doi:10.1104/pp.15.00793
[55] Li Y, Zhu J, Wu H, Liu C, Huang C, Lan J, et al. Precise base editing of non-allelic acetolactate synthase genes confers sulfonylurea herbicide resistance in maize. *Crop J.* 2020;8(3):449–456. doi:10.1016/j.cj.2019.10.001

[56] Jiang YY, Chai YP, Lu MH, Han XL, Lin Q, Zhang Y, et al. Prime editing efficiently generates W542L and S621I double mutations in two ALS genes in maize. *Genome Biol.* 2020;21(1). doi:10.1186/s13059-020-02170-5
[57] Kuang Y, Li S, Ren B, Yan F, Spetz C, Li X, et al. Base-editing-mediated artificial evolution of OsALS1 in planta to develop novel herbicide-tolerant rice germplasms. *Mol Plant.* 2020;13(4):565–572. doi:10.1016/j.molp.2020.01.010
[58] Li S, Li J, He Y, Xu M, Zhang J, Du W, et al. Precise gene replacement in rice by RNA transcript-templated homologous recombination. *Nat Biotechnol.* 2019;37(4):445–450. doi:10.1038/s41587-019-0065-7
[59] Sun Y, Zhang X, Wu C, He Y, Ma Y, Hou H, et al. Engineering herbicide-resistant rice plants through CRISPR/Cas9-mediated homologous recombination of acetolactate synthase. *Mol Plant.* 2016;9(4):628–631. doi:10.1016/j.molp.2016.01.001
[60] Wang F, Xu Y, Li W, Chen Z, Wang J, Fan F, et al. Creating a novel herbicide-tolerance OsALS allele using CRISPR/Cas9-mediated gene editing. *Crop J.* 2021;9(2):305–312. doi:10.1016/j.cj.2020.06.001
[61] Wang T, Deng Z, Zhang X, Wang H, Wang Y, Liu X, et al. Tomato DCL2b is required for the biosynthesis of 22-nt small RNAs, the resulting secondary siRNAs, and the host defense against ToMV. *Hortic Res.* 2018;5(1). doi:10.1038/s41438-018-0073-7
[62] Butt H, Rao GS, Sedeek K, Aman R, Kamel R, Mahfouz M. Engineering herbicide resistance via prime editing in rice. *Plant Biotechnol J.* 2020;18(12):2370–2372. doi:10.1111/pbi.13399
[63] Ghorbani Faal P, Farsi M, Seifi A, Mirshamsi Kakhki A. Virus-induced CRISPR-Cas9 system improved resistance against tomato yellow leaf curl virus. *Mol Biol Rep.* 2020;47(5):3369–3376. doi:10.1007/s11033-020-05409-3
[64] Ortigosa A, Gimenez-Ibanez S, Leonhardt N, Solano R. Design of a bacterial speck resistant tomato by CRISPR/Cas9-mediated editing of SlJAZ2. *Plant Biotechnol J.* 2019;17(3):665–673. doi:10.1111/pbi.13006
[65] de Toledo Thomazella D, Brail Q, Dahlbeck D, Staskawicz B. CRISPR-Cas9 mediated mutagenesis of a DMR6 ortholog in tomato confers broad-spectrum disease resistance. *BioRxiv.* 2016. doi:10.1101/064824
[66] Tang L, Mao B, Li Y, Lv Q, Zhang L, Chen C, et al. Knockout of OsNramp5 using the CRISPR/Cas9 system produces low Cd-accumulating indica rice without compromising yield. *Sci Rep.* 2017;7(1). doi:10.1038/s41598-017-14832-9
[67] Songmei L, Jie J, Yang L, Jun M, Shouling X, Yuanyuan T, et al. Characterization and evaluation of OsLCT1 and OsNramp5 mutants generated through CRISPR/Cas9-mediated mutagenesis for breeding low Cd rice. *Rice Sci.* 2019;26(2):88–97. doi:10.1016/j.rsci.2019.01.002
[68] Nekrasov V, Wang C, Win J, Lanz C, Weigel D, Kamoun S. Rapid generation of a transgene-free powdery mildew resistant tomato by genome deletion. *Sci Rep.* 2017;7(1). doi:10.1038/s41598-017-00578-x
[69] Zhang R, Chen S, Meng X, Chai Z, Wang D, Yuan Y, et al. Generating broad-spectrum tolerance to ALS-inhibiting herbicides in rice by base editing. *Sci China Life Sci.* 2021;64(10):1624–1633. doi:10.1007/s11427-020-1800-5
[70] Tashkandi M, Ali Z, Aljedaani F, Shami A, Mahfouz MM. Engineering resistance against *Tomato yellow leaf curl virus* via the CRISPR/Cas9 system in tomato. *Plant Signal Behav.* 2018;13(10). doi:10.1080/15592324.2018.1525996
[71] Shu P, Li Z, Min D, Zhang X, Ai W, Li J, et al. CRISPR/Cas9-mediated *SlMYC2* mutagenesis adverse to tomato plant growth and MeJA-induced fruit resistance to *Botrytis cinerea*. *J Agric Food Chem.* 2020;68(20):5529–5538. doi:10.1021/acs.jafc.9b08069
[72] Danilo B, Perrot L, Mara K, Botton E, Nogué F, Mazier M. Efficient and transgene-free gene targeting using Agrobacterium-mediated delivery of the CRISPR/Cas9 system in tomato. *Plant Cell Rep.* 2019;38(4):459–462. doi:10.1007/s00299-019-02373-6
[73] Koseoglou E. *The Study of SlPMR4 CRISPR/Cas9- Mediated Tomato Allelic Series for Resistance Against Powdery Mildew.* Wageningen University, Wageningen, The Netherlands; 2017.
[74] Wu J, Chen C, Xian G, Liu D, Lin L, Yin S, et al. Engineering herbicide-resistant oilseed rape by CRISPR/Cas9-mediated cytosine base-editing. *Plant Biotechnol J.* 2020;18(9):1857–1859. doi:10.1111/pbi.13368
[75] Wang Z, Hardcastle TJ, Pastor AC, Yip WH, Tang S, Baulcombe DC. A novel DCL2-dependent miRNA pathway in tomato affects susceptibility to RNA viruses. *Genes Dev.* 2018;32(17–18):1155–1160. doi:10.1101/gad.313601.118

[76] Yoon YJ, Venkatesh J, Lee JH, Kim J, Lee HE, Kim DS, et al. Genome editing of eIF4E1 in tomato confers resistance to pepper mottle virus. *Front Plant Sci.* 2020;11. doi:10.3389/fpls.2020.01098
[77] Zhang P, Du H, Wang J, Pu Y, Yang C, Yan R, et al. Multiplex CRISPR/Cas9-mediated metabolic engineering increases soya bean isoflavone content and resistance to soya bean mosaic virus. *Plant Biotechnol J.* 2020;18(6):1384–1395. doi:10.1111/pbi.13302
[78] Chandrasekaran J, Brumin M, Wolf D, Leibman D, Klap C, Pearlsman M, et al. Development of broad virus resistance in non-transgenic cucumber using CRISPR/Cas9 technology. *Mol Plant Pathol.* 2016;17(7):1140–1153. doi:10.1111/mpp.12375
[79] Zafar K, Khan MZ, Amin I, Mukhtar Z, Yasmin S, Arif M, et al. Precise CRISPR-Cas9 mediated genome editing in super basmati rice for resistance against bacterial blight by targeting the major susceptibility gene. *Front Plant Sci.* 2020;11. doi:10.3389/fpls.2020.00575
[80] Li S, Shen L, Hu P, Liu Q, Zhu X, Qian Q, et al. Developing disease-resistant thermosensitive male sterile rice by multiplex gene editing. *J Integr Plant Biol.* 2019;61(12):1201–1205. doi:10.1111/jipb.12774

7 Functional Genomics for Abiotic Stress Tolerance in Crops

Muhammad Khuram Razzaq, Guangnan Xing, Muhammad Basit Shehzad, Ghulam Raza, and Reena Rani

7.1 INTRODUCTION

Food security is threatened by climate change. Abiotic stresses, such as salinity, drought, temperature, and heavy metals, provide a significant barrier to agricultural production and significantly lower crop yields globally. For crop production to be sustainable, germplasm development with resistance to biotic and abiotic influences is crucial. Genetic variations causing both Mendelian and complex disorders may be studied and mapped using a framework developed by genomic research [1]. The molecular concept "omics" denotes a thorough evaluation of many molecules. Multi-omics technologies represent a holistic assessment of the molecules that make up an organism or cell to detect genes, mRNA, proteins, metabolites, etc. in a biological context [2,3]. Expression quantitative trait loci (eQTL), which have proven essential in the analysis of genome-wide association studies (GWAS) and the simulation of biological networks, were also mapped using array technologies in the early 2000s [4]. Since then, a wide variety of new omics tools have been created that can examine complete collections of metabolites, transcripts, and proteins in addition to the genome. Together, these insights offered a justification for the creation of systems biology tools that combine several omics data types to find molecular patterns linked to biotic and abiotic stresses [5]. The discovery of stress-associated genes in crops has been aided by functional genomics [6,7]. Publicly available web-based databases are a valuable tool for plant genomics, particularly for identifying genes that respond to stress. At the genome level, a method for finding stress-responsive genes should soon be available, thanks to recent advancements in genomic methodologies and the availability of the complete genome sequences of many plant species [8]. GWAS have found genes that respond to stress and their advantageous alleles for complex trait loci. The development of genomic databases has made it possible for bioinformatics tools to use homology and synteny to detect stress-resistant gene families in a variety of plant species [9,10]. Metabolomics is one of the most general omics technologies and can be applied to a wide range of organisms. It has been effectively used to investigate the genetic phenotypes that plants exhibit in reaction to abiotic stress and identify specific patterns linked to stress tolerance. These investigations have emphasized the critical role of primary metabolites, including amino acids, sugars, and intermediates of the Krebs cycle, as both direct indicators of photosynthetic failure and agents of osmotic adaptation [11]. Moreover, under stressful conditions, transcriptomics can reveal the entire transcriptional level of the whole genome, which is useful in detecting the complex controlling system associated with the adaptability of plants and stress tolerance [12]. The identification of essential proteins that confer abiotic stress tolerance was greatly aided by proteomics, particularly those involved in photosynthesis, defense response, signal transduction, carbon metabolism, protein synthesis, energy metabolism, and redox homeostasis

 DOI: 10.1201/9781003387060-7

[13]. Ionomics offers novel systems for knowing how the plant's ionome functions under stress as well as for locating the genes and regulatory pathways involved in transportation, and mineral accumulation that participates in various molecular mechanisms under either normal or stressful circumstances [14]. Hence, multi-omics approaches are composed to detect molecular insight under stress and stress tolerance genes that will be helpful to improve economically important traits in crop plants.

7.2 MULTI-OMICS APPROACHES

7.2.1 Genomics

In contrast to genetics, which investigated individual variations or single genes, genomics, the first omics field to emerge, concentrated on the study of complete genomes [15]. Functional genomics has discovered multiple genes that regulate biotic and abiotic stress responses in plants. Crop plants have been genetically modified to develop tolerance to biotic and abiotic stresses [16]. From the genomics of wild crop relatives, several novel candidate genes have been identified for the tolerance of biotic and abiotic stresses in crops [11]. For example, a chip-seq genome-wide investigation revealed 21 ABA TFs and their extensive regulatory network [17]. The genomic region of *Solanum americanum* that imparts resistance to late blight was identified using combined genetic mapping and RenSeq [18]. A population comprised of 199 recombinant inbred lines was generated from a narrow-leaf wild parent Changling and a broad-leaf Yiqianli to examine leaflet width (LW), length (LL), and length to width (LLW). In total, 10 LW, 13LL, and 9 LLW QTLs were detected, including 17 new QTLs. Six genes were predicted as the most suitable candidate genes, and three PCR markers, YC-16-3, BARCSOYSSR_16_0796, and Gm16PAV0653, were proposed for marker-assisted selection of JQS-16, LLW, and LW [19]. Soybean resistance to salt stress was also increased by overexpression of GmNHX6. The salt tolerance was correlated with an SNP in the promoter region of NHX6. These findings showed that the soybean *NHX6* gene is crucial for the plant's ability to withstand alkaline salt stress [20]. A PAM2 domain is present in the protein that GsERD15B encodes, enabling it to interact with poly(A)-binding proteins to activate transcription. We suggest that GsERD15B overexpression likely improved salt tolerance by enhancing the expression levels of genes associated with catalase-peroxidase, cation transport, proline content, dehydration response, and ABA signaling. They proposed that natural variation in the GsERD15B promoter affects soybean salt tolerance. The combination of functional genomics and high-throughput phenotyping results in novel crop enhancement strategies. Figure 7.1 summarizes functional genomics approaches and their mechanism of action.

7.2.2 Transcriptomics

The methods used to investigate an organism's transcriptome and the collection of all its RNA transcripts are known as "transcriptomic technology". Multiple stress tolerance-associated genes may be found utilizing recently developed RNA sequencing, digital profiling, microarrays, and serial analysis of gene expression to find key gene functions [21]. Genes associated with pollen abortion were identified through transcriptome analysis in cytoplasmic male-sterile soybeans. Moreover, several DEGs are linked to pollen wall growth, sugar transport, carbohydrate metabolism, transcription factor, and reactive oxygen species (ROS) metabolism [22]. Transcriptome profiling of GXS87-16 was done during severe drought stress. Hence, the DEGs were allocated to 47 GO classes and 93 KEGG pathway analyses, respectively. These pathways were involved in numerous processes such as plant-pathogen interaction, mRNA surveillance, RNA transport, and plant hormone signal transduction [23]. There aren't many global gene expression profile datasets available to study

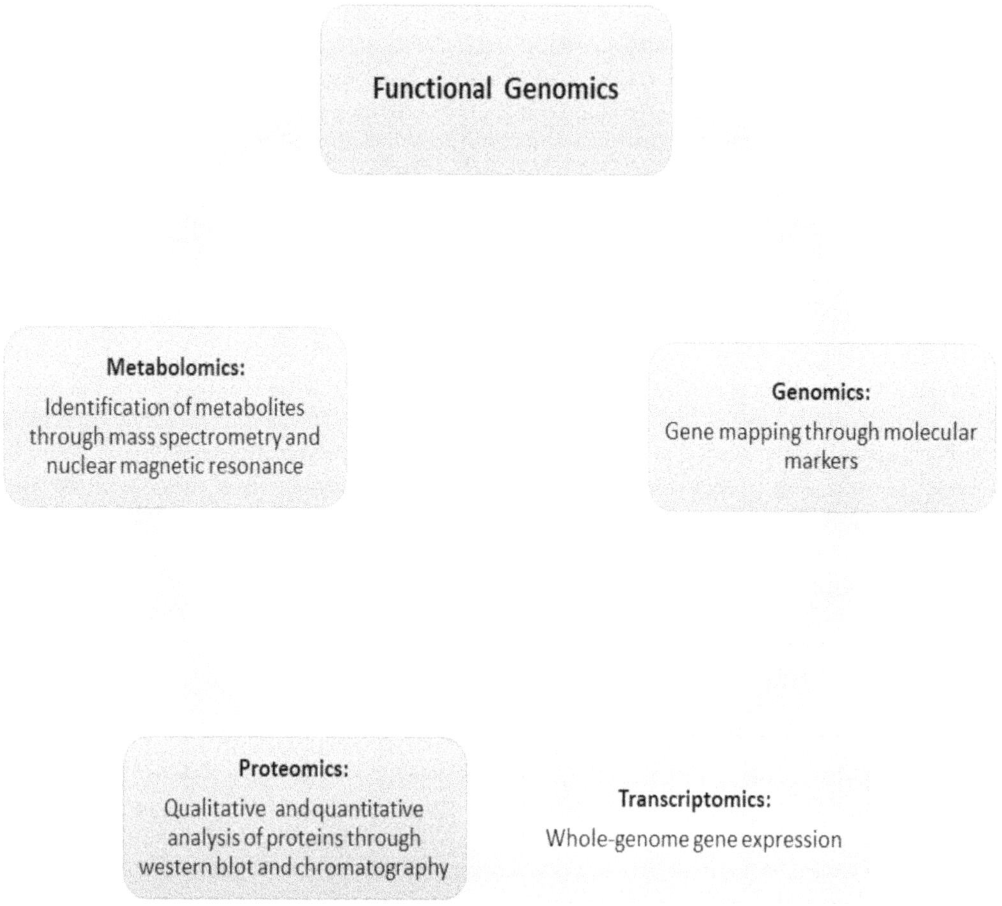
Functional Genomics
Metabolomics:
Identification of metabolites through mass spectrometry and nuclear magnetic resonance
Genomics:
Gene mapping through molecular markers
Proteomics:
Qualitative and quantitative analysis of proteins through western blot and chromatography
Transcriptomics:
Whole-genome gene expression

FIGURE 7.1 Functional genomics approaches and how these work.

peppers' mechanisms for coping with environmental stress. This study documented the RNA-seq of peppers after treatment with salinity, osmotic stress, cold, and heat at six diverse growth stages. This study provided molecular insight into numerous stimuli to help the development of stress-tolerant cultivars [24]. Compounds, live microbes, or their components are called "plant biostimulants", and they can change how plants develop. Biostimulants' impact is revealed in numerous ways: via transcriptomic changes, epigenomic, physiological, proteomic, morphological, and biochemical. However, a criterion that permits biostimulants to be categorized based on how they affect biological systems may be developed using a method that focuses on the influence of a specific field, such as transcriptomics [25].

7.2.3 Proteomics

Proteomics is the quantitative and qualitative analysis of all proteins expressed in an organism, tissue, or cell [26]. Recently, the proteome's reactivity to abiotic stressors was profiled using a variety of mass spectrometry techniques [27]. When utilized in plant stress and genome-wide research, mass spectrometry proteomics gives substantial proteome information. To determine the role of specific proteins in abiotic stress-induced transmission as well as differentially regulated stress-resistant proteins, proteome profiles can be examined [28]. Wheat leaf phosphoproteome research was conducted [29]. S-adenosylmethionine has been found in several isoforms in soybeans grown in both flood and drought conditions [30]. In tomato plants under drought stress, chloroplast proteins with nuclear proteins have been found [31]. Plant reactions under salt stress have been investigated in halophytes and glycophytes through proteomics. At the proteomics level, most work has been done on model plants such as *Arabidopsis thaliana* and *Nicotiana tobaccum* [32]. In rice, using Pro-Q Diamond stain to analyze the changes in roots phosphoproteome under NaCl (150mM), they identified 20 upregulated and 18 downregulated proteins. Proteins including GALP hydrogenase, ATP synthase beta, and protein kinase, while proteins such as Hsp70, GST, and mannose-binding lectin were upregulated [33]. The 20-kDa chaperonin is upregulated, which effectively protects soybean proteins against salinization. Moreover, proteomics was applied to 34 distinct plant species that were exposed to salinity. They documented 2171 salt-responsive proteins and organized them based on transcription, translation, CO_2 assimilation, defense and stress interaction, energy metabolism, protein transport, carbohydrate synthesis, cell division, cell structure, and several others with indefinite functions [34]. Furthermore, the scientists found the *Glycine max* root proteome under drought stress and found a total of 45 proteins, of which two proteins were novel. Five proteins' expression was upregulated, and the expression of 21 proteins was downregulated [34]. When the roots of soybeans were examined using proteomic techniques under drought stress, an increase in dehydrin and ferritin was discovered. Dehydrins and Late-Embryogenesis Abundant (LEA) proteins that lessen the negative effects of ROS can significantly increase plant development under stress [35]. Heat shock proteins, low (15–45 kDa) and high (60–110 kDa) are synthesized in plants when the temperature is high [36]. Researchers looked at the proteome analysis of heat-stressed rice leaf tissue. On exposure to high temperatures for 12 to 24 hours compared to controls, they found 48 proteins. Eighteen of the discovered proteins were HSPs, including the BiP molecular chaperone, Cpn60, HSP100, and smHSPs [36]. A novel technology known as proteomics makes it possible to identify pathways and proteins linked to plant stress and physiology. The stress-related proteins' understanding and use in biology for crop improvement are further improved by proteomics.

7.2.4 Metabolomics

The study of metabolites, or the tiny molecules that serve as the substrates, intermediaries, and end products of cellular metabolism, is known as metabolomics [37,38]. A large number of hub genes based on environment and their gene-metabolite regulation networks are discovered using integrated transcriptomic and metabolomics assay. Two hub genes, ZmGLK44 and Bx12, play key roles in controlling the production of metabolites in maize and its ability to withstand drought, according to extensive molecular and genetic investigations [39]. Under drought stress (DS), the number of sugars (including ribose, xylonate, sucrose, glucose, raffinose, fructose, erythronate, and raffinose) and the glycolysis intermediary pyruvate increased in the roots [40]. Arabidopsis experienced reduced myoinositol levels as a result of dehydration stress [41]. Urea cycles and glutathione, as well as lipid and carbohydrate metabolism, are crucial for *Zea mays*' osmoprotection, preservation of the membrane, and antioxidant defense during DS [42]. 4-hydroxy-2-oxoglutaric acid was undetectable in the leaves of *N. tabacum* but rose 20-fold and 70-fold in the roots throughout the first 1 to 2 hours of drought stress, respectively. This shows that in the presence of DS, N. tabacum develops 4-hydroxy-2-oxoglutaric acid, which is then converted into glyoxylate and pyruvate when water is available [43]. It was discovered that under chronic and acute (100 mM) salinity stress (SS), saponins were overexpressed in the shoots of *M. sativa* whereas *M. arborea* did so in the roots, which is the plant's section where the majority of saponins are overexpressed and produced [44]. Taxol, glycyrrhetinate terpenoids, and oleanolate 3-d-glucuronoside-28-glucoside, were found in higher concentrations in the SS-treated roots of S. brachiata, whereas the production of sesquiterpenoids such as desoxyhemigossypol-6-methyl ether, heliespirone C, costunolide, and 15-hydroxysolavetivone was found in the stressed leaves [45]. Contemporary metabolomics, including the creation of metabolomics data, uses two fundamental techniques: MS and NMR [46]. In reaction to heat stress (HS), leaves generate a variety of diverse metabolites (DPMs), including cofactors, nucleotides, lipids, amino acids, peptides, carbohydrates, nucleotides, and lipids. Several DPMs, including methionine, xylose, ribose, isoleucine, gluconate, lysine, xylitol alanine, and deoxyribose are involved in the pathways of cellular processes, for example, glycolysis, the TCA cycle, starch biosynthesis, and the pentose phosphate pathway were affected by heat stress [47]. By combining the metabolic profile and transcriptomics, regulatory networks were predicted for the Myb28 and Myb29 transcription factor genes in Arabidopsis. These genes are particularly important for the generation of aliphatic GSL and the expression of biosynthetic genes [48]. Understanding secondary metabolism can now take a new path from functional genomics, proteomics, metabolomics, and transcriptomics. The application of omics technologies to some model crops for abiotic stress tolerance is summarized in Table 7.1.

TABLE 7.1
Some Examples of Omics Technologies for Stress Tolerance Molecular Insight in Crops

Crop	Omics Approach	Stress	Main Findings	Ref.
Thale cress (*Arabidopsis thaliana*)	Transcriptomics, Metabolomics, and Proteomics	Nitrate, Sulfate	Increased expression of immunity genes, low-N compounds that worsen glucosinolate degradation, and low-S compounds that inhibit glucosinolate formation	[49]
Thale cress (*Arabidopsis thaliana*)	Proteomics	Selenium	Differentially expressed protein species were observed after the addition of selenium in transgenic and non-transgenic plants. These proteins were involved in metabolism, signal transduction, the glycolysis pathway, and defense	[50]
Thale cress (*Arabidopsis thaliana*)	Proteomics	Salt stress	AtCBL1 is a protein that belongs to the calcineurin B-like (CBL) protein family and is strongly inducible by a variety of stress signals. It works in the salt stress signal pathway and favorably regulates the plant's tolerance to salt	[51]
Thale cress (*Arabidopsis thaliana*)	Proteomics	Salt and osmotic stress	Change in cellular proteins 6 h after NaCl and sorbitol treatment. 2949 proteins were observed, of which 266 significantly changed in quantity, and 75 responded to salt and sorbitol	[52]
Thale cress (*Arabidopsis thaliana*)	Proteomics	Cold Stress	Plants were subjected to cold stress for 1 week at two distinct temperatures (6°C and 8°C). 22 differentially expressed proteins were found and identified by MALDI-TOF and ESI-MS/MS techniques	[53]
Thale cress (*Arabidopsis thaliana*)	Metabolomics	Cadmium	The impact of the heavy metal cadmium on the metabolome of the Arabidopsis thaliana plant was studied. Various statistical comparisons of the metabolic fingerprints seemed useful for isolating and identifying some distinguishing metabolites	[54]
Thale cress (*Arabidopsis thaliana*)	Metabolomics	Temperature stress	In response to both heat and cold stress, metabolites increased. Cold shock had a far greater impact on the metabolome than heat shock. In response to heat and cold shocks, 143 and 311 metabolites were changed, respectively	[55]

(Continued)

TABLE 7.1 (*Continued*)
Some Examples of Omics Technologies for Stress Tolerance Molecular Insight in Crops

Crop	Omics Approach	Stress	Main Findings	Ref.
Thale cress (*Arabidopsis thaliana*)	Transcriptomics	salt, osmotic, cold, heat, and abscisic acid	Using the whole-genome tiling array technique, the impact of abiotic stresses was investigated, and hundreds of newly stress-induced transcribed regions were discovered	[56]
Thale cress (*Arabidopsis thaliana*)	Transcriptomics	Clubroot disease	Find interactions between the pathogen and the host plant	[57]
Tomato (*Solanum lycopersicum*)	Microarray gene expression	Biotic and abiotic stresses	Meta-analytical approach was used to characterize genes	[58]
Tomato (*Solanum lycopersicum*)	Transcriptomics, and Metabolomics	Nitrogen, Phosphorous, and Sulfur	Interactions between transcription factors, biosynthetic genes, and metabolomics reactions were discovered	[59]
Tomato (*Solanum lycopersicum*)	RNA-Seq	Abscisic acid	Differentially expressed genes were associated with different pathways	[60]
Tomato (*Solanum lycopersicum*)	Proteomics	Salinity	Salt stress applied to the plant for 14 days, showed significant abundance variation in 26 proteins that were evolved in energy metabolism, especially in the salt-tolerant genotype	[61]
Tomato (*Solanum lycopersicum*)	Transcriptomics	Exogenous abscisic acid	Exogenous abscisic acid influences the ABA signaling pathway and regulates the genes involved in stress tolerance	[62]
Tomato (*Solanum lycopersicum*)	RNA sequencing	Salt and oxidative stress	Tomato transcriptional reactions and cytokinin levels are regulated by salt and oxidative stress	[63]

7.3 CONCLUSION AND FUTURE OUTLOOK

Abiotic stresses, such as high or low temperature, drought, metalloids, exposure to heavy metals, and flooding are the main obstacles in agriculture and continue to be the main cause of low agricultural production and output in the world. To understand gene activities in plants, high-throughput testing of research platforms is necessary, especially novel initiatives for plant phenotyping (http://www.lemnatec.com) systems. To predict gene function, we needed more data, better data analysis, access to data and tools, and high-throughput confirmation through experiments. Multi-omics approaches mainly access knowledge about genes, proteins, master regulators, biosynthetic pathways, cross-talk, biological networks, and specifically about plant stress reactions. The post-genomic period has provided many new opportunities to properly understand various functioning characteristics of plants. The complex interactions of such biological systems have been substantially clarified through genomes, metabolomics, and phenomics. Functional genomics is the use of omics-based methods to characterize the role of a gene or genes. The advent of functional genomics methodologies has made it easier to analyze several genes and their products linked to many plant developmental and defensive processes. The study of abiotic stress tolerance is now thought to be possible with the help of functional genomics, which enables the analysis of the perception of stress, signaling, and tolerance responses. This analysis can be done by comparing the metabolite profiles of stressed and control tissues. With this background, plant breeders are attempting to improve field crops' tolerance and productivity by applying the knowledge learned from model plants. Omics has delivered a polaroid at the tissue and cell level by illustrating and evaluating biomolecules to improve organism interactions and functions.

REFERENCES

1. Tam V, Patel N, Turcotte M, et al. Benefits and limitations of genome-wide association studies. *Nature Reviews Genetics.* 2019;20(8):467–484.
2. Hasin Y, Seldin M, Lusis A. Multi-omics approaches to disease. *Genome Biology.* 2017;18(1):1–15.
3. Razzaq MK, Aleem M, Mansoor S, et al. Omics and CRISPR-Cas9 approaches for molecular insight, functional gene analysis, and stress tolerance development in crops. *International Journal of Molecular Sciences.* 2021;22(3):1292.
4. Joehanes R, Zhang X, Huan T, et al. Integrated genome-wide analysis of expression quantitative trait loci aids interpretation of genomic association studies. *Genome Biology.* 2017;18(1):1–24.
5. Manzoni C, Kia DA, Vandrovcova J, et al. Genome, transcriptome and proteome: The rise of omics data and their integration in biomedical sciences. *Briefings in Bioinformatics.* 2018;19(2):286–302.
6. Shi T, Iqbal S, Ayaz A, et al. Analyzing differentially expressed genes and pathways associated with pistil abortion in Japanese apricot via RNA-Seq. *Genes.* 2020;11(9):1079.
7. Li M, Chen L, Zeng J, et al. Identification of additive-epistatic QTLs conferring seed traits in soybean using recombinant inbred lines. *Frontiers in Plant Science.* 2020;11:566056.
8. Segelbacher G, Bosse M, Burger P, et al. New developments in the field of genomic technologies and their relevance to conservation management. *Conservation Genetics.* 2022;23:217–242.
9. Yang L, Cao H, Zhang X, et al. Genome-wide identification and expression analysis of tomato ADK gene family during development and stress. *International Journal of Molecular Sciences.* 2021;22(14):7708.
10. Mubarik MS, Wang X, Khan SH, et al. Engineering broad-spectrum resistance to cotton leaf curl disease by CRISPR-Cas9 based multiplex editing in plants. *GM Crops & Food.* 2021;12(2):647–658.
11. Arbona V, Manzi M, de Ollas C, et al. Metabolomics as a tool to investigate abiotic stress tolerance in plants. *International Journal of Molecular Sciences.* 2013;14(3):4885–4911.
12. Wang X, Li N, Li W, et al. Advances in transcriptomics in the response to stress in plants. *Global Medical Genetics.* 2020;7(2):30–34.
13. Halder T, Choudhary M, Liu H, et al. Wheat proteomics for abiotic stress tolerance and root system architecture: Current status and future prospects. *Proteomes.* 2022;10(2):17.
14. Ali S, Tyagi A, Bae H. Ionomic approaches for discovery of novel stress-resilient genes in plants. *International Journal of Molecular Sciences.* 2021;22(13):7182.
15. Razzaq MK, Akhter M, Ahmad RM, et al. CRISPR-Cas9 based stress tolerance: New hope for abiotic stress tolerance in chickpea (*Cicer arietinum*). *Molecular Biology Reports.* 2022;49:8977–8985.
16. Parmar N, Singh KH, Sharma D, et al. Genetic engineering strategies for biotic and abiotic stress tolerance and quality enhancement in horticultural crops: A comprehensive review. *3 Biotech.* 2017;7(4):1–35.
17. Song L, Huang S-S C, Wise A, et al. A transcription factor hierarchy defines an environmental stress response network. *Science.* 2016;354(6312):aag1550.
18. Witek K, Jupe F, Witek AI, et al. Accelerated cloning of a potato late blight-resistance gene using RenSeq and SMRT sequencing. *Nature Biotechnology.* 2016;34(6):656–660.
19. Zeng J, Li M, Qiu H, et al. Identification of QTLs and joint QTL segments of leaflet traits at different canopy layers in an interspecific RIL population of soybean. *Theoretical and Applied Genetics.* 2022;135:4261–4275.
20. Jin T, An J, Xu H, et al. A soybean sodium/hydrogen exchanger GmNHX6 confers plant alkaline salt tolerance by regulating Na+/K+ homeostasis. *Frontiers in Plant Science.* 2022;13:938635.
21. Moustafa K, Cross JM. Genetic approaches to study plant responses to environmental stresses: An overview. *Biology.* 2016;5(2):20.
22. Bai Z, Ding X, Zhang R, et al. Transcriptome analysis reveals the genes related to pollen abortion in a cytoplasmic male-sterile soybean (*Glycine max* (L.) Merr.). *International Journal of Molecular Sciences.* 2022;23(20):12227.
23. Li C, Wang Z, Nong Q, et al. Physiological changes and transcriptome profiling in *Saccharum spontaneum* L. leaf under water stress and re-watering conditions. *Scientific Reports.* 2021;11(1):1–14.
24. Kang W-H, Sim YM, Koo N, et al. Transcriptome profiling of abiotic responses to heat, cold, salt, and osmotic stress of *Capsicum annuum* L. *Scientific Data.* 2020;7(1):1–7.
25. González-Morales S, Solís-Gaona S, Valdés-Caballero MV, et al. Transcriptomics of biostimulation of plants under abiotic stress. *Frontiers in Genetics.* 2021;12:583888.
26. Graves PR, Haystead TA. Molecular biologist's guide to proteomics. *Microbiology and Molecular Biology reviews.* 2002;66(1):39–63.

27. Luan H, Shen H, Pan Y, et al. Elucidating the hypoxic stress response in barley (*Hordeum vulgare* L.) during waterlogging: A proteomics approach. *Scientific Reports.* 2018;8(1):1–13.
28. Barkla BJ. Identification of abiotic stress protein biomarkers by proteomic screening of crop cultivar diversity. *Proteomes.* 2016;4(3):26.
29. Zhang M, Lv D, Ge P, et al. Phosphoproteome analysis reveals new drought response and defense mechanisms of seedling leaves in bread wheat (*Triticum aestivum* L.). *Journal of Proteomics.* 2014;109:290–308.
30. Wang YS, Yao HY, Xue HW. Lipidomic profiling analysis reveals the dynamics of phospholipid molecules in Arabidopsis thaliana seedling growth. *Journal of Integrative Plant Biology.* 2016;58(11):890–902.
31. Tamburino R, Vitale M, Ruggiero A, et al. Chloroplast proteome response to drought stress and recovery in tomato (*Solanum lycopersicum* L.). *BMC Plant Biology.* 2017;17(1):1–14.
32. Flannery SE, Pastorelli F, Wood WH, et al. Comparative proteomics of thylakoids from Arabidopsis grown in laboratory and field conditions. *Plant Direct.* 2021;5(10):e355.
33. Ahmad P, Abdel Latef AA, Rasool S, et al. Role of proteomics in crop stress tolerance. *Frontiers in Plant Science.* 2016;7:1336.
34. Zhang H, Han B, Wang T, et al. Mechanisms of plant salt response: Insights from proteomics. *Journal of Proteome research.* 2012;11(1):49–67.
35. Hossain Z, Khatoon A, Komatsu S. Soybean proteomics for unraveling abiotic stress response mechanism. *Journal of Proteome Research.* 2013;12(11):4670–4684.
36. Lee DG, Ahsan N, Lee SH, et al. A proteomic approach in analyzing heat-responsive proteins in rice leaves. *Proteomics.* 2007;7(18):3369–3383.
37. Lee SJ, Trostel A, Le P, et al. Cellular stress created by intermediary metabolite imbalances. *Proceedings of the National Academy of Sciences.* 2009;106(46):19515–19520.
38. Shahid A, Khurshid M, Aslam B, et al. Cyanobacteria derived compounds: Emerging drugs for cancer management. *Journal of Basic Microbiology.* 2022;62(9):1125–1142.
39. Zhang F, Wu J, Sade N, et al. Genomic basis underlying the metabolome-mediated drought adaptation of maize. *Genome Biology.* 2021;22(1):1–26.
40. de Miguel M, Guevara MÁ, Sánchez-Gómez D, et al. Organ-specific metabolic responses to drought in *Pinus pinaster* Ait. *Plant Physiology and Biochemistry.* 2016;102:17–26.
41. Urano K, Maruyama K, Ogata Y, et al. Characterization of the ABA-regulated global responses to dehydration in Arabidopsis by metabolomics. *The Plant Journal.* 2009;57(6):1065–1078.
42. Yang L, Fountain JC, Ji P, et al. Deciphering drought-induced metabolic responses and regulation in developing maize kernels. *Plant Biotechnology Journal.* 2018;16(9):1616–1628.
43. Rabara RC, Tripathi P, Reese RN, et al. Tobacco drought stress responses reveal new targets for Solanaceae crop improvement. *BMC Genomics.* 2015;16(1):1–23.
44. Sarri E, Termentzi A, Abraham EM, et al. Salinity stress alters the secondary metabolic profile of *M. sativa*, *M. arborea* and their hybrid (Alborea). *International Journal of Molecular Sciences.* 2021;22(9):4882.
45. Benjamin JJ, Lucini L, Jothiramshekar S, et al. Metabolomic insights into the mechanisms underlying tolerance to salinity in different halophytes. *Plant Physiology and Biochemistry.* 2019;135:528–545.
46. Kim HK, Verpoorte R. Sample preparation for plant metabolomics. *Phytochemical Analysis: An International Journal of Plant Chemical and Biochemical Techniques.* 2010;21(1):4–13.
47. Das A, Rushton PJ, Rohila JS. Metabolomic profiling of soybeans (*Glycine max* L.) reveals the importance of sugar and nitrogen metabolism under drought and heat stress. *Plants.* 2017;6(2):21.
48. Hirai MY, Sugiyama K, Sawada Y, et al. Omics-based identification of *Arabidopsis* Myb transcription factors regulating aliphatic glucosinolate biosynthesis. *Proceedings of the National Academy of Sciences.* 2007;104(15):6478–6483.
49. Luo J, Havé M, Clement G, et al. Integrating multiple omics to identify common and specific molecular changes occurring in Arabidopsis under chronic nitrate and sulfate limitations. *Journal of Experimental Botany.* 2020;71(20):6471–6490.
50. Maciel BC, Barbosa HS, Pessôa GS, et al. Comparative proteomics and metallomics studies in *Arabidopsis thaliana* leaf tissues: Evaluation of the selenium addition in transgenic and nontransgenic plants using two-dimensional difference gel electrophoresis and laser ablation imaging. *Proteomics.* 2014;14(7–8):904–912.
51. Shi S, Chen W, Sun W. Comparative proteomic analysis of the *Arabidopsis cbl1* mutant in response to salt stress. *Proteomics.* 2011;11(24):4712–4725.
52. Ndimba BK, Chivasa S, Simon WJ, et al. Identification of *Arabidopsis* salt and osmotic stress responsive proteins using two-dimensional difference gel electrophoresis and mass spectrometry. *Proteomics.* 2005;5(16):4185–4196.

53. Amme S, Matros A, Schlesier B, et al. Proteome analysis of cold stress response in *Arabidopsis thaliana* using DIGE-technology. *Journal of Experimental Botany*. 2006;57(7):1537–1546.
54. Ducruix C, Vailhen D, Werner E, et al. Metabolomic investigation of the response of the model plant *Arabidopsis thaliana* to cadmium exposure: Evaluation of data pretreatment methods for further statistical analyses. *Chemometrics and Intelligent Laboratory Systems*. 2008;91(1):67–77.
55. Kaplan F, Kopka J, Haskell DW, et al. Exploring the temperature-stress metabolome of Arabidopsis. *Plant Physiology*. 2004;136(4):4159–4168.
56. Zeller G, Henz SR, Widmer CK, et al. Stress-induced changes in the *Arabidopsis thaliana* transcriptome analyzed using whole-genome tiling arrays. *The Plant Journal*. 2009;58(6):1068–1082.
57. Wu F, Chen Y, Tian X, et al. Genome-wide identification and characterization of phased small interfering RNA genes in response to *Botrytis cinerea* infection in *Solanum lycopersicum*. *Scientific Reports*. 2017;7(1):3019.
58. Bielecka M, Watanabe M, Morcuende R, et al. Transcriptome and metabolome analysis of plant sulfate starvation and resupply provides novel information on transcriptional regulation of metabolism associated with sulfur, nitrogen and phosphorus nutritional responses in Arabidopsis. *Frontiers in Plant Science*. 2015;5:805.
59. Ashrafi-Dehkordi E, Alemzadeh A, Tanaka N, et al. Meta-analysis of transcriptomic responses to biotic and abiotic stress in tomato. *PeerJ*. 2018;6:e4631.
60. Zhao Y, Bi K, Gao Z, et al. Transcriptome analysis of *Arabidopsis thaliana* in response to *Plasmodiophora brassicae* during early infection. *Frontiers in Microbiology*. 2017;8:673.
61. Manaa A, Mimouni H, Wasti S, et al. Comparative proteomic analysis of tomato (*Solanum lycopersicum*) leaves under salinity stress. *Plant Omics*. 2013;6(4):268–277.
62. Wang Y, Tao X, Tang X-M, et al. Comparative transcriptome analysis of tomato (*Solanum lycopersicum*) in response to exogenous abscisic acid. *BMC Genomics*. 2013;14(1):1–14.
63. Keshishian EA, Hallmark HT, Ramaraj T, et al. Salt and oxidative stresses uniquely regulate tomato cytokinin levels and transcriptomic response. *Plant Direct*. 2018;2(7):e00071.

8 CRISPR/Cas9 for Enhancing Resilience of Crops to Abiotic Stresses

Gyanendra Kumar Rai, Danish Mushtaq Khanday, Gayatri Jamwal, and Monika Singh

8.1 INTRODUCTION

Stress is a physiologically changed situation brought on by events that tend to upset the equilibrium. Various stressors, such as drought, excessive heat or cold, salt, floods, oxidative stress, and heavy metal toxicity, are frequently experienced by plants. Stress is the norm; normalcy is the exception. Injury, illness, or abnormal physiology are brought on by constraints like stress or extremely unexpected changes placed on normal metabolic rhythms. (Kumar et al., 2013). By 2050, crop yields will have decreased by at least 20% due to climate change, according to the World Bank. Traditional breeding and transgenic have made some progress in producing stress-tolerant crop plants and increasing crop production. The rapid development of innovative methods for breeding and development has resulted in a strong emphasis on genetic engineering applications. However, traditional genetic engineering has encountered various challenges and constraints, including ethical considerations, a detrimental influence on the environment, and most notably, the difficulty in manipulating vast genomes of higher plants (Nemudryi et al., 2014). One of the most important methods for crop improvement in contemporary agriculture is the creation of better cultivars of crops by mutation, hybridization, and rDNA technology. Conventional breeding, in contrast, entirely depends on existing genetic differences and crossing strategies, which will continue to impede agricultural growth and need a significant investment of time and money to introduce desirable traits into crops. Furthermore, genetic variety has declined considerably, reducing the possibility of improving traits through cross-breeding (Zhang et al., 2020). Furthermore, the commercialization of GM/biotech crops is hampered by social concerns as well as time-consuming and expensive regulatory review procedures (Chen et al., 2019). Through the modification of several regulatory elements, genes, and chromosomes in the finest cultivars, genome engineering is revolutionizing plant biology. The nucleases utilized in genome editing are called engineered nucleases, and they are sequence-specific. It consists of two components: a DNA targeting element that directs the nuclease to a specific region of the genome, and a nuclease element that cuts the targeted genomic regions. ZFN, TALENs, and the CRISPR-Cas system are the three types of designed nucleases found so far. These sequence-specific nucleases generate double-stranded breaks at specific genomic sites, which are later restored either through HDR (homology-directed repair) or through NHEJ (non-homologous end joining) (Figure 8.1) (El-Mounadi et al., 2020). So far, several genome editing approaches have been applied to improve various crop plant properties (Figure 8.2) (Sedeek et al., 2019). The CRISPR/Cas9 approach, which is presently utilized worldwide for plant gene editing, has expedited crop breeding beyond what was previously thought possible. Technical complexity and low efficiency plague ZFNs and TALENs, whereas the CRISPR/Cas9 system is simple and very efficient. CRISPR/Cas9 has emerged as a viable technique for improving the genetic architecture of various plant genomes

DOI: 10.1201/9781003387060-8

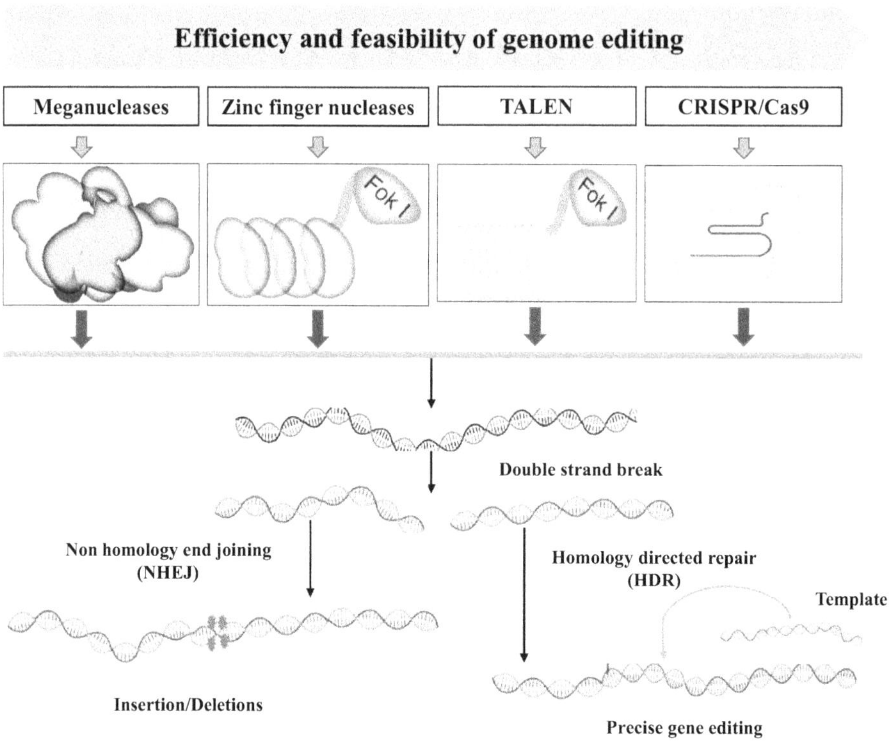

FIGURE 8.1 Several generations of nucleases utilized for genetic manipulation, as well as the DNA repair mechanisms employed to change target DNA.

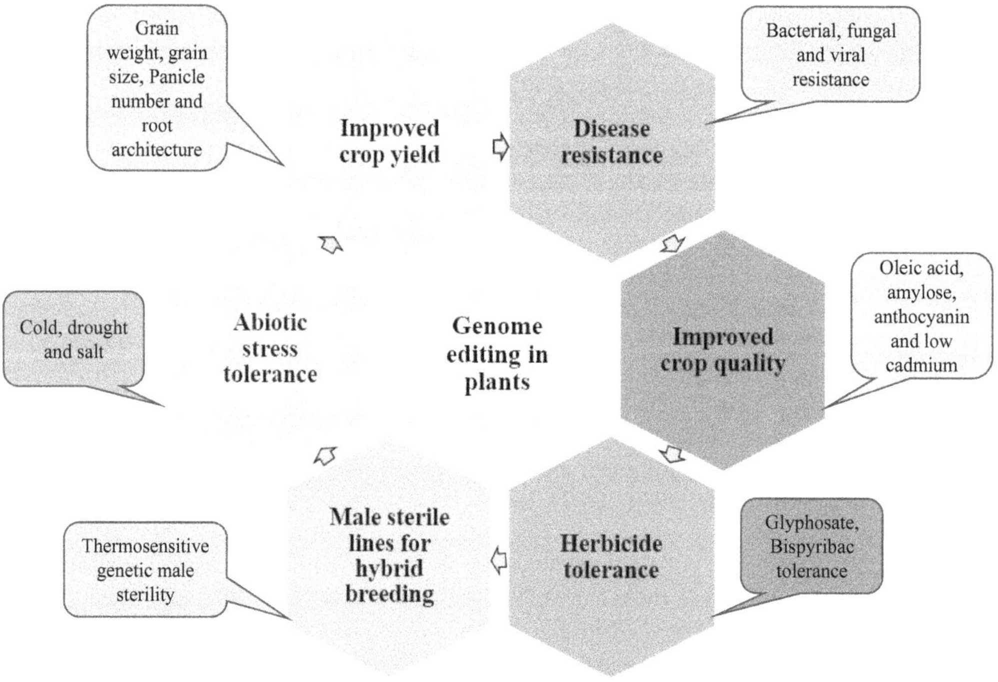

FIGURE 8.2 Genome editing applications in field crops. (Mushtaq et al., 2021b.)

TABLE 8.1
Comparison of Various Gene Editing Tools

S.no	Properties	ZFN	TALEN	Crisper/Cas9
1	Restriction site	20-35bp	20-40bp	20-23bp
2	Nuclease Component	fokI	fokI	Cas9
3	Effectiveness	High	High	High
4	Identification molecule	Protein-DNA	Protein-DNA	Protein-DNA
5	Cost	High	Moderate	Low
6	limitations	Time intensive	Time intensive	Off -targeting

Source: (Mushtaq et al., 2021b)

especially agriculturally important crops, due to its greater effectiveness, lower cost, and ease of implementation (Zhang et al., 2020; Bortesi and Fischer, 2015). Figure 8.2 depicts various applications of CRISPR/Cas9 in plant breeding. A comparison table of three designed nucleases has been depicted (Table 8.1).

CRISPR/Cas, which has currently replaced traditional resistance breeding methods as the industry standard, may be used to develop crops in a specific direction and significantly decrease the breeding life.

8.2 ORIGIN AND WORKING OF CRISPER/Cas9

Clustered regularly interspaced short palindromic repeats (CRISPR) are short, direct repeat sequences found in the intergenic regions of the prokaryotic genome. These repeats are interspersed with spacer regions and harbor several hundred base pair leader sequences on one side of the repeat cluster. CRISPRs were first detected in *E. coli* by Yoshizumi Ishino while analyzing the gene involved in the conversion of alkaline phosphatase (Ishino et al., 1987). Subsequently, they were reported in archaea *Haloferax mediterranei* (Mojica et al., 1993) and then in many other bacterial and archaeal species. Comparative genomic analyses between the CRISPR spacer regions and plasmid, bacteriophage, and archaeal virus sequences suggested CRISPR to be a part of the microorganism's immune system. Further analysis by Jansen et al., in 2002 found four conserved genes to be present adjacent to the CRISPR regions in many prokaryotes, two of which showed structural similarities with helicases and RecB exonucleases. These were designated as CRISPR-associated genes 1 to 4 (cas 1 to cas 4). Incidentally, the same group coined the term CRISPR for the repeat sequences with which cas genes were associated.

Since their initial discovery, CRISPR/Cas systems have been recognized in several bacterial and archeal genomes. However, their presence in eukaryotic and viral genomes has not been reported. The CRISPR/Cas system has garnered much attention due to the ability of Cas proteins to induce directional cuts in foreign DNA and in certain cases RNA, which can be exploited to cleave target molecules at desired sites. When it comes to genome editing in eukaryotic organisms, the most widely studied and used system is CRISPR/Cas 9 of *Streptococcus pyogenes*. The Cas 9 protein was first identified by Bolotin et al. (2005) (initially named Cas 5), which was later characterized as a type II prokaryotic defense system against intruding bacteriophages (Jinek et al., 2012). Together with the discovery of CRISPR-associated RNA (crRNA) (Gasiunas et al., 2012) and trans-activating CRISPR RNA (tracrRNA) (Deltcheva et al., 2011) which can be combined into single guide RNA (sg-RNA) for target DNA recognition (Jinek et al., 2012), the CRISPR/Cas9 is now being used extensively for genome engineering in humans, animals, and plants. As a component of the immune system, it relies on double-strand breaks (DSBs) generated by invasive viral DNA at certain locations. As a result, DSBs trigger a DNA repair mechanism in host cells through HDR or NHEJ (Figure 8.1), resulting in INDELS (insertions or deletions) in the invading viral DNA, rendering it ineffective against the host bacterium (Zaidi et al., 2020).

In general, CRISPR/Cas9-driven genome editing in plant genomes consists of four steps. First, utilizing a number of web-based sgRNA generation tools, users produce sgRNAs, particularly for a gene of interest. These programs ask users to provide the location, name, or DNA sequence of each relevant gene as well as information about the species in question. Plants express sgRNA more when small nuclear RNA promoters viz., U3 and U6 are present. Massive data acquisition and comprehensive studies of sgRNA in plant cells are needed to increase the precision of computational sgRNA identification. The second stage requires employing a transient transformation system, such as hairy root or protoplast transformation using CRISPR before utilizing it for genetic manipulation, which is the best way to confirm the functionality of sgRNA. The third stage involves the stable integration of sgRNA and Cas9 into the plant genome, as well as the delivery of a genome editing construct into plant cells, usually by particle bombardment or Agrobacterium-mediated transformation. Last but not least, RE/PCR genotyping, which is verified by sequencing, is used to check for mutations in transformed or regenerated plants (Figure 8.3).

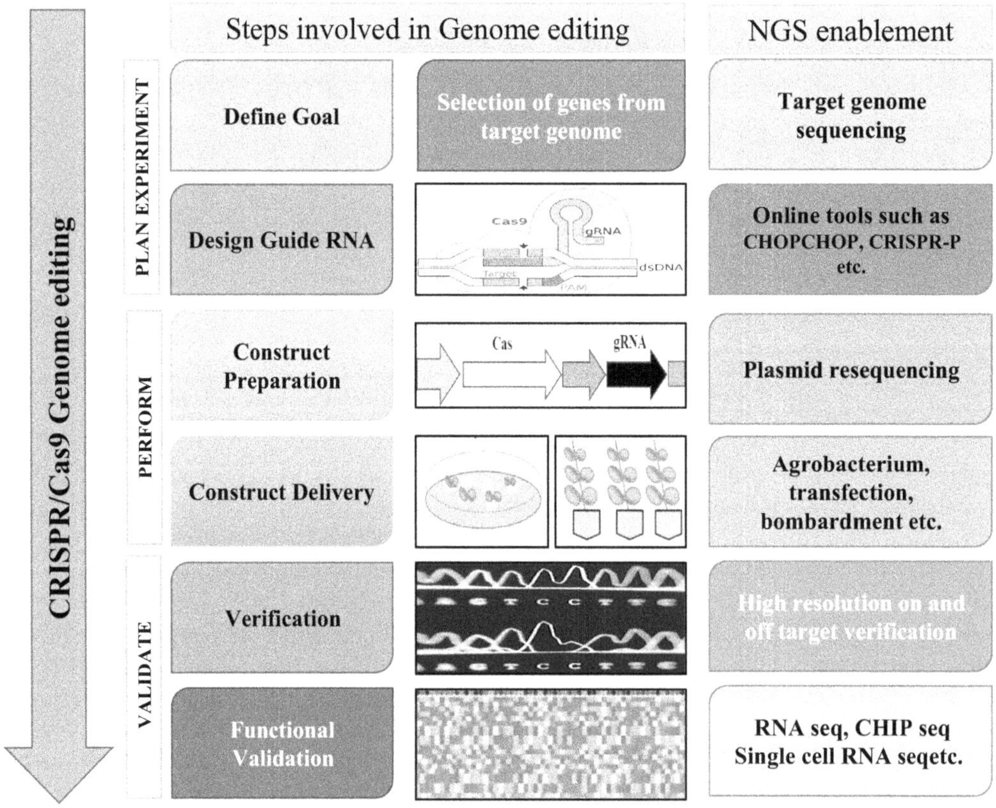

FIGURE 8.3 The CRISPR/cas9 working concept for plant genome editing. (Mushtaq et al., 2021c.)

8.3 CRISPR/Cas9 TECHNIQUE FOR ENHANCING PLANT ABIOTIC STRESS RESISTANCE

CRISPR/Cas9, a robust genetic manipulation technology, has indeed been successfully applied in a variety of taxa, notably bacteria, mammals, and plants. Due to its effectiveness for precise gene editing, it has been employed in crop development which would undoubtedly open up many new avenues. The CRISPR/Cas9 platform has been utilized in model species such as Arabidopsis, Oryza, Nicotiana, and many others, along with crops including wheat, sorghum, maize, soybean, and tomato, respectively (Osakabe and Osakabe, 2017). It is particularly helpful in investigating gene functions by generating loss-of-function plant mutants. Aside from that, some researchers are currently interested in employing CRISPR/Cas9 to improve plant abiotic stress tolerance. Abiotic stress responses in plants involve numerous genes engaged in metabolic, signaling, and regulatory processes. CRISPR/Cas9 technique can be employed for targeting single or many genes in order to improve plant stress tolerance. This approach can be used to create point mutations, insertions, or deletions as well as transcriptional modulation for target-based gene editing (Jain, 2015). The primary considerations for enhancing abiotic stress resistance in plants through CRISPR/cas9 technique include choosing an appropriate promoter for cas9 expression, structure of guide RNA, creation of new allelic variants for abiotic stress-responsive genes, and engineering a suitable and efficient gene delivery method (Osakabe and Osakabe, 2017). Paixão et al. (2019) targeted the *ABRE1* gene to improve drought stress tolerance in Arabidopsis since the drought stress reaction is controlled by abscisic acid-responsive element binding proteins 1 and 2 (ABRE1/ABF2). CRISPR/Cas9 gene editing system was utilized to activate endogenous ABRE1 promoter. Furthermore, a catalytically inactive Cas9 (dCas9) was coupled with histone acetyltransferase (HAT) both of which are crucial for gene transcription. CRISPR/dCas9/HAT in Arabidopsis thaliana increases drought stress tolerance by positively regulating ABRE1. Alfatih et al. (2020) utilized CRISPR/Cas9 technology to develop rice mutants resistant to Paraquat by knocking out a gene, *OsPQT3*. Further, these mutants were reported to have more resistance toward salt as well as oxidative stress. Osakabe et al. (2016) employed the CRISPR/Cas9 technique in Arabidopsis to create novel variants for the *OST2* gene, which plays a crucial role in stomatal conductance which in turn boosts the drought stress resilience. Truncated guide RNA/ Tru-gRNA was utilized for site-specific mutagenesis, and thus no off-target effects were noticed. Because of elevated OST2 gene expression in germ line cells coupled with altered stress responses, Arabidopsis exhibits a heritable stress tolerance.

Barrero et al. (2022) used the CRISPR/Cas9 system for editing the *SPDT* gene (*SULTR3; 4*) (*SULTR-like Phosphorus Distribution Transporter*), chosen to improve phosphorus utilization efficiency and grain quality in Colombian *tropical japonica* upland rice cultivars. The edited rice protoplast was to be used for creating lines with improved efficiency in phosphorus uptake from acidic soils. Alam et al. (2022) generated an adenine base deletion at position 91 in the *OsbHLH024* gene in rice and found it to improve plant survival under salt stress. To improve rice survivability under drought conditions, Park et al. (2022) used CRISPR/Cas9 to identify *OsSAP* (*Oryza sativa Senescence-associated protein*) as a candidate drought response gene for the short breeding cycle in rice and found that the regenerated plants from callus were homozygous for *OsSAP* deletion and exhibited stable expression of CRISPR/Cas9. Wang et al. (2022) identified and characterized the *OsNAC092* gene encoding an NAC transcription factor in rice as a negative regulator of drought tolerance, which when knocked out improved drought resistance in rice. A similar observation was reported by Kim et al. (2023) by knocking out the *OsPUB7* (Plant U-box gene in *Oryza sativa*) gene, which increased drought stress tolerance in rice. CRISPR/ Cas9 system was employed to carry gene knockouts in rice by targeting the PYL1/4/6 (abscisic acid receptor) gene subfamily, which resulted in the production of rice cultivars having remarkable tolerance toward high temperatures (Miao et al., 2018).

CRISPR/cas9 generated mutants in maize for the *ARGOS8* gene resulted in increased yield in drought-prone areas (Shi et al., 2017). CRISPR/cas9 technology was implemented to target the OST2/AHA1 gene as well as AtEF1, which acts as a tissue-specific promoter for cas9 translation in Arabidopsis, which led to greater stomatal responsiveness amid drought conditions (Osakabe and Osakabe, 2017). These results indicate that the CRISPR/cas9 system seems to have the capability to produce drought-tolerant field crops through different allelic variants. Various researchers have attempted to reduce mineral toxicity using CRISPR/cas9 technology, with promising results. Using CRISPR/Cas9 technology, a group of researchers was successful in developing low cesium rice seedlings by inactivating the *OsHAK1* gene which is responsible for the transport of K^+ ions across cellular membranes (Nieves-Cordones et al., 2017). OsARM1 knock-out mutant rice plants were developed using CRISPR/Cas9 and demonstrated improved arsenic tolerance (Wang et al., 2017). CRISPR/Cas9 system was utilized to create rice plants with lower cadmium accumulation by knocking down gene(OsNramp5) which is responsible for the transportation of metals across membranes (Tang et al., 2017). Herbicide-resistant rice was created with the application of Acetolactate synthase (ALS) homologous recombination (Sun et al., 2016). Furthermore, targeted base editing was carried out on the watermelon ALS gene through the CRISPR/Cas9 approach to modify single amino acid in polypeptide and resulting mutants demonstrated improved herbicide resistance (Tian et al., 2018). This method of targeting ALS1/ALS2 resulted in the production of herbicide-resistant maize plants (Svitashev et al., 2015). The implementation of this highly effective genome editing technique to ameliorate the issues produced by diverse abiotic stress conditions is now being investigated. Chen et al. (2021) used CRISPR/Cas 9 technology to enhance tomato plant resistance to drought and rehydration ability by creating a loss of function of SlARF4 (*Solanum lycopersicum* Auxin Response Factor 4) mutant lines. CRISPR/Cas9 edited tomato plants for SlHyPRP1 exhibited higher salt tolerance during germination and vegetative conditions. This gene editing technology was used to specifically eliminate the negative response domains in the SlHyPRP1 (*Solanum lycopersicum* hybrid proline-rich protein 1) showing that this technique can also be used for engineering multidomain proteins (Tran et al., 2021).

Salinity stress tolerance in soybean plants was shown to improve after knocking out six abscisic acid (ABA)-induced transcription repressors (*AITRs*) genes whose products are nuclear proteins with transcription repression activity, using the CRISPR-Cas9 system (Wang et al., 2021). Drought tolerance in soybean (*Glycine max* L.) was shown to improve by CRISPR/Cas9 mediated gene editing of the *GmHdz4* transcription factor (Zhong et al., 2022). In another study, a CRISPR/Cas9 gene-editing system with a single gRNA was used to generate soybean double and triple mutants with the deletion of PYL genes i.e. *gmpyl17/19 and gmpyl17/18/19* respectively. It was observed that the double mutant was less susceptible to abscisic acid than the wild type at the germination stage (Zhang et al., 2022). These genes could be used as potential targets for molecular breeding to generate soybean plants with improved productivity under stressful conditions.

Decades of research on CRISPR/Cas9 technology have revealed that it is a versatile genetic engineering tool used for targeting promoters, repressing genes, creating knockouts for molecular breeding, etc. Being simple and highly efficient, this technique has opened new avenues for functional genetics research for improving crop yield, nutritional efficiency, and response to different kinds of biotic and abiotic stresses. Most importantly it produces transgene-free genetically engineered plants which fall within the domains of GMO regulations and guidelines. In a nutshell, CRISPR/ Cas9 technology is valuable for developing sustainable agriculture systems. Table 8.2 summarizes the recent advancements in developing abiotic resistance in plants through the CRISPR/Cas9 system as discussed above.

TABLE 8.2
CRISPR/Cas9 Driven Gene Editing in Crops to Boost Resilience to Abiotic Stress

Plant	Gene of Interest	Modified Trait	References
Arabidopsis thaliana	ABRE1	Drought stress resistance	Paixão et al. (2019)
	OST2		Osakabe et al. (2016)
	OST2/AHA1, AtEF1		Osakabe and Osakabe (2017)
Citrullus lanatus	ALS	Improved Herbicide resistance	Tian et al. (2018)
Oryza sativa	OsPQT3	Salt and oxidative stress resilience	Alfatih et al. (2020)
	PYL1/4/6	Heat stress tolerance	Miao et al. (2018)
	OsHAK1	Low cesium	Nieves-Cordones et al. (2017)
	OsARM1	Arsenic tolerance	Wang et al. (2017)
	OsNramp5	Low cadmium	Tang et al. (2017)
	ALS	Herbicide resistance	Sun et al. (2016)
	SPDT	Phosphorus uptake in acidic soil conditions	Barrero et al. (2022)
	HLH024	Salt stress resistance	Alam et al. (2022)
	SAP	Drought resistance	Park et al. (2022)
	NAC092	Drought resistance	Wang et al. (2022)
	PUB7	Drought resistance	Kim et al. (2023)
Zea mays	ARGOS8	Drought stress resistance	Svitashev et al. (2015)
	ALS 1, ALS 2	Herbicide resistance	
Solanum lycopersicum	SlARF4	Drought resistance	Tran et al. (2021)
	SlHyPRP1	Salt stress	Chen et al. (2021)
Glycine max	AITR	Salt stress	Wang et al. (2021)
	Hdz4	Drought stress	Zhong et al. (2022)
	PLY17/19	Response to ABA	Zhang et al. (2022)

8.4 CRISPR/Cas9 AND ITS EFFICACY IN GENOME EDITING FOR IMPROVING CROP ABIOTIC STRESS TOLERANCE

Many factors influence the efficiency of CRISPR/Cas9 genome editing. The protein level of the Cas9 structure strongly increases overall editing effectiveness, thus adequate codon optimization for the crop of interest is required. (Zhang et al., 2016). According to the experiments, the desired codon-optimized Cas9 appears to provide improved editing efficiency (Johnson et al., 2015; Xu et al., 2015). Since the 35S promoter has lesser expression in germline cells which makes it more susceptible to transgenic gene silencing, the Ubiquitin promoter is occasionally employed instead of the standard 35S promoter (Jiang et al., 2013). For each plant species, RNA Polymerase III must be identified to facilitate guide RNA expression. Promoter TaU3 outperforms OsU3 and AtU6–26 in maize, but OsU6 promoter outperforms OsU6 in rice (Mikami et al., 2015; Xing et al., 2014). Sometimes it is necessary to co-express multiple guide RNAs and unique U3/U6 Ubiquitin promoters, and the same can be achieved via poly-cistronic tRNA/guide RNA (Ma et al., 2015; Xie et al., 2015). Gene editing may be more successful if guide RNA is structurally altered by intermolecular duplex length extension and mutation in a significantly weaker transcription termination signal (Dang et al., 2015). To carry out effective gene editing in this scenario, it is necessary to select a suitable vector design that assures high Cas9 and guides RNA expression.

8.5 CONCLUSION AND FUTURE THRUSTS

CRISPR/Cas9 is a simplistic, efficient, highly specific, and multiplexable technology. It has upgraded crop breeding practices and boosted hopes for future generation breeding. Until recently, few studies have reported on the resilience and adaptability of the CRISPR/Cas9 system for generating climate-smart crops. Such research suggests that this strategy could be employed in molecular breeding programs to improve crop tolerance toward varying abiotic stresses. Naturally existing elements like genes, promoters, cis-regulatory elements, and epigenetic modifications may permit the editing of regulatory and metabolic networks to govern abiotic stress tolerance in field crops. Thus, CRISPR-Cas-based genome editing has immense promise for producing food plants with long-term climatic resistance (Mushtaq et al., 2021a). The CRISPR/Cas9 gene-editing method has completely changed the way that basic research is done in the fields of agriculture and plants. It now allows for gene knockout, knockin, replacement, point mutation, gene regulation, base editing, and other locus-specific modifications. Further due to its adaptability and endurance, it can also be utilized to develop high-throughput mutant libraries (Chen et al., 2019). Particular sequences of promoters, structural genes, transcriptional factors, and so forth might be taken out from the library and integrated into plants to adjust specific features, resulting in desired new kinds of plants. CRISPR/Cas9 enables the imitation of domestication of wild as well as semi-domesticated plants by changing monogenic traits in wild relatives of crops with polygenic features of interest, as well as providing a source of varied germplasm essential for breeding. For instance, wild-type tomato plants are resilient to some abiotic stresses, unlike currently produced tomato varieties. The use of CRISPR/Cas9 technology enhances the nutritional value and growth habits of domesticated wild tomatoes while maintaining their inherent stress tolerance (Chen et al., 2019; Jain, 2015). Using this approach, future domestication of wild types of many crops may be enhanced to facilitate their growth under challenging circumstances like extreme temperatures, poor nutrient availability in soil, water shortages, and so forth without compromising yield.

REFERENCES

Alam, M.S., Kong, J., Tao, R., Ahmed, T., Alamin, M., Alotaibi, S.S., Abdelsalam, N.R. and Xu, J.H., 2022. CRISPR/Cas9 mediated knockout of the OsbHLH024 transcription factor improves salt stress resistance in rice (*Oryza sativa* L.). *Plants*, *11*(9), p. 1184.

Alfatih, A., Wu, J., Jan, S.U., Zhang, Z., Xia, J. and Xiang, C., 2020. Loss of rice *PARAQUAT TOLERANCE 3* confers enhanced resistance to abiotic stresses and increases grain yield in field. *Plant, Cell & Environment*, *43*(11), pp. 2743–2754. https://doi.org/10.1111/pce.13856

Barrero, L.S., Willmann, M.R., Craft, E.J., Akther, K.M., Harrington, S.E., Garzon-Martinez, G.A., Glahn, R.P., Piñeros, M.A. and McCouch, S.R., 2022. Identifying genes associated with abiotic stress tolerance suitable for CRISPR/Cas9 editing in upland rice cultivars adapted to acid soils. *Plant Direct*, 6(12), p. e469.

Bolotin, A., Quinquis, B., Sorokin, A. and Dusko Ehrlich, S., 2005. Clustered regularly interspaced short palindrome repeats (CRISPRs) have spacers of extrachromosomal origin, *Microbiology*, *151*, pp. 2551–2561.

Bortesi, L. and Fischer, R., 2015. The CRISPR/Cas9 system for plant genome editing and beyond. *Biotechnology Advances*, *33*, pp. 41–52.

Chen, K., Wang, Y., Zhang, R., Zhang, H. and Gao, C., 2019. CRISPR/Cas genome editing and precision plant breeding in agriculture. *Annual Review of Plant Biology*, *70*, pp. 667–697. https://doi.org/10.1146/annurev-arplant-050718-100049.

Chen, M., Zhu, X., Liu, X., Wu, C., Yu, C., Hu, G., Chen, L., Chen, R., Bouzayen, M., Zouine, M. and Hao, Y., 2021. Knockout of auxin response factor SlARF4 improves tomato resistance to water deficit. *International Journal of Molecular Sciences*, *22*(7), p. 3347.

Dang, Y., Jia, G., Choi, J., Ma, H., Anaya, E., Ye, C., Shankar, P. and Wu, H., 2015. Optimizing sgRNA structure to improve CRISPR-Cas9 knockout efficiency. *Genome Biology*, *16*, p. 280.

Deltcheva, E., Chylinski, K., Sharma, C. M., Gonzales, K., Chao, Y., Pirzada, Z.A., Eckert, M.R., Vogel, J. and Charpentier, E., 2011. CRISPR RNA maturation by trans-encoded small RNA and host factor RNase III. *Nature*, *471*, pp. 602–607.

El-Mounadi, K., Morales-Floriano, M.L. and Garcia-Ruiz, H., 2020. Principles, applications, and biosafety of plant genome editing using CRISPR-Cas9. *Frontiers in Plant Science*, *11*, p. 56.

Gasiunas, G., Barrangou, R., Horvath, P. and Siksnys, V., 2012. Cas9-crRNA ribonucleoprotein complex mediates specific DNA cleavage for adaptive immunity in bacteria. *Proceedings of the National Academy of Sciences of the United States of America*, *109*, pp. E2579–E2586.

Ishino, Y., Shinagawa, H., Makino, K., Amemura, M. and Nakata, A., 1987. Nucleotide sequence of the iap gene, responsible for alkaline phosphatase isozyme conversion in Escherichia coli, and identification of the gene product. *Journal of Bacteriology*, *169*, pp. 5429–5433.

Jain, M., 2015. Function genomics of abiotic stress tolerance in plants: A CRISPR approach. *Frontiers in Plant Science*, *6*, pp. 2011–2014. https://doi.org/10.3389/fpls.2015.00375.

Jansen, R., Embden, J.D., Gaastra, W. and Schouls, L.W., 2002. Identification of genes that are associated with DNA repeats in prokaryotes. *Molecular Microbiology*, *43*, pp. 1565–1575.

Jiang, W., Zhou, H., Bi, H., Fromm, M., Yang, B. and Weeks, D.P., 2013. Demonstration of CRISPR/Cas9/sgRNA-mediated targeted gene modification in Arabidopsis, tobacco, sorghum and rice. *Nucleic Acids Research*, *41*, p. e188. https://doi.org/10.1093/nar/gkt780.

Jinek, M., Chylinski, K., Fonfara, I., Hauer, M., Doudna, J. A. and Charpentier, E., 2012. A programmable dual-RNA-guided DNA endonuclease in adaptive bacterial immunity. *Science*, *337*, pp. 816–821. https://doi.org/10.1126/ science.1225829.

Johnson, R.A., Gurevich, V., Filler, S., Samach, A. and Levy, A.A., 2015. Comparative assessments of CRISPR-Cas nucleases' cleavage efficiency *in planta*. *Plant Molecular Biology*, *87*, pp. 143–156. https://doi.org/10.1007/s11103-014-0266-x.

Kim, M.S., Ko, S.R., Jung, Y.J., Kang, K.K., Lee, Y.J. and Cho, Y.G., 2023. Knockout mutants of OsPUB7 generated using CRISPR/Cas9 revealed abiotic stress tolerance in rice. *International Journal of Molecular Sciences*, *24*(6), p. 5338.

Kumar, R.R., Goswami, S., Singh, K., Rai, G.K. and Rai, R.D., 2013. Modulation of redox signal transduction in plant system through induction of free radical/ROS scavenging redox-sensitive enzymes and metabolites. *Australian Journal of Crop Science*, *7*(11), pp. 1744–1751.

Ma, X., Zhang, Q., Zhu, Q., Liu, W., Chen, Y., Qiu, R., Wang, B., Yang, Z., Li, H., Lin, Y. and Xie, Y., 2015. A robust CRISPR/Cas9 system for convenient, high-efficiencymultiplex genomeediting in monocot and dicot plants. Molecular Plant. https://doi.org/ 10.1016/j.molp.2015.04.007.

Miao, C., Xiao, L., Hua, K., Zou, C., Zhao, Y., Bressan, R.A. and Zhu, J.K., 2018. Mutations in a subfamily of abscisic acid recepto genes promote rice growth and productivity. *Proceedings of the National Academy of Sciences of the United States of America*, *115*, 6058–6063. https://doi.org/10.1073/pnas.1804774115.

Mikami, M., Toki, S. and Endo, M., 2015. Parameters affecting frequency of CRISPR/Cas9 mediated targeted mutagenesis in rice. *Plant Cell Reports*, *34*, pp. 1–9. https://doi.org/10.1007/s00299-015-1826-5.

Mojica, M.J., Juez, G. and Rodríguez-Valera, F. 1993. Transcription at different salinities of *Haloferax mediterranei* sequences adjacent to partially modified *Pst*I sites. *Molecular Microbiology*, *9*(3), pp. 613–621.

Mushtaq, M., Ahmad Dar, A., Skalicky, M., Tyagi, A., Bhagat, N., Basu, U., Bhat, B.A., Zaid, A., Ali, S., Dar, T.U.H. and Rai, G.K., 2021a. CRISPR-based genome editing tools: Insights into technological breakthroughs and future challenges. *Genes*, *12*(6), p. 797.

Mushtaq, M., Dar, A.A., Basu, U., Bhat, B.A., Mir, R.A., Vats, S., Dar, M.S., Tyagi, A., Ali, S., Bansal, M. and Rai, G.K., 2021b. Integrating CRISPR-Cas and next generation sequencing in plant virology. *Frontiers in Genetics*, *12*, p. 735489.

Mushtaq, M., Rai, G.K., Kumar, R.R. and Basu, U., 2021c. CRISPR/Cas genome editing to modify abiotic stress responses in plants. In *Plant Abiotic Stress Tolerance: Physiochemical and Molecular Avenues* (edited by Gyanendra Kumar Rai and RR Kumar, ISBN: 9788193243657). Deepika Book Agency, Delhi, India.

Nemudryi, A.A., Valetdinova, K.R., Medvedev, S.P. and Zakian, M., 2014. TALEN and CRISPR/Cas genome editing systems: Tool of discovery. *Acta Naturae*, *6*(22), pp. 19–40.

Nieves-Cordones, M., Mohamed, S., Tanoi, K., Kobayashi, N.I., Takagi, K., Vernet, A., Guiderdoni, E., Périn, C., Sentenac, H. and Véry, A.A., 2017. Production of low-Cs+ rice plants by inactivation of the K+ transporter OsHAK1 with the CRISPR-Cas system. *The Plant Journal*, *92*, pp. 43–56. https://doi.org/10.1111/tpj.13632.

Osakabe, Y. and Osakabe, K., 2017. Genome editing to improve abiotic stress responses in plants. *Progress in Molecular Biology and Translational Science*, *149*, pp. 99–109. https://doi.org/10.1016/bs.pmbts.2017.03.007.

Osakabe, Y., Watanabe, T., Sugano, S.S., Ueta, R. and Ishihara, R., 2016. Optimization of CRISPR/Cas9 genome editing to modify abiotic stress responses in plants. *Scientific Reports*, *6*(1), p. 26685. https://doi.org/10.1038/srep26685.

Paixão, J.F.R., Gillet, F., Ribeiro, T.P., Bournaud, C., Lourenço-tess, I.T., Noriega, D.D., De Melo, B.P., De Almeida-engler, J. and Grossi-de-sa, M.F., 2019. Improved drought stress tolerance in Arabidopsis by CRISPR/dCas9 fusion with a Histone AcetylTransferase. *Scientific Reports*, *9*(1), p. 8080. https://doi.org/10.1038/s41598-019-44571-y.

Park, J.R., Kim, E.G., Jang, Y.H., Jan, R., Farooq, M., Ubaidillah, M. and Kim, K.M., 2022. Applications of CRISPR/Cas9 as new strategies for short breeding to drought gene in rice. Frontiers in Plant Science, *13*, p. 850441.

Sedeek, K.E.M., Mahas, A. and Mahfouz, M. 2019. Plant genome engineering for targeted improvement of crop traits. Frontiers in Plant Science, *10*, p. 114.

Shi J, Gao H, Wang H, Lafitte HR, Archibald RL, Yang M, Hakimi SM, Mo H, Habben JE. 2017. ARGOS8 variants generated by CRISPR-Cas9 improve maize grain yield under field drought stress conditions. *Plant Biotechnology Journal*, *15*(2), pp. 207–216.

Sun, Y., Zhang, X., Wu, C., He, Y., Ma, Y., Hou, H., Guo, X., Du, W., Zhao, Y. and Xia, L., 2016. Engineering herbicide-resistant Rice plants through CRISPR/Cas9-mediated homologous recombination of acetolactate synthase. *Molecular Plant*, *9*, pp. 628–631. https:// doi.org/10.1016/j.molp.2016.01.001.

Svitashev, S., Young, J.K., Schwartz, C., Gao, H., Falco, S.C. and Cigan, A.M., 2015. Targeted mutagenesis, precise gene editing, and site-specific gene insertion in maize using Cas9 and guide RNA. *Plant Physiology*, *169*, pp. 931–945. https://doi.org/10.1104/ pp.15.00793.

Tang, L., Mao, B., Li, Y., Lv, Q., Zhang, L., Chen, C., He, H., Wang, W., Zeng, X., Shao, Y., Pan, Y., Hu, Y., Peng, Y., Fu, X., Li, H., Xia, S. and Zhao, B., 2017. Knockout of OsNramp5 using the CRISPR/Cas9 system produces low cd-accumulating indicarice without compromising yield. *Scientific Reports*, 7, pp. 1–12. https://doi.org/10.1038/s41598-017-14832-9.

Tian, S., Jiang, L., Cui, X., Zhang, J., Guo, S., Li, M., Zhang, H., Ren, Y., Gong, G., Zong, M., Liu, F., Chen, Q. and Xu, Y., 2018. Engineering herbicide-resistant watermelon variety through CRISPR/Cas9-mediated base-editing. *Plant Cell Reports*, *37*, pp. 1353–1356. https://doi.org/10.1007/s00299-018-2299-0.

Tran, M.T., Doan, D.T.H., Kim, J., Song, Y.J., Sung, Y.W., Das, S., Kim, E.J., Son, G.H., Kim, S.H., Van Vu, T. and Kim, J.Y., 2021. CRISPR/Cas9-based precise excision of SlHyPRP1 domain (s) to obtain salt stress-tolerant tomato. *Plant Cell Reports*, *40*, pp. 999–1011.

Wang, B., Wang, Y., Yu, W., Wang, L., Lan, Q., Wang, Y., Chen, C. and Zhang, Y., 2022. Knocking out the transcription factor OsNAC092 promoted rice drought tolerance. *Biology*, *11*(12), p. 1830.

Wang, F.Z., Chen, M.X., Yu, L.J., Xie, L.J., Yuan, L.B., Qi, H., Xiao, M., Guo, W., Chen, Z., Yi, K., Zhang, J., Qiu, R., Shu, W., Xiao, S. and Chen, Q.F., 2017. OsARM1, an R2R3 MYB transcription factor, is involved in regulation of the response to arsenic stress in rice. *Frontiers in Plant Science*, *8*, p. 1868. https://doi.org/10.3389/fpls.2017.01868.

Wang, T., Xun, H., Wang, W., Ding, X., Tian, H., Hussain, S., Dong, Q., Li, Y., Cheng, Y., Wang, C. and Lin, R., 2021. Mutation of GmAITR genes by CRISPR/Cas9 genome editing results in enhanced salinity stress tolerance in soybean. Frontiers in Plant Science, *12*, p. 2752.

Xie, K., Minkenberg, B., Yang, Y., 2015. Boosting CRISPR/Cas9 multiplex editing capability with the endogenous tRNA-processing system. *Proceedings of the National Academy of Sciences of the United States of America*, *112*, pp. 3570–3575. https://doi.org/10.1073/pnas.1420294112.

Xing, H.L., Dong, L., Wang, Z.P., Zhang, H.Y., Han, C.Y., Liu, B., Wang, X.C. and Chen, Q.J., 2014. A CRISPR/Cas9 toolkit for multiplex genome editing in plants. *BMC Plant Biology*, 14, pp. 1–12. https://doi.org/10.1186/s12870-014-0327-y.

Xu, R.F., Li, H., Qin, R.Y., Li, J., Qiu, C.H., Yang, Y.C., Ma, H., Li, L., Wei, P.C. and Yang, J.B., 2015. Generation of inheritable and "transgene clean" targeted genome-modified rice in later generations using the CRISPR/Cas9 system. *Scientific Reports*, *5*, p. 11491. https://doi.org/10.1038/srep11491.

Zaidi, S. S., A., Mahas, A., Vanderschuren, H. and Mahfouz, M. M., 2020. Engineering crops of the future: CRISPR approaches to develop climateresilient and disease-resistant plants. *Genome Biology*, *21*, p. 289. https://doi.org/10.1186/s13059-020-02204-y.

Zhang, C., Xu, W., Wang, F., Kang, G., Yuan, S., Lv, X., Li, L., Liu, Y. and Yang, J., 2020. Expanding the base editing scope to GA and relaxed NG PAM sites by improved xCas9 system. *Plant Biotechnology Journal*, *18*, p. 884.

Zhang, D., Li, Z. and Li, J., 2016. Targeted gene manipulation in plants using the CRISPR/Cas technology. *Journal of Genetics and Genomics*, *43*, pp. 251–262. https://doi.org/10.1016/j. jgg.2016.03.001.

Zhang, Z., Wang, W., Ali, S., Luo, X. and Xie, L., 2022. CRISPR/Cas9-mediated multiple knockouts in abscisic acid receptor genes reduced the sensitivity to ABA during soybean seed germination. *International Journal of Molecular Sciences*, *23*(24), p. 16173.

Zhong, X., Hong, W., Shu, Y., Li, J., Liu, L., Chen, X., Islam, F., Zhou, W. and Tang, G., 2022. CRISPR/Cas9 mediated gene-editing of GmHdz4 transcription factor enhances drought tolerance in soybean (*Glycine max* [L.] Merr.). *Frontiers in Plant Science*, 13, p. 988505.

9 CRISPR-Based Genome Editing for Improving Nutrient Use Efficiency and Functional Genomics of Nutrient Stress Adaptation in Plants

Lekshmy Sathee, Sinto Antoo,
Gadpayale Durgeshwari Prabhakar,
Arpitha S R, Sudhir Kumar, Archana Watts,
Viswanathan Chinnusamy, Balaji Balamurugan,
and Asif Ali Vadakkethil

9.1 INTRODUCTION

Since the dawn of agriculture, constant efforts have been made to feed the ever-expanding population. To meet the projected food demands, a 50% increase by 2030 and a cent percent increase by 2050 needed to be achieved in food production (Pastor et al., 2019). For sustained food security, a higher nutrient use efficiency, especially that of nitrogen (N) is of major importance. By 2050, the world population is estimated to surpass 9.7 billion (Billen et al., 2015). It is essential to ensure food and nutritional security with essential micronutrients provided in the diet for a healthy life. Mineral malnutrition is a critical worldwide challenge as nearly one in every two children is micronutrient deficient in developing countries (Connorton et al., 2017). The inadequacy in intake of key micronutrients termed "Hidden Hunger" affects 2 billion people worldwide with 30%–40% of it solely due to iron (Fe) deficiency anemia (Trijatmiko et al., 2016).

The use of genome editing can be to (1) create a knockout mutant of "sensitivity genes", (2) knockdown a negative regulator in the plant's stress response pathway (Feng et al., 2013), (3) enhance the expression, (4) enhance enzyme activity, and (5) improve the stability of positive regulators in the stress response pathways (Sathee et al., 2022). The advent of genome editing after the discovery of meganucleases (Cermak et al., 2011), Zn finger nucleases (ZFNs) (Voytas, 2013), transcription activator-like effector nucleases (TALENs) (Gaj et al., 2013), and clustered regularly interspaced short palindromic repeat (CRISPR) systems (Jinek et al., 2012) has now boosted the pace of crop improvement. CRISPR systems have become the gold standard of genome editing and have outlived ZFNs and TALENs as the former use RNA-DNA interaction to guide the double-strand break (DSB) instead of the latter using protein-DNA interaction to determine the target specificity (Sander and Joung, 2014), making them a simple and efficient tool for genetic manipulation. The Cas nuclease drives specific DSB due to the presence of a nuclease and a DNA-binding domain (Bhullar et al., 2010; Lieber, 2010). The DSB induced in DNA is repaired by either Non-Homologous End-Joining (NHEJ) or Homologous Recombination (HR) in the cell. Sequence changes are caused by error-prone NHEJ or by introducing a modified repair template in the Homology Directed Repair

 DOI: 10.1201/9781003387060-9

(HDR) (Davis and Chen, 2013). The initial discovery of CRISPR/Cas was as an adaptive strategy evolved by bacteria to combat infectious DNA viruses (Ishino et al., 1987). In the natural system, a functional guide RNA (gRNA) is synthesized by the bacteria upon base pairing of CRISPR-RNA (crRNA) and trans-activating CRISPR-RNA (tracrRNA) by the CRISPR loci. In in-vitro, synthetic single-guide RNA (sgRNA) is synthesized by transcription of CRISPR loci. A functional complex is formed between the gRNA/sgRNA and CRISPR-associated nuclease (Cas9) to guide the nuclease to the specific genomic loci with the help of a 20-bp complementation of spacer sequence in the sgRNA and cleave upstream of Protospacer Adjacent Motif (PAM) sequence (Xie and Yang, 2013). Pathak et al. (2008) and Lightfoot (2009) reported that transgenic plants overexpressing genes for nutrient assimilation did not significantly improve nutrient utilization efficiency (NtUE). This article discusses the potential of CRISPR/Cas technology for improving nutrient stress tolerance and NtUE.

9.2 GENOME EDITING FOR IMPROVING NITROGEN USE EFFICIENCY IN PLANTS

Nitrogen plays a crucial role in plants through its involvement in numerous physiological processes like plant growth, foliar expansion, and yield production. Nitrogen is also an inseparable part of important biological molecules like amino acids, nucleic acids, ATP, chlorophyll, and phytohormones which are involved in photosynthesis, carbon and N metabolisms, and the synthesis of proteins (Crawford and Forde, 2002; Frink et al., 1999). Thus, better nitrogen use efficiency (NUE) can aid in better plant performance and crop yields. Generally, NUE is described as the ratio of economic yield to the total amount of N supplied (Masclaux-Daubresse et al., 2010) and can be divided into N uptake efficiency (NUpE) and N utilization efficiency (NUtE). NUpE is the amount of N taken up by plants from all the N sources available in the soil, whereas NUtE is the ability of plants to assimilate and remobilize N in the plant (Good et al., 2004; Moll et al., 1982; Moose and Below, 2009) and NUE can also be described as the product of NUpE and NUtE. Thus, exploiting NUE is an important research area with a scope of improvement in NUpE, NUtE, the interaction of source-sink tissues for N transport, N signaling, and other regulating pathways contributing to plant N status (Moose and Below, 2009).

Prominence has been given to improve NUE in cereals and several targets have been edited by the CRISPR system, especially in rice. NUE can be affected by soil N fluctuations despite the presence of efficient N uptake and signaling mechanisms in plants. It is established that the external nitrate signal is perceived by a functional homolog of AtNRT1.1 in rice called OsNRT1.1B/OsNPF6.5 (Hu et al., 2015). Li et al. (2018a) replaced the *NRT1.1B* gene from japonica rice with an elite and mutated corresponding allele from indica rice by CRISPR/Cas9- donor repair template strategy which uses an HDR mechanism for precise gene replacement. In rice, members of the amino acid transporter family called amino acid permeases (AAPs) localized in the cellular membrane are known to transport certain amino acids from source to sink (Tegeder, 2012). OsAAP3 transports basic amino acids like Lys and Arg (Taylor et al., 2015), and its expression levels were negatively correlated to the number of tillers in rice. When *OsAAP3* was knocked out in *Japonica* ZH11 and KY13 by CRISPR, there was a significant increase in the number of tillers, shoot biomass, grain yield, and NUE as compared to their wild-type controls. It was proposed that OsAAP3 is involved in bud outgrowth elongation and enhances yield by increasing tiller number through alterations in concentrations of different amino acids (Lu et al., 2018). Similarly, a significant increase in the number of tillers, the number of grains per plant, grain yield, and NUE was observed when *OsAAP5* was knocked out in the ZH11 background (Wang et al., 2019). The reduction in transcript abundance of *OsAAP5* increased tiller number by maintaining higher cytokinin levels via a reduction in concentrations of Lys and Arg (basic amino acids) and Val and Ala (neutral amino acids). In rice, N-mediated heading date-1 (Nhd1) also regulates NUE (Zhang et al., 2021b). *Nhd1* knockout mutants by CRISPR had extended phenology leading to higher total N uptake (NUpE) and yield under long-day conditions. The mutants maintained higher and lower levels of N in seeds and vegetative tissues, respectively leading to significantly higher NUtE. Loss-of-function mutation in *Nhd1* increases the expression and enzyme activity of Fd-GOGAT and may reduce the expression of a few amino acid transporters like *OsLHT1* involved in the transport of many N amino acids for N allocation. Thus, both ammonium assimilation and distribution of amino acids from source to sink are enhanced by the inactivation of Nhd1 leading to higher physiological NUtE. Recently, additional factors involved in NUE were identified in rice. The *ABNORMAL CYTOKININ RESPONSE1* or *ABC1* genes encoding for Fd-GOGAT were identified in rice (Yang et al., 2016b). The *abc1-1* is a weak mutant allele with a recessive mutation and the plants showed typical nitrogen-deficiency-like symptoms. These mutants showed retarded growth usually 11–13 days after germination and had reduced height and number of tillers (only one to two) with the symptoms aggravating at later growth stages. Functional characterization of a genetic suppressor of *abc1-1* led to the identification of *are1* (Abnormal Cytokinin Response 1 Repressor 1, *OsARE1*), which partially rescued the nitrogen deficient syndrome of *abc1-1*. It is a highly conserved, chloroplast-localized protein that was later identified in rice as a negative regulator of Fd-GOGAT (Wang et al., 2018). Under low N conditions, the loss of function of *OsARE1* delayed senescence, improved NUE, and increased rice yields by 10%–20% (Wang et al., 2018). The prominent phenotypes of *are1* mutants include delayed heading date by 3–5 days, stay green trait at the grain filling stage, higher accumulation of chlorophyll, higher root-to-shoot ratio, and higher root biomass. Analysis of a panel of 2,155 Asian cultivated rice accessions from the 3,000 Rice Genomes Project revealed a small six base pair insertions in 18% indica and 48% aus accessions in the promoter region of the *OsARE1*. The existing natural variations in the promoter region of *ARE1* were correlated to its expression levels and grain yield proposing ARE1 to be a key mediator of NUE and a promising target for the genetic improvement of NUE in rice. The orthologs of *ARE1* in major cereals like wheat and barley play a similar role. In a hydroponics study, *Taare1* mutants in wheat had enhanced tolerance to N deficiency (Zhang et al., 2021a). The mutant lines of AABBdd and aabbDD showed a delay in foliar senescence, and a significant increase in 1,000-grain weights and grain yield without growth defects compared to the control plants. It also improved grain size by increasing grain length and width in mutants. Wheat *are1* mutants also showed enhanced higher NUE and grain yield as compared to the wild type (Guo et al., 2021). Barley *Hvare1* mutants had increased plant height, delayed senescence, higher tiller number, increased grain protein, and enhanced grain yield (Karunarathne et al., 2022). In *Osare1* mutants, a frameshift mutation in *ARE1* enhanced the grain yield of the Indica variety (Wang et al., 2021). The heading date of *are1–3, are1–4*, and *are1–5* mutants developed in the background of Huanghuazhan (HHZ, indica variety) was delayed by 2–3 days (Wang et al., 2021). Although the grain weight was slightly reduced, the grain yields were increased under both HN and LN conditions by 10.4%–21.8% in *are1–3, are1–4*, and *are1–5* mutants as compared to the HHZ wild-type control.

Several miRNAs are known to control grain yield in cereals. Recently, miR396 was shown to regulate grain yield by targeting growth-regulating factors (GRFs) transcription factors (Kim and Tsukaya, 2015). OsGRF4 controls the development of grains through the regulation of brassinosteroid-induced genes, and grain size and yield were reported to be increased by disruption of the target site of miR396 in *OsGRF4* (Che et al., 2015; Duan et al., 2015). Among eight miR396 gene family members, miR396e and miR396f (miR396ef) members were involved in regulation panicle and seed development. In a hydroponic culture with or without N supply, it was observed that *mir396ef* mutant plants generated by the CRISPR system had significantly higher N than control wild-type plants (Zhang et al., 2020a,b). Under field conditions, these *mir396ef* mutants had a ~15% increase in grain yield and a 25% increase in above-ground dry biomass under N-deficit conditions (0 kg/ha of external N applied). (Zhang et al., 2021) generated DULL NITROGEN RESPONSE1 (*DNR1*) mutants, a quantitative trait locus of NUE which is also involved in auxin

TABLE 9.1
Functions and Effect of Genome Editing/Genetic Manipulation for Nitrogen Use Efficiency

Gene	Functions and Effects of Genome Editing/Genetic Manipulation	Reference
NRT1.1B	Replacement of the japonica rice allele with indica allele using by CRISPR/Cas9- donor repair template strategy created Nitrate transporter NRT1.1B with mutant sites and improved the NUE	Li et al. (2019)
OsAAP3	When OsAAP3 was knocked out in Japonica ZH11 and KY13 by CRISPR, there was a significant increase in the number of tillers, shoot biomass, grain yield, and NUE as compared to their wild-type controls.	Lu et al. (2018)
OsAAP5	The reduction in transcript abundance of OsAAP5 increased tiller number by maintaining higher cytokinin levels via a reduction in concentrations of Lys and Arg (basic amino acids) and Val and Ala (neutral amino acids).	Wang et al. (2019)
OsDNR1	DULL NITROGEN RESPONSE1 (DNR1) mutants showed enhanced nitrate uptake and NR activity in larger biomass, increased grains per panicle, higher yield, and NUE.	Zhang et al. (2021)
Os ARE1/TaARE1/HvARE1	The loss of function of ARE1 delayed senescence, improved NUE, and increased yield.	Wang et al. (2018)
OsNPF6.1^{HapB}	OsNPF6.1HapB is an elite haplotype of nitrate transporter that improves nitrate uptake and yield under low N. Variation in the coding and promoter sequence confers uptake efficacy and improved transactivation by OsNAC42. Making it a good candidate for allele replacement for high NUE.	Tang et al. (2019)
OsGS2/OSGRF4	CRISPR/Cas9 mediated editing of miR396 binding site in GS2 generated a gain of function GS2^{E} alleles with improved yield and nitrogen use efficiency.	Wang et al. (2022)
SEEDLING BIOMASS 1	These mutants had significantly larger biomass, increased grains per panicle, higher yield, and NUE.	Xu et al. (2021)

homeostasis. The mutants had a disrupted DNR1 due to a 2-bp deletion in the second exon. The *dnr1* mutants had increased plant height, reduced the number of tillers, and increased IAA content. The absence of functional DNR1 enhanced nitrate uptake and NR activity as compared to ZH11 wild-type plants. *SEEDLING BIOMASS 1* QTL on chromosome 1 control rice seedling biomass. Xu et al., (2021) generated *sbm1* CRISPR knockout mutants in the background of Nipponbare. These mutants had significantly larger biomass, increased grains per panicle, higher yield, and NUE.

9.3 GENOME EDITING FOR IMPROVING PHOSPHORUS UE AND FUNCTIONAL GENOMICS

Phosphorus (P) plays a key role in the growth and development of plants by contributing to various metabolic activities in different forms. Moreover, P governs modern biochemical functions and controls the growth of ecosystems (López-Arredondo et al., 2014). Phosphorus as a macro phytonutrient forms an integral part of cell walls and aids important functions such as photosynthesis, metabolism, translocation of sugars, and the transformation of various metabolites (López-Arredondo et al., 2014). Phosphorus takes part in various biochemical pathways and is used in the biosynthesis of many vital biomolecules (Gojon et al., 2009). Increased use of P fertilizers has contributed to enhancements in crop production and food security (Kirkby and Johnston, 2008). Only inorganic P (Pi) can be readily taken up and utilized by plants, mostly from topsoil. It leads to a major disposal of applied fertilizers where the plants are unable to consume them effectively. This ultimately results in economic loss and environmental pollution. Phosphorus resources are getting depleted globally due to the extensive application, and the deficiency can significantly affect yield potential.

To address these issues, the P acquisition efficiency (PAE) and P utilization efficiency (PUE) of plants need to be improved, which are inherently linked to the uptake, transport, and metabolism of Pi in plants with proper recycling and homeostasis balance (Prathap et al., 2022; Zeeshan et al., 2020). The permanent solution is to perform genome editing strategies to enhance PUE, and it also aids in deciphering the functional genomics of phosphorus-related genes and pathways. Genome editing strategies are nowadays widely utilized for improving various traits in different plants. There are mainly two strategies for improving PUE, either by the enhancement of intake or by the optimization of usage (Jha et al., 2023; Svistoonoff et al., 2007).

To cope with the limited P-induced stress, plants have developed various strategies. Their cells have been shown to have undergone major changes in how they function, including changes in their metabolic activities, energy consumption, and levels of hormone production. Phosphorus uptake primarily occurs through the roots. Plants can directly acquire P from phytate, although plants have a very limited ability to obtain P from other organic sources outside the roots (Richardson et al., 2011). In situations where phosphorus deficiency is present, modifications in root structure become vital to optimize efficient absorption and metabolic functions ((Pang et al., 2018; Sathee et al., 2022). Certain plant species have evolved a mechanism to increase the accessibility of Pi by excreting organic acids. These organic acids aid in the release of Pi from various Pi-containing complexes in the soil that otherwise remain untouched by plants (Bello, 2021). For example, the Al-activated Malate Transporters (ALMTs) family of transporters has a significant impact on phosphorus uptake in soybeans and contributes to the maintenance of K and Fe homeostasis in lupin through the regulation of malate exudation in acidic soil conditions (Li et al., 2017a; Xu et al., 2020). Additionally, the induction of Pi transporters becomes necessary to enable the utilization of P in the form of phospholipids (Devaiah et al., 2009). Furthermore, under conditions of P deficiency, certain genes that are known to be responsive to such conditions, including IPS1, RNS1, PLDZ2, Pht1;4, and At4, exhibit significant upregulation (Hillwig et al., 2008; Meng et al., 2012; Nagarajan et al., 2011; Park et al., 2014).

Phosphate transporters present a promising opportunity for the development of genetically modified crops with higher PAE. PHT and PHO are gene families that encode Pi transporters which enable the acquisition of P and maintain homeostasis in plants. PHT1 and AtPHT1 are transporter proteins that primarily function under low concentrations of Pi (Guo et al., 2008, 2013; Mudge et al., 2002). AtPHT1;5 is involved in transporting Pi from the source to the receiving organ based on developmental signals and P levels (Nagarajan et al., 2011). AtPHT1;9, on the other hand, plays a crucial role in Pi absorption at the root-soil interface and facilitates the movement of Pi from the root to the shoot (Bari et al., 2006). Another transporter, AtPHT2;1, is located in the inner shell membrane of the chloroplast and serves as a low-affinity Pi transporter, ensuring the proper distribution of Pi within plants (Wang et al., 2017). Additionally, AtPHT2;1 is involved in transferring Pi to the xylem vessels in the roots (Stefanovic et al., 2007). The PHO1 transporter, belonging to the SPX-EXS subfamily, regulates root-to-shoot Pi transport, while VPT1 and OsSPX-MFS1-3, from the SPX-MFS subfamily, regulate Pi transport within the vascular vacuoles (Ruan et al., 2018). AtPHO2 is responsible for encoding E2-bound ubiquitin, which plays a crucial role in regulating P levels in shoots, and preventing excessive accumulation of phosphate ions (Bari et al., 2006). Furthermore, it was discovered that the transcriptional expressions of these genes in Arabidopsis vary depending on the internal and external P states, suggesting that their activity is influenced by the plant's P status (López-Arredondo et al., 2014) The knockout of the SlPHO1.1 gene in tomatoes using CRISPR/Cas9 led to a Pi starvation response characterized by reduced shoot weight, enhanced root biomass, increased root-to-shoot ratio, increased anthocyanin accumulation, and higher soluble Pi content (Zhao et al., 2019). Further exploration through CRISPR editing of the six PHO1 genes (SlPHO1;1-SlPHO1;6) in tomatoes could potentially provide insights into their specific roles. In Arabidopsis and rice, the development of root hairs and homeostasis maintenance are found to be regulated by the transcription factors PHR1 and PHR2. SPX1/2 and SPX4 were found to inhibit AtPHR1/OsPHR2, which is a positive regulator of Pi metabolic signal pathways and acquisition.

The NIGT1-SPX-PHR cascade regulates Pi uptake and signaling in response to nitrogen availability (Ueda et al., 2020). Under low Pi conditions, the CRISPR/Cas9 generated *nigtl.1 nigtl.2* double mutant showed lower P uptake and higher N uptake (Wang et al., 2020). Downregulation of different negative regulators of Pi signaling, such as AtWRKY6, AtWRKY43, IPS1/2, PHO2, miR827, etc., and upregulation of various positive regulators, such as AtPHR1, At PHR2, AVP1/AVP1D, Cm-PAP10.1, Cm-PAP10.2, Cm-RNS1, MFSs, OsSPX, PAP, PHT1, PHO1, SQD2, TaALMT1, VPT, etc., can effectively improve PUE in plants (Hasan et al., 2016; Ruan et al., 2018).

The initial transcription factor identified in connection with P deficiency is PHR1. PHR1 is an MYB-type transcription factor that specifically reacts to P deficiency at the posttranscriptional level. It plays a crucial role in regulating genes that are involved in the response to P deficiency (Rubio et al., 2001) PHRs are involved in the control of miR399 expression, leading to a decrease in the transcript levels of PHO2, a protein that negatively regulates signaling pathways associated with P deficiency (Baek et al., 2013). Rice mutants created through CRISPR/Cas9 technology, specifically targeting the genes phrl, phr2, and phr3, exhibited notable decreases in plant growth (Ruan et al., 2019). The overexpression of OsPHR3 resulted in enhanced growth in environments with low Pi levels suggesting that OsPHR3 plays a positive role in Pi signaling and maintaining homeostasis (Sun et al., 2018). Apart from the PHR transcription factors, several others play a role in Pi signaling, including AtMYB2, AtMYB62, OsMYB1, OsMYB2, OsMYB2P-1, OsMYB4, OsMYB4P, OsMYB5, OsPTF1, OsARF1, OsWRK, GmERF1 and GmWRKY6 (Baek et al., 2013; Devaiah et al., 2009; Wang et al., 2014). Mutation of the MYB1 transcription factor using the CRISPR/Cas9 system resulted in an enhancement in Pi uptake and accumulation. Additionally, the mutation of MYB1 also had an impact on the expression of several genes associated with Pi starvation signaling and Pi transporters, including IPS, miR399j, PHT1;2, PHT1;9, PHT1;10, and PHO2.1 (Gu et al., 2017). Double mutant of *Osvpel Osvpe2,* vacuolar Pi efflux transporters developed using CRISPR/Cas9 exhibited elevated levels of vascular Pi content when subjected to low Pi stress, in comparison to the wild-type plants (Xu et al., 2019). A transcription factor, Zn finger C2H2, was discovered that responds to low P levels by acting at the transcription level within the nucleus (Devaiah et al., 2007). The main helix transcription factor OsPTF1 in the bHLH gene family was identified as being responsible for conferring resistance to P stress in rice (Yang et al., 2016a). The Purple Acid Phosphatase (PAP) enzymes play a crucial role in hydrolyzing phosphate monoesters to produce Pi and the genes responsible for encoding these enzymes exhibit a robust induction in response to P deficiency. Pi absorption is improved, and biomass growth and production are increased by overexpressing PAP. The gene Phosphorus Starvation Tolerance 1 (PSTOL1) encodes a protein kinase that is specific to P. This gene has demonstrated significance in enhancing grain yield under low-P soil conditions and stimulating early root growth in different rice cultivars, thereby conferring tolerance to P deficiency (Gamuyao et al., 2012). So far, several target genes associated with P acquisition and utilization are available, including those responsible for encoding transport proteins. These genes, along with those involved in organic acid production, have shown potential for enhancing P uptake and assimilation. However, it is important to note that excessive P levels resulting from the manipulation of Pi transporter genes can lead to growth retardation (Aung et al., 2006). As more and more genetic factors governing the PUE are being revealed, biotechnology strategies can be effectively deployed for further crop improvement and understanding their functional characterization.

9.4 GENOME EDITING FOR IMPROVING POTASSIUM UE AND FUNCTIONAL GENOMICS

Potassium (K) is an essential macronutrient that plays a vital role in plant cell physiology for enzymatic activity, production of turgor, cellular growth, osmotic regulation, control of membrane electric potential, and pH homeostasis. K^+ concentration in soil solution or at the interface of roots and soil varies between 0.1 and 20 mM while high K^+ concentration is maintained in the living plant cells which is around 80–100 mM. This high concentration of cytoplasmic K^+ is relatively stable and utilized to carry out cytoplasmic enzymatic activities. K is taken up from the soil against the concentration gradient with the help of cortical and root epidermal cells. K+ can be stored in vacuoles for osmoregulation or can be transferred from the root symplast to the shoot via the xylem. The transport of K ions between the cytoplasm and vacuoles plays a crucial role in maintaining K^+ homeostasis within plant cells. K^+ transporters and K^+ channels are involved in the transport or uptake of K from the external environment to living plant cells (Wang and Wu, 2015, 2013).

The regulation of K use efficiency (KUE) involves an intricate gene network associated with the efficiency of uptake and utilization. The uptake of K (is primarily influenced by root morphology and distribution in soils, as well as the net influx of K^+ into the roots. On the other hand, utilization efficiency is connected to the translocation of K^+ into various organs, the ability to maintain optimal cytosolic K^+ concentration, and an increased capacity to substitute sodium (Na+) with K^+. *Arabidopsis* thaliana has been found to possess a total of 71 K^+ channels and transporters, categorized into six gene families (Gierth et al., 2005; Wang and Wu, 2013). The transporter family HAC/KUP/KT is important for K^+ transport in plants. They comprise three transporter families (KUP/HAK/KT, HKT, and CPA families) and three channel families (Shaker, TPK, and Kir-like families) where AKT1 has been established as the key inward-rectifying Shaker K+ channel involved in plant K acquisition. It is primarily expressed in the epidermis of *Arabidopsis* roots and plays a crucial role in facilitating K^+ uptake from the soil into root cells. Three K^+ transporters, namely OsHAK1, OsHAK5, and OsHAK21, have been identified and characterized in rice for their role in facilitating the transport of K+ into the root. In addition, tonoplast-located NHX transporters, namely AtNHX1 and AtNHX2, function as $Na^+(K^+)/H^+$ antiporters responsible for facilitating K^+ uptake into the vacuole. These NHX antiporters play a crucial role in establishing the vacuolar K^+ pool and are involved in regulating plant tolerance to salt stress, flower development, and stomatal movement (Gierth et al., 2005). Therefore, manipulating the expression of K+ transporters and channels using CRISPR/Cas9 can improve the K use efficiency in plants. This can involve either the upregulation of high-affinity K transporters or the downregulation of low-affinity transporters.

9.5 GENOME EDITING AND POTENTIAL TARGETS TO ENHANCE KUE

Potassium use efficiency can be improved by manipulating the expression of transporters, transcription factors responsible for regulating K^+ homeostasis, and regulatory gene networks to achieve sustainable yield with enhanced quality and stress tolerance. Genome editing using CRISPR/Cas9 is one of the most promising tools to achieve this goal. Genome editing holds the potential to target negative regulators of nutrient signaling, offering a promising avenue for enhancing nutrient uptake and stress signaling in resource-poor conditions. Researchers propose that the AKT1 K channel plays a role in perceiving low K levels in *Arabidopsis* roots, leading to the regulation of root growth. As the external K concentration decreases, the growth of the primary root in wild-type plants is progressively hindered. AKT1 is a crucial K^+ channel that undergoes modulation by several regulators. AtKC1, a subunit of the K^+ channel, interacts with AKT1 to create a heteromeric K^+ channel, which effectively inhibits AKT1-mediated K^+ uptake. Furthermore, the activity of the AKT1/AtKC1 heteromeric K^+ channel is regulated by the interaction of two SNARE (soluble N-ethyl maleimide sensitive factor attachment protein receptor) proteins, SYP121 and VAMP721, with AtKC1. A study by Li et al. (2017b) confirmed the involvement of AKT1 for low K+ stress tolerance by auxin-mediated root growth inhibition in Arabidopsis. In this respect, the role of ATK1 in root inhibition can be explored using the CRISPR/Cas9 editing approach to understand its exact molecular mechanism. In addition, in the same pathway, KC1, a subunit of the K^+ channel can be manipulated using a genome editing approach to unravel the mechanism of ATK1 regulation and root growth. The suppression of AtAKT1-AtKC1 heteromeric channel protein is mediated by Regulatory SNARE protein VAMP721 by targeting vesicles to the plasma membrane (Li et al., 2017b). SNARE proteins can also act as a potential target for genome editing. The transcription factor AtARF2 acts as a repressor for AtHAK5. Under favorable K^+ conditions, it binds to the promoter of AtHAK5 and inhibits its expression. However, in the case of K^+ deficiency, AtARF2 undergoes phosphorylation, causing it to dissociate from the promoter of AtHAK5. This enables the AtHAK5 expression. ARF2 can be a potential candidate for genome editing for altering the expression of HAK5 to enhance K acquisition and utilization in plants. OsHAK3, a K transporter, is required for maintaining K^+ homeostasis in rice under low K and high salinity conditions. Knockout of OsHAK3 by the CRISPR system reduced K+ uptake and increased sensitivity to low K and salinity stress. In another study, CRISPR/Cas9 mediated mutation of K^+ transporter OsHAK1 resulted in reduced Cs (+) uptake by rice plants. Reports suggest K+ homeostasis in Arabidopsis and rice is regulated by activation of AtHAK5 accumulation in the plasma membrane by Raf-like MAPKK kinase (AtILK1) and rice receptor-like kinase OsRUPO. CRISPR/Cas9 approach used for developing OsRUPO knockout lines in rice demonstrated higher levels of K^+ in pollen tubes than the wild-type plants by activating the K^+ transporter, OsHAK1 (Liu et al., 2016).

Xiaohui Mao, along with a team of researchers from Fujian Agriculture and Forestry University in China, spearheaded an investigation into OsPRX2, the rice homolog of At2-CysPrxB, and its impact on rice's tolerance to K deficiency. Their study revealed that OsPRX2 was primarily located in the chloroplast. By overexpressing OsPRX2, the researchers observed an enhanced tolerance to K deficiency. Conversely, when OsPRX2 was knocked out using CRISPR, significant leaf defects and impaired stomatal opening occurred under K-deficient conditions. These findings confirm that OsPRX2 holds promise as a potential target for genetic modification aimed at improving plants' resilience to K deficiency (Mao et al., 2018). Thus, the above-mentioned transporters, transcription factors, posttranscriptional and post-translational regulators, etc., act as regulatory checkpoints from K+ uptake to transport and serve as crucial targets for genome editing for enhancement of crop yield under multiple environmental stress and K+ limited conditions.

To date, there have been no reports of genome editing of any K transporters, channels, or TFs to improve the K use efficiency in plants. OsAKT2 is an exclusively K^+ permeable channel, osakt2 CRISPR mutants were sensitive to salt stress and accumulated more Na^+ than wild type (Tian et al., 2021). CRISPR activation of OsAKT2 can thus improve K nutrition under salinity. Genome editing technologies, such as CRISPR/Cas9, have opened new possibilities for enhancing the traits of plants, including improving their K use efficiency (Mao et al., 2018). K is an essential macronutrient for plants, and enhancing their ability to efficiently utilize K can lead to improved crop yields and reduced fertilizer requirements. Here are some potential targets for genome editing to enhance K use efficiency in plants:

1. **K transporters**: Genes encoding K transporters play a crucial role in the uptake, distribution, and homeostasis of K within plants. Modifying these genes through genome editing can potentially enhance the efficiency of K uptake from the soil and its transport to different plant tissues. For example, editing genes responsible for high-affinity K transporters can increase the uptake of K from low-K environments (Gierth et al., 2005). Genome editing can be used to modify genes encoding K channels and pumps to enhance their activity or specificity. This can lead to improved K uptake, efflux, and compartmentalization within different plant tissues (Adams and Shin, 2014; Ragel et al., 2019).
2. **K sensors and signaling components**: Plants possess mechanisms to sense and respond to K availability in their surroundings. Modifying genes involved in K sensing and signaling pathways can help improve plant responses to varying K levels. This can include genes encoding K sensors, transcription factors, and kinases involved in K signaling cascades (Shin et al., 2005).
3. **Root architecture**: Altering the root system architecture can impact the plant's ability to explore and acquire K from the soil. By modifying genes involved in root growth, branching, or gravitropism, it is possible to enhance the plant's ability to efficiently access K-rich regions of the soil (Templalexis et al., 2021).

4. **K storage and remobilization**: Plants store K in various tissues, including leaves, stems, and roots. Modifying genes involved in K storage and remobilization processes can help improve the plant's ability to retain and utilize K efficiently during periods of limited availability. For example, enhancing the expression of genes involved in K remobilization from senescing leaves to developing tissues can improve K use efficiency (Sardans and Peñuelas, 2012).

To enhance crop KUE, understanding the genetic foundations is crucial for targeted improvements. Numerous studies across different species have been conducted, resulting in the identification of quantitative trait loci (QTLs) linked to plant responses to K^+ deficiency. High-resolution QTLs related to KUE have been discovered, revealing that substituting K^+ with Na^+ is likely a key component in low-K^+ conditions. Although multiple Na+ and K+ transporters are believed to be involved in this process, one of the primary candidates responsible for increased Na+ uptake is OsHKT2.1. Thus, breeding programs can target the OsHKT2.1 transporter and other identified candidates for improving crop performance under low-K+ conditions. It's important to note that the successful implementation of genome editing strategies to enhance K use efficiency in plants requires a comprehensive understanding of the target genes and their functions. Additionally, rigorous testing and evaluation of edited plant lines should be conducted to ensure their safety, efficacy, and compliance with regulatory requirements.

9.6 GENOME EDITING FOR IMPROVING SULFUR UE AND FUNCTIONAL GENOMICS

Sulfur (S) is one of the crucial macroelements in plants. After N, P, and K, it is regarded as the fourth essential element for crop plants. The amino acids like cystine and methionine, glutathione, proteins, vitamins, co-enzymes, prosthetic groups, and secondary metabolites like glucosinolate and sulfoflavonoids all require S for biosynthesis (Hu et al., 2019; Kopriva et al., 2009). In recent years, S deficiency has become a significant issue in plant nutrition; As a result of reduced S dioxide (SO_2) emissions and decreased supply of S through mineral fertilization. To maintain the homeostasis of S and S-derived compounds, plants have developed a network of transporters. To store the sulfur or to route it through metabolic pathways allowing for the synthesis of important S-containing metabolites, specific intra-, and intercellular transporters are needed. Some of the potential targets and strategies for the improvement of sulfur use efficiency (SUE), S uptake efficiency (SUpE; S uptake per S externally applied), and S assimilation efficiency (SUaE; S assimilated to amino acids and proteins per S externally applied) by genome editing tools are discussed below.

Sulfate influx in plants is primarily facilitated by proton/sulfate co-transport systems. The discovery of genes encoding sulfate transporters originated from the functional complementation of yeast strains with sulfate transporter mutations, leading to the identification of the SULTR gene family (Gigolashvili and Kopriva, 2014). These sulfate transporters exhibit structural similarities to solute transporters that are membrane-bound with 12 transmembrane domains. Additionally, at their C termini, they possess STAS (sulfate transporters and antisigma factor antagonists) domains which likely play regulatory roles in controlling transporter activity and membrane localization (Gigolashvili and Kopriva, 2014). Arabidopsis harbors 12 *SULTR* genes, while rice, poplar, *Selaginella moellendorffii*, and *Physcomitrella patens* possess 11, 13, and 5 *SULTR* genes, respectively, within their sequenced genomes (Gigolashvili and Kopriva, 2014; Takahashi et al., 2012). Mutant studies have demonstrated that SULTR1.1 and SULTR1.2, the high-affinity sulfate transporters in Arabidopsis are predominantly expressed in root epidermis, and cortex, root hairs, and their expression is upregulated under sulfate deficiency. Plants lacking both transporters exhibit impaired sulfate uptake at low concentrations and experience significant growth defects. SULTR1;3, a novel sulfate transporter gene, belongs to the family of phloem-specific sulfate transporter found in the roots and participates in the inter-organ movement of S nutrients in vascular plants in

Arabidopsis through the expression of the fusion protein of *SULTR1;3* and green fluorescent protein in transgenic plants (Yoshimoto et al., 2003). Moreover, there are two low-affinity group 2 sulfate transporters located in the vasculature, responsible for inter-tissue sulfate transport. In Arabidopsis, *SULTR2.1* is predominantly expressed in xylem parenchyma and pericycle cells, and its expression is upregulated under S-deficient conditions (Li et al., 2020). According to Kirschner et al., (2018), the positive regulator upstream of SULTR2;2 plays a critical role in transporting SO_4^{2-} into companion cells and could be employed as a synthetic module to control or design activities in Arabidopsis bundle sheath cells and vein cells.

Among the sulfate transporters, SULTR3;5 was the first group 3 transporter to be characterized which modulates the function of SULTR2;1 during the transport of SO_4^{2-} from roots to shoots, but is not involved in sulfate transport itself (Gigolashvili and Kopriva, 2014). In Arabidopsis, SULTR4.1 and SULTR4.2 are present in the vacuole membrane used for transferring stored SO_4^{2-} from the vacuole. Under S-deficient conditions, the expression of SULTR4.1 and SULTR4.2 is upregulated, facilitating the release of stored SO_4^{2-} from the vacuole to meet the plant's S demand. On the other hand, SULTR5.2, classified as a member of group 5, has been identified as a molybdenum transporter and exhibits distinct characteristics compared to other members of the sulfate transporter family (Li et al., 2020).

In comparison to research on other essential plant nutrients, the study of regulatory mechanisms governing S assimilation and uptake has been relatively limited. However, recent advancements in genomic, transcriptomic, and metabolomic research have significantly contributed to our understanding of the physiological and molecular regulatory processes involved in plant responses to S deficiency stress. S assimilation in plants and algae is governed by a combination of transcriptional and posttranscriptional mechanisms. Sulfate assimilation is mainly governed at the stages of sulfate uptake and APS (adenosine 5′-phosphosulfate) reduction, where the expression levels of sulfate transporters and APR (APS reductase) are tightly controlled in relation to sulfate availability, closely revealing the process of sulfate import and reduction (Kusano et al., 2011; Yoshimoto et al., 2007). Furthermore, the regulation of sulfate uptake involves intricate posttranscriptional mechanisms. A key factor in controlling plant S nutrition is the regulation by the cysteine synthase complexes (CSC), which serve as precursors for cysteine biosynthesis and sense the availability of sulfide. Under S-deprived conditions, the sulfide level decreases while OAS (O-acetyl serine) accumulates, triggering the dissociation of the CSC and a reduction in SAT activity. Numerous experimental studies have consistently revealed a strong relationship between sulfate uptake regulation and transporter mRNA levels, especially those belonging to group 1 (Smith et al., 1997; Yoshimoto et al., 2002). Sulfate deprivation, which leads to increased uptake of sulfate, triggers upregulation of transcript abundance for Arabidopsis *SULTR1;1, 1;2, 2;1, 4;1,* and *4;2*. Certain genes responsive to S deficiency in Arabidopsis, such as APR3, are known to be regulated by systemic signals and are thought to be associated with the decrease in OAS (O-acetyl serine) content (Yoshimoto et al., 2003). Transcriptome analysis has revealed that P deficiency induces the upregulation of S transporters in Arabidopsis. The transcription factor PHR1 (Phosphate Response 1), a key player in the P deficiency signaling pathway, is anticipated to play a crucial role in the interaction between P and S nutrient pathways in Arabidopsis. By utilizing *phr* mutants, it has been demonstrated that the regulation of SULTR1.3 by P deficiency is dependent on PHR1, which also inhibits *SULTR2.1* and *SULTR3.4* expression (Rouached et al., 2011).

SLIM1, a member of the ethylene insensitive3-like (EIL) family of transcription factors, is the first identified transcription factor involved in up-regulating S transporters under S-deficient conditions. The slim1 mutant exhibits impaired induction of GFP fluorescence and native *SULTR1;2* transcripts, resulting in reduced sulfate uptake and growth under S-limited conditions. The two major metabolic processes influenced by sulfur nutrition are sulfur assimilation and glucosinolate synthesis (Maruyama-Nakashita, 2017). MYB transcription factors are positive regulators of glucosinolate biosynthesis and were discovered by a genetic screen for Arabidopsis dominant mutants. The first *myb28* knockout mutant created in *B.oleracea* using CRISPR/Cas9 gene editing technology leads to reduced transcript abundance of A-GSL biosynthesis genes and decreased accumulation of glucoraphanin, a methionine-derived glucosinolate in the leaves and flowers of field-grown *myb28* mutant broccoli plants (Neequaye et al., 2021). The 5′-promoter region of *SULTR1;1* contains an auxin response factor (ARF) binding site specific to S response, known as the S-responsive element (SURE). It confers transcriptional regulation specific to S limitation. Although *SLIM1* mRNA remains unaffected by S starvation, the precise mechanisms of its action and its actual binding to the *SURE* element are not yet fully understood (Maruyama-Nakashita, 2017; Maruyama-Nakashita et al., 2005).

miR395, on the other hand, plays a role in regulating S deficiency signals in plants, specifically targeting the sulfate transporter SULTR2;1 for posttranscriptional degradation of target gene transcripts. It accumulates under S starvation conditions and targets three ATP Sulfurylase homologs (*APS1, APS3,* and *APS4*) encoding the APS enzyme, and this process is dependent on SLIM1 (Maruyama-Nakashita, 2017). *Pri-miR408* in Arabidopsis encodes a microRNA-encoded peptide called *miPEP408* that controls the expression of *miR408* and its targets, and the corresponding phenotype. *MiR408* knockout mutants developed by the CRISPR system confirmed tolerance by alteration of phenotype and modulation in the gene expression of the S reduction pathway affecting sulfate and glutathione accumulation. Plants overexpressing *miR408* showed severe sensitivity to biotic and abiotic stresses under low S conditions (Kumar et al., 2023).

9.7 GENOME EDITING FOR IMPROVING MANGANESE UE AND FUNCTIONAL GENOMICS

Manganese (Mn) is one of the inevitable micronutrients that have a role in metabolism including photosynthesis, respiration, hormone signaling, abiotic stress, and biotic stress. Mn availability depends majorly on the pH of the soil. At low pH (acidic) Mn is highly available, resulting in accumulation in plants causing toxicity. In contrast, at high pH (basic), the Mn availability is less due to Mn oxide conversion causing deficiency. Thus Mn deficiency can easily be found in calcareous soil. Contrary to that, Mn toxicity highly occurs in acidic soil. The optimum pH for Mn availability for plants is 5 to 6.5. Mn is required for the catalytic activity of Photosystem II (PSII) and Oxygen evolving complex (OEC) of light reaction and co-factors for several other enzymes such as Mn superoxide dismutase, nucleotide polymerase, phosphoenolpyruvate carboxykinase, oxalate oxidase and isocitrate dehydrogenase (Marschner, 1995). Mn is an environmentally important element as it is a co-factor of an enzyme involved in the photosynthetic oxygen evolution (OEC) which is a prime source in the oxygen cycle. Change in Mn amount positively or negatively affects the plant's growth directly. Mn-deficient plants show interveinal chlorosis, necrotic spots, distorted leaves, and stunted growth. In addition to that, flower development is also impaired under Mn deficiency (Schmidt et al., 2016). High Mn causes toxicity to plants due to the competition with the Fe ions. These two are divalent cations and often involve the same transporter system. A higher concentration of one metal will cause the secondary deficiency of another. Hence Mn uptake and sequestration should be homeostatic in the plants for proper growth. Barley crop shows high affinity and low-affinity uptake of Mn

NRAMP (Natural Resistance-Associated Macrophage protein) involved in the Mn uptake which is a divalent metal carrier protein, transports Fe^{2+}, Mn^{2+}, Zn^{2+}, Cd^{2+}, Co^{2+}, Cu^{2+}, Ni^{2+}, and Pb^{2+}. NRAMP exists in most organisms as a protein family including bacteria, fungi, algae, plants, and animals. The homologous protein to NRMAP1 in humans is DMT1. The selectivity and binding affinity of protein members in the NRAMP family vary based on the conformation. For instance, *Arabidopsis thaliana* contains six members in the *AtNRAMP* family. Of these, AtNRAMP1 shows a high affinity for Mn and its role under Mn deficiency is essential for growth (Cailliatte et al., 2010). Ca signaling plays a role in Mn homeostasis. Ca^{2+}-dependent phosphorylation by CPK21 and CPK23 on NRAMP protein enhances Mn uptake in the Arabidopsis. This CPK21 and CPK23 also activate iron transporter protein IRT1. Under Mn-deficient condition, AtNRMAP3 and AtNRAMP4 are involved in the export of Mn from the vacuoles thus helping to normalize the deficient condition (Lanquar et al., 2010). Overexpression of *AtNRAMP2* enhances plant growth under Mn-deficient conditions (Gao et al., 2018).

Zn-regulated transporter/Fe-regulated transporter-like protein (ZIP/IRT) is a transporter protein, that plays a major role in Fe, Zn, and Cd transport. It also involves the Mn transport in plants to some extent. Arabidopsis AtIRT1 is involved in the transport of Fe, Mn, and Zn. AtIRT2 has more role in Mn transport especially the transport from root to shoot. The knockout line of AtIRT2 shows more tolerance to high Mn conditions (Milner et al., 2013). Barley crop, HvIRT1 plays a role in the transport of Mn and its distribution in the plant system (Long et al., 2018). And this IRT1 in barley is highly expressed in Mn-efficient genotypes (Pedas et al., 2008). *OsZIP6* shows more expression under Mn-deficient conditions (Kavitha et al., 2015).

Arabidopsis contains 11 CAX-like transporters and plays a role in the transport of Ca^{2+}, Mn^{2+}, and Cd^{2+} into the vacuole. In Arabidopsis, YSL mutants show sensitivity to high levels of Mn. The mutants include single mutants *ysl4-2*, *ysl6-4*, *ysl6-5*, and double mutant *ysl4ysl6* (Conte et al., 2013). The sensitivity to high Mn^{2+} denotes their minor role in Mn transport and regulation. OsYSL2 of rice transport the Mn to shoot through phloem transport (Ishimaru et al., 2010). Whereas OsYSL6 has a role in the Mn detoxification in rice under high Mn (Sasaki et al., 2011). ECA3, a member of the P_{2A}-ATPase family in Arabidopsis has an important function in Ca and Mn transport (Mills et al., 2008).

Mn toxicity can be tolerated by the plant through compartmentation through sequestering in the vacuoles and Golgi bodies. Mn-CDF transporter, a member of the cation diffusion facilitator (CDF) family has a role in the detoxification of Mn. CDF reverse H^+ transporter thereby exports the Mn^{2+} from the cytoplasm to vacuoles or outside of the cell. Metal tolerance protein 8 (MTP8) classified as Mn-CDF involved in Mn tolerance. Cloning of *Stylosanthes hamata* ShMTP8 in Arabidopsis increases the tolerance to Mn through sequestering in the vacuoles (Delhaize et al., 2003; Pittman, 2005). The homologs to ShMTP8 in Arabidopsis are AtMTP8 and AtMTP11 participate in Mn detoxification (Delhaize et al., 2007; Eroglu et al., 2016). In rice, OsMTP8.1 and OsMTP9 localized on the tonoplast and plasma membrane, respectively, have a role in the Mn accumulation to vacuole and export from the cell. OsMTP11 overexpression in rice increases the tolerance to Mn (Ma et al., 2018). When the plant is under high Mn, the Ca signaling activates this MTP transporter through Ca-dependent protein kinase (CPK). This signaling protein also phosphorylates the serine group at the 20th, 22nd, and 24th positions in the N terminal of NRAMP1(Castaings et al., 2021). Zhang et al., 2023 reported the phosphorylation is mediated by two calcineurin B-like proteins CBL 1/9 and CIPK23 kinase at Ser20/22 of NRAMP1. This induces the clathrin-mediated endocytosis of NRMAP1 and reduces the distribution on the plasma membrane, thereby reducing the effect of Mn toxicity. The OsNRAMP3 under high Mn conditions degrade within a short time to prevent the Mn toxic effects on the young leaves by preventing transport. Similarly, cation exchanger, antiporter protein AtCAX2 and AtCAX4 Ca^{2+}/H^+ in Arabidopsis involved in accumulating Mn in vacuoles. An increase in Mn toxicity can be found by overexpressing the P2A type ATPase, AtECA1 (Wu et al., 2002).

The selectivity of transporter for the proteins can vary by change in single amino. For example, CAX2 and CAX1 in Arabidopsis and rice, respectively, can eliminate the Mn^{2+} transport retaining the Ca^{2+} by a change in an amino acid sequence (Shigaki et al., 2003). The above-mentioned transporter can be Mn^{2+} selective by amino acid change (Pittman et al., 2004). At low Mn conditions, NRAMP1 of Arabidopsis plays a role in the Mn uptake (Cailliatte et al., 2010). Phosphatidyl inositol 3-phosphate binding protein, AtPH1 determines the localization of this NRAMP (Agorio et al., 2017). MntH, an NRAMP protein in *S.typhimurium* and *E. coli* involved in the selective uptake of Mn transport. Analyzing several site-directed mutagenic studies of the NRAMP protein family shows the transmembrane domains 4 and 6 are highly essential for transport activity. Mutation at the TM6 domain of *E.coli* NRAMP affects Mn transport activity. DraNramp of, a high affinity Mn uptake protein found in *Deinococcus radiodurans* whose post-translation modification

TABLE 9.2
Targets for Genome Editing/Genetic Manipulation for Improving Manganese Content

Plant	Protein Family and Number of Members	Protein	Localization	Function	Reference
Arabidopsis thaliana	Natural Resistance Associated Macrophage protein (AtNRAMP) (6)	AtNRAMP1	Root plasma membrane	High-affinity Mn transporter and low-affinity Fe uptake. Expressed normally in roots and increased under Mn-deficient condition	Curie, (2000); Cailliatte et al. (2010); Zhang et al. (2023)
		AtNRAMP3	Vacuole	Essential for seed germination under low Fe	Lanquar et al. (2010)
		AtNRAMP4	Vacuole	Essential for seed germination under low Fe	Lanquar et al. (2010)
		AtIRT1	Root epidermal plasma membrane	Transport diverse cations including Mn2+	Korshunova et al. (1999); Krausko et al. (2021)
	Zinc-regulated transporter/iron regulated transporter-like Protein (AtZIP/IRT)	AtZIP1	Roots and shoots	Expression increased in shoots at Mn deficiency	Milner et al. (2013)
		AtZIP2	Roots and shoots	Expression decreases in shoots under Mn deficiency. Knockout lines show more tolerance to high Mn	Milner et al. (2013)
	P2A-type ATPase genes (4)	AtECA3	Golgi	Involves in Mn uptake and transport	Mills et al. (2008); Wu et al. (2002)
	Metal tolerance protein (AtMTP)-12	AtMTP8	Tonoplast	Accumulate Mn and Fe in the seeds and prevent the seed from Mn toxicity. Sequestrate Mn in the root vacuoles under high Mn	Höller et al. (2022)
		AtMTP11	Trans Golgi and pre vacuolar compartment	Transfer Mn and play a role in excess Mn	Delhaize et al. (2007)
	Cation exchanger (AtCAX) (28)	AtCAX2	Vacuole	Transport Fe, Mn, and Cd	Schaaf et al. (2002)
		AtCAX4, AtCAX5	Vacuole	Transport Mn into vacuoles	Schaaf et al. (2002)

(Continued)

TABLE 9.2 (*Continued*)
Targets for Genome Editing/Genetic Manipulation for Improving Manganese Content

Plant	Protein Family and Number of Members	Protein	Localization	Function	Reference
Rice	OsNRAMP (7)	OsNRAMP3	Vascular bundles	Distributes Mn to young leaves and panicles under Mn	Yamaji et al. (2013); Yang et al. (2013)
		OsNRAMP5	Exodermis and endodermis of root	Transport both Mn and Fe but not Zn	Sasaki et al. (2012)
		OsZIP6	Plasma membrane	Shows more expression under Mn deficiency	Kavitha et al. (2015)
	Yellow stripe-like protein (OsYSL) (18)	OsYSL2	Phloem cells	Transport Mn and Fe in phloem	Ishimaru et al. (2010); Koike et al. (2004)
		OsYSL6	Plasma membrane	Play a role in Mn toxicity tolerance	Sasaki et al. (2011)
	OsMTP (11)	OsMTP8.1, OsMTP8.2	Tonoplast	Play a role in Mn toxicity tolerance by accumulating Mn in vacuoles	Chen et al. (2013); Takemoto et al. (2017)
		OsMTP9	Root	Involve in Mn uptake	
		OsMTP11	Golgi	Mn transporter in rice and increases tolerance to Mn toxicity. Compartmentalize Mn and ensure the fertility in rice	Tsunemitsu et al. (2018)
Maize	ZmNRAMP (8)	ZmNRAMP2	Xylem parenchyma and root stele	Transport Mn from root to shoot	Guo et al. (2022)
Barley	HvZIP (16)	HvIRT1	roots	Have a role in Mn uptake and its distribution in the shoot	Long et al. (2018)
Vigna radiata	VrNRAMP (6)	VrNramp5	Exodermis and endodermis	Involves in Mn and Cd transport	Qian et al. (2022)
Camellia sinensis	CsMTP (13)	MTP8, MTP8.2	Plasma membrane	Transport excess Mn^{2+} out of the cell	Li et al. (2017c); Zhang et al. (2020c)
Medicago truncatula	MtMTP (7)	MtZIP4, MtZIP7	-	Transport Mn^{2+}	Lopez-Millan et al. (2004)
Pear	PcMTP (7)	PbMTP8.1	Vacuole and shoot of plants	Increase metal tolerance on expression in Arabidopsis	Li et al. (2021)

affects the conformation and functionality. A point mutation in the alanine in the TM6 domain of DRaNramp disables the Mn transport activity. The His 232 and His 237 of DraNrmp act like a switch between the inward open state and outward open state on the metal binding and release. Among the Mn-uptaking proteins found in the roots, IRT1 is responsible for most of the Mn uptake. The M248 residue determines the selectivity of the substrate in AtNRAMP3. Substitution of Met 248 by Ala showed a decrease in Cd transport and less efficiency in Fe and Mn transport and serine substitution on Met 248 did not transport the Mn and reduction in Fe and Cd transport. Mutation in the G171 of AtNRAMP3 affects the Fe transport without disturbing Mn and Cd (Li et al., 2018a). Out of 19 mutations reported by Li et al (2018b) H250A single substitution affects the Mn transport but not Fe and this similar change in amino acid can be utilized to develop tolerance to Mn toxicity. Thus this amino acid sequence can be modified into required amino acids by making non-synonyms change in the codon using dCas9 base editors and prime editors. MicroRNA also involve in regulating the Mn transporter for example MiR167 in *B. napus* regulates the expression of Nramp1 under Cd stress (Meng). Cloning and overexpression of TaCNR5 in rice and Arabidopsis have increased Mn and Zn tolerance and accumulates Mn and Zn in the rice grains which can be utilized for biofortification (Qiao et al., 2019). In AtCAX2, Mn specificity is conferred by the amino acids 177–186 at the T4 domain (Shigaki et al., 2003).

Regulating Mn uptake and export is a critical system and often gets affected by the Fe concentration. Increasing MnUE needs more selectivity which can be made possible through genome editing and other related techniques. However, increasing the Mn^{2+} uptake and accumulation results in toxicity. Hence there should be an equilibrium between these two which is possible only through adopting proper regulating tools. Changing the expression of genes involved in Mn^{2+} uptake and improving the protein for selectivity by making the changes in the conformation and binding affinity through base editing and prime editing dCas9 coupled base editors and prime editors make them more precise. Further studying the epigenomic profiles of these proteins helps us to understand and modify regulation.

9.8 GENOME EDITING FOR FUNCTIONAL GENOMICS OF CALCIUM SIGNALING

Calcium (Ca) is one of the extremely significant intracellular secondary messenger molecules ubiquitously involved in several signal transduction pathways in plants. An increase in intracellular Ca concentration in plant cells acts as a response to a stimulus perceived by plants. A rise in the cytosolic free Ca^{2+} concentration is observed in response to a variety of physiological stimuli including light, touch, plant hormones, pathogenic elicitors, and abiotic conditions such as cold, drought, and excessive salt. In most cases, Ca^{2+} inflow through channels or Ca^{2+} outflow through pumps causes these Ca^{2+} spikes (Dodd et al., 2010) Genome editing tools like CRISPR/Cas9 a game-changing tool that aids in selectively modifying genes and regulatory elements involved in Ca signaling, thereby unraveling their roles in cellular processes. Several studies have successfully utilized CRISPR/Cas9 to investigate Ca channels, transporters, and modulators, shedding light on their functional genomics and paving the way for finding solutions to overcome environmental challenges. Ca^{2+} signaling encompasses Ca channels (Ca exporters and exchangers), modulators, and regulators. The signaling components include Calmodulins (CaMs), Ca^{2+} dependent protein kinases (CDPKs), Calneurin B-like proteins (CBLs), Ca^{2+}/CaM-dependent protein kinases (CCaMKs) which acts as sensor and responders, the transporters include Cyclic nucleotide-gated channels (CNGCs), Ca^{2+}-ATPases, Ca^{2+}/H^+ exchangers (CAXs), Glutamate-like receptors (GLRs) and mechanosensitive channels (Demidchik and Maathuis, 2007; Dodd et al., 2010; Kudla, 2018).

Ca sensors consist of EF hands, which are highly conserved helix-turn-helix structures. The EF hands are targeted by increased expression of miR1432 which activates signal transduction pathways (Budak et al., 2015). Calmodulin is a key player and sensor of Ca signaling. It is an acidic protein encoded by intron-rich genes and harbors EF-hand. CaM transmits the information to transcription factors that interact with the relevant genes, such as calmodulin-binding transcription activators (CAMTA). CAMTAs (Calmodulin-Binding transcription activators), bZIPs, WRKY IIDs, CBP60s, MYBs, MADs box, and NAC proteins are only a few of the more than 90 TFs that have been classified as Calmodulin-Binding Proteins (CBPs) (Edel, 2017). A study by (Akama et al., 2020) showed truncation of the calmodulin-binding domain of C-terminal OsGAD3 (Oryza sativa Glutamate decarboxylase 3) done using CRISPR/Cas9 resulted in increased concentration (7-fold) of non-protein amino acid GABA (γ-Aminobutyric acid) production. As a signaling molecule, GABA is essential for many aspects of plant growth and development in addition to stress responses. According to Xu et al. (2017), a Ca^{2+}-binding protein called calmodulin-like 41 (CML41) localized in plasmodesmata mediates the closure of plasmodesmata in response to immunogenic bacterial flagellin peptide (flg22). (Kim et al., 2002) proposed mildew resistant locus (MLO) encoded plasma membrane-associated protein in barley has a calmodulin-binding C-terminal domain which is a requisite for powdery mildew infection susceptibility. Loss of function mutants confers resistance to powdery mildew diseases. Aranjuelo et al. (2009) produced transgenic tobacco plants tolerant to high salinity by expressing catalytic and regulatory components of yeast calcineurin. The calcineurin protein, which regulates gene expression and ion transporter activity to assist ion homeostasis in NaCl-stressed cells by changing the Na/K/Ca balance, may be responsible for the improvement in salt tolerance. NtCBP4 (*Nicotiana tabacum* calmodulin-binding protein) can modulate plant tolerance to heavy metals. transgenics expressing NtCBP4 showed hypersensitivity to Pb^{2+} and improved tolerance to Ni^{2+} (Arazi et al., 2002). Ca^{2+}-CaM-regulated kinases (CCaMKs) are involved in the regulation of the formation of root nodules and arbuscular mycorrhiza (Kistner and Parniske, 2002). Ca-dependent protein kinases (CDPKs) are found throughout the plant kingdom and in some protozoans. The structure of CDPKs consists of a Variable N-terminal domain, and functional domains (protein kinase domain, an autoinhibitory region, and a calmodulin-like domain). When Ca binds to a calmodulin-like domain the CDPKs get activated and regulate downstream components of Ca signaling (Schulz et al., 2013).

By using CRISPR/Cas9 to engineer phosphorylation pathways, it may be possible to increase both agricultural productivity and immunity. Threonine (T505) and serine (S512) residues in the C-terminal regions of rice OsCPK18 and its paralog OsCPK4 are phosphorylated by OsMPK5. OsCPK18 and OsCPK4 are effective regulators of plant height and yield-related traits. Further research reveals that OsMPK5 and OsCPK18 regulate genes associated with development independently but collaborate to control genes related to defense (Li et al., 2022). CBL senses Ca and interacts with CIPKs to decode the signals. CBLs have been linked to ABA signaling during stress responses and developmental programming in plants. In the presence of ABA and under conditions of drought, salt stress, or cold stress, the complex formed by the proteins CBL/SCaBP (calcineurin B-like protein/SOS3-like Ca-binding protein) and its interaction partner CIPK/PKS, which is like SnRK3, is recognized to be an essential regulator of signal transduction (Umezawa et al., 2006). Additionally, the CBL-CIPK complex controls low K+ uptake as well as ABA-mediated stomatal responses in Arabidopsis. ABA responses are controlled by the CBL9-CIPK3-ABA repressor 1 (ABR1) pathway both during seed germination and during adult developmental stages. CIPK23 negatively effects nitrate response (Guo et al., 2021). *OsCIPK12, OsCIPK15,* and *OsCIPK03* expressed in rice were shown to be osmotic stress-responsive, overexpression of *OsCIPK03, OsCIPK15,* and *OsCIPK12* improved tolerance to cold, salt, and to drought (Xiang et al., 2007). According to reports, a variety of abiotic stress signals is actively mediated by CMLs (CaM-like proteins). CML9, CML24, and CML37 expression have been linked to phytopathogenic bacteria *Pseudomonas syringae* and the African cotton leafworm *Spodoptera littoralis*. Reports claim that *AtCML24* and *AtCML9*, two CMLs, are also involved in ABA responses and ionic stress, whereas CLM36 regulates ACA8 (Ca^{2+} ATPase) activity. AtCML2 has also been demonstrated to regulate pollen tube development in addition to seed and pollen germination. Like CML42, it has been demonstrated that CML41 affects how plasmodesmata close during plant immunological responses, but CML42 gene deletion led to aberrant trichomes.

The [Ca^{2+}] cyt maintenance is controlled by (1) Ca^{2+}-permeable cation channels, which allow a passive input of Ca^{2+} in time with the electrochemical gradient; (2) Ca^{2+}-ATPases, which regulate the Ca^{2+}/H+ ratio; (3) intracellular substances that buffer [Ca^{2+}] cyt; and iv) exchangers that transfer Ca^{2+}against the electrochemical gradient. CNGCs are regulated through the binding of cyclic nucleotides. CNGCs allow the transport of both monovalent and divalent cations. They are involved in nutrition, transport of divalent heavy metal ions, Ca^{2+} homeostasis, growth of pollen tubes (CNGC 18), tolerance to salinity, cell death, and cold and drought stresses. They are also associated with symbiotic responses in *Medicago truncatula*. And relate to developing rhizobial as well as mycorrhizal symbiotic associations. The P-type Ca^{2+} ATPases use ATP as the energy source to transport Ca^{2+} across the membrane. Low capacity but high-affinity transporters. Ca^{2+} fluxes are produced by Ca^{2+} pumps and channels to mediate Ca^{2+}-dependent ABA signaling in plants. The plasma membrane-localized Ca^{2+}-ATPases and the vacuolar channel activity of TPC1 (Two pore channels) participate in ABA signaling and its effect on ABA-induced stomatal regulation (Peiter et al., 2005).

Ca^{2+}/H^+ exchangers (CAXs) are low-affinity but high-capacity transporters. The cax1 mutant exhibited a reduction in Ca^{2+}/H^+ antiporter activity in tonoplast, V-type H^+ translocating ATPase activity and enhanced tonoplast Ca^{2+} ATPase activity and showed enhanced growth under Mn and magnesium stress (Cheng et al., 2003). In Arabidopsis and tobacco ligand-gated ion channels like GLRs control Ca^{2+} influx across the plasma membrane and the apical [Ca^{2+}] cyt gradient, affecting the development of the pollen tube elucidated through knockout mutants *Atglrl.2* and *Atglr3.7* (Michard et al., 2011). In plants, mechanosensitive channels control Ca^{2+} signals in response to mechanical osmotic stress and membrane tension. These include the two-pore K (TPK) and mechanosensitive-like channel (MSL) families as well as the Mid1-complementing activity channel (MCA) family (Gosh et al., 2021).

The identification of genes and gene families encoding a variety of Ca^{2+} sensors and transporters, as well as their functional analysis in various environments and crop systems, opens opportunities to control how plants react to stress. The application of genome-wide CRISPR screens further expanded our knowledge of Ca signaling networks. Despite challenges, genome editing continues to be a transformative tool for investigating the complexities of Ca signaling and holds immense potential for future discoveries.

9.9 GENOME EDITING FOR IMPROVING IRON UE AND FUNCTIONAL GENOMICS FOR BIOFORTIFICATION

Iron (Fe) being an essential micronutrient is a part of hemoglobin and myoglobin biosynthesis, for the formation of heme-containing enzymes, and other Fe-containing enzymes which are needed for electron transfer and redox reactions (McDowell, 2003). Deficiency of Fe can also lower immunity, reduce work productivity due to tiredness, and impair cognitive development (Murata et al., 2010). On average, a daily Fe intake of 10 mg/day for children (4–8 years), 8 mg/day for adult men (31–50 years), and 18 mg/day for adult women (19–50 years) is recommended (Trumbo et al., 2001). One of the widely considered, long-term, and sustainable approaches to ameliorate micronutrient deficiency is biofortification of staple crops by reducing the concentrations of anti-nutritional factors and other secondary metabolites as cereals are generally rich in mineral ions which are otherwise limited by the presence of these anti-nutritional factors (Raboy, 2009). One of the major factors of such kind in cereals (wheat and rice) is phytic acid, also known as Myo-inositol hexakisphosphate or IP6. In maize, mutants developed for low phytic acid (*lpa*) either had a defect in ABC transporter called multidrug resistance-associated protein (MRP) with unknown function (Shi et al., 2007) or had a mutation in an inositol phosphate kinase gene (Shi et al., 2003) or myoinositol kinase enzyme

(Shi et al., 2005). The *lpa* mutants in barley had higher free phosphate (Pi) and reduced phytic acid with loss in activity of inositol phosphate kinase (Dorsch et al., 2003). It was also observed that the silencing of a putative transporter of phytic acid in wheat called ABCC13 reduced phytic acid levels without changing Fe levels in grains. Previously, wheat inositol pentakisphosphate kinase (*TaIPK1*) was reported to catalyze the last step of phytic acid biosynthesis (Bhati et al., 2014). Targeting *TaIPK1* by RNAi-mediated gene silencing in C306 hexaploid wheat significantly reduced phytate levels, and increased Fe and Zn in mature grains (Aggarwal et al., 2018). Previously, the knockdown of *IPK1* by RNAi in rice also increased seed Fe and Zn concentrations significantly (Ali et al., 2013). Recently, attempts have been made to knock out *TaIPK1* by CRISPR/Cas9 system. The *TaIPK1* knockout lines were generated by targeting the first and second exons of *TaIPK1* and led to the accumulation of Fe and Zn in edited lines as compared to non-edited controls (Ibrahim et al., 2022). There was a 60%–70% reduction in *TaIPK1* transcript abundance and a 1.5–2.1-fold increase in Fe and a 1.6–1.9-fold increase in Zn content in grains suggesting a key role of *TaIPK1* in wheat biofortification. In tomatoes, the SQUAMOSA PROMOTER-BINDING PROTEIN-LIKE (SPL) transcription factor SlSPL-CNR is a negative regulator of Fe-deficiency responses. SlSPL-CNR knockout lines exhibited enhanced Fe-deficiency response (Zhu et al., 2022). Although the success stories of genome editing tools to improve Fe content are currently scarce, identification of various targets related to Fe homeostasis opens new arenas to improve Fe use efficiency to produce Fe-rich crops.

9.10 GENOME EDITING FOR IMPROVING ZN UE AND FUNCTIONAL GENOMICS FOR ZN BIOFORTIFICATION

With the present COVID-19 pandemic, Zn (Zn) shortage can increase infection rates (Trijatmiko et al., 2016), underlining the need for Zn biofortification. Because rice is a staple food for more than half of the world's population, biofortification of Zn in rice (Oryza sativa L.) has the potential to considerably improve global Zn nutritional status (Boonyaves et al., 2017) CRISPR/Cas9 editing of TaIPK1 in wheat significantly reduced phytate levels while boosting grain Fe and Zn content (Ibrahim et al., 2022). The CRISPR-Cas mediated disruption of Triticum aestivum inositol pentakisphosphate 2-kinase 1 (TaIPK1) in wheat produces an increase in Zn accumulation in grains (Ramireddy et al., 2018) transgenic barley with decreased root cytokinin and an increase in root biomass. Transgenic barley plants overexpressing CYTOKININ OXIDASE/DEHYDROGENASE were rich in micro and macronutrients while exhibiting limited soil mobility (P, Mn, and Zn). Brown rice Zn levels were reduced when the OsCKX4 gene was silenced or the OsCKX2 homologs were removed. However, knocking down CYP735A, which is involved in the synthesis of trans-zeatin (tZ-type) cytokinins, using CRISPR/Cas9 raises Zn concentrations (Gao et al., 2019). In Zn-deficient conditions, overexpression of AtZIP1, a Zn transporter, boosted plant growth and Zn transport rates in barley (Ramesh et al., 2004). Similarly, Arabidopsis AtMTP1 and AtZIP1 gene overexpression in cassava increased Zn accumulation in the edible section of the storage root compared to control plants in Zn deprivation (Gaitán et al., 2015). Several other Zn carriers (OsZIP4, AtMHX1, AtMPT1, AhHMA4, ZmIRT1, ZmZIP3, and NcTZN1) have also been over-expressed in rice, Arabidopsis, and Nicotiana tabacum (tobacco) to increase Zn uptake during Zn deficit.

1. QTL analysis was used to identify two alleles of the OsHMA7 transporter gene in rice. Overexpression of allele 284 in transgenic rice lines resulted in increased Fe and Zn accumulation as well as tolerance to iron- and Zn-deficient soils (Kappara et al., 2018). OsHMA7 transcript levels were five times greater in these transgenic strains. Transgenic rice with nicotianamine and 2-deoxymugenic acid (DMA) overexpression accumulates more Fe and Zn in the rice (Banakar et al., 2017). In rice endosperm, these transgenic rice plants could

collect up to 4-fold more Fe and 2-fold more Zn (Zn). A meta-analysis of grain Fe and Zn QTLs in rice and maize found 22 MQTLs in rice, 4 syntenic MQTL-related areas, and 3 MQTL-containing candidate genes in maize. Rice corn orthologs GRMZM2G366919 and GRMZM2G178190 have been identified as naturally occurring macrophage (NRAMP) genes related to resistance. The phylogenetic study indicated these genes as biofortification candidates since they are responsible for the natural fluctuation in Fe and Zn concentrations. The researchers discovered numerous potential genes involved in Fe and Zn homeostasis. These include OsNAS2, OsYSL2, OsIRT2, OsDMAS1, and OsYSL15, which play critical roles in two strategies: Strategy I (acquiring Fe2+ from the soil) and Strategy II (acquiring Fe3+ from the soil), as well as long-distance transport in shoots and loading Fe in seeds. Because of their high accuracy and specificity, editing genes related to Fe and Zn accumulation in rice grain for biofortification offers a significant advantage over existing approaches or technologies (Agarwal et al., 2014; Anuradha et al., 2012).

2. Trijatmiko et al. (2016) discovered NASFer-274, a transgenic event comprising rice nicotianamine synthase (OsNAS2) and soybean ferritin (SferH-1) genes with a single locus insertion that resulted in no yield penalty or altered grain quality. In polished rice grain, the transgenic line had 15 g/g Fe and 45.7 g/g Zn. The Caco-2 cell assay revealed that the grains were bioavailable, with no hazardous heavy metals found (Trijatmiko et al., 2016). Overexpression of genes involved in Zn translocation and mobilization, with enhanced bioavailability of Zn and no yield penalty, is an important strategy to improve grain Zn. The genes of the CDF family are important in Zn absorption and translocation. The ZIP family protein IRT1 plays an important role in Zn absorption in A. thaliana root cells (Korshunova et al., 1999). Overexpression of NA synthase (NAS) via 35S enhancer elements resulted in 2–3-fold increases in Zn content in paddy (Wang et al., 2013). Similarly, transgenic rice expressing the rice actin1 promoter-controlled barley nicotianamine synthase gene HvNAS1 accumulated 2–3-fold more Zn in polished rice grains (Aung et al., 2019). Biofortification of grains with NAS alone or in combination with ferritin has a high potential for fighting global human mineral deficiency (Singh et al., 2017). Two primary QTLs for Zn concentration in grain 1B and 2B were discovered; of these, QTL over 2B was colorized with QTL for grain Fe concentration (Tiwari et al., 2016). A substantial QTL for grain Zn on 2B was also found to be co-located with the QTL for grain Fe concentration. Transgenic techniques have played an important role in Fe and Zn enhancement in cereal crops, although they are limited by poor acceptability and stringent regulatory processes. Precise editing with CRISPR/Cas9 will aid in increasing the Fe and Zn content of cereals while avoiding linkage drag and biosafety concerns. Fe and Zn fortification and bioavailability in grains using next-generation methods. High-throughput phenotyping technologies, such as inductively coupled plasma-mass spectrometry (ICP-MS), X-ray fluorescence spectrometry, and newly developed synchrotron-based X-ray fluorescence microscopy, are also advancing quickly, complementing deep genetic analysis and increasing the efficiency of Fe/Zn gene discovery in hexaploid wheat (Hisano et al., 2021; Qiao et al., 2019).
3. Due to wheat's hexaploidy, the massive genome size (16 Gb), and the abundance of repetitive sequences (>85%), it proved extremely challenging to find causative genes through map-based cloning. Gpc-B1, obtained from *T. turgidum* ssp. microcodes and described as NAM-B1, a NAC family TF, was the first Fe/Zn-related gene extracted prior to the release of the Chinese Spring wheat reference genome sequence. It hastens senescence and boosts the remobilization of Fe, Zn, and nitrogen molecules from source organs to grain. The functional allele is uncommon in recent Australian cultivars and absent in Chinese wheat. The functional Gpc-B1 allele boosted wheat grain Fe and Zn content but caused certain unfavorable features such as senescence, reduced ear number, and low grain weight (Tabbita et al., 2017; Uauy et al., 2006).

4. Comparative genomics is an effective method for identifying and cloning orthologous genes with conserved activities (Valluru et al., 2011). Six genes representing 14 loci have been identified as influencing wheat grain Fe/Zn. Two VIT paralogs, TaVIT1 and TaVIT2, were identified in wheat by homologous cloning and were found in three copies on wheat group 2 and 5 chromosomes, respectively. TaVIT2 regulates vacuolar Fe transport in the endosperm and is effective in wheat biofortification (Connorton et al., 2017). Cell number regulator 2 (TaCNR2) and TaCNR5 found in wheat have conserved domains comparable to plant cadmium resistance proteins (PCR). Furthermore, TaCNR2 and TaCNR5 were identified as factors influencing heavy metal transport and accumulation in plant seeds (Qiao et al., 2019). Furthermore, the wheat ABCC transporters TaABCC13 and TaIPK1, which have copies in each sub-genome, have been implicated in PA biosynthesis. RNAi-mediated TaABCC13 downregulation lowered PA without changing Fe and Zn concentrations in wheat grains, whereas TaIPK1 silencing decreased PA while increasing grain Fe and Zn contents (Aggarwal et al., 2018). Gene editing technologies, particularly developing plant prime editing (PPE), allow for precision gene alteration, such as single base substitution and multiple base insertion/deletion, and so have the potential to enable wheat Fe/Zn biofortification (Lin et al., 2020). Genomic selection (GS) could be a promising strategy for Fe/Zn biofortification, however, one critical element influencing GS accuracy is whether key QTL/gene markers are included in the model as a fixed effect (Heffner et al., 2009).

9.11 CONCLUSIONS

The knowledge of regulators that participate in the signaling, acquisition, export, and redistribution of nutrients in plants is vast. Generally, heterologous systems like yeast and *Xenopus* are used to examine the transport function of plant nutrient transporters. CRISPR/Cas system has now assumed center stage for gene functional validation and characterization. By modifying nutrient transporters and modifying motifs in promoters and transcription factors, we may gain insights into the potential application of the CRISPR/Cas system for enhancing nutrient use efficiency in plants. Cas9 nuclease is directed at a particular genomic region using a specially designed short sgRNA. Three to four bp upstream from the PAM site at the target site where the Cas9 makes a double-stranded break. Double-strand breaks are then repaired by the cell's inherent repair mechanisms, HDR or NHEJ (Miglani, 2017). Frameshift mutations and targeted gene knockouts are caused by random insertions/deletions and indels that are caused by error-prone NHEJ repair (Jinek et al., 2012). In contrast to other genome editing techniques, such as TALENs and ZFNs, CRISPR/Cas9 provides precise and effective targeted modification in the genome of any organism in a straightforward manner. Tables 9.1 and 9.2 both list the various genome editing targets for enhancing NtUE and nutrient stress tolerance. Figure 9.1 depicts the genome editing techniques employable for enhancing NtUE. The CRISPR/Cas9 system has advanced to the point where it can now edit multiple genes simultaneously (Donohoue et al., 2018), and mutations can be targeted to specific areas of coding genes, such as their untranslated regions (Mao et al., 2018), promoter regions (Seth, 2016), microRNAs (Chang et al., 2016), and non-coding RNAs (ncRNAs). The method known as CRISPRi and CRISPRa uses transcription factors coupled to dCas9 to either up or down-regulate the expression of a gene or genes of interest by inhibiting or enhancing transcription by RNA polymerase. With the help of sophisticated genome editing tools and knowledge of the genes controlling nutrient balance in plants, scientists can successfully create plants with specific characteristics. Through the use of CRISPR-Cas technology, high NtUE, resource-efficient plants will be developed, speeding up genetic yield enhancement. The development of non-transgenic crops with improved nutrient use efficiency would increase food security while reducing the use of synthetic fertilizers.

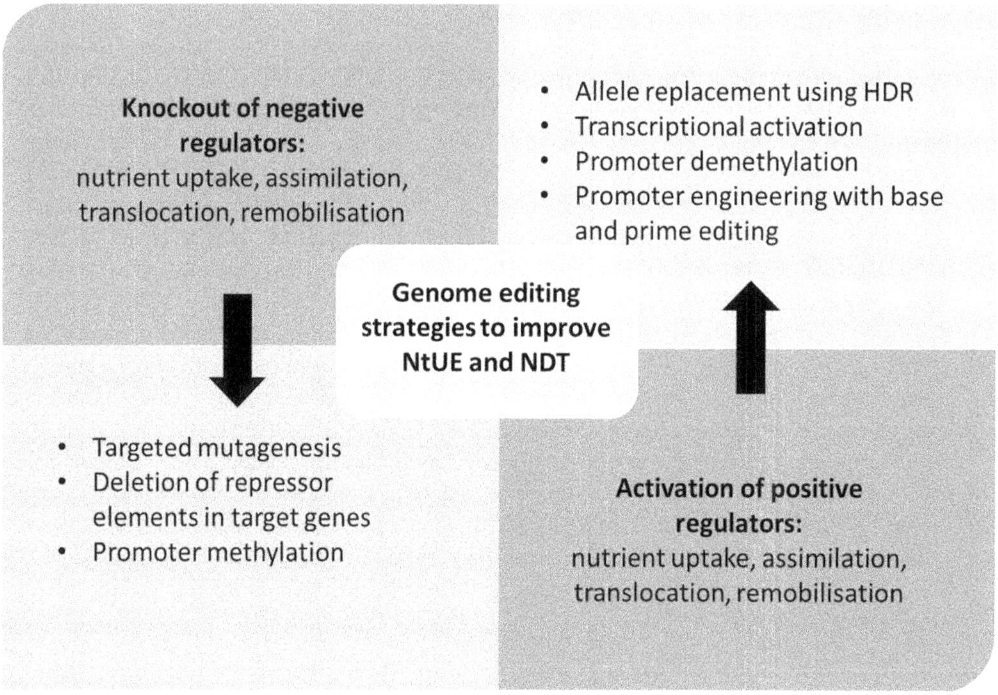

FIGURE 9.1 Genome editing strategies to improve nutrient use efficiency (NtUE) and nutrient deficiency tolerance (NDT).

REFERENCES

Adams, E., Shin, R., 2014. Transport, signaling, and homeostasis of potassium and sodium in plants. *J. Integr. Plant Biol.* 56, 231–249. https://doi.org/10.1111/JIPB.12159

Agarwal, S., Venkata, T.V.G.N., Kotla, A., Mangrauthia, S.K., Neelamraju, S., 2014. Expression patterns of QTL based and other candidate genes in Madhukar×Swarna RILs with contrasting levels of iron and zinc in unpolished rice grains. *Gene* 546, 430–436. https://doi.org/10.1016/j.gene.2014.05.069

Aggarwal, S., Kumar, A., Bhati, K.K., Kaur, G., Shukla, V., Tiwari, S., Pandey, A.K., 2018. RNAi-mediated downregulation of inositol pentakisphosphate kinase (IPK1) in wheat grains decreases phytic acid levels and increases Fe and Zn accumulation. *Front. Plant Sci.* 9, 259.

Agorio, A., Giraudat, J., Bianchi, M.W., Marion, J., Espagne, C., Castaings, L., Lelièvre, F., Curie, C., Thomine, S., Merlot, S., 2017. Phosphatidylinositol 3-phosphate-binding protein AtPH1 controls the localization of the metal transporter NRAMP1 in Arabidopsis. *Proc. Natl. Acad. Sci. U. S. A.* 114, E3354–E3363.

Akama, K., Akter, N., Endo, H., Kanesaki, M., Endo, M., Toki, S., 2020. An in vivo targeted deletion of the calmodulin-binding domain from rice glutamate decarboxylase 3 (OsGAD3) increases γ-aminobutyric acid content in grains. *Rice* 13. https://doi.org/10.1186/S12284-020-00380-W

Ali, N., Paul, S., Gayen, D., Sarkar, S.N., Datta, K., Datta, S.K., 2013. Development of low phytate rice by RNAi mediated seed-specific silencing of inositol 1, 3, 4, 5, 6-pentakisphosphate 2-kinase gene (IPK1). *PLoS One* 8, e68161.

Anuradha, K., Agarwal, S., Rao, Y.V., Rao, K. V., Viraktamath, B.C., Sarla, N., 2012. Mapping QTLs and candidate genes for iron and zinc concentrations in unpolished rice of Madhukar×Swarna RILs. *Gene* 508, 233–240. https://doi.org/10.1016/j.gene.2012.07.054

Aranjuelo, I., Pardo, A., Biel, C., Savé, R., Azcón-bieto, J., Nogués, S., 2009. Leaf carbon management in slow-growing plants exposed to elevated CO2. *Glob. Chang. Biol.* 15, 97–109. https://doi.org/10.1111/j.1365-2486.2008.01829.x

Arazi T, Sunkar R, Kaplan B, Fromm H., 1999. A tobacco plasma membrane calmodulin-binding transporter confers Ni2+ tolerance and Pb2+ hypersensitivity in transgenic plants. *Plant J.* 20(2), 171–82. doi: 10.1046/j.1365-313x.1999.00588.x.

Aung, K., Lin, S.I., Wu, C.C., Huang, Y.T., Su, C.L., Chiou, T.J., 2006. pho2, a phosphate overaccumulator, is caused by a nonsense mutation in a microRNA399 target gene. *Plant Physiol.* 141, 1000–1011. https://doi.org/10.1104/pp.106.078063

Aung, M.S., Masuda, H., Nozoye, T., Kobayashi, T., Jeon, J.S., An, G., Nishizawa, N.K., 2019. Nicotianamine synthesis by OsNAS3 is important for mitigating iron excess stress in rice. *Front. Plant Sci.* 10. https://doi.org/10.3389/FPLS.2019.00660

Baek, D., Kim, M.C., Chun, H.J., Kang, S., Park, H.C., Shin, G., Park, J., Shen, M., Hong, H., Kim, W.Y., Kim, D.H., Lee, S.Y., Bressan, R.A., Bohnert, H.J., Yun, D.J., 2013. Regulation of miR399f transcription by AtMYB2 affects phosphate starvation responses in Arabidopsis. *Plant Physiol.* 161, 362–373. https://doi.org/10.1104/pp.112.205922

Banakar, R., Alvarez Fernandez, A., Díaz-Benito, P., Abadia, J., Capell, T., Christou, P., 2017. Phytosiderophores determine thresholds for iron and zinc accumulation in biofortified rice endosperm while inhibiting the accumulation of cadmium. *J Exp Bot* 68, 4983–4995.

Bari, R., Pant, B.D., Stitt, M., Scheible, W.R., 2006. PHO2, microRNA399, and PHR1 define a phosphate-signaling pathway in plants. *Plant Physiol.* 141, 988–999. https://doi.org/10.1104/pp.106.079707

Bello, S.K., 2021. An overview of the morphological, genetic and metabolic mechanisms regulating phosphorus efficiency via root traits in soybean. *J. Soil Sci. Plant Nutr.* 21, 1013–1029. https://doi.org/10.1007/S42729-021-00418-Y

Bhati, K.K., Aggarwal, S., Sharma, S., Mantri, S., Singh, S.P., Bhalla, S., Kaur, J., Tiwari, S., Roy, J.K., Tuli, R., others, 2014. Differential expression of structural genes for the late phase of phytic acid biosynthesis in developing seeds of wheat (*Triticum aestivum* L.). *Plant Sci.* 224, 74–85.

Bhullar, N.K., Zhang, Z., Wicker, T., Keller, B., 2010. Wheat gene bank accessions as a source of new alleles of the powdery mildew resistance gene *Pm3*: a large scale allele mining project. *BMC Plant Biol.* 10, 1–13.

Billen, G., Lassaletta, L., Garnier, J., 2015. A vast range of opportunities for feeding the world in 2050: trade-off between diet, N contamination and international trade. *Environ. Res. Lett.* 10, 025001.

Boonyaves, K., Wu, T.Y., Gruissem, W., Bhullar, N.K., 2017. Enhanced grain iron levels in iron-regulated metal transporter, nicotianamine synthase, and ferritin gene cassette. *Front. Plant Sci.* 8. https://doi.org/10.3389/FPLS.2017.00130

Budak, H., Kantar, M., Bulut, R., Akpinar, B.A., 2015. Stress responsive miRNAs and isomiRs in cereals. *Plant Sci.* https://doi.org/10.1016/j.plantsci.2015.02.008

Cailliatte, R., Schikora, A., Briat, J.-F., Mari, S., Curie, C., 2010. High-affinity manganese uptake by the metal transporter NRAMP1 is essential for *Arabidopsis* growth in low manganese conditions. *Plant Cell* 22, 904–917.

Castaings, L., Alcon, C., Kosuth, T., Correia, D., Curie, C., 2021. Manganese triggers phosphorylation-mediated endocytosis of the Arabidopsis metal transporter NRAMP1. *Plant J.* 106, 1328–1337.

Cermak, T., Doyle, E.L., Christian, M., Wang, L., Zhang, Y., Schmidt, C., Baller, J.A., Somia, N. V., Bogdanove, A.J., Voytas, D.F., 2011. Efficient design and assembly of custom TALEN and other TAL effector-based constructs for DNA targeting. *Nucleic Acids Res.* 39. https://doi.org/10.1093/NAR/GKR218

Chang, H., Yi, B., Ma, R., Zhang, X., Zhao, H., Xi, Y., 2016. CRISPR/cas9, a novel genomic tool to knock down microRNA in vitro and in vivo. *Sci. Rep.* 6, 1–12.

Che, R., Tong, H., Shi, B., Liu, Y., Fang, S., Liu, D., Xiao, Y., Hu, B., Liu, L., Wang, H., others, 2015. Control of grain size and rice yield by GL2-mediated brassinosteroid responses. *Nat. Plants* 2, 1–8.

Chen, Z., Fujii, Y., Yamaji, N., Masuda, S., Takemoto, Y., Kamiya, T., Yusuyin, Y., Iwasaki, K., Kato, S., Maeshima, M., 2013. Mn tolerance in rice is mediated by MTP8. 1, a member of the cation diffusion facilitator family. *J. Exp. Bot.* 64, 4375–4387.

Cheng, N.H., Pittman, J.K., Barkla, B.J., Shigaki, T. and Hirschi, K.D., 2003. The Arabidopsis cax1 mutant exhibits impaired ion homeostasis, development, and hormonal responses and reveals interplay among vacuolar transporters. *The Plant Cell* 15(2),347–364.

Connorton, J.M., Jones, E.R., Rodríguez-Ramiro, I., Fairweather-Tait, S., Uauy, C., Balk, J., 2017. Wheat vacuolar iron transporter TaVIT2 transports Fe and Mn and is effective for biofortification. *Plant Physiol.* 174, 2434–2444. https://doi.org/10.1104/PP.17.00672

Conte, S.S., Chu, H.H., Chan-Rodriguez, D., Punshon, T., Vasques, K.A., Salt, D.E., Walker, E.L., 2013. *Arabidopsis thaliana* Yellow Stripe1-Like4 and Yellow Stripe1-Like6 localize to internal cellular membranes and are involved in metal ion homeostasis. *Front. Plant Sci.* 4, 283.

Crawford, N.M., Forde, B.G., 2002. Molecular and developmental biology of inorganic nitrogen nutrition. *Arab. B.* 1, e0011. https://doi.org/10.1199/tab.0011

Curie, C., Alonso, J.M., JEAN, M.L., Ecker, J.R., Briat, J.F., 2000. Involvement of NRAMP1 from *Arabidopsis thaliana* in iron transport. *Biochem. J.* 347, 749–755.

Davis, A.J., Chen, D.J., 2013. DNA double strand break repair via non-homologous end-joining. *Transl. Cancer Res.* 2, 130–143. https://doi.org/10.3978/j.issn.2218-676x.2013.04.02

Delhaize, E., Gruber, B.D., Pittman, J.K., White, R.G., Leung, H., Miao, Y., Jiang, L., Ryan, P.R., Richardson, A.E., 2007. A role for the *AtMTP11* gene of Arabidopsis in manganese transport and tolerance. *Plant J.* 51, 198–210.

Delhaize, E., Kataoka, T., Hebb, D.M., White, R.G., Ryan, P.R., 2003. Genes encoding proteins of the cation diffusion facilitator family that confer manganese tolerance. *Plant Cell* 15, 1131–1142.

Demidchik, V., Maathuis, F.J.M., 2007. Physiological roles of nonselective cation channels in plants: from salt stress to signalling and development. *New Phytol.* 175, 387–404. https://doi.org/10.1111/J.1469-8137.2007.02128.X

Devaiah, B.N., Madhuvanthi, R., Karthikeyan, A.S., Raghothama, K.G., 2009. Phosphate starvation responses and gibberellic acid biosynthesis are regulated by the *MYB62* transcription factor in *Arabidopsis*. *Mol. Plant* 2, 43–58.

Devaiah, B.N., Nagarajan, V.K., Raghothama, K.G., 2007. Phosphate homeostasis and root development in Arabidopsis are synchronized by the zinc finger transcription factor ZAT6. *Plant Physiol.* 145, 147–159. https://doi.org/10.1104/pp.107.101691

Dodd, A.N., Kudla, J., Sanders, D., 2010. The language of calcium signaling. *Annu. Rev. Plant Biol.* 61, 593–620. https://doi.org/10.1146/ANNUREV-ARPLANT-070109-104628

Donohoue, P.D., Barrangou, R., May, A.P., 2018. Advances in industrial biotechnology using CRISPR-Cas systems. *Trends Biotechnol.* 36, 134–146.

Dorsch, J.A., Cook, A., Young, K.A., Anderson, J.M., Bauman, A.T., Volkmann, C.J., Murthy, P.P.N., Raboy, V., 2003. Seed phosphorus and inositol phosphate phenotype of barley low phytic acid genotypes. *Phytochemistry* 62, 691–706.

Duan, P., Ni, S., Wang, J., Zhang, B., Xu, R., Wang, Y., Chen, H., Zhu, X., Li, Y., 2015. Regulation of OsGRF4 by OsmiR396 controls grain size and yield in rice. *Nat. Plants* 1. https://doi.org/10.1038/NPLANTS.2015.203

Edel, K.H., Marchadier, E., Brownlee, C., Kudla, J., Hetherington, A.M., 2017. The evolution of calcium-based signalling in plants. *Curr. Biol.* 27, R667–R679.

Eroglu, S., Meier, B., von Wirén, N., Peiter, E., 2016. The vacuolar manganese transporter MTP8 determines tolerance to iron deficiency-induced chlorosis in Arabidopsis. *Plant Physiol.* 170, 1030–1045.

Feng, Z., Zhang, B., Ding, W. et al., 2013. Efficient genome editing in plants using a CRISPR/Cas system. *Cell Res.* 23, 1229–1232. https://doi.org/10.1038/cr.2013.114

Frink, C.R., Waggoner, P.E., Ausubel, J.H., 1999. Nitrogen fertilizer: retrospect and prospect. *Proc. Natl. Acad. Sci. U. S. A.* https://doi.org/10.1073/pnas.96.4.1175

Gaj, T., Gersbach, C.A., Barbas, C.F., 2013. ZFN, TALEN, and CRISPR/Cas-based methods for genome engineering. *Trends Biotechnol.* 31, 397–405. https://doi.org/10.1016/J.TIBTECH.2013.04.004

Gaitán-Solís, E., Taylor, N.J., Siritunga, D., Stevens, W., Schachtman, D.P., 2015. Overexpression of the transporters *AtZIP1* and *AtMTP1* in cassava changes zinc accumulation and partitioning. *Front. Plant Sci.* 6, 492.

Gamuyao, R., Chin, J.H., Pariasca-Tanaka, J., Pesaresi, P., Catausan, S., Dalid, C., Slamet-Loedin, I., Tecson-Mendoza, E.M., Wissuwa, M., Heuer, S., 2012. The protein kinase Pstol1 from traditional rice confers tolerance of phosphorus deficiency. *Nature* 488, 535–539. https://doi.org/10.1038/nature11346

Gao, H., Xie, W., Yang, C., Xu, J., Li, J., Wang, H., Chen, X., Huang, C., 2018. NRAMP2, a trans-Golgi network-localized manganese transporter, is required for *Arabidopsis* root growth under manganese deficiency. *New Phytol.* 217, 179–193.

Gao, S., Xiao, Y., Xu, F., Gao, X., Cao, S., Zhang, F., Wang, G., Sanders, D., Chu, C., 2019. Cytokinin-dependent regulatory module underlies the maintenance of zinc nutrition in rice. *New Phytol.* 224, 202–215. https://doi.org/10.1111/NPH.15962

Gierth, M., Mäser, P., Schroeder, J.I., 2005. The potassium transporter *AtHAK5* functions in K+ deprivation-induced high-affinity K+ uptake and *AKT1* K+ channel contribution to K+ uptake kinetics in Arabidopsis roots. *Plant Physiol.* 137, 1105–1114.

Gigolashvili, T., Kopriva, S., 2014. Transporters in plant sulfur metabolism. *Front. Plant Sci.* 5, 442.

Gojon, A., Nacry, P., Davidian, J.C., 2009. Root uptake regulation: a central process for NPS homeostasis in plants. *Curr. Opin. Plant Biol.* https://doi.org/10.1016/j.pbi.2009.04.015

Good, A.G., Shrawat, A.K., Muench, D.G., 2004. Can less yield more? Is reducing nutrient input into the environment compatible with maintaining crop production? *Trends Plant Sci.* 9, 597–605. https://doi.org/10.1016/j.tplants.2004.10.008

Gu, M., Zhang, J., Li, H., Meng, D., Li, R., Dai, X., Wang, S., Liu, W., Qu, H., Xu, G., 2017. Maintenance of phosphate homeostasis and root development are coordinately regulated by MYB1, an R2R3-type MYB transcription factor in rice. *J. Exp. Bot.* 68, 3603–3615. https://doi.org/10.1093/JXB/ERX174

Guo, B., Jin, Y., Wussler, C., Blancaflor, E.B., Motes, C.M., Versaw, W.K., 2008. Functional analysis of the Arabidopsis PHT4 family of intracellular phosphate transporters. *New Phytol.* 177, 889–898. https://doi.org/10.1111/j.1469-8137.2007.02331.x

Guo, C., Zhao, X., Liu, X., Zhang, L., Gu, J., Li, X., Lu, W., Xiao, K., 2013. Function of wheat phosphate transporter gene TaPHT2;1 in Pi translocation and plant growth regulation under replete and limited Pi supply conditions. *Planta* 237, 1163–1178. https://doi.org/10.1007/s00425-012-1836-2

Guo, J., Long, L., Chen, A., Dong, X., Liu, Z., Chen, L., Wang, J., Yuan, L., 2022. Tonoplast-localized transporter ZmNRAMP2 confers root-to-shoot translocation of manganese in maize. *Plant Physiol.* 190, 2601–2616.

Guo, M., Wang, Q., Zong, Y., Nian, J., Li, H., Li, J., Wang, T., Gao, C., Zuo, J., 2021. Genetic manipulations of TaARE1 boost nitrogen utilization and grain yield in wheat. *J. Genet. Genomics* 48, 950–953. https://doi.org/10.1016/J.JGG.2021.07.003

Hasan, Md Mahmudul, Hasan, Md Mainul, da Silva, J.A.T., Li, X., 2016. Regulation of phosphorus uptake and utilization: transitioning from current knowledge to practical strategies. *Cell. Mol. Biol. Lett.* 21, 1–19.

Heffner, E.L., Sorrells, M.E., Jannink, J.L., 2009. Genomic selection for crop improvement. *Crop Sci.* 49, 1–12. https://doi.org/10.2135/cropsci2008.08.0512

Hillwig, M.S., LeBrasseur, N.D., Green, P.J., MacIntosh, G.C., 2008. Impact of transcriptional, ABA-dependent, and ABA-independent pathways on wounding regulation of RNS1 expression. *Mol. Genet. Genomics* 280, 249–261. https://doi.org/10.1007/s00438-008-0360-3

Hisano, H., Abe, F., Hoffie, R.E., Kumlehn, J., 2021. Targeted genome modifications in cereal crops. *Breed. Sci.* 71, 405–416. https://doi.org/10.1270/jsbbs.21019

Höller, S., Küpper, H., Brückner, D., Garrevoet, J., Spiers, K., Falkenberg, G., Andresen, E., Peiter, E., 2022. Overexpression of *METAL TOLERANCE PROTEIN8* reveals new aspects of metal transport in *Arabidopsis thaliana* seeds. *Plant Biol.* 24, 23–29.

Hu, B., Jiang, Z., Wang, W., Qiu, Y., Zhang, Z., Liu, Y., Li, A., Gao, X., Liu, L., Qian, Y., Huang, X., Yu, F., Kang, S., Wang, Yiqin, Xie, J., Cao, S., Zhang, L., Wang, Yingchun, W., Xie, Q., Kopriva, S., Chu, C., 2019. Nitrate-NRT1.1B-SPX4 cascade integrates nitrogen and phosphorus signalling networks in plants. *Nat. Plants* 5, 401–413. https://doi.org/10.1038/S41477-019-0384-1

Hu, B., Wang, W., Ou, S., Tang, J., Li, H., Che, R., Zhang, Z., Chai, X., Wang, H., Wang, Y., others, 2015. Variation in NRT1. 1B contributes to nitrate-use divergence between rice subspecies. *Nat. Genet.* 47, 834.

Ibrahim, S., Saleem, B., Rehman, N., Zafar, S.A., Naeem, M.K., Khan, M.R., 2022. CRISPR/Cas9 mediated disruption of Inositol Pentakisphosphate 2-Kinase 1 (TaIPK1) reduces phytic acid and improves iron and zinc accumulation in wheat grains. *J. Adv. Res.* 37, 33–41. https://doi.org/10.1016/J.JARE.2021.07.006

Ishimaru, Y., Masuda, H., Bashir, K., Inoue, H., Tsukamoto, T., Takahashi, M., Nakanishi, H., Aoki, N., Hirose, T., Ohsugi, R., 2010. Rice metal-nicotianamine transporter, OsYSL2, is required for the long-distance transport of iron and manganese. *Plant J.* 62, 379–390.

Ishino, Y., Shinagawa, H., Makino, K., Amemura, M., Nakatura, A., 1987. Nucleotide sequence of the iap gene, responsible for alkaline phosphatase isoenzyme conversion in Escherichia coli, and identification of the gene product. *J. Bacteriol.* 169, 5429–5433. https://doi.org/10.1128/JB.169.12.5429-5433.1987

Jha, U.C., Nayyar, H., Parida, S.K., Beena, R., Pang, J., Siddique, K.H.M., 2023. Breeding and genomics approaches for improving phosphorus-use efficiency in grain legumes. *Environ. Exp. Bot.* 205, 105120. https://doi.org/10.1016/J.ENVEXPBOT.2022.105120

Jinek, M., Chylinski, K., Fonfara, I., Hauer, M., Doudna, J.A., Charpentier, E., 2012. A programmable dual-RNA--guided DNA endonuclease in adaptive bacterial immunity. *Science* 337, 816–821.

Kappara, S., Neelamraju, S., Ramanan, R., 2018. Down regulation of a heavy metal transporter gene influences several domestication traits and grain Fe-Zn content in rice. *Plant Sci.* 276, 208–219. https://doi.org/10.1016/J.PLANTSCI.2018.09.003

Karunarathne, S.D., Han, Y., Zhang, X.Q., Li, C., 2022. CRISPR/Cas9 gene editing and natural variation analysis demonstrate the potential for HvARE1 in improvement of nitrogen use efficiency in barley. *J. Integr. Plant Biol.* 64, 756–770. https://doi.org/10.1111/JIPB.13214/SUPPINFO

Kavitha, P.G., Kuruvilla, S., Mathew, M.K., 2015. Functional characterization of a transition metal ion transporter, OsZIP6 from rice (*Oryza sativa* L.). *Plant Physiol. Biochem.* 97, 165–174.

Kim, J.H., Tsukaya, H., 2015. Regulation of plant growth and development by the GROWTH-REGULATING FACTOR and GRF-INTERACTING FACTOR duo. *J. Exp. Bot.* 66, 6093–6107.

Kim, M.C., Panstruga, R., Elliott, C., Möller, J., Devoto, A., Yoon, H.W., Park, H.C., Cho, M.J., Schulze-Lefert, P., 2002. Calmodulin interacts with MLO protein to regulate defence against mildew in barley. *Nature* 416, 447–450. https://doi.org/10.1038/416447A

Kirkby, E.A., Johnston, A.E., 2008. Soil and fertilizer phosphorus in relation to crop nutrition. In: White, P.J., Hammond, J.P. (eds) *The Ecophysiology of Plant-Phosphorus Interactions. Plant Ecophysiology*, vol. 7, pp. 177–223. Springer, Dordrecht. https://doi.org/10.1007/978-1-4020-8435-5_9

Kirschner, S., Woodfield, H., Prusko, K., Koczor, M., Gowik, U., Hibberd, J.M., Westhoff, P., 2018. Expression of *SULTR2*; 2, encoding a low-affinity sulphur transporter, in the Arabidopsis bundle sheath and vein cells is mediated by a positive regulator. *J. Exp. Bot.* 69, 4897–4906.

Kistner C, Parniske M., 2002. Evolution of signal transduction in intracellular symbiosis. *Trends Plant Sci.* 7, 511–518.

Koike, S., Inoue, H., Mizuno, D., Takahashi, M., Nakanishi, H., Mori, S., Nishizawa, N.K., 2004. OsYSL2 is a rice metal-nicotianamine transporter that is regulated by iron and expressed in the phloem. *Plant J.* 39, 415–424.

Kopriva, S., Mugford, S.G., Matthewman, C., Koprivova, A., 2009. Plant sulfate assimilation genes: redundancy versus specialization. *Plant Cell Rep.* 28, 1769–1780.

Korshunova, Y.O., Eide, D., Gregg Clark, W., Lou Guerinot, M., Pakrasi, H.B., 1999. The IRT1 protein from Arabidopsis thaliana is a metal transporter with a broad substrate range. *Plant Mol. Biol.* 40, 37–44. https://doi.org/10.1023/A:1026438615520

Krausko, M., Labajová, M., Peterková, D., Jásik, J., 2021. Specific expression of AtIRT1 in phloem companion cells suggests its role in iron translocation in aboveground plant organs. *Plant Signal. Behav.* 16, 1925020.

Kudla, J., 2018. Advances and current challenges in calcium signaling. *New Phytol.* 218, 414–431.

Kumar, R.S., Sinha, H., Datta, T., Asif, M.H., Trivedi, P.K., 2023. microRNA408 and its encoded peptide regulate sulfur assimilation and arsenic stress response in Arabidopsis. *Plant Physiol.* 192, 837–856.

Kusano, M., Tabuchi, M., Fukushima, A., Funayama, K., Diaz, C., Kobayashi, M., Hayashi, N., Tsuchiya, Y.N., Takahashi, H., Kamata, A., Yamaya, T., Saito, K., 2011. Metabolomics data reveal a crucial role of cytosolic glutamine synthetase 1;1 in coordinating metabolic balance in rice. *Plant J.* 66, 456–466. https://doi.org/10.1111/j.1365-313X.2011.04506.x

Lanquar, V., Ramos, M.S., Lelièvre, F., Barbier-Brygoo, H., Krieger-Liszkay, A., Krämer, U., Thomine, S., 2010. Export of vacuolar manganese by AtNRAMP3 and AtNRAMP4 is required for optimal photosynthesis and growth under manganese deficiency. *Plant Physiol.* 152, 1986–1999.

Li, B., Tester, M., Gilliham, M., 2017a. Chloride on the move. *Trends Plant Sci.* https://doi.org/10.1016/j.tplants.2016.12.004

Li, J., Wang, L., Zheng, L., Wang, Y., Chen, X., Zhang, W., 2018a. A functional study identifying critical residues involving metal transport activity and selectivity in natural resistance-associated macrophage protein 3 in *Arabidopsis thaliana. Int. J. Mol. Sci.* 19, 1430.

Li, J., Wu, W.-H., Wang, Y., 2017b. Potassium channel AKT1 is involved in the auxin-mediated root growth inhibition in *Arabidopsis* response to low K+ stress. *J. Integr. Plant Biol.* 59, 895–909.

Li, J., Zhang, X., Sun, Y., Zhang, J., Du, W., Guo, X., Li, S., Zhao, Y., Xia, L., 2018b. Efficient allelic replacement in rice by gene editing: a case study of the NRT1. 1B gene. *J. Integr. Plant Biol.* 60, 536–540.

Li, J., Zheng, L., Fan, Y., Wang, Y., Ma, Y., Gu, D., Lu, Y., Zhang, S., Chen, X., Zhang, W., 2021. Pear metal transport protein PbMTP8. 1 confers manganese tolerance when expressed in yeast and *Arabidopsis thaliana. Ecotoxicol. Environ. Saf.* 208, 111687.

Li, Q., Gao, Y., Yang, A., 2020. Sulfur homeostasis in plants. *Int. J. Mol. Sci.* 21, 8926.

Li J, Zhang X, Sun Y, Zhang J, Du W, Guo X, Li S, Zhao Y, Xia L., 2018. Efficient allelic replacement in rice by gene editing: A case study of the NRT1.1B gene. *J Integr Plant Biol.* 60(7), 536–540. doi: 10.1111/jipb.12650.

Li, H., Zhang, Y., Wu, C., Bi, J., Chen, Y., Jiang, C., Cui, M., Chen, Y., Hou, X., Yuan, M., Xiong, L., 2022. Fine-tuning OsCPK18/OsCPK4 activity via genome editing of phosphorylation motif improves rice yield and immunity. *Plant Biotechnol J.* 20(12), 2258–2271.

Li, Q., Li, Y., Wu, X., Zhou, L., Zhu, X., Fang, W., 2017c. Metal transport protein 8 in *Camellia sinensis* confers superior manganese tolerance when expressed in yeast and *Arabidopsis thaliana. Sci. Rep.* 7, 39915.

Lieber, M.R., 2010. The mechanism of double-strand DNA break repair by the nonhomologous DNA end-joining pathway. *Annu Rev Biochem* 79, 181–211. https://doi.org/10.1146/annurev.biochem.052308.093131

Lightfoot, D.A., 2009. Genes for use in improving nitrate use efficiency in crops. In: Jenks, Matthew A. and Wood, Andrew J. (eds.) *Genes for Plant Abiotic Stress.* Wiley, NY, p. 240.

Lin, Q., Zong, Y., Xue, C., Wang, S., Jin, S., Zhu, Z., Wang, Y., Anzalone, A. V., Raguram, A., Doman, J.L., Liu, D.R., Gao, C., 2020. Prime genome editing in rice and wheat. *Nat. Biotechnol.* 38, 582–585. https://doi.org/10.1038/s41587-020-0455-x

Liu, L., Zheng, C., Kuang, B., Wei, L., Yan, L., Wang, T., 2016. Receptor-like kinase RUPO interacts with potassium transporters to regulate pollen tube growth and integrity in rice. *PLoS Genet.* 12, e1006085.

Long, L., Persson, D.P., Duan, F., Jørgensen, K., Yuan, L., Schjoerring, J.K., Pedas, P.R., 2018. The iron-regulated transporter 1 plays an essential role in uptake, translocation and grain-loading of manganese, but not iron, in barley. *New Phytol.* 217, 1640–1653.

López-Arredondo, D.L., Leyva-González, M.A., González-Morales, S.I., López-Bucio, J., Herrera-Estrella, L., 2014. Phosphate nutrition: improving low-phosphate tolerance in crops. *Annu. Rev. Plant Biol.* 65, 95–123. https://doi.org/10.1146/annurev-arplant-050213-035949

Lopez-Millan, A.-F., Ellis, D.R., Grusak, M.A., 2004. Identification and characterization of several new members of the ZIP family of metal ion transporters in *Medicago truncatula. Plant Mol. Biol.* 54, 583.

Lu, K., Wu, B., Wang, J., Zhu, W., Nie, H., Qian, J., Huang, W., Fang, Z., 2018. Blocking amino acid transporter OsAAP3 improves grain yield by promoting outgrowth buds and increasing tiller number in rice. *Plant Biotechnol. J.* 16, 1710–1722. https://doi.org/10.1111/PBI.12907

Ma, G., Li, Jiyu, Li, Jingjun, Li, Y., Gu, D., Chen, C., Cui, J., Chen, X., Zhang, W., 2018. OsMTP11, a trans-Golgi network localized transporter, is involved in manganese tolerance in rice. *Plant Sci.* 274, 59–69.

Mao, X., Zheng, Y., Xiao, K., Wei, Y., Zhu, Y., Cai, Q., Chen, L., Xie, H., Zhang, J., 2018. OsPRX2 contributes to stomatal closure and improves potassium deficiency tolerance in rice. *Biochem. Biophys. Res. Commun.* 495, 461–467. https://doi.org/10.1016/J.BBRC.2017.11.045

Mao, Y., Yang, X., Zhou, Y., Zhang, Z., Botella, J.R., Zhu, J.-K., 2018. Manipulating plant RNA-silencing pathways to improve the gene editing efficiency of CRISPR/Cas9 systems. *Genome Biol.* 19, 1–15.

Marschner, H., 1995. *Mineral Nutrition of Higher Plants*, 2nd edn. Institute of Plant Nutrition University of Hohenheim, Germany.

Maruyama-Nakashita, A., 2017. Metabolic changes sustain the plant life in low-sulfur environments. *Curr. Opin. Plant Biol.* 39, 144–151.

Maruyama-Nakashita, A., Nakamura, Y., Watanabe-Takahashi, A., Inoue, E., Yamaya, T., Takahashi, H., 2005. Identification of a novel *cis*-acting element conferring sulfur deficiency response in Arabidopsis roots. *Plant J.* 42, 305–314.

Masclaux-Daubresse, C., Daniel-Vedele, F., Dechorgnat, J., Chardon, F., Gaufichon, L., Suzuki, A., 2010. Nitrogen uptake, assimilation and remobilization in plants: challenges for sustainable and productive agriculture. *Ann. Bot.* 105, 1141–1157. https://doi.org/10.1093/AOB/MCQ028

McDowell, L.R., 2003. *Minerals in Animal and Human Nutrition*. Elsevier, Amsterdam, the Netherlands.

Meng, Y., Shao, C., Wang, H., Jin, Y., 2012. Target mimics: an embedded layer of microRNA-involved gene regulatory networks in plants. *BMC Genomics* 13. https://doi.org/10.1186/1471-2164-13-197

Michard, E., Lima, P.T., Borges, F., Silva, A.C., Portes, M.T., Carvalho, J.E., Gilliham, M., Liu, L.H., Obermeyer, G. and Feijó, J.A., 2011. Glutamate receptor–like genes form Ca2+ channels in pollen tubes and are regulated by pistil D-serine. *Science* 332(6028), 434–437.

Miglani, G.S., 2017. Genome editing in crop improvement: present scenario and future prospects. *J. Crop Improv.* 31, 453–559.

Mills, R.F., Doherty, M.L., López-Marqués, R.L., Weimar, T., Dupree, P., Palmgren, M.G., Pittman, J.K., Williams, L.E., 2008. ECA3, a Golgi-localized P2A-type ATPase, plays a crucial role in manganese nutrition in Arabidopsis. *Plant Physiol.* 146, 116–128.

Milner, M.J., Seamon, J., Craft, E., Kochian, L. V, 2013. Transport properties of members of the ZIP family in plants and their role in Zn and Mn homeostasis. *J. Exp. Bot.* 64, 369–381.

Moll, R.H., Kamprath, E.J., Jackson, W.A., 1982. Analysis and interpretation of factors which contribute to efficiency of nitrogen utilization 1. *Agron. J.* 74, 562–564. https://doi.org/10.2134/agronj1982.0002196 2007400030037x

Moose, S., Below, F.E. (2009). Biotechnology approaches to improving maize nitrogen use efficiency. In: Kriz, A.L., Larkins, B.A. (eds) *Molecular Genetic Approaches to Maize Improvement. Biotechnology in Agriculture and Forestry*, vol 63. Springer, Berlin, Heidelberg. https://doi.org/10.1007/978-3-540-68922-5_6

Mudge, S.R., Rae, A.L., Diatloff, E., Smith, F.W., 2002. Expression analysis suggests novel roles for members of the Pht1 family of phosphate transporters in *Arabidopsis. Plant J.* 31, 341–353. https://doi.org/10.1046/J.1365-313X.2002.01356.X

Murata, Y., Yamamoto, K., Yamaguchi, Y., Morishita, H., 2010. The expression method of the spacecraft operations procedure, In: *SpaceOps 2010 Conference Delivering on the Dream Hosted by NASA Marshall Space Flight Center and Organized by AIAA*, p. 2232.

Nagarajan, V.K., Jain, A., Poling, M.D., Lewis, A.J., Raghothama, K.G., Smith, A.P., 2011. Arabidopsis Pht1;5 mobilizes phosphate between source and sink organs and influences the interaction between phosphate homeostasis and ethylene signaling. *Plant Physiol.* 156, 1149–1163. https://doi.org/10.1104/PP.111.174805

Neequaye, M., Stavnstrup, S., Harwood, W., Lawrenson, T., Hundleby, P., Irwin, J., Troncoso-Rey, P., Saha, S., Traka, M.H., Mithen, R., 2021. CRISPR/Cas9-mediated gene editing of *MYB28* genes impair glucoraphanin accumulation of *Brassica oleracea* in the field. *Cris. J.* 4, 416–426.

Pang, J., Ryan, M.H., Lambers, H., Siddique, K.H., 2018. Phosphorus acquisition and utilisation in crop legumes under global change. *Curr. Opin. Plant Biol.* 45, 248–254. https://doi.org/10.1016/j.pbi.2018.05.012

Park, B.S., Seo, J.S., Chua, N.H., 2014. NITROGEN LIMITATION ADAPTATION Recruits PHOSPHATE2 to target the phosphate transporter PT2 for degradation during the regulation of *Arabidopsis* phosphate homeostasis. *Plant Cell* 26, 454–464. https://doi.org/10.1105/TPC.113.120311

Pastor, A.V, Palazzo, A., Havlik, P., Biemans, H., Wada, Y., Obersteiner, M., Kabat, P., Ludwig, F., 2019. The global nexus of food--trade--water sustaining environmental flows by 2050. *Nat. Sustain.* 2, 499–507.

Pathak, R.R., Ahmad, A., Lochab, S., Raghuram, N., 2008. Molecular physiology of plant nitrogen use efficiency and biotechnological options for its enhancement. *Curr. Sci.* 94, 1394–1403.

Pedas, P., Ytting, C.K., Fuglsang, A.T., Jahn, T.P., Schjoerring, J.K., Husted, S., 2008. Manganese efficiency in barley: identification and characterization of the metal ion transporter HvIRT1. *Plant Physiol.* 148, 455–466.

Peiter, E., Maathuis, F., Mills, L. *et al.*, 2005. The vacuolar Ca^{2+}-activated channel TPC1 regulates germination and stomatal movement. *Nature* 434, 404–408. https://doi.org/10.1038/nature03381

Pittman, J.K., 2005. Managing the manganese: molecular mechanisms of manganese transport and homeostasis. *New Phytol.* 167, 733–742.

Pittman, J.K., Cheng, N., Shigaki, T., Kunta, M., Hirschi, K.D., 2004. Functional dependence on calcineurin by variants of the *Saccharomyces cerevisiae* vacuolar Ca2+/H+ exchanger Vcx1p. *Mol. Microbiol.* 54, 1104–1116.

Prathap, V., Kumar, A., Maheshwari, C., Tyagi, A., 2022. Phosphorus homeostasis: acquisition, sensing, and long-distance signaling in plants. *Mol. Biol. Reports* 2022 1, 1–16. https://doi.org/10.1007/S11033-022-07354-9

Qian, M., Li, X., Tang, L., Peng, Y., Huang, X., Wu, T., Liu, Y., Liu, X., Xia, Y., Peng, K., 2022. VrNramp5 is responsible for cadmium and manganese uptake in *Vigna radiata* roots. *Environ. Exp. Bot.* 199, 104867.

Qiao, K., Wang, F., Liang, S., Wang, H., Hu, Z., Chai, T., 2019. New biofortification tool: wheat TaCNR5 enhances zinc and manganese tolerance and increases zinc and manganese accumulation in rice grains. *J. Agric. Food Chem.* 67, 9877–9884.

Raboy, V., 2009. Approaches and challenges to engineering seed phytate and total phosphorus. *Plant Sci.* 177, 281–296.

Ragel, P., Raddatz, N., Leidi, E.O., Quintero, F.J., Pardo, J.M., 2019. Regulation of K+ nutrition in plants. *Front. Plant Sci.* 10. https://doi.org/10.3389/FPLS.2019.00281

Ramireddy, E., Hosseini, S.A., Eggert, K., Gillandt, S., Gnad, H., von Wirén, N., Schmülling, T., 2018. Root engineering in barley: increasing cytokinin degradation produces a larger root system, mineral enrichment in the shoot and improved drought tolerance. *Plant Physiol.* 177, 1078–1095. https://doi.org/10.1104/PP.18.00199

Ramesh, S. A., Choimes, S., and Schachtman, D. P., 2004. Over-expression of an *Arabidopsis* zinc transporter in *Hordeum vulgare* increases short-term zinc uptake after zinc deprivation and seed zinc content. *Plant Mol. Biol.* 54, 373–385. doi: 10.1023/B:PLAN.0000036370.70912.34

Richardson, A.E., Lynch, J.P., Ryan, P.R., Delhaize, E., Smith, F.A., Smith, S.E., Harvey, P.R., Ryan, M.H., Veneklaas, E.J., Lambers, H., Oberson, A., Culvenor, R.A., Simpson, R.J., 2011. Plant and microbial strategies to improve the phosphorus efficiency of agriculture. *Plant Soil* 349, 121–156.

Rouached, H., Secco, D., Arpat, B., Poirier, Y., 2011. The transcription factor PHR1 plays a key role in the regulation of sulfate shoot-to-root flux upon phosphate starvation in Arabidopsis. *BMC Plant Biol.* 11, 1–10.

Ruan, W., Guo, M., Wang, X., Guo, Z., Xu, Z., Xu, L., Zhao, H., Sun, H., Yan, C., Yi, K., 2019. Two RING-finger ubiquitin E3 ligases regulate the degradation of SPX4, an internal phosphate sensor, for phosphate homeostasis and signaling in rice. *Mol. Plant* 12, 1060–1074. https://doi.org/10.1016/j.molp.2019.04.003

Ruan, W., Guo, M., Xu, L., Wang, X., Zhao, H., Wang, J., Yi, K., 2018. An SPX-RLI1 module regulates leaf inclination in response to phosphate availability in rice. *Plant Cell* 30, 853–870. https://doi.org/10.1105/TPC.17.00738

Rubio, V., Linhares, F., Solano, R., Martín, A.C., Iglesias, J., Leyva, A., Paz-Ares, J., 2001. A conserved MYB transcription factor involved in phosphate starvation signaling both in vascular plants and in unicellular algae. *Genes Dev.* 15, 2122–2133. https://doi.org/10.1101/gad.204401

Sander, J.D., Joung, J.K., 2014. CRISPR-Cas systems for editing, regulating and targeting genomes. *Nat. Biotechnol.* 32, 347–350. https://doi.org/10.1038/NBT.2842

Sardans, J., Peñuelas, J., 2012. The role of plants in the effects of global change on nutrient availability and stoichiometry in the plant-soil system. *Plant Physiol.* 160, 1741–1761. https://doi.org/10.1104/pp.112.208785

Sasaki, A., Yamaji, N., Xia, J., Ma, J.F., 2011. OsYSL6 is involved in the detoxification of excess manganese in rice. *Plant Physiol.* 157, 1832–1840.

Sasaki, A., Yamaji, N., Yokosho, K., Ma, J.F., 2012. Nramp5 is a major transporter responsible for manganese and cadmium uptake in rice. *Plant Cell* 24, 2155–2167.

Sathee, L., Jagadhesan, B., Pandesha, P.H., Barman, D., Adavi B.S., Nagar, S., Krishna, G.K., Tripathi, S., Jha, S.K., Chinnusamy, V., 2022. Genome editing targets for improving nutrient use efficiency and nutrient stress adaptation. *Front. Genet.* 13, 900897. https://doi.org/10.3389/FGENE.2022.900897

Schaaf, G., Catoni, E., Fitz, M., Schwacke, R., Schneider, A., Von Wiren, N., Frommer, W.B., 2002. A putative role for the vacuolar calcium/manganese proton antiporter AtCAX2 in heavy metal detoxification. *Plant Biol.* 4, 612–618.

Schmidt, S.B., Jensen, P.E., Husted, S., 2016. Manganese deficiency in plants: the impact on photosystem II. *Trends Plant Sci.* 21, 622–632.

Schulz P, Herde M, Romeis T., 2013. Calcium-dependent protein kinases: hubs in plant stress signaling and development. *Plant Physiol.* 163(2), 523–30. doi: 10.1104/pp.113.222539.

Seth, K., 2016. Current status of potential applications of repurposed Cas9 for structural and functional genomics of plants. *Biochem. Biophys. Res. Commun.* 480, 499–507.

Shi, J., Wang, H., Hazebroek, J., Ertl, D.S., Harp, T., 2005. The maize low-phytic acid 3 encodes a myo-inositol kinase that plays a role in phytic acid biosynthesis in developing seeds. *Plant J.* 42, 708–719.

Shi, J., Wang, H., Schellin, K., Li, B., Faller, M., Stoop, J.M., Meeley, R.B., Ertl, D.S., Ranch, J.P., Glassman, K., 2007. Embryo-specific silencing of a transporter reduces phytic acid content of maize and soybean seeds. *Nat. Biotechnol.* 25, 930–937.

Shi, J., Wang, H., Wu, Y., Hazebroek, J., Meeley, R.B., Ertl, D.S., 2003. The maize low-phytic acid mutant lpa2 is caused by mutation in an inositol phosphate kinase gene. *Plant Physiol.* 131, 507–515.

Shigaki, T., Pittman, J.K., Hirschi, K.D., 2003. Manganese specificity determinants in the *Arabidopsis*Metal/H+ Antiporter CAX2. *J. Biol. Chem.* 278, 6610–6617.

Shin, R., Berg, R.H., Schachtman, D.P., 2005. Reactive oxygen species and root hairs in Arabidopsis root response to nitrogen, phosphorus and potassium deficiency. *Plant Cell Physiol.* 46, 1350–1357. https://doi.org/10.1093/pcp/pci145

Singh, S.P., Keller, B., Gruissem, W., Bhullar, N.K., 2017. Rice *NICOTIANAMINE SYNTHASE 2* expression improves dietary iron and zinc levels in wheat. *Theor. Appl. Genet.* 130, 283–292. https://doi.org/10.1007/S00122-016-2808-X

Smith, F.W., Hawkesford, M.J., Ealing, P.M., Clarkson, D.T., Vanden Berg, P.J., Belcher, A.R., Warrilow, A.G.S., 1997. Regulation of expression of a cDNA from barley roots encoding a high affinity sulphate transporter. *Plant J.* 12, 875–884.

Stefanovic A, Ribot C, Rouached H, Wang Y, Chong J, Belbahri L, Delessert S, Poirier Y., 2007. Members of the PHO1 gene family show limited functional redundancy in phosphate transfer to the shoot, and are regulated by phosphate deficiency via distinct pathways. *Plant J.* 50(6), 982–94. doi: 10.1111/j.1365-313X.2007.03108.x.

Sun, Y., Luo, W., Jain, A., Liu, L., Ai, H., Liu, X., Feng, B., Zhang, L., Zhang, Z., Guohua, X., Sun, S., 2018. OsPHR3 affects the traits governing nitrogen homeostasis in rice. *BMC Plant Biol.* 18. https://doi.org/10.1186/s12870-018-1462-7

Svistoonoff, S., Creff, A., Reymond, M., Sigoillot-Claude, C., Ricaud, L., Blanchet, A., Nussaume, L., Desnos, T., 2007. Root tip contact with low-phosphate media reprograms plant root architecture. *Nat. Genet.* 39, 792–796. https://doi.org/10.1038/ng2041

Tabbita, F., Pearce, S., Barneix, A.J., 2017. Breeding for increased grain protein and micronutrient content in wheat: ten years of the GPC-B1 gene. *J. Cereal Sci.* 73, 183–191. https://doi.org/10.1016/j.jcs.2017.01.003

Takahashi, H., Buchner, P., Yoshimoto, N., Hawkesford, M.J., Shiu, S.-H., 2012. Evolutionary relationships and functional diversity of plant sulfate transporters. *Front. Plant Sci.* 2, 119.

Takemoto, Y., Tsunemitsu, Y., Fujii-Kashino, M., Mitani-Ueno, N., Yamaji, N., Ma, J.F., Kato, S.-I., Iwasaki, K., Ueno, D., 2017. The tonoplast-localized transporter MTP8. 2 contributes to manganese detoxification in the shoots and roots of *Oryza sativa* L. *Plant Cell Physiol.* 58, 1573–1582.

Tang, W., Ye, J., Yao, X. *et al.*, 2019, Genome-wide associated study identifies NAC42-activated nitrate transporter conferring high nitrogen use efficiency in rice. *Nat Commun.* 10, 5279. https://doi.org/10.1038/s41467-019-13187-1

Taylor, M.R., Reinders, A., Ward, J.M., 2015. Transport function of rice amino acid permeases (AAPs). *Plant Cell Physiol.* 56, 1355–1363. https://doi.org/10.1093/pcp/pcv053

Tegeder, M., 2012. Transporters for amino acids in plant cells: some functions and many unknowns. *Curr. Opin. Plant Biol.* 15, 315–321.

Templalexis, D., Tsitsekian, D., Liu, C., Daras, G., Šimura, J., Moschou, P., Ljung, K., Hatzopoulos, P., Rigas, S., 2021. Potassium transporter TRH1/KUP4 contributes to distinct auxin-mediated root system architecture responses. *Plant Physiol.* https://doi.org/10.1093/PLPHYS/KIAB472

Tian, Q., Shen, L., Luan, J., Zhou, Z., Guo, D., Shen, Y., Jing, W., Zhang, B., Zhang, Q., Zhang, W., 2021. Rice shaker potassium channel OsAKT2 positively regulates salt tolerance and grain yield by mediating K+ redistribution. *Plant. Cell Environ.* 44, 2951–2965. https://doi.org/10.1111/PCE.14101

Tiwari, C., Wallwork, H., Arun, B., Mishra, V., Velu, G., Stangoulis, J., 2016. Molecular mapping of quantitative trait loci for zinc, iron and protein content in the grains of hexaploid wheat. *Euphytica* 207, 563–570.

Trijatmiko, K.R., Duenãs, C., Tsakirpaloglou, N., Torrizo, L., Arines, F.M., Adeva, C., Balindong, J., Oliva, N., Sapasap, M. V., Borrero, J., Rey, J., Francisco, P., Nelson, A., Nakanishi, H., Lombi, E., Tako, E., Glahn, R.P., Stangoulis, J., Chadha-Mohanty, P., Johnson, A.A.T., Tohme, J., Barry, G., Slamet-Loedin, I.H., 2016. Biofortified indica rice attains iron and zinc nutrition dietary targets in the field. *Sci. Rep.* 6. https://doi.org/10.1038/SREP19792

Trumbo, P., Yates, A.A., Schlicker, S., Poos, M., 2001. Dietary reference intakes. *J. Am. Diet. Assoc.* 101, 294.

Tsunemitsu, Y., Genga, M., Okada, T., Yamaji, N., Ma, J.F., Miyazaki, A., Kato, S., Iwasaki, K., Ueno, D., 2018. A member of cation diffusion facilitator family, MTP11, is required for manganese tolerance and high fertility in rice. *Planta* 248, 231–241.

Uauy, C., Brevis, J.C., Dubcovsky, J., 2006. The high grain protein content gene Gpc-B1 accelerates senescence and has pleiotropic effects on protein content in wheat. *J. Exp. Bot.* 57, 2785–2794. https://doi.org/10.1093/JXB/ERL047

Ueda, Y., Kiba, T., Yanagisawa, S., 2020. Nitrate-inducible NIGT1 proteins modulate phosphate uptake and starvation signalling via transcriptional regulation of SPX genes. *Plant J.* 102, 448–466. https://doi.org/10.1111/tpj.14637

Umezawa T, Fujita M, Fujita Y, Yamaguchi-Shinozaki K, Shinozaki K., 2006. Engineering drought tolerance in plants: discovering and tailoring genes to unlock the future. *Curr Opin Biotechnol.* 17(2), 113–22. doi: 10.1016/j.copbio.2006.02.002.

Valluru, R., Link, J., Claupein, W., 2011. Natural variation and morpho-physiological traits associated with water-soluble carbohydrate concentration in wheat under different nitrogen levels. *F. Crop. Res.* 124, 104–113. https://doi.org/10.1016/j.fcr.2011.06.008

Voytas, D.F., 2013. Plant genome engineering with sequence-specific nucleases. *Annu. Rev. Plant Biol.* 64, 327–350. https://doi.org/10.1146/ANNUREV-ARPLANT-042811-105552

Wang X, Wang HF, Chen Y, Sun MM, Wang Y, Chen YF., 2020. The transcription factor NIGT1.2 modulates both phosphate uptake and nitrate influx during phosphate starvation in Arabidopsis and Maize. *Plant Cell* 32(11), 3519–3534. doi: 10.1105/tpc.20.00361.

Wang, D., Lv, S., Jiang, P., Li, Y., 2017. Roles, regulation, and agricultural application of plant phosphate transporters. *Front. Plant Sci.* 8. https://doi.org/10.3389/FPLS.2017.00817

Wang, H., Xu, Q., Kong, Y.H., Chen, Y., Duan, J.Y., Wu, W.H., Chen, Y.F., 2014. Arabidopsis WRKY45 transcription factor activates *PHOSPHATE TRANSPORTER1;1* expression in response to phosphate starvation. *Plant Physiol.* 164, 2020–2029. https://doi.org/10.1104/pp.113.235077

Wang, J., Wu, B., Lu, K., Wei, Q., Qian, J., Chen, Y., Fang, Z., 2019. The amino acid permease 5 (Osaap5) regulates tiller number and grain yield in rice. *Plant Physiol.* 180, 1031–1045. https://doi.org/10.1104/PP.19.00034

Wang, M., Gruissem, W., Bhullar, N.K., 2013. Nicotianamine synthase overexpression positively modulates iron homeostasis-related genes in high iron rice. *Front. Plant Sci.* 4. https://doi.org/10.3389/fpls.2013.00156

Wang, W., Wang, W., Pan, Y., Tan, C., Li, H., Chen, Y., Liu, X., Wei, J., Xu, N., Han, Y. and Gu, H., 2022. A new gain-of-function OsGS2/GRF4 allele generated by CRISPR/Cas9 genome editing increases rice grain size and yield. *The Crop J.* 10(4), 1207–1212.

Wang, Q., Nian, J., Xie, X., Yu, H., Zhang, J., Bai, J., Dong, G., Hu, J., Bai, B., Chen, L., others, 2018. Genetic variations in ARE1 mediate grain yield by modulating nitrogen utilization in rice. *Nat. Commun.* 9, 1–10.

Wang, Q., Su, Q., Nian, J., Zhang, J., Guo, M., Dong, G., Hu, J., Wang, R., Wei, C., Li, G., others, 2021. The Ghd7 transcription factor represses ARE1 expression to enhance nitrogen utilization and grain yield in rice. *Mol. Plant* 14, 1012–1023.

Wang, Y., Wu, W.H., 2013. Potassium transport and signaling in higher plants. *Annu. Rev. Plant Biol.* https://doi.org/10.1146/annurev-arplant-050312-120153

Wang, Y., Wu, W.H., 2015. Genetic approaches for improvement of the crop potassium acquisition and utilization efficiency. *Curr. Opin. Plant Biol.* 25, 46–52. https://doi.org/10.1016/J.PBI.2015.04.007

Wu, Z., Liang, F., Hong, B., Young, J.C., Sussman, M.R., Harper, J.F., Sze, H., 2002. An endoplasmic reticulum-bound Ca2+/Mn2+ pump, ECA1, supports plant growth and confers tolerance to Mn2+ stress. *Plant Physiol.* 130, 128–137.

Xiang, Y., Huang, Y. and Xiong, L., 2007. Characterization of stress-responsive CIPK genes in rice for stress tolerance improvement. *Plant physiology*, *144*(3), 1416–1428.

Xie, K., Yang, Y., 2013. RNA-Guided genome editing in plants using a CRISPR-Cas system. *Mol. Plant* 6, 1975–1983. https://doi.org/10.1093/mp/sst119

Xu, J., Shang, L., Wang, J., Chen, M., Fu, X., He, H., Wang, Z., Zeng, D., Zhu, L., Hu, J., others, 2021. The SEEDLING BIOMASS 1 allele from indica rice enhances yield performance under low-nitrogen environments. *Plant Biotechnol. J.* 19, 1681.

Xu, J.M., Wang, Z.Q., Wang, J.Y., Li, P.F., Jin, J.F., Chen, W.W., Fan, W., Kochian, L. V., Zheng, S.J., Yang, J.L., 2020. Low phosphate represses histone deacetylase complex1 to regulate root system architecture remodeling in *Arabidopsis*. *New Phytol.* 225, 1732–1745. https://doi.org/10.1111/NPH.16264

Xu, L., Zhao, H., Wan, R., Liu, Y., Xu, Z., Tian, W., Ruan, W., Wang, F., Deng, M., Wang, J., others, 2019. Identification of vacuolar phosphate efflux transporters in land plants. *Nat. Plants* 5, 84–94.

Xu B, Cheval C, Laohavisit A, Hocking B, Chiasson D, Olsson TSG, Shirasu K, Faulkner C, Gilliham M., 2017. A calmodulin-like protein regulates plasmodesmal closure during bacterial immune responses. *New Phytol.* 215(1), 77–84. doi: 10.1111/nph.14599.

Yamaji, N., Sasaki, A., Xia, J.X., Yokosho, K., Ma, J.F., 2013. A node-based switch for preferential distribution of manganese in rice. *Nat. Commun.* 4, 2442.

Yang, M., Zhang, W., Dong, H., Zhang, Y., Lv, K., Wang, D., Lian, X., 2013. OsNRAMP3 is a vascular bundles-specific manganese transporter that is responsible for manganese distribution in rice. *PLoS One* 8, e83990.

Yang, T., Hao, L., Yao, S., Zhao, Y., Lu, W., Xiao, K., 2016a. TabHLH1, a bHLH-type transcription factor gene in wheat, improves plant tolerance to Pi and N deprivation via regulation of nutrient transporter gene transcription and ROS homeostasis. *Plant Physiol. Biochem.* 104, 99–113. https://doi.org/10.1016/j.plaphy.2016.03.023

Yang, X., Nian, J., Xie, Q., Feng, J., Zhang, F., Jing, H., Zhang, J., Dong, G., Liang, Y., Peng, J., Wang, G., Qian, Q., Zuo, J., 2016b. Rice ferredoxin-dependent glutamate synthase regulates nitrogen-carbon metabolomes and is genetically differentiated between *japonica* and *indica* subspecies. *Mol. Plant* 9, 1520–1534. https://doi.org/10.1016/J.MOLP.2016.09.004

Yoshimoto, N., Inoue, E., Saito, K., Yamaya, T., Takahashi, H., 2003. Phloem-localizing sulfate transporter, Sultr1; 3, mediates re-distribution of sulfur from source to sink organs in Arabidopsis. *Plant Physiol.* 131, 1511–1517.

Yoshimoto, N., Inoue, E., Watanabe-Takahashi, A., Saito, K., Takahashi, H., 2007. Posttranscriptional regulation of high-affinity sulfate transporters in Arabidopsis by sulfur nutrition. *Plant Physiol.* 145, 378–388.

Yoshimoto, N., Takahashi, H., Smith, F.W., Yamaya, T., Saito, K., 2002. Two distinct high-affinity sulfate transporters with different inducibilities mediate uptake of sulfate in Arabidopsis roots. *Plant J.* 29, 465–473.

Zeeshan, M., Lu, M., Sehar, S., Holford, P., Wu, F., 2020. Comparison of biochemical, anatomical, morphological, and physiological responses to salinity stress in wheat and barley genotypes deferring in salinity tolerance. *Agronomy* 10, 127.

Zhang S, Zhu L, Shen C, Ji Z, Zhang H, Zhang T, Li Y, Yu J, Yang N, He Y, Tian Y, Wu K, Wu J, Harberd NP, Zhao Y, Fu X, Wang S, Li S. Natural allelic variation in a modulator of auxin homeostasis improves grain yield and nitrogen use efficiency in rice. Plant Cell. 2021 May 5;33(3):566-580. doi: 10.1093/plcell/koaa037. PMID: 33955496; PMCID: PMC8136903.

Zhang, J., Zhang, H., Li, S., Li, J., Yan, L., Xia, L., 2021a. Increasing yield potential through manipulating of an ARE1 ortholog related to nitrogen use efficiency in wheat by CRISPR/Cas9. *J. Integr. Plant Biol.* 63, 1649–1663. https://doi.org/10.1111/JIPB.13151/SUPPINFO

Zhang, J., Zhou, Z., Bai, J., Tao, X., Wang, L., Zhang, H., Zhu, J.-K., 2020a. Disruption of MIR396e and MIR396f improves rice yield under nitrogen-deficient conditions. *Natl. Sci. Rev.* 7, 102–112. https://doi.org/10.1093/nsr/nwz142

Zhang, S., Zhang, Y., Li, K., Yan, M., Zhang, J., Yu, M., Tang, S., Wang, L., Qu, H., Luo, L., Xuan, W., Xu, G., 2021b. Nitrogen mediates flowering time and nitrogen use efficiency via floral regulators in rice. *Curr. Biol.* 31, 671–683.e5.

Zhang, X., Li, Q., Xu, W., Zhao, H., Guo, F., Wang, P., Wang, Y., Ni, D., Wang, M., Wei, C., 2020c. Identification of MTP gene family in tea plant (*Camellia sinensis* L.) and characterization of CsMTP8. 2 in manganese toxicity. *Ecotoxicol. Environ. Saf.* 202, 110904.

Zhang, Z., Fu, D., Xie, D., Wang, Z., Zhao, Y., Ma, X., Huang, P., Ju, C., Wang, C., 2023. CBL1/9-CIPK23-NRAMP1 axis regulates manganese toxicity. *New Phytol.*

Zhao, P., You, Q., Lei, M., 2019. A CRISPR/Cas9 deletion into the phosphate transporter SlPHO1; 1 reveals its role in phosphate nutrition of tomato seedlings. *Physiol. Plant.* 167, 556–563.

Zhu H, Wang J, Jiang D, Hong Y, Xu J, Zheng S, Yang J, Chen W., 2022. The miR157-SPL-CNR module acts upstream of bHLH101 to negatively regulate iron deficiency responses in tomato. *J Integr Plant Biol.* 64(5), 1059–1075. doi: 10.1111/jipb.13251.

10 CRISPR-Cas Genome Modification for Non-Transgenic Disease-Resistant, High Yielding and High-Nutritional Quality Wheat

S. M. Hisam Al Rabbi, Israt Nadia, and Tofazzal Islam

10.1 INTRODUCTION

Wheat (*Triticum aestivum* L.) is the primary diet, supplying not only 20% of the total calories needed but also a quarter of protein consumption (FAO, http://faostat.fao.org). Various abiotic and biotic stresses significantly impact wheat productivity. Climate change makes these stresses more frequent in many wheat-growing countries. For example, the recent intercontinental spread and epidemic outbreaks of a destructive wheat blast disease in Bangladesh (Islam et al., 2016) and Zambia (Tembo et al., 2020) jeopardize the global production of wheat and food sufficiency (Islam et al., 2020; Islam et al., 2019). Through genomic research, a South American species called *Magnaporthe oryzae* (Latorre et al., 2022) was found to be widespread in both countries. There are some other diseases, such as Fusarium head blight (FHB), rust, powdery mildew, etc., that also cause yield reduction of wheat. On the other hand, the presence of gluten and asparagine in the grains of wheat is problematic for human health.

Wheat is a hexaploid plant that complicates its genetic improvement. It is likely abandoned by the genetic engineers. Conventional breeding of wheat takes 10–15 years for the improvement of a trait. Recently, the annotation of the whole genome of wheat has been published. The Clustered regularly interspaced short palindromic repeats (CRISPR)-Cas genome editing technology has emerged as a strong toolkit for editing the genome of crop plants since 2013 (Xie & Yang, 2013). Commercial products of CRISPR-edited plants and other organisms are now available in some countries. It is progressively used to change various plants' genomes, including wheat. The methodological improvement (Xie et al., 2015; Kim et al., 2018; Zhang et al., 2019b; Liu et al., 2020), and the utilization of CRISPR technology for the improvement of wheat has significantly progressed in recent years (Taj et al., 2022; Sánchez-León et al., 2018; Jouanin et al., 2020; Zhang et al., 2021a; Okada et al., 2019; Li et al., 2021c). This technology has been used in wheat improvement for disease resistance (Taj et al., 2022), grain quality (Sánchez-León et al., 2018), abiotic stress tolerance (Debbarma et al., 2019; Bhat et al., 2021), hybrid seed production (Longin et al., 2013; Mühleisen et al., 2014), functional genomics (Li et al., 2021a) and disease diagnosis at the point-of-care (Sánchez et al., 2022). Multiplex genome editing, base editing, prime editing, and epigenetic editing are the recent advancements of the CRISPR knowledge that offer new opportunity for the rapid

DOI: 10.1201/9781003387060-10

development of transgene-free stress-tolerant, high-quality, and high-yielding wheat. Nevertheless, some challenges exist in the uses of the CRISPR-Cas techniques of hexaploid wheat and their practical application in the field.

Some good articles on CRISPR-Cas genome modification of wheat and some other economically important crop plants have been described in some research and review articles (Sánchez-León et al., 2018; Haque et al. 2018; Islam 2019; Li et al. 2022). This chapter reviews the updates of CRISPR-Cas genome modification in wheat and discusses the promise of this revolutionary technology for developing transgene-free disease-resistant, high-yielding, and high-nutritional quality wheat. The knowledge gap and challenges of the CRISPR application techniques in the editing of wheat are also articulated.

10.2 CRISPR-Cas SYSTEM IN NATURE

CRISPR and Cas proteins are components of the bacterial and archaeal protective mechanisms against invaders like plasmids and viruses (Bhaya et al., 2011; Terns & Terns, 2011; Wiedenheft et al., 2012) using small RNAs to identify invading nucleic acids by matching and subsequently inactivating them by cleavage. Cas genes are arranged in operons in CRISPR-Cas systems. CRISPR arrays are made up of sequences that target invading genomes (spacers) interspersed with palindromic repeats (Bhaya et al., 2011; Terns & Terns, 2011; Wiedenheft et al., 2012).

Immunity mediated by CRISPR-Cas takes place in three stages. Firstly, the virus and plasmid kept invading bacteria and archaea. At one point, some foreign sequences from the attackers got introgressed near the CRISPR array's proximal end of the host chromosome. They remain there afterward as protospacers (Bhaya et al., 2011; Terns & Terns, 2011; Wiedenheft et al., 2012). Secondly, the precursor CRISPR RNA (pre-crRNA) molecules are produced from the CRISPR loci upon invasion. The endogenous nuclease then separates them from one another (Deltcheva et al., 2011). Thirdly, the resulting short CRISPR RNAs (crRNAs) identify and duo with the corresponding protospacer nucleotide sequences of the attacking organisms and lead Cas proteins to cleave and inactivate (Figure 10.1) (Brouns et al., 2008).

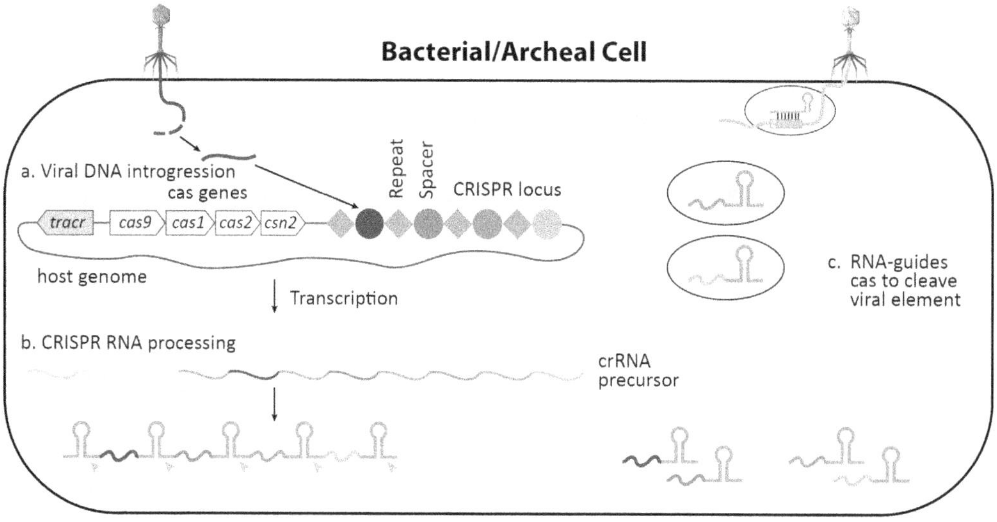

FIGURE 10.1 The basic concept of CRISPR-based immune system of bacteria and archea.

10.3 CRISPR-Cas AS A TECHNOLOGY OF GENOME EDITING

After Jinek et al. (2012) engineered the natural CRISPR genome editing, it became programmable for the beneficial use of mankind. Trans-activating crRNA (tracrRNA) and mature crRNAs naturally couple up to form a dual RNA structure. This structure joins with Cas9 enzymes and guides them to find the target location to make the double-stranded cleavage. The nuclease has two domains. The Cas9 RuvC domain breaks the non-complementary strand, and the Cas9 HNH breaks the complementary strand to the crRNA-guide sequence.

The natural dual-tracrRNA:crRNA structure was fixed up in a single RNA chimera, later referred to as single guide (sg) RNA, to control sequence-specific breakage making the genome editing process much simpler (Figure 10.2) (Jinek et al., 2012). The sgRNA's guide sequence (usually 20 nucleotides) directs CRISPR-Cas9 to edit any desired DNA sequence whenever it is close to a protospacer adjacent motif (PAM). Therefore, now the CRISPR-Cas9 method just needs the guide RNA sequence for the modification, unlike ZFNs or TALENs. This is why CRISPR-Cas9 technology has advanced so quickly. Because of this, the scientific community has quickly and broadly embraced the CRISPR-Cas9 technology to manipulate or alter the targeted genomic fragments of various cells and species (Doudna & Charpentier, 2014).

The initial single-target editing system was modified to spawn many gRNAs from one polycistronic gene (Xie et al., 2015). The endogenous tRNA-processing mechanism was utilized to introduce multiplex editing. Precisely cleaving habit at both ends of the tRNA precursor enables the multiplexing to be straightforward. Synthetic genes arranged in tandem with tRNA-gRNA structure were successfully and accurately translated *in vivo* into gRNAs containing the necessary 5′ targeting sequences, which instructed Cas9 to alter several targets on the chromosome. With high effectiveness (up to 100%), multiplex genome editing and deletion of chromosomal fragments were easily accomplished in transgenic rice plants. Because tRNA's processings are essentially constant across all creatures, it has received a lot of attention and has been altered in several ways for the convenience of researchers.

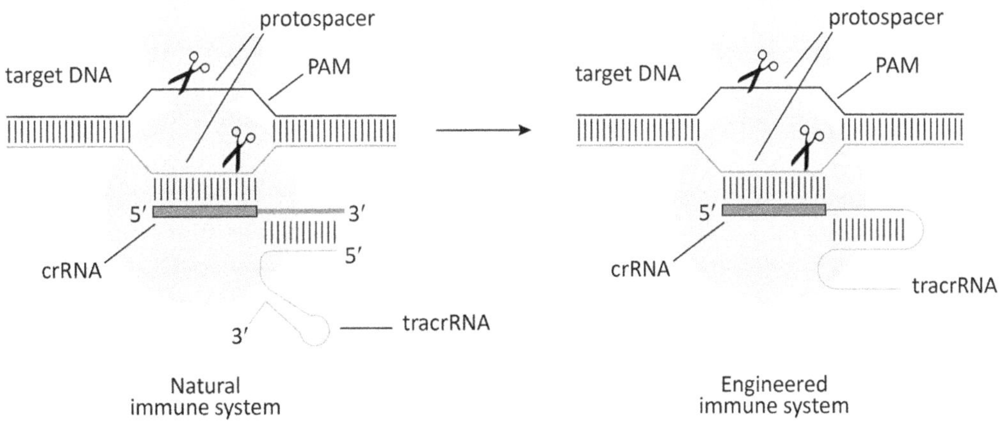

FIGURE 10.2 The engineering of sgRNA from the dual RNA system.

10.4 CRISPR-Cas GENOME EDITING OF WHEAT

The biolistics-based transformation technique is used most commonly for delivering the CRISPR-Cas9 complex in wheat. Lower efficiency to recover sought-after mutants remained a major hurdle for many genome editing techniques in wheat. These techniques were eventually required to maintain a large plant population. In this light, an *Agrobacterium*-delivered CRISPR-Cas9 system has been introduced with a single binary vector, where a ubiquitin gene and a maize promoter were utilized to express Cas9 and wheat U6 promoters were used to express the guide RNA cassette (Zhang et al., 2019b). The system in wheat has created 68 mutants for the TaCKX2-1, TaGW2, TaGLW7, and TaGW8 grain-regulatory genes in the plants of T0, T1, and T2 generation creating 10% mutations. In a single generation, homozygous stable mutations can be achieved from a sizable population. Unlike the majority of plant species, more than 10 bp deletions are the most common kinds of mutation in wheat. TaCKX2-D1 homozygous plants dramatically enhanced the number of grains per spikelet. Thus, the *Agrobacterium*-mediated delivery mechanism with CRISPR-Cas9 can be offered as a superior alternative strategy for editing the wheat genome (Zhang et al., 2019b).

Multiplexing in wheat (Liu et al., 2020) with the promoters (OsU6a, TaU3, and TaU6) was proven to be a better option. Here *Agrobacterium* was used to introduce sgRNAs into wheat. They found TaU3 was a superior option to TaU6 and OsU6a. In addition, two sgRNAs were more successful at editing than one because they produced mutants with large-scale section deletions between the two sgRNAs.

Kim et al. (2018) used the CRISPR-Cas9 genome editing method in the wheat protoplast to carry out targeted editing of the genes for two stress-responsive transcription factors, wheat dehydration-responsive element binding protein 2 (TaDREB2), and wheat ethylene-responsive factor 3 (TaERF3). Using the temporary expression of sgRNA and the Cas9 protein within wheat protoplasts, genome alteration of TaDREB2 and TaERF3 was accomplished. With the help of sequencing, T7 endonuclease assay, and restriction enzyme digestion assay, it was confirmed that wheat protoplast mutagenesis was more successful. Also, many constructed sgRNA off-target sites were examined, and amplicon sequencing proved the uniqueness of genome editing. General conclusions indicated that the CRISPR-Cas9 genome editing might be set up on the protoplasts of wheat with great ease (Jouanin et al., 2020).

10.5 CRISPR IN CREATING LONG-LASTING DISEASE-RESISTANT WHEAT

CRISPR genome editing paves new ways to design disease-resistant crops. It addresses the key barriers to crop development by modifying the genome to precisely yield non-transgenic plants (Zaidi et al., 2020; Taj et al., 2022). Therefore, it may effectively be utilized to target wheat genes that are vulnerable to biotic stress.

To uncover harmful susceptible genes (S genes) in bread wheat, published research papers revealing the genetic factors related to different diseases and various insect pests were thoroughly analyzed. Thirty-one genetic variables (S genes) were discovered for powdery mildew, stripe rust, FHB, and leaf rust (Table 10.1) to be negative regulators of these diseases out of all known genetic factors, implying that their suppression, deletion, or down-regulation are crucial. In the case of powdery mildew, five S genes have been reported so far. Using CRISPR only two of them were inactivated. Virus-induced gene silencing (VIGS) was utilized to inactivate the remaining genes (Table 10.1). Functional genomics through VIGS discovered 16 S alleles for stripe rust in several studies. FHB has nine reported S genes. CRISPR-based knockout was done only in two cases of them. By RANi or VIGS, the others are silenced. By using VIGS, the only S gene linked to leaf rust was silenced. (Table 10.1).

A milestone was achieved in retaining yield along with wheat disease resistance (Li et al., 2022). Previously it was known that mildew resistance locus (TaMLO) loss-of-function mutations confer wheat with long-term and overall resistance in powdery mildew (Jaganathan et al.,

TABLE 10.1
Discovered S Genes for the Major Wheat Diseases, and Their Editing Nature

Disease	S Gene/Genes	Editing Nature	References
Powdery mildew	TaMLO	CRISPR mediated knockout	Jaganathan et al. (2018)
	TaEDR1	CRISPR mediated knockout	Zaidi et al. (2018)
	TaHDA6, TaHDT701, TaHOS15	Knockdown by virus-induced gene silencing (VIGS)	Zhi et al. (2020)
Stripe rust	TaNAC2	Knockdown by VIGS	Zhang et al. (2018)
	TaRop10 and TaTrxh9	Knockdown by VIGS	Shi et al. (2021)
	TaNAC30	Knockdown by VIGS	Wang et al. (2018)
	TaNAC1	Knockdown by VIGS	Wang et al. (2015)
	TaNAC21/22	Knockdown by VIGS	Feng et al. (2014a)
	TaSTP13	Knockdown by VIGS	Huai et al. (2020)
	TaWRKY49	Knockdown by VIGS	Wang et al. (2017)
	TaLSD1	Knockdown by VIGS	Guo et al. (2013)
	TaMDHAR4	Knockdown by VIGS	Feng et al. (2014b)
	TaDIR1-2	Knockdown by VIGS	Ahmed et al. (2017)
	TaULP5	Knockdown by VIGS	Feng et al. (2016)
	TaNUDX23	Knockdown by VIGS	Yang et al. (2020)
	TaMDAR6	Knockdown by VIGS	Abou-Attia et al. (2016)
	TaEIL1	Knockdown by VIGS	Duan et al. (2013)
	TaBln1	Knockdown by VIGS	Guo et al. (2022)
Fusarium head blight (FHB)	Ta_PYL4AS_A	Knockdown by VIGS	Gordon et al. (2016); Fabre et al. (2020)
	TaSSI2	Knockdown by VIGS	Hu et al. (2018); Fabre et al. (2020)
	TaNAC032	Knockdown by VIGS	Soni et al. (2021)
	TaTIR1	RNAi-mediated knockdown	Su et al. (2021)
	TaLpx-1	RNAi-mediated knockdown	Nalam et al. (2015); Jaganathan et al. (2018)
	TaHRC	CRISPR mediated knockout	Su et al. (2019); Li et al. (2021b)
	TaABCC6, TaNFXL1, TansLTP9.4	CRISPR mediated knockout	Zaman et al. (2019)
Leaf rust	TaNAC35	VIGS	Zhang et al. (2021b)

2018). However, the resistance was related to longer growth days and yield reduction (Büschges et al., 1997; Consonni et al., 2006). Therefore, this resistance was not very much appealing to the farmers. The knocking down of the MLO-B1 locus activates Tonoplast monosaccharide transporter 3 (TaTMT3B), which alleviates the growth and yield losses linked with MLO disruption (Li et al., 2022).

10.6 CRISPR FOR POINT-OF-CARE WHEAT DISEASE DIAGNOSIS

Efficient diagnosis and genotyping are prerequisites for any effective disease management or breeding scheme. For plant genotyping, pathogen detection, and GMO identification in crops, traditional techniques like visual inspection, pure culture isolation, microscopy, polymerase chain reaction (PCR), immunological tests, matrix-assisted laser desorption ionization-time of flight mass spectrometry (MALDI-TOF MS), DNA microarrays (Khakimov et al., 2022), DNA hybridization and restriction fragment length polymorphism (RFLP), are frequently used (Henry, 2014). Nucleic acids are the primary biomarkers for allele tracking, infections, and biological pollutants because of their stability, repeatable amplification, and ease of interfacing with various reporter systems (Zheng et al., 2021). However, even PCR, the currently most widely used technique for detecting nucleic acids, has significant flaws. It is lengthy and needs specialized know-how and lab facilities (Zheng et al., 2021; Zhu et al., 2020). Together, the traditional detection techniques are difficult and time-consuming, requiring specialized laboratory equipment. Hence, point-of-care quick identification systems that are straightforward, accurate, and deployable in the field are urgently needed to replace lab-based time-consuming phenotyping and genotyping.

For field detection, isothermal amplification techniques are suitable as they are fast and single-stepped, and appropriate for point-of-care testing (POCT) and point-of-need testing (PONT) applications. Among these techniques, recombinase polymerase amplification (RPA) and loop-mediated isothermal amplification (LAMP) are most suitable for POCT. With the potential for a visual readout on lateral flow assay (LFA) strips or colorimetric assays, integrating isothermal amplification with CRISPR systems, in particular, makes the tools very sensitive to precisely detect mutations, transgenes, and phytopathogens (Aman et al., 2020; Kaminski et al., 2021; Zheng et al., 2021; Zou et al., 2020; Sánchez et al., 2022).

Some other CRISPR-based detections have been developed like RT-LAMP and DETECTR to produce an LFA readout (Broughton et al., 2020; Sánchez et al., 2022). Another technique, AIOD-CRISPR, needs only a single pot for fluorescent detections (Ding et al., 2020). Again, Bio-SCAN, an emerging biotin-CRISPR coupling-based extraordinarily perceptive, economical, and user-friendly pathogen detection platform has recently been developed (Ali et al., 2022). It has been successfully deployed to screen wheat germplasm for the Lr34 and Lr67 alleles rendering overall stripe rust resistance. Also, it was very efficient to detect two disastrous fungal pathogens (*Puccinia striiformis* f. sp. *tritici* and *Magnaporthe oryzae Triticum*) in wheat (Sánchez et al., 2022).

10.7 IMPROVEMENT OF NUTRITIONAL QUALITY OF GRAIN

Coeliac disease, an autoimmune illness, develops in people who are genetically prone to it when wheat, rye, and barley gluten proteins are consumed. The causal peptide for the disease lies in the a-gliadin gene family. Two sgRNAs were created (Sánchez-León et al., 2018) to specifically target a conserved area next to the 33-mer coding sequence in the a-gliadin genes. Twenty-one mutant lines were produced, and all of them displayed a significant decrease in a-gliadins. A line was achieved where 35 genes were mutated out of 45 identified genes related to the disease showing 85% less immunoreactivity. Transgene-free lines were found, and none of the lines showed any off-target alterations. The transgene-free wheat lines with lesser gluten might be used to create safer food and act as a starting point for introducing this feature into superior wheat cultivars (Sánchez-León et al., 2018). Also, Jouanin et al. (2020) inactivated gliadins lessening the exposure of patients to the disease.

Again, amylose and resistant starch (RS) rich foods can considerably improve the health of people and reduce the likelihood of non-infectious, life-threatening illnesses. But in contemporary wheat varieties, this crucial RS is still scarce. Li et al. (2021a) reported that CRISPR-Cas9-mediated modification of TaSBEIIa in the winter wheat cv Zhengmai 7698 (ZM) and spring wheat cv Bobwhite resulted in the production of high-amylose wheat. They created several non-transgenic mutant lines in ZM and Bobwhite, with either partial or triple-null TasbeIIa alleles. Triple-null lines showed more substantial effects on the composition of starch, amylopectin structures, nutrient content, and physiochemical qualities. The triple-null lines' flours had many more essential compounds for public health like soluble pentosan, RS, amylose, and protein. Together, they produced transgene-free high-amylose and high RS wheat.

10.8 CRISPR FOR MORE PRODUCTIVE WHEAT

The aberrant cytokinin response1 repressor1 (ARE1) orthologs may be altered using CRISPR to develop high-yielding wheat with high nitrogen use efficiency (NUE) (Zhang et al., 2021a). Fertilizers containing nitrogen (N) are widely used in agriculture to increase wheat output to fulfill the rising global demand for food. The low NUE of current wheat varieties and excessive N fertilizer application, however, are accelerating ecological decline and pollution. To enhance these yield-related traits, the premium Chinese winter wheat cultivar Zheng Mai 7698 was then taken for editing TaARE1. Several mutant lines without transgenes were created with either triple-null or partial TaARE1 alleles using CRISPR-Cas9. Every mutant line without transgenes showed improved resistance to N deficiency, delayed senescence, and greater grain production under field circumstances. Particularly, the mutant lines AABBdd and aabbDD displayed normal growing late senescent plants producing more grains than the wild control. Together, the findings highlight the possibility of modifying ARE1 orthologs using gene editing for high-yielding wheat breeding with enhanced NUE (Zhang et al., 2021a).

10.9 HYBRID SEED PRODUCTION AND FUNCTIONAL GENOMICS OF WHEAT

Hybrid breeding technology has achieved enormous yield gains in different crops across various production environments since it offers more than 10% yield heterosis (Longin et al., 2013; Mühleisen et al., 2014). To utilize the advantage, the hybrid seed has a long history of being widely used, especially when it comes to rice and maize. Wheat lacks the privilege due to its pronounced inbreeding nature, and the lack of an easy and efficient hybrid seed production system. Strong inbreeding nature is the primary obstacle since cross-pollination between inbred lines with different genetic backgrounds is a prerequisite for any commercial hybrid seed production. Female parent self-pollination must be avoided. Unluckily, we are lacking simple methods to separate sexes and force them for outcrossing in wheat (Whitford et al., 2013). To overcome this hurdle, CRISPR-Cas9 was used. The CRISPR-Cas9 system created Ms1 inactivated wheat with complete male sterility. The introduction of recessive ms1 alleles for male sterility makes a wheat hybridization platform commercially viable (Okada et al., 2019).

CRISPR also breaks the deadlock and brings momentum to the functional genomics of wheat. Being an allohexaploid ($2n$ = 6 9 = 42, AABBDD), wheat has three similar subgenomes derived from three kin parents (Petersen et al., 2006). Therefore, it has three similar, functionally redundant, and complementary copies of each gene. As a result, it is very unlikely that the genes in the A, B, and D genomes would mutate simultaneously by spontaneous or induced mutagenesis to make a mutant phenotype successfully. Mainly this complicated polyploidy in wheat limits its functional genomics and breeding to flourish, particularly in comparison to other grains like rice and corn (Borrill et al., 2015). The utilization of CRISPR-Cas9 made it possible to target several homoeoalleles of the wheat genome at the same time and subsequent mutant phenotype generation to hasten the advancement of functional genomics. Already a review has been published of wheat functional genomics where the outcomes of mutation with CRISPR for forty-four genes in wheat were studied (Li et al., 2021a).

10.10 TISSUE CULTURE IS NO LONGER ESSENTIAL IN CRISPR

An effective delivery system is needed for large-scale plant genome modification (Atkins & Voytas, 2020; Gao, 2021). The current plant genome modification system necessitates tissue culture, which is a lengthy and laborious process with a limited regeneration ability that confines its application to a narrow variety of genotypes (Li et al., 2021c). Wheat is also less responsive to tissue culture. However, in wheat, hopefully, transformation is taking place within an active meristem (Hamada et al., 2017). In this approach, somatic cells of entire plants get developmental controllers and gene-editing tools. This causes meristems to produce shoots with specified DNA alterations and gene edits are passed down to the next level. The *de novo* gene-edited meristem formation bypasses tissue culture and is expected to solve the bottlenecking in gene-editing activities (Maher et al., 2020).

A sgRNA delivery vector based on the Barley Stripe Mosaic Virus (BSMV) has been proven efficient for Cas9-based editing in wheat. Three wheat varieties had mutant progenies in the following generation ranging from 12.9% to 100%, and 53.8%–100%. Also, modifications can be multiplexed designing the vector containing various sgRNAs. Additionally, Cas9-edited wheat pollen was used for crossing with wild-type wheat to develop an edited wheat plant without transgenes (Li et al., 2021c).

10.11 CONCLUSION

Wheat provides about 20% of calories worldwide (Rabbi et al., 2021) and needs further improvement in terms of yields, and nutrition for food security. Although climate change is threatening us with the potential advent of new wheat diseases, and the world is already overburdened with the population which tends to be ever-increasing, CRISPR beckons the scientific world with hope. It makes us optimistic to take care of a lot of issues that appeared to be almost irreparable just a decade ago.

CRISPR is the archaeal and bacterial adaptive immune system to combat the attacking viruses and plasmids was a two-RNA system initially. Later, it was devised as a single RNA system. This event opened a new arena of CRISPR utilization. Since then, CRISPR become programmable. People started genome editing using CRISPR, especially for silencing negative regulators. Finally, the engineering process is further super-tuned by multiplexing, by which more than one genomic site can be sliced accommodating more target sequences into a single construct. Endogenous nucleases are utilized here for cleavage of the RNA transcript to make each guide RNA independent so that they can associate with Cas protein and break down different targets. These multiplexing systems are getting exceedingly popular since they can make bigger deletions in the genome creating higher mutation frequencies.

Although Arabidopsis and rice were extensively genome edited for so many traits, wheat has a slow adaptation rate. Mainly because of the large and complex wheat genome and also less responsiveness of wheat to tissue culture. The initial biolistic plasmid delivery system also has some shortfalls. Later *Agrobacterium*-mediated plasmid delivery system was optimized and proved to be more efficient. Also, protoplast was successfully used as a starting material to work on for genome editing.

A coeliac disease patient does not necessarily curtail consuming wheat commodities now, since the disease-provoking gene in the wheat genome can be edited through CRISPR. Also, more nutritious, amylose, and RS-rich wheat, is now available. This genome editing technology opens up the improvisation of several POCTs for field deployable pathogen detection of wheat. Production of wheat hybrid seeds on an industrial scale gets handy through the introduction of male sterile alleles through genome editing. Also, functional genomics in wheat could not flourish since the fact that, being allohexaploid wheat has three similar copies for each gene. Traditional mutations were unlikely to mutate all simultaneously. Mutation in a single copy could not stop the expression completely impeding mutant phenotypes. The potentiality of CRISPR was explored in wheat to target mutate in many sites at a time to create mutant phenotypes for functional genomics study on many occasions. Less responsiveness of wheat for tissue culture can be bypassed now through the in-planta application of CRISPR. Also, BSMV is now introduced for the same purpose. Genome-edited pollen can be directly used for fertilization to avoid tissue culture.

Functional genomics provided thirty-one negative regulators for making wheat more disease-resistant. The majority of them were silenced either by VIGS or RANi. Only a few plants were knocked out with the help of CRISPR for wheat functional genomics. Therefore, there is ample opportunity to make wheat disease-resistant with CRISPR to create more appealing non-transgenic plants. Also, different alleles can be knocked out altogether to get pyramided knocked out alleles

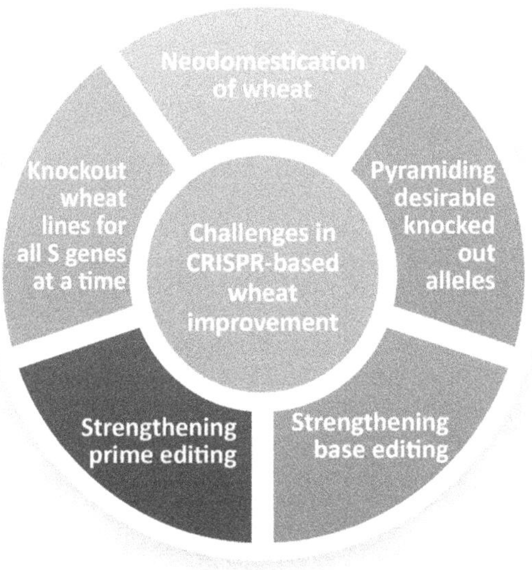

FIGURE 10.3 The challenges of CRISPR for the development of wheat.

for the S gene making more stable disease resistance. Base editing could be used more frequently (Komor et al., 2016) in wheat (Zhang et al., 2019a) to increase the mutation rate. Also, prime editing (Anzalone et al., 2019) is now obtainable to incorporate a correct nucleotide sequence replacing an erroneous one in a gene that is yet to be utilized fully for wheat development (Lin et al., 2020). Neodomestication is also another concept that largely can utilize the adaptation quality of wild wheat germplasm to the stresses by domesticating it using CRISPR (Figure 10.3) (Hickey et al., 2019).

All this compiled information in a single manuscript will help wheat researchers in their exploration. Also, it will help to find out research gaps in this area to work on. Together, they will hopefully make a mark for the food security of the world.

REFERENCES

Abou-Attia, M. A., Wang, X., Nashaat Al-Attala, M., Xu, Q., Zhan, G., & Kang, Z. (2016). TaMDAR6 acts as a negative regulator of plant cell death and participates indirectly in stomatal regulation during the wheat stripe rust-fungus interaction. *Physiologia Plantarum*, *156*(3), 262–277. https://doi.org/10.1111/ppl.12355

Ahmed, S. M., Liu, P., Xue, Q., Ji, C., Qi, T., Guo, J., Guo, J., & Kang, Z. (2017). TaDIR1-2, a wheat ortholog of lipid transfer protein AtDIR1 contributes to negative regulation of wheat resistance against Puccinia striiformis f. sp. *tritici*. *Frontiers in Plant Science*, *8*(April). https://doi.org/10.3389/fpls.2017.00521

Ali, Z., Sánchez, E., Tehseen, M., Mahas, A., Marsic, T., Aman, R., Sivakrishna Rao, G., Alhamlan, F. S., Alsanea, M. S., Al-Qahtani, A. A., Hamdan, S., & Mahfouz, M. (2022). Bio-SCAN: A CRISPR/dCas9-based lateral flow assay for rapid, specific, and sensitive detection of SARS-CoV-2. *ACS Synthetic Biology*, *11*(1), 406–419. https://doi.org/10.1021/acssynbio.1c00499

Aman, R., Mahas, A., & Mahfouz, M. (2020). Nucleic acid detection using CRISPR/Cas biosensing technologies. *ACS Synthetic Biology*, *9*(6), 1226–1233. https://doi.org/10.1021/acssynbio.9b00507

Anzalone, A. V., Randolph, P. B., Davis, J. R., Sousa, A. A., Koblan, L. W., Levy, J. M., Chen, P. J., Wilson, C., Newby, G. A., Raguram, A., & Liu, D. R. (2019). Search-and-replace genome editing without double-strand breaks or donor DNA. *Nature*, 576(7785), 149–157. https://doi.org/10.1038/s41586-019-1711-4

Atkins, P. A., & Voytas, D. F. (2020). Overcoming bottlenecks in plant gene editing. *Current Opinion in Plant Biology*, *54*, 79–84. https://doi.org/10.1016/j.pbi.2020.01.002

Bhat, M. A., Mir, R. A., Kumar, V., Shah, A. A., Zargar, S. M., Rahman, S., & Jan, A. T. (2021). Mechanistic insights of CRISPR/Cas-mediated genome editing towards enhancing abiotic stress tolerance in plants. *Physiologia Plantarum*, *172*(2), 1255–1268. https://doi.org/10.1111/ppl.13359

Bhaya, D., Davison, M., & Barrangou, R. (2011). CRISPR/Cas systems in bacteria and archaea: Versatile small RNAs for adaptive defense and regulation. *Annual Review of Genetics*, *45*, 273–297. https://doi.org/10.1146/annurev-genet-110410-132430

Borrill, P., Adamski, N., & Uauy, C. (2015). Genomics as the key to unlocking the polyploid potential of wheat. *New Phytologist*, 208(4), 1008–1022.

Broughton, J. P., Deng, X., Yu, G., Fasching, C. L., Servellita, V., Singh, J., Miao, X., Streithorst, J. A., Granados, A., Sotomayor-Gonzalez, A., Zorn, K., Gopez, A., Hsu, E., Gu, W., Miller, S., Pan, C. Y., Guevara, H., Wadford, D. A., Chen, J. S., & Chiu, C. Y. (2020). CRISPR-Cas12-based detection of SARS-CoV-2. *Nature Biotechnology*, *38*(7), 870–874. https://doi.org/10.1038/s41587-020-0513-4

Brouns, S. J. J., Jore, M. M., Lundgren, M., Westra, E. R., Slijkhuis, R. J. H., Snijders, A. P. L., Dickman, M. J., Makarova, K. S., Koonin, E. V., & Van Der Oost, J. (2008). Small CRISPR RNAs guide antiviral defense in prokaryotes. *Science*, *321*(5891), 960–964. https://doi.org/10.1126/science.1159689

Büschges, R., Hollricher, K., Panstruga, R., Simons, G., Wolter, M., Frijters, A., Van Daelen, R., Van der Lee, T., Diergaarde, P., Groenendijk, J., Töpsch, S., Vos, P., Salamini, F., & Schulze-Lefert, P. (1997). The barley Mlo gene: A novel control element of plant pathogen resistance. *Cell*, *88*(5), 695–705. https://doi.org/10.1016/S0092-8674(00)81912-1

Consonni, C., Humphry, M. E., Hartmann, H. A., Livaja, M., Durner, J., Westphal, L., Vogel, J., Lipka, V., Kemmerling, B., Schulze-Lefert, P., Somerville, S. C., & Panstruga, R. (2006). Conserved requirement for a plant host cell protein in powdery mildew pathogenesis. *Nature Genetics*, *38*(6), 716–720. https://doi.org/10.1038/ng1806

Debbarma, J., Sarki, Y. N., Saikia, B., Boruah, H. P. D., Singha, D. L., & Chikkaputtaiah, C. (2019). Ethylene response factor (ERF) family proteins in abiotic stresses and CRISPR-Cas9 genome editing of ERFs for multiple abiotic stress tolerance in crop plants: A review. *Molecular Biotechnology*, *61*(2), 153–172. https://doi.org/10.1007/s12033-018-0144-x

Deltcheva, E., Chylinski, K., Sharma, C. M., Gonzales, K., Chao, Y., Pirzada, Z. A., Eckert, M. R., Vogel, J., & Charpentier, E. (2011). CRISPR RNA maturation by trans-encoded small RNA and host factor RNase III. *Nature*, *471*(7340), 602–607. https://doi.org/10.1038/nature09886

Ding, X., Yin, K., Li, Z., Lalla, R. V., Ballesteros, E., Sfeir, M. M., & Liu, C. (2020). Ultrasensitive and visual detection of SARS-CoV-2 using all-in-one dual CRISPR/Cas12a assay. *Nature Communications*, *11*(1), 1–10. https://doi.org/10.1038/s41467-020-18575-6

Doudna, J. A., & Charpentier, E. (2014). The new frontier of genome engineering with CRISPR/Cas9. *Science*, *346*(6213). https://doi.org/10.1126/science.1258096

Duan, X., Wang, X., Fu, Y., Tang, C., Li, X., Cheng, Y., Feng, H., Huang, L., & Kang, Z. (2013). TaEIL1, a wheat homologue of AtEIN3, acts as a negative regulator in the wheat-stripe rust fungus interaction. *Molecular Plant Pathology*, *14*(7), 728–739. https://doi.org/10.1111/mpp.12044

Fabre, F., Rocher, F., Alouane, T., Langin, T., & Bonhomme, L. (2020). Searching for FHB resistances in bread wheat: susceptibility at the crossroad. *Frontiers in Plant Science*, *11*(June), 731. https://doi.org/10.3389/fpls.2020.00731

Feng, H., Duan, X., Zhang, Q., Li, X., Wang, B., Huang, L., Wang, X., & Kang, Z. (2014a). The target gene of tae-miR164, a novel NAC transcription factor from the NAM subfamily, negatively regulates resistance of wheat to stripe rust. *Molecular Plant Pathology*, *15*(3), 284–296. https://doi.org/10.1111/mpp.12089

Feng, H., Liu, W., Zhang, Q., Wang, X., Wang, X., Duan, X., Li, F., Huang, L., & Kang, Z. (2014b). TaMDHAR4, a monodehydroascorbate reductase gene participates in the interactions between wheat and *Puccinia striiformis* f. sp. *tritici*. *Plant Physiology and Biochemistry*, *76*, 7–16.

Feng, H., Wang, Q., Zhao, X., Han, L., Wang, X., & Kang, Z. (2016). TaULP5 contributes to the compatible interaction of adult plant resistance wheat seedlings-stripe rust pathogen. *Physiological and Molecular Plant Pathology*, *96*, 29–35. https://doi.org/10.1016/j.pmpp.2016.06.008

Gao, C. (2021). Genome engineering for crop improvement and future agriculture. *Cell*, *184*(6), 1621–1635. https://doi.org/10.1016/j.cell.2021.01.005

Gordon, C. S., Rajagopalan, N., Risseeuw, E. P., Surpin, M., Ball, F. J., Barber, C. J., Buhrow, L. M., Clark, S. M., Page, J. E., Todd, C. D., Abrams, S. R., & Loewen, M. C. (2016). Characterization of triticum aestivum abscisic acid receptors and a possible role for these in mediating fusairum head blight susceptibility in wheat. *PLoS One*, *11*(10), 1–23. https://doi.org/10.1371/journal.pone.0164996

Guo, J., Bai, P., Yang, Q., Liu, F., Wang, X., Huang, L., & Kang, Z. (2013). Wheat zinc finger protein TaLSD1, a negative regulator of programmed cell death, is involved in wheat resistance against stripe rust fungus. *Plant Physiology and Biochemistry*, *71*, 164–172. https://doi.org/10.1016/j.plaphy.2013.07.009

Guo, S., Zhang, Y., Li, M., Zeng, P., Zhang, Q., Li, X., Xu, Q., Li, T., Wang, X., Kang, Z., & Zhang, X. (2022). TaBln1, a member of the Blufensin family, negatively regulates wheat resistance to stripe rust by reducing Ca^{2+} influx. Plant Physiology, *189*(3), 1380–1396. https://doi.org/10.1093/plphys/kiac112

Hamada, H., Linghu, Q., Nagira, Y., Miki, R., Taoka, N., & Imai, R. (2017). An in planta biolistic method for stable wheat transformation. *Scientific Reports*, *7*(1), 2–9. https://doi.org/10.1038/s41598-017-11936-0

Haque, E., Taniguchi, H., Hassan, M.M., Bhowmik, P., Karim, M.R., Smiech, M. et al. (2018). Application of CRISPR/Cas9 genome editing technology for the improvement of crops cultivated in tropical climates: recent progress, prospects, and challenges. *Frontiers in Plant Science 9*, 617.

Henry, R. J. (2014). Genomics strategies for germplasm characterization and the development of climate resilient crops. *Frontiers in Plant Science*, *5*(FEB), 1–4. https://doi.org/10.3389/fpls.2014.00068

Hickey, L. T., N. Hafeez, A., Robinson, H., Jackson, S. A., Leal-Bertioli, S. C. M., Tester, M., Gao, C., Godwin, I. D., Hayes, B. J., & Wulff, B. B. H. (2019). Breeding crops to feed 10 billion. *Nature Biotechnology*, *37*(7), 744–754. https://doi.org/10.1038/s41587-019-0152-9

Hu, L., Mu, J., Su, P., Wu, H., Yu, G., Wang, G., Wang, L., Ma, X., Li, A., Wang, H., Zhao, L, & Kong, L. (2018). Multi-functional roles of *TaSSI2* involved in Fusarium head blight and powdery mildew resistance and drought tolerance. *Journal of Integrative Agriculture*, *17*(2), 368–380. https://doi.org/10.1016/S2095-3119(17)61680-0

Huai, B., Yang, Q., Wei, X., Pan, Q., Kang, Z., & Liu, J. (2020). TaSTP13 contributes to wheat susceptibility to stripe rust possibly by increasing cytoplasmic hexose concentration. BMC *Plant Biology*, *20*(1), 1–17. https://doi.org/10.1186/s12870-020-2248-2

Islam, M. T., Croll, D., Gladieux, P., Soanes, D. M., Persoons, A., Bhattacharjee, P., Hossain, M. S., Gupta, D. R., Rahman, M. M., Mahboob, M. G., Cook, N., Salam, M. U., Surovy, M. Z., Sancho, V. B., Maciel, J. L. N., NhaniJúnior, A., Castroagudín, V. L., Reges, J. T. D. A., Ceresini, P. C., … Kamoun, S. (2016). Emergence of wheat blast in Bangladesh was caused by a South American lineage of Magnaporthe oryzae. *BMC Biology*, *14*(1), 1–11. https://doi.org/10.1186/s12915-016-0309-7

Islam, T. (2019). CRISPR-Cas technology in modifying food crops. *CAB Reviews* 14, 1–16.

Islam, M. T., Gupta, D. R., Hossain, A., Roy, K. K., He, X., Kabir, M. R., Singh, P. K., Khan, M. A. R., Rahman, M., & Wang, G. L. (2020). Wheat blast: A new threat to food security. *Phytopathology Research*, *2*(1). https://doi.org/10.1186/s42483-020-00067-6

Islam, M. T., Kim, K. H., & Choi, J. (2019). Wheat blast in Bangladesh: The current situation and future impacts. *Plant Pathology Journal*, *35*(1), 1–10. https://doi.org/10.5423/PPJ.RW.08.2018.0168

Jaganathan, D., Ramasamy, K., Sellamuthu, G., Jayabalan, S., & Venkataraman, G. (2018). CRISPR for crop improvement: An update review. *Frontiers in Plant Science*, *9*(July), 1–17. https://doi.org/10.3389/fpls.2018.00985

Jinek, M., Chylinski, K., Fonfara, I., Hauer, M., Doudna, J. A., & Charpentier, E. (2012). A programmable dual-RNA-guided DNA endonuclease in adaptive bacterial immunity. *Science*. *337*(August), 816–821. https://www.sciencemag.org/content/337/6096/816/suppl/DC1

Jouanin, A., Gilissen, L. J. W. J., Schaart, J. G., Leigh, F. J., Cockram, J., Wallington, E. J., Boyd, L. A., van den Broeck, H. C., van der Meer, I. M., America, A. H. P., Visser, R. G. F., & Smulders, M. J. M. (2020). CRISPR/Cas9 gene editing of gluten in wheat to reduce gluten content and exposure-reviewing methods to screen for coeliac safety. *Frontiers in Nutrition*, *7*(April). https://doi.org/10.3389/fnut.2020.00051

Kaminski, M. M., Abudayyeh, O. O., Gootenberg, J. S., Zhang, F., & Collins, J. J. (2021). CRISPR-based diagnostics. *Nature Biomedical Engineering*, *5*(7), 643–656. https://doi.org/10.1038/s41551-021-00760-7

Khakimov, A., Salakhutdinov, I., Omolikov, A., & Utaganov, S. (2022). Traditional and current-prospective methods of agricultural plant diseases detection: A review. *IOP Conference Series: Earth and Environmental Science*, *951*(1). https://doi.org/10.1088/1755-1315/951/1/012002

Kim, D., Alptekin, B., & Budak, H. (2018). CRISPR/Cas9 genome editing in wheat. *Functional and Integrative Genomics*, *18*(1), 31–41. https://doi.org/10.1007/s10142-017-0572-x

Komor, A. C., Kim, Y. B., Packer, M. S., Zuris, J. A., & Liu, D. R. (2016). Programmable editing of a target base in genomic DNA without double-stranded DNA cleavage. *Nature*, *533*, 420–424. https://doi.org/10.1038/nature17946

Latorre, S. M., Were, V. M., Foster, A. J., Langner, T., Malmgren, A., Harant, A., Asuke, S., Reyes-Avila, S., Gupta, D. R., Jensen, C., Ma, W., Mahmud, N. U., Mehebub, M. S., Mulenga, R. M., Muzahid, A. N. M., Paul, S. K., Rabby, S. M. F., Raha, A. A. M., Ryder, L., ... Kamoun, S. (2022). A pandemic clonal lineage of the wheat blast fungus. *BioRxiv*, 2022.06.06.494979. https://doi.org/10.1371/journal.pbio.3002052

Li, J., Li, Y., & Ma, L. (2021a). Recent advances in CRISPR/Cas9 and applications for wheat functional genomics and breeding. *aBIOTECH*, *2*(4), 375–385. https://doi.org/10.1007/s42994-021-00042-5

Li, S., Zhang, C., Li, J., Yan, L., Wang, N., & Xia, L. (2021b). Present and future prospects for wheat improvement through genome editing and advanced technologies. *Plant Communications*, *2*(4), 100211. https://doi.org/10.1016/j.xplc.2021.100211

Li, S., Lin, D., Zhang, Y., Deng, M., Chen, Y., Lv, B., Li, B., Lei, Y., Wang, Y., Zhao, L., Liang, Y., Liu, J., Chen, K., Liu, Z., Xiao, J., Qiu, J. L., & Gao, C. (2022). Genome-edited powdery mildew resistance in wheat without growth penalties. *Nature*, *602*(7897), 455–460. https://doi.org/10.1038/s41586-022-04395-9

Li, T., Hu, J., Sun, Y., Li, B., Zhang, D., Li, W., Liu, J., Li, D., Gao, C., Zhang, Y., & Wang, Y. (2021c). Highly efficient heritable genome editing in wheat using an RNA virus and bypassing tissue culture. *Molecular Plant*, *14*(11), 1787–1798. https://doi.org/10.1016/j.molp.2021.07.010

Lin, Q., Zong, Y., Xue, C., Wang, S., Jin, S., Zhu, Z., Wang, Y., Anzalone, A. V., Raguram, A., Doman, J. L., Liu, D. R., & Gao, C. (2020). Prime genome editing in rice and wheat. *Nature Biotechnology*, *38*(5), 582–585. https://doi.org/10.1038/s41587-020-0455-x

Liu, H., Wang, K., Jia, Z., Gong, Q., Lin, Z., Du, L., Pei, X., & Ye, X. (2020). Efficient induction of haploid plants in wheat by editing of *TaMTL* using an optimized *Agrobacterium*-mediated CRISPR system. *Journal of Experimental Botany*, *71*(4), 1337–1349. https://doi.org/10.1093/jxb/erz529

Longin, C. F. H., Gowda, M., Mühleisen, J., Ebmeyer, E., Kazman, E., Schachschneider, R., Schacht, J., Kirchhoff, M., Zhao, Y., & Reif, J. C. (2013). Hybrid wheat: Quantitative genetic parameters and consequences for the design of breeding programs. *Theoretical and Applied Genetics*, *126*(11), 2791–2801. https://doi.org/10.1007/s00122-013-2172-z

Maher, M. F., Nasti, R. A., Vollbrecht, M., Starker, C. G., Clark, M. D., & Voytas, D. F. (2020). Plant gene editing through de novo induction of meristems. *Nature Biotechnology*, *38*(1), 84–89. https://doi.org/10.1038/s41587-019-0337-2

Mühleisen, J., Piepho, H. P., Maurer, H. P., Longin, C. F. H., & Reif, J. C. (2014). Yield stability of hybrids versus lines in wheat, barley, and triticale. *Theoretical and Applied Genetics*, *127*(2), 309–316. https://doi.org/10.1007/s00122-013-2219-1

Nalam, V. J., Alam, S., Keereetaweep, J., Venables, B., Burdan, D., Lee, H., Trick, H. N., Sarowar, S., Makandar, R., & Shah, J. (2015). Facilitation of *Fusarium graminearum* infection by 9-lipoxygenases in Arabidopsis and wheat. *Molecular Plant-Microbe Interactions*, *28*(10), 1142–1152. https://doi.org/10.1094/MPMI-04-15-0096-R

Okada, A., Arndell, T., Borisjuk, N., Sharma, N., Watson-Haigh, N. S., Tucker, E. J., Baumann, U., Langridge, P., & Whitford, R. (2019). CRISPR/Cas9-mediated knockout of Ms1 enables the rapid generation of male-sterile hexaploid wheat lines for use in hybrid seed production. *Plant Biotechnology Journal*, *17*(10), 1905–1913. https://doi.org/10.1111/pbi.13106

Petersen, G., Seberg, O., Yde, M., & Berthelsen, K. (2006). Phylogenetic relationships of *Triticum* and *Aegilops* and evidence for the origin of the **A**, **B**, and **D** genomes of common wheat (*Triticum aestivum*). *Molecular Phylogenetics and Evolution*, *39*(1), 70–82. https://doi.org/10.1016/j.ympev.2006.01.023

Rabbi, S. M. H. A., Kumar, A., Mohajeri Naraghi, S., Simsek, S., Sapkota, S., Solanki, S., Alamri, M. S., Elias, E. M., Kianian, S., Missaoui, A., & Mergoum, M. (2021). Genome-wide association mapping for yield and related traits under drought stressed and non-stressed environments in wheat. *Frontiers in Genetics*, *12*(June), 1–13. https://doi.org/10.3389/fgene.2021.649988

Sánchez-León, S., Gil-Humanes, J., Ozuna, C. V., Giménez, M. J., Sousa, C., Voytas, D. F., & Barro, F. (2018). Low-gluten, nontransgenic wheat engineered with CRISPR/Cas9. *Plant Biotechnology Journal*, *16*(4), 902–910. https://doi.org/10.1111/pbi.12837

Sánchez, E., Ali, Z., Islam, T., & Mahfouz, M. (2022). A CRISPR-based lateral flow assay for plant genotyping and pathogen diagnostics. *Plant Biotechnology Journal*, *20*(12), 2418–2429. https://doi.org/10.1111/pbi.13924

Shi, B., Wang, J., Gao, H., Yang, Q., Wang, Y., Day, B., & Ma, Q. (2021). The small GTP-binding protein TaRop10 interacts with TaTrxh9 and functions as a negative regulator of wheat resistance against the stripe rust. Plant Science, 309. https://doi.org/10.1016/j.plantsci.2021.110937

Soni, N., Altartouri, B., Hegde, N., Duggavathi, R., Nazarian-Firouzabadi, F., & Kushalappa, A. C. (2021). TaNAC032 transcription factor regulates lignin-biosynthetic genes to combat Fusarium head blight in wheat. Plant Science, 304(December 2020), 110820. https://doi.org/10.1016/j.plantsci.2021.110820

Su, P., Zhao, L., Li, W., Zhao, J., Yan, J., Ma, X., Li, A., Wang, H., & Kong, L. (2021). Integrated metabolo-transcriptomics and functional characterization reveals that the wheat auxin receptor TIR1 negatively regulates defense against *Fusarium graminearum*. *Journal of Integrative Plant Biology*, *63*(2), 340–352. https://doi.org/10.1111/jipb.12992

Su, Z., Bernardo, A., Tian, B., Chen, H., Wang, S., Ma, H., Cai, S., Liu, D., Zhang, D., Li, T., Trick, H., St. Amand, P., Yu, J., Zhang, Z., & Bai, G. (2019). A deletion mutation in TaHRC confers Fhb1 resistance to Fusarium head blight in wheat. *Nature Genetics*, *51*(7), 1099–1105. https://doi.org/10.1038/s41588-019-0425-8

Taj, M., Sajjad, M., Li, M., Yasmeen, A., Mubarik, M. S., Kaniganti, S., & He, C. (2022). Potential targets for CRISPR/Cas knockdowns to enhance genetic resistance against some diseases in wheat (*Triticum aestivum* L.). *Frontiers in Genetics*, *13*(June), 1–5. https://doi.org/10.3389/fgene.2022.926955

Tembo, B., Mulenga, R. M., Sichilima, S., M'siska, K. K., Mwale, M., Chikoti, P. C., Singh, P. K., He, X., Pedley, K. F., Peterson, G. L., Singh, R. P., & Braun, H. J. (2020). Detection and characterization of fungus (*Magnaporthe oryzae pathotype Triticum*) causing wheat blast disease on rain-fed grown wheat (*Triticum aestivum* L.) in Zambia. *PLoS One*, 15(9 September), 1–10. https://doi.org/10.1371/journal.pone.0238724

Terns, M. P., & Terns, R. M. (2011). CRISPR-based adaptive immune systems. *Current Opinion in Microbiology*, *14*(3), 321–327. https://doi.org/10.1016/j.mib.2011.03.005

Wang, B., Wei, J., Song, N., Wang, N., Zhao, J., & Kang, Z. (2018). A novel wheat NAC transcription factor, TaNAC30, negatively regulates resistance of wheat to stripe rust. *Journal of Integrative Plant Biology*, *60*(5), 432–443. https://doi.org/10.1111/jipb.12627

Wang, F., Lin, R., Feng, J., Chen, W., Qiu, D., & Xu, S. (2015). TaNAC1 acts as a negative regulator of stripe rust resistance in wheat, enhances susceptibility to Pseudomonas syringae, and promotes lateral root development in transgenic Arabidopsis thaliana. *Frontiers in Plant Science*, *6*(FEB), 1–17. https://doi.org/10.3389/fpls.2015.00108

Wang, J., Tao, F., Tian, W., Guo, Z., Chen, X., Xu, X., Shang, H., & Hu, X. (2017). The wheat WRKY transcription factors TaWRKY49 and TaWRKY62 confer differential high-temperature seedling-plant resistance to *Puccinia striiformis* f. sp. *tritici*. *PLoS One*, *12*(7), 1–23. https://doi.org/10.1371/journal.pone.0181963

Whitford, R., Fleury, D., Reif, J. C., Garcia, M., Okada, T., Korzun, V., & Langridge, P. (2013). Hybrid breeding in wheat: Technologies to improve hybrid wheat seed production. *Journal of Experimental Botany*, *64*(18), 5411–5428. https://doi.org/10.1093/jxb/ert333

Wiedenheft, B., Sternberg, S. H., & Doudna, J. A. (2012). RNA-guided genetic silencing systems in bacteria and archaea. *Nature*, *482*(7385), 331–338. https://doi.org/10.1038/nature10886

Xie, K., Minkenberg, B., & Yang, Y. (2015). Boosting CRISPR/Cas9 multiplex editing capability with the endogenous tRNA-processing system. *Proceedings of the National Academy of Sciences of the United States of America*, *112*(11), 3570–3575. https://doi.org/10.1073/pnas.1420294112

Xie, K., & Yang, Y. (2013). RNA-Guided genome editing in plants using a CRISPR/Cas system. *Molecular Plant*, *6*(6), 1975–1983. https://doi.org/10.1093/mp/sst119

Yang, Q., Huai, B., Lu, Y., Cai, K., Guo, J., Zhu, X., Kang, Z., & Guo, J. (2020). A stripe rust effector Pst18363 targets and stabilises TaNUDX23 that promotes stripe rust disease. *New Phytologist*, *225*(2), 880–895. https://doi.org/10.1111/nph.16199

Zaidi, S. S. E. A., Mahas, A., Vanderschuren, H., & Mahfouz, M. M. (2020). Engineering crops of the future: CRISPR approaches to develop climate-resilient and disease-resistant plants. *Genome Biology*, *21*(1), 1–19. https://doi.org/10.1186/s13059-020-02204-y

Zaidi, S. S. E. A., Mukhtar, M. S., & Mansoor, S. (2018). Genome editing: Targeting susceptibility genes for plant disease resistance. *Trends in Biotechnology*, *36*(9), 898–906. https://doi.org/10.1016/j.tibtech.2018.04.005

Zaman, Q. U., Li, C., Cheng, H., & Hu, Q. (2019). Genome editing opens a new era of genetic improvement in polyploid crops. *Crop Journal*, *7*(2), 141–150. https://doi.org/10.1016/j.cj.2018.07.004

Zhang, J., Zhang, H., Li, S., Li, J., Yan, L., & Xia, L. (2021a). Increasing yield potential through manipulating of an ARE1 ortholog related to nitrogen use efficiency in wheat by CRISPR/Cas9. *Journal of Integrative Plant Biology*, *63*(9), 1649–1663. https://doi.org/10.1111/jipb.13151

Zhang, N., Yuan, S., Zhao, C., Park, R. F., Wen, X., Yang, W., Zhang, N., & Liu, D. (2021b). TaNAC35 acts as a negative regulator for leaf rust resistance in a compatible interaction between common wheat and *Puccinia triticina*. *Molecular Genetics and Genomics*, *296*(2), 279–287. https://doi.org/10.1007/s00438-020-01746-x

Zhang, R., Liu, J., Chai, Z., Chen, S., Bai, Y., Zong, Y., Chen, K., Li, J., Jiang, L., & Gao, C. (2019a). Generation of herbicide tolerance traits and a new selectable marker in wheat using base editing. *Nature Plants*, *5*(5), 480–485. https://doi.org/10.1038/s41477-019-0405-0

Zhang, X. M., Zhang, Q., Pei, C. L., Li, X., Huang, X. L., Chang, C. Y., Wang, X. J., Huang, L. L. & Kang, Z. S. (2018). TaNAC2 is a negative regulator in the wheat-stripe rust fungus interaction at the early stage. *Physiological and Molecular Plant Pathology*, *102*, 144–153. https://doi.org/10.1016/j.pmpp.2018.02.002

Zhang, Z., Hua, L., Gupta, A., Tricoli, D., Edwards, K. J., Yang, B., & Li, W. (2019b). Development of an *Agrobacterium*-delivered CRISPR/Cas9 system for wheat genome editing. *Plant Biotechnology Journal*, *17*(8), 1623–1635. https://doi.org/10.1111/pbi.13088

Zheng, C., Wang, K., Zheng, W., Cheng, Y., Li, T., Cao, B., Jin, Q., & Cui, D. (2021). Rapid developments in lateral flow immunoassay for nucleic acid detection. *Analyst*, *146*(5), 1514–1528. https://doi.org/10.1039/d0an02150d

Zhi, P., Kong, L., Liu, J., Zhang, X., Wang, X., Li, H., Sun, M., Li, Y., & Chang, C. (2020). Histone deacetylase TaHDT701 functions in TaHDA6-TaHOS15 complex to regulate wheat defense responses to *Blumeria graminis* f.sp. *tritici*. *International Journal of Molecular Sciences*, *21*(7). https://doi.org/10.3390/ijms21072640

Zhu, H., Zhang, H., Xu, Y., Laššáková, S., Korabečná, M., & Neužil, P. (2020). PCR past, present and future. *BioTechniques*, *69*(4), 317–325. https://doi.org/10.2144/BTN-2020-0057

Zou, Y., Mason, M. G., & Botella, J. R. (2020). Evaluation and improvement of isothermal amplification methods for point-of-need plant disease diagnostics. *PLoS One*, 15(6 June), 1–19. https://doi.org/10.1371/journal.pone.0235216

11 Using CRISPR to Unlock the Full Potential of Plant Immunity

Mumin Ibrahim Tek, Abdul Razak Ahmed, Ozer Calis, and Hakan Fidan

11.1 INTRODUCTION

Our understanding of plant–pathogen interactions has undergone significant advancements since the establishment of the gene-for-gene theory proposed by Harold Henry Flor [1,2]. Flor's studies on the genetic interactions between host resistance genes and pathogen avirulence genes paved the way for further investigations in plant pathology. Over time, the field has witnessed remarkable progress with the advent of molecular biology techniques, such as RNA interference (RNAi) and transgene technologies, enabling precise manipulation of organism genomes, including plants. Among the notable developments in molecular biology, the emergence of site-specific nucleases has played a pivotal role in unraveling the intricacies of plant–pathogen interactions and facilitating the generation of disease-resistant plants. Particularly, the optimization of CRISPR/Cas9 (Clustered Regularly Interspaced Short Palindromic Repeats/CRISPR-associated protein 9) technology has revolutionized our ability to manipulate the genomes of model organisms as well as plants, offering a rapid and efficient means to generate mutant lines [3,4]. This powerful tool has opened new avenues for studying the regulatory elements involved in plant–pathogen interactions by creating loss-of-function mutants and expanding our knowledge in this field. In this chapter, we will delve into the utilization of CRISPR-based approaches to advance our understanding of plants and their interactions with phytopathogens, drawing upon the wealth of previous research findings. We will explore how CRISPR technology has enabled precise manipulation of host genes, allowing researchers to dissect the complex molecular mechanisms underlying plant defense responses. Furthermore, we will address the potential of this technology in enhancing plant resistance by precisely modifying host genes and developing commercially viable disease-resistant plants.

11.2 THE SIGNIFICANCE OF PLANT–PATHOGEN INTERACTIONS IN ENHANCING PLANT IMMUNITY

The first step of engineering disease resistance in plants is understanding host immunity within host–pathogen interactions. Throughout evolution, plants have developed sophisticated defense mechanisms to counteract the onslaught of phytopathogens. The arms race between hosts and microbes resembles a biochemical battle, where pathogens employ diverse strategies, including the deployment of novel effectors, to overcome the plant's defense systems. However, most pathogens are intercepted by the plant's robust multilayered immune system, while only a few succeed in overcoming the defense system and infecting the host [5,6].

Plant immunity comprises two branches: Pathogen Triggered Immunity (PTI) and Effector Triggered Immunity (ETI) [7]. PTI is initiated when the host recognizes Microbe/Pathogen-Associated Molecular Patterns (M/PAMPs) through pattern recognition receptors (PRRs) [8]. This recognition triggers a series of defense responses that hinder pathogen growth and survival. These responses include the reinforcement of the cell wall through the production of cellulose,

DOI: 10.1201/9781003387060-11

pectin, and lignin; secretion of antimicrobials; production of reactive oxygen species (ROS); and induction of non-specific necrosis. PTI provides broad-spectrum resistance and is considered the plant's basal defense [9]. However, successful pathogens can evade PTI by employing effectors that disrupt PTI-mediated defenses, leading to effector-triggered susceptibility (ETS).

However, plants are not defenseless, and their immune system is not limited to this first layer. ETI is triggered when pathogen effectors are recognized by host resistance (R) proteins, particularly intracellular nucleotide-binding leucine-rich-repeats (NB-LRRs) proteins. This recognition event activates distinct defense mechanisms that specifically target pathogens and elicit an immune response known as race-specific resistance. ETI often culminates in the hypersensitive response (HR), a programmed cell death that serves as a decisive defense strategy following pathogen recognition [10]. ETI and ETS are dynamic, as pathogens can use their various effectors to evade or suppress ETI. In response, plants acquire new resistance genes to counter these evolving pathogen strains. This constant adaptation of defense strategies highlights the evolutionary arms race between plants and pathogens, necessitating ongoing research and exploration of novel approaches to enhance plant resistance.

Understanding the complex mechanisms underlying plant–pathogen interactions is crucial for the development of effective disease-resistant strategies. Harnessing the power of CRISPR technology, with its precise genome editing capabilities, offers promising opportunities to study and manipulate key elements involved in plant defense [11,12]. Targeting defense-related genes is a valuable approach to unraveling the intricate interactions between plants and pathogens, allowing us to decipher the molecular mechanisms underlying plant immunity. Through the identification and characterization of the genes involved in plant defense, we can engineer resistance traits in crops, bolstering their ability to fend off pathogens and reducing yield losses [13,14]. The knowledge gained from studying plant–pathogen interactions provides invaluable insights that can be harnessed to develop cultivars with enhanced disease resistance.

11.3 UNRAVELING PLANT IMMUNITY IN THE CRISPR ERA

The utilization of CRISPR technology to generate loss-of-function mutants offers significant benefits in investigating the function of genes involved in plant immunity. By precisely targeting specific genes with guide RNA (gRNA), CRISPR/Cas9 enables the disruption or knockout (KO) of their function through the non-homologous end joining (NHEJ) repair mechanism after the occurrence of a DNA double-strand break (DSB) caused by Cas9 endonuclease activity [15], allowing researchers to decipher their roles in defense mechanisms (Figure 11.1). This approach provides valuable insights into the functional significance of individual genes, unveiling their contributions to plant immunity and disease resistance [16]. In addition, the ability of CRISPR technology has facilitated multiplex gene-editing and the possibility of investigating the complex interplay between multiple genes involved in plant immunity. This capability is particularly advantageous in plants, as compared to other organisms where achieving multiplex gene targeting is more challenging. For instance, CRISPR/Cas9 editing in mammalian cells, especially in the context of multiplexed editing, is the potential activation of DNA damage-response mechanisms regulated by the transcription factor *p53*, which can result in cellular senescence or apoptosis [17–19] By targeting multiple genes simultaneously, researchers can investigate the synergistic effects and functional relationships between these genes, providing deeper insights into the intricate mechanisms of plant defense and disease resistance.

On the other hand, the generation of resistant plants against plant pathogens by manipulating host *susceptibility* (*S*) genes has become easier and more efficient thanks to CRISPR/Cas9 [20]. The proteins encoded by *S*-genes cause host recognition by the pathogen, defense suppression, and facilitate penetration. *S* genes encode immune suppressors [21]. These proteins are critical for diseases caused by biotrophic pathogens such as powdery mildew (PM) and downy mildew (DM) because their vitality is directly associated with host interactions [22]. Therefore, disruption of

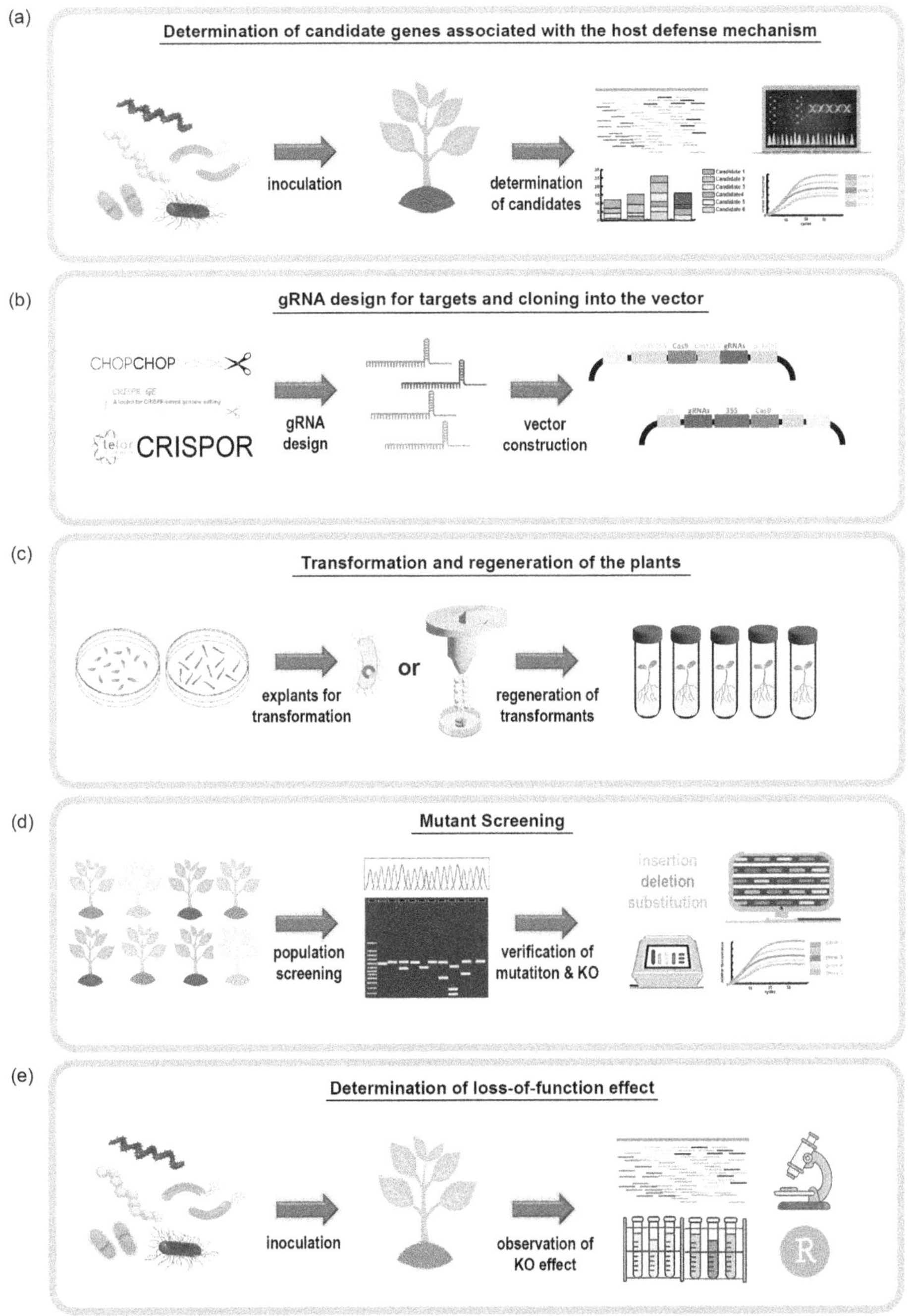

FIGURE 11.1 Workflow for generating loss-of-function mutants using CRISPR/Cas9 to investigate the role of genes in plant defense response against pathogens. (a) Identification of candidate genes through transcriptomic analysis, followed by validation using qPCR. (b) Selection of gRNAs by using different tools [24–26] and transfer into a Cas9 vector for the targeted plant. (c) Transformation of plant explants (cotyledon, hypocotyl) via Agrobacterium or particle bombardment and regeneration of T0 plants. T1 seeds can be harvested after acclimatization and maturation of T0 plants. (d) Screening of the T1 population using PCR-based techniques or sequencing to identify mutant plants; and confirmation of KO at the target site using bioinformatic and transcriptomic analyses. (e) Assessment of loss-of-function effects in T2 plants through advanced microscopy, biochemistry, and bioinformatic approaches.

S genes confers a broad spectrum and durable resistance in the host against plant pathogens. As mentioned above CRISPR is a versatile tool to generate loss-of-function mutants and hence CRISPR is useful for generating mutant lines which have a mutation that causes KO on the target *S* gene [23].

11.4 ENHANCED RESISTANCE: STRATEGIES AGAINST OOMYCETES, FUNGAL AND BACTERIAL PATHOGENS

CRISPR/Cas9 technology has enabled the targeting of genes involved in host–pathogen interactions to enhance plant immunity against a range of pathogens, including oomycetes, fungal pathogens, and bacterial pathogens. By disrupting these genes, we can elucidate their functions in defense mechanisms and develop strategies to improve plant resistance. A good example is the targeted disruption of the TALE-binding elements in the host. KO of rice (*Oryza sativa*) sugar transporter genes *OsSWEET11* and *OsSWEET14* resulted in broad-spectrum resistance against *Xanthomonas oryzae* pv. *oryzae* (*Xoo*), the causal agent of bacterial blight, and the engineered rice line displayed resistance to most *Xoo* strains (LN18 and PXO61), indicating successful disruption of ETS [27]. Also, the disruption of the *OsSWEET14* gene in rice conferred significant resistance against both African (AXO1947) and Asian (PXO86) strains of *Xoo*. The engineered line exhibited normal growth and increased plant height, highlighting the potential of *OsSWEET14* in developing breeding strategies for resistance against *Xoo* [28]. On the other hand, disruption of the TALE-binding elements in the host is not limited to *OsSWEETs* in rice. Researchers have used CRISPR/Cas9 to target *CsLOB1* promoter, effector binding element (EBE_{PthA4}) in the *CsLOB1* promoter, which is recognized by the main effector PthA4 of *Xanthomonas citri* subsp. *citri* (*Xcc*), in Wanjincheng orange (*Citrus sinensis* Osbeck). After the editing loss-of-*CsLOB1*-function mutant lines showed enhanced resistance against *Xcc* [29]. Similarly, the same approach was applied to editing *DLOB2*, *DLOB3*, *DLOB9*, *DLOB10*, *DLOB11*, and *DLOB12* in Duncan grapefruit (*Citrus*×*paradisi*). The mutant lines *DLOB9*, *DLOB10*, *DLOB11*, and *DLOB12* showed no typical canker symptoms caused by *Xcc*, and even lower numbers of pustules were observed compared to the wild-type (WT) during later stages of inoculation [30].

In addition to targeting effector binding elements, manipulating host genes involved in the regulation of responses against environmental stimuli by using CRISPR/Cas9 has emerged as an alternative and powerful strategy to enhance plant defense against pathogens. KO of the Phospholipase C2 gene (*SlPLC2*) in tomato plants led to increased resistance against the fungal pathogen *Botrytis cinerea*. The KO plants displayed reduced ROS production and showed changes in the expression of marker genes associated with JA and SA responses [31]. Similarly, the functional role of *JAZ2* was characterized as a major COR/JA-Ile co-receptor in *Arabidopsis thaliana* (*AtJAZ2*) controlling stomatal dynamics during the bacterial invasion. In another study, KO of *SlJAZ2* in tomato enabled the disruption of the antagonistic interaction between the SA and JA defense pathways, leading to enhanced resistance against *B. cinerea* [32]. Manipulation of the phytohormone signaling pathways that are involved in response against plant pathogens is one of the highly focused CRISPR/Cas9 editing studies in plants. Two orthologs of the *A. thaliana S* gene *DMR6*, namely *SlDMR6-1* and *SlDMR6-2*, were inactivated in tomato, and loss-of-function mutants were generated. The absence of *SlDMR6-1*, but not *SlDMR6-2*, in mutant lines, resulted in enhanced resistance against bacterial *Pseudomonas syringae pv. tomato*, *X. gardneri*, and *X. perforans*, the oomycete pathogen *Phytophthora capsici*, and the fungal PM pathogen *Oidium neolycopersici*. This resistance was accompanied by increased SA levels and heightened activation of plant immune responses. Disease resistance assays confirmed the improved resistance of *Sldmr6-1* mutants against these pathogens, with elevated SA levels and enhanced immune responses observed in the mutant lines. In addition, both *Sldmr6-1* and *Sldmr6-2* were found to possess SA-5 hydroxylase activity. These findings highlight the broad-spectrum resistance of the mutants, demonstrating their effectiveness against three different genetically distant pathogens: bacteria, oomycetes, and fungi [33].

Transcription factors (TFs) have emerged as promising targets for CRISPR/Cas9 due to their pivotal role in plant immunity, regulating genes involved in PTI/ETI, defense response pathways such as hormone or phytoalexin synthesis, and transcriptional reprogramming [34]. In rice, the induced mutations in the *Bsr-d1* (C2H2-type TF), *Pi21*, and *ERF922* (APETELA2/ethylene response factor type TF) genes resulted in increased resistance against *Magnaporthe oryzae*, with the *pi21* or *erf922* single mutants also exhibiting inhibition of *Xoo* infection. The disruption of *Bsr-d1* function led to broad-spectrum resistance by modulating the expression of peroxidase genes and elevating H_2O_2 levels, while KO mutants of *Bsr-d1*, *Pi21*, and *ERF922* showed enhanced expression of defense-related genes associated with immune signaling pathways, indicating improved resistance mechanisms [35]. In addition, previous studies utilizing CRISPR/Cas9 and multi-gRNAs to target *ERF922* demonstrated enhanced resistance against *M. oryzae* in edited T2 plants [36]. The *VvWRKY52* gene, which has been implicated in biotic stress responses [37], was targeted using the CRISPR/Cas9 in grape (*Vitis vinifera*). Four guide RNAs were designed for *VvWRKY52,* and 22 mutants were identified, including lines with biallelic and heterozygous mutations after *Agrobacterium*-mediated transformation. Enhanced resistance to *B. cinerea* was observed in *VvWRKY52* KO lines, with the biallelic mutants exhibiting the highest level of resistance [38].

On the other hand, the Mildew Locus (*MLO*) is the most popular target for using CRISPR/Cas9 to generate PM-resistant cultivars and investigate *mlo*-based resistance in various plants. The loss of *mlo*-mediated broad-spectrum and durable resistance was first identified in barley (*Hordeum vulgare*) against resistance against PM caused by *Blumeria graminis* f.sp *hordei* (*Bgh*) [39]. Subsequently, *MLO*s were identified in various plant species, including *A. thaliana* [40], wheat (*Triticum aestivum*) [41], tomato (*Solanum lycopersicum*) [42], pea (*Pisum sativum*) [43], pepper (*Capsicum annuum*) [42] and cucumber *Cucumis sativus* [44–46]. It has also been reported that mutations on Clade IV and V confer broad-spectrum and durable resistance against PM in monocots and dicots, respectively. The application of CRISPR/Cas9 has greatly facilitated the development of PM-resistant cultivars by targeting mutated Clade IV and V *MLO*s in various plants such as tomato, cucumber, grapevine, tobacco (*Nicotiana tabacum*), and wheat. The loss-of-function allele of *MLO* represents one of the most extensively studied recessive inherited resistance against PMs [47].

PM-resistant *S. lycopersicum* cv. Tomelo was generated with CRISPR/Cas9. Loss-of-function mutants that have 48 bp deletion on *SlMlo1* were generated in less than 10 months. Moreover, the elimination of T-DNA has allowed the generation of transgene-free mutants that are indistinguishable from naturally occurring mutants [48]. Similarly, *NtMLO1* and *NtMLO2* were targeted in *Nicotiana tabacum* by using the CRISPR/Cas9 system. Generated double-mutants have shown enhanced resistance against *Golovinomyces cichoracearum* [49]. PM resistance was successfully engineered in susceptible *C. sativus* cv. 'Ilan' using CRISPR/Cas9 by targeting *CsaMLO8*. Two transgene-free lines, *Csamlo-cr-1* (with a 5 bp deletion on exon 1) and *Csamlo-cr-2* (with a 1280 bp deletion/10 bp insertion from exon 1 to 5) exhibited high resistance against PM under semi-commercial conditions [50]. However, studies that focused on *MLO*s are not only focused on generating PM-resistant cultivars. Various studies have focused on *mlo*-based defense response against PM pathogens.

*MLO*s encode plant-specific proteins localized to the plasma membrane with seven transmembrane domains, acting as susceptibility factors for PM [51]. *Mlo*-based resistance involves multiple pathways and interacts with other defense mechanisms such as programmed cell death (PCD) and papilla formation [52]. In *A. thaliana*, papilla formation, controlled by PENETRATION (PEN) and SNAREs VESICLE-ASSOCIATED MEMBRANE PROTEINs (VAMPs), represents a robust response against PM pathogens [53–55]. Recent studies classified *mlo*-mediated resistance as part of non-host resistance (NHR) in *A. thaliana*, with PEN proteins playing a crucial role in *AtMLO2*-mediated resistance [56,57]. However, the exact mechanism of *mlo*-based resistance remains ambiguous and still needs to be demystified. Notably, the calmodulin-binding domain of barley MLO increases susceptibility to PM, highlighting the importance of Ca^{+2} signaling in the early stage of infection for host recognition by PM pathogens [58]. *VvMLO3* and *VvMLO4* were targeted with CRISPR/Cas9 in *V. vinifera*, and various mutational patterns were observed for targets. Edited four lines for *VvMLO3* displayed enhanced resistance to PM. Their response against *Erysiphe necator* was associated with host cell death, cell wall apposition (CWA), and the accumulation of H_2O_2 [59]. In another study, *CsaMLO1*, *CsaMLO8*, and *CsaMLO11* were targeted in *C. sativus* simultaneously by using CRISPR/Cas9, and different loss-of-function mutants were generated. Altered defense responses were observed in these mutants against *P. xanthii* at different stages of infection. *P. xanthii* could not germinate on leaves of *CsaMLO1*^{A137N}/*CsaMLO8*/11^{del} and *CsaMLO8/11*^{del} due to pre-invasive defense response in the host, while hypersensitive reaction (HR) restricted invasion of *P. xanthii* in *CsaMLO1*^{del} and *CsaMLO11*^{del} mutants. Decreased H_2O_2 accumulation was also detected in *CsaMLO1*^{A137N}/*CsaMLO8*/11^{del} and *CsaMLO8/11*^{del} when they were compared with WT [60].

Besides *MLO*s, other genes have also been considered in conferring resistance against PM in plants. One such gene is *Powdery Mildew Resistance 4* (*PMR4*), which plays a crucial role in callose synthesis and defense against PM. *A. thaliana* mutants *pmr4* and *gsl5*, exhibit reduced callose production but still form papillae, raising questions about their resistance mechanism [61,62]. The *pmr4* mutant displays a HR and upregulation of salicylic acid (SA) signaling genes, indicating complex interactions between callose deposition, SA signaling, and defense responses [63,64]. CRISPR/Cas9 was used to target the *PMR4* in *S. lycopersicum* to investigate the loss-of-function effect of *PMR4* against *O. neolycopersici*. Mutants exhibited reduced susceptibility to PM, as evidenced by disease symptom scoring and fungal biomass quantification. Histological observations revealed increased hypersensitive response-like cell death and impaired pathogen growth in the mutants [65]. CRISPR/Cas9 technology was also employed to modify homologs of *enhanced disease resistance1* (*EDR*) in bread wheat (*T. aestivum*), resulting in *Taedr1* mutant plants. Notably, no off-target mutations were observed in these mutants. The Taedr1 plants displayed resistance to *Blumeria graminis* f. sp. *tritici* without experiencing mildew-induced cell death [66].

Utilizing CRISPR/Cas9 technology, researchers have successfully generated mutants with enhanced resistance and altered defense responses, demonstrating the potential of genome editing in improving plant immunity against various pathogens. These findings contribute to the development of novel strategies for disease control in agriculture.

11.5 DISRUPTION PLANT–VIRUS INTERACTIONS: CHALLENGES AND OPPORTUNITIES

Manipulation of plant–virus interactions has contributed immensely to our understanding of virus infection mechanisms and host-defense responses. By targeting and disrupting various components of the host–virus interaction, we can uncover novel insights into viral replication or infection, host-defense mechanisms, and the interplay between the two. This knowledge opens a world of opportunities to develop virus-resistance cultivars by using CRISPR-based tools to target *S* genes or antiviral strategies in the host [67]. Moreover, various plants that have resistance against viruses have already been developed by targeting *S* genes such as *Eukaryotic translation initiation factors* (*eIFs*) with CRISPR-based technologies in the host.

eIFs, particularly *eIF4E*, have emerged as crucial players in plant–virus interactions. In addition to its role in cellular translation regulation, *eIF4E*'s association with viral proteins (VPg) binding to the host's mRNA 5′-terminal cap is essential for potyvirus infection [68–71]. Loss of *eIF4E* function leads to recessive resistance against potyviruses, and naturally occurring mutants as well as controlled mutations suppressing *eIF4E* function have been identified [72]. Notably, various recessive genes encoding *eIF4E* variants, such as *pot-1* in *S. lycopersicum*, *rym4*, *rym5*, and *rym6* in *H. vulgare*, and *mol* in *Lactuca sativa*, have been extensively studied for their roles in plant–virus interactions [72–74]. Given *eIF4E*'s pivotal role in determining host susceptibility or resistance to viral pathogens, it represents a promising target for managing plant viruses in agricultural production [75]. Therefore, after the optimization of the CRISPR/Cas9 for gene editing in plants, *eIF*s have become one of the popular targets to apply this novel technology in various plants.

In *A. thaliana, eIF(iso)4E* was targeted by using CRISPR/Cas9 and generated T3 homozygous mutants showed complete resistance against turnip mosaic virus (TuMV). Assessment of T3 lines also demonstrated that homozygous mutations in *eIF(iso)4E* did not adversely affect plant vigor, as indicated by comparable dry weights and flowering times to WT plants under standard growth conditions [76]. In another study, *eIF4E* was targeted in *C. sativus*, and T3 generation mutants were inoculated with cucumber vein yellowing virus (CVYV), zucchini yellow mosaic virus (ZYMV), and papaya ring spot mosaic virus-W (PRSV-W). Homozygous loss-of-function mutants exhibit resistance against inoculated viruses, while heterozygous mutants and WT plants remain susceptible [77]. The allelic and positional effects of *eIF4E* mutations in cucumber were investigated by targeting different positions of *eIF4E* in *C. sativus* using CRISPR/Cas9. Loss-of-function mutant lines G27 and G247 were generated, and their progeny was extensively produced. Among the F1 plants, only those with a homozygous mutation at the 1st exon exhibited complete resistance against PRSV, ZYMV, and watermelon mosaic virus (WMV), while F1 plants with a homozygous mutation at the third exon showed reduced symptoms and lower viral accumulation compared to non-edited F1 plants [78]. Moreover, CRISPR/Cas9 was utilized to induce mutations at different residues (Y1059, V1060, and V1061) in *eIF4 gamma (eIF4G)* to create new sources of resistance against rice tungro spherical virus (RTSV) and rice tungro bacilliform virus. This study focused on the RTSV-susceptible rice variety IR64, which is widely cultivated in tropical Asia. Among the various mutated *eIF4G* alleles investigated, only those resulting in in-frame mutations in the "SVLFPNLAGKS" residues, particularly NL adjacent to the "YVV" residues, conferred resistance to RTSV. It was observed that *eIF4G* plays a critical role in normal development, as alleles leading to truncated *eIF4G* were unable to be maintained in a homozygous state. The resulting plants exhibited RTSV resistance and improved yield under controlled greenhouse conditions [79]. In another study, five *eIF4E*s, called eIF4E, eIF(iso)4E-1, *eIF(iso)4E-2*, novel cap-binding protein-*1* (*nCBP-1*), and *nCBP-2* were targeted in cassava (*Manihot esculenta* Crantz) to engineer resistance against Cassava brown streak virus (CBSV) and Ugandan cassava brown streak virus (UCBSV). When inoculated with CBSV, the ncbp-1/ncbp-2 mutants exhibited delayed and attenuated CBSD aerial symptoms, as well as reduced severity and incidence of storage root necrosis. The suppression of disease symptoms correlated with decreased virus levels in storage roots compared to WT controls. This study demonstrates the feasibility of simultaneously modifying multiple genes in cassava to achieve tolerance against CBSD [80].

In addition to targeting *eIF*s, CRISPR/Cas9 has been used to inactivate different host genes to engineer virus resistance in plants. One notable development is the KO of *TOBAMOVIRUS MULTIPLICATION1* (*TOM1*), which is essential for tobamovirus multiplication [81], including Tomato brown rugose fruit virus (ToBRFV). ToBRFV poses a significant threat to tomato production worldwide, as it has overcome *Tm-2*^{2}-mediated resistance and is spreading rapidly [82–84]. However, the disruption of the four *TOM1* homologs (*TOM1a*, *TOM1b*, *TOM1c*, and *TOM1d*) in tomatoes resulted in resistance against ToBRFV, with no detectable accumulation of ToBRFV coat protein (CP) and no noticeable growth or fruit production defects observed in the quadruple mutants. This advancement holds promise for safeguarding tomato cultivation against ToBRFV [85]. Resistance against Tomato yellow leaf curl virus (TYLCV) was achieved in tomatoes by disrupting the *Ty5* (*SlPelo*) using CRISPR/Cas9. *Ty-5* locus encodes a protein that is responsible for synthesizing the messenger RNA surveillance factor known as Pelota (PELO). PELO proteins play a critical role in ribosome recycling during protein synthesis. The edited loss-of-SlPelo-function mutant lines for lo G1-41-40, G1-41-42, and G1-41-44 exhibited enhanced resistance to TYLCV, showing no symptoms, and their Disease Severity Indexes (DSI) were comparable to mock-inoculated plants. In contrast, the DSI of the WT plants after inoculation was 2.5 out of 4. Furthermore, TYLCV accumulation was not detected in G1-41-40 following infection with Agrobacterium carrying the virulent plasmid (pCAM-TYLCV-1.7mer), confirming the effectiveness of the CRISPR/Cas9-mediated disruption in preventing virus accumulation [86].

CRISPR-based approaches have also been employed to directly target viral genomes, offering a promising strategy for combating viral infections. For instance, various RNA-guided Cas endonucleases such as Cas9 or Cas13a have been utilized to suppress viral propagation through transient or stable expression of CRISPR cassettes in plants [20]. This targeted genome editing approach offers a potential strategy for developing antiviral methods and mitigating the impact of severe plant viral diseases. Several studies have successfully employed CRISPR technology to target viral genomes of plant pathogenic viruses, with CRISPR/Cas9-mediated targeting being widely used against Geminiviruses. One notable application involved the degradation of the intergenic region (IR), Rep, and RepA regions of bean yellow dwarf virus (BeYDV) by transient expression of Cas9 and sgRNA in planta, resulting in a reduction of virus copy number [87]. Following the same strategy, TYLCV, beet curly top virus (BCTV), and Merremia mosaic virus (MeMV) were interfered with in *N benthamiana* at the same time [88]. Besides the transient expression, CRISPR/Cas9-overexpressed *A. thaliana* that targeted beet severe curly top virus (BSCTV) was generated and there was no detectable BSCTV accumulation in C3, even if eight obvious off-target alterations out of potential ten candidates in C3 [89].

In other studies, Banana streak virus (BSV) [90], Wheat dwarf virus (WDV) [91], Cotton Leaf Curl Multan virus (CLCuMuV) [92], and TYLCV [93] genomes were directly targeted by using CRISPR/Cas9 to suppress propagation of viruses in the hosts. Also, CRISPR/Cas13a, an RNA-targeting CRISPR system, was used for directly targeting CP and Hc-Pro of Turnip Mosaic Virus (TuMV) [94] and Potato Virus Y (PVY) [95], while CRISPR/Cas13a was used to target Cucumber Mosaic Virus (CMV) and Tobacco mosaic virus (TMV) [96]. Since the targeted mutations occur directly in the viral genome, they can potentially disrupt conserved regions or essential functions shared among different strains, rendering them ineffective in infecting the host plant. Overall, the targeted modification of viral genomes using CRISPR/Cas9 holds great potential for developing durable and broad-spectrum resistance against plant viral pathogens. Further research and technological advancements in this field are necessary to overcome challenges and fully harness the power of CRISPR/Cas9 as a tool for viral genome editing and crop protection.

Disruption of plant–virus interactions using gene-editing technologies such as CRISPR/Cas9 presents both challenges and opportunities. While the studies previously showcase the potential of CRISPR/Cas9 in conferring resistance against various plant pathogens, several challenges need to be addressed. Firstly, off-target effects and unintended mutations can occur, raising concerns about potential negative impacts on plant growth and genome integrity. In addition, the rapid evolution and emergence of new viral strains can render engineered resistance ineffective over time. Furthermore, the delivery of CRISPR components and efficient targeting of viral genomes in different plant species pose technical hurdles. Despite these challenges, the application of CRISPR/Cas9 technology in disrupting plant–virus interactions provides opportunities for developing more resilient and disease-resistant crop varieties. Further research and advancements in delivery methods, target selection, and understanding of plant–virus interactions will help overcome these challenges and unlock the full potential of genome editing for sustainable agriculture.

11.6 OVERCOMING CHALLENGES IN DEVELOPING DISEASE-RESISTANT CULTIVARS

Overcoming challenges associated with developing disease-resistant cultivars is crucial for sustainable agriculture. One of the key challenges is the rapid evolution of plant pathogens, which can render resistance genes ineffective over time. To address this, continuous monitoring and surveillance of pathogen populations is necessary to identify new strains and develop strategies to overcome their virulence. *S*-mutated mediated resistance represents a promising alternative to *R*-mediated resistance in the development of disease-resistant cultivars [97]. While *R*-mediated resistance relies on specific recognition of pathogen effectors by plant resistance genes, *S*-mutated mediated resistance operates by introducing targeted mutations in the susceptibility (*S*) genes of the host plant. This approach disrupts the interaction between the pathogen and the plant, rendering the plant less susceptible to infection [98]. One of the major advantages of *S*-mutated mediated resistance is its potential for broad-spectrum resistance. Unlike *R*-mediated resistance, which often targets specific strains or races of pathogens, *S*-mutated mediated resistance can protect against a wide range of pathogen variants [99]. This is particularly valuable in the face of rapidly evolving pathogens that can quickly overcome *R*-mediated resistance. Another advantage is the potential for durability. Since *S*-mutated mediated resistance targets essential plant genes involved in pathogen susceptibility, it can impose strong selective pressure on the pathogen population, making it less likely for resistance-breaking mutants to emerge. This offers long-lasting protection against diseases and reduces the need for frequent deployment of new resistance genes [20]. Therefore, CRISPR-based gene-editing tools are functional for generating disease-resistant cultivars by creating mutations that lead to KO of *S*-genes.

However, the generation of disease-resistant cultivars through CRISPR-based genome editing faces several challenges. These include selecting the appropriate targets, delivering CRISPR reagents into plants, regenerating edited plants, screening for desired mutations, minimizing off-target effects, and achieving large-scale production of resistant cultivars. For example, in the case of PM resistance in wheat, mutations in the *MLO* gene have been found to confer resistance. However, *mlo* mutants can also have pleiotropic effects, leading to growth penalties and yield losses in addition to PM resistance. To address this, researchers targeted a 304-kilobase pair deletion in the *MLO-B1* locus, resulting in the ectopic activation of *Tonoplast monosaccharide transporter 3* (*TaTMT3B*). This targeted approach helped reduce the growth and yield penalties associated with *MLO* disruption in wheat. As a result, they successfully generated the *Tamlo-R32* mutant, which exhibits PM resistance without growth penalty [100]. Therefore, when selecting candidates to reduce susceptibility, the function of the genes in plant growth must be considered.

The next challenge in developing disease-resistant cultivars is the delivery of CRISPR reagents into plants through transformation techniques. *Agrobacterium*-mediated transformation is used when a CRISPR/Cas9 expression cassette is to be delivered into the plant genome for integration. Particle bombardment is another commonly used technique, and various optimized protocols are available for both methods [101,102]. These techniques rely on integrating randomly CRISPR reagents into the plant genome, enabling their continuous expression. Different plant parts, such as cotyledons, hairy roots, hypocotyls, or true leaves, can serve as explants for *Agrobacterium*-mediated transformation [16,103]. In addition, DNA-free delivery methods, such as in vitro transcripts (IVTs), transient expression of plasmid DNA constructs, or Ribonucleoprotein (RNP) complexes, have emerged as viable options in recent years for genome editing in plants [104–106]. These advancements in transformation techniques offer several additional benefits in generating disease-resistant plants. They allow for the efficient delivery of CRISPR reagents, enabling targeted genome editing and the generation of desired mutations. Moreover, the use of DNA-free delivery methods reduces the risk of unwanted integration events and allows for precise targeting of viral genomes, minimizing the interference of virus propagation. Overall, these advancements provide researchers with valuable tools to develop disease-resistant cultivars with improved efficacy (Figure 11.2).

Another challenge in plant gene editing is the screening of mutants and addressing off-target effects. After the transformation, the regenerated plants need to be screened and verified for mutations. The screening workflow depends on the number of targets and regenerated plants. PCR-based approaches can be useful for screening a large number of regenerated plants and targets. Several PCR-based methods have been developed for detecting homozygous mutants, using designed primers specific to the guide RNA (gRNA) targets and upstream of a protospacer-adjacent motif (PAM). Mutation Sites Based Specific Primers PCR (MSBSP-PCR) and Annealing at Critical Temperature PCR (ACT-PCR) are cost-effective and simple methods for screening homozygous/biallelic mutants from regenerated plants [107,108]. In addition, quantitative real-time PCR with high-resolution

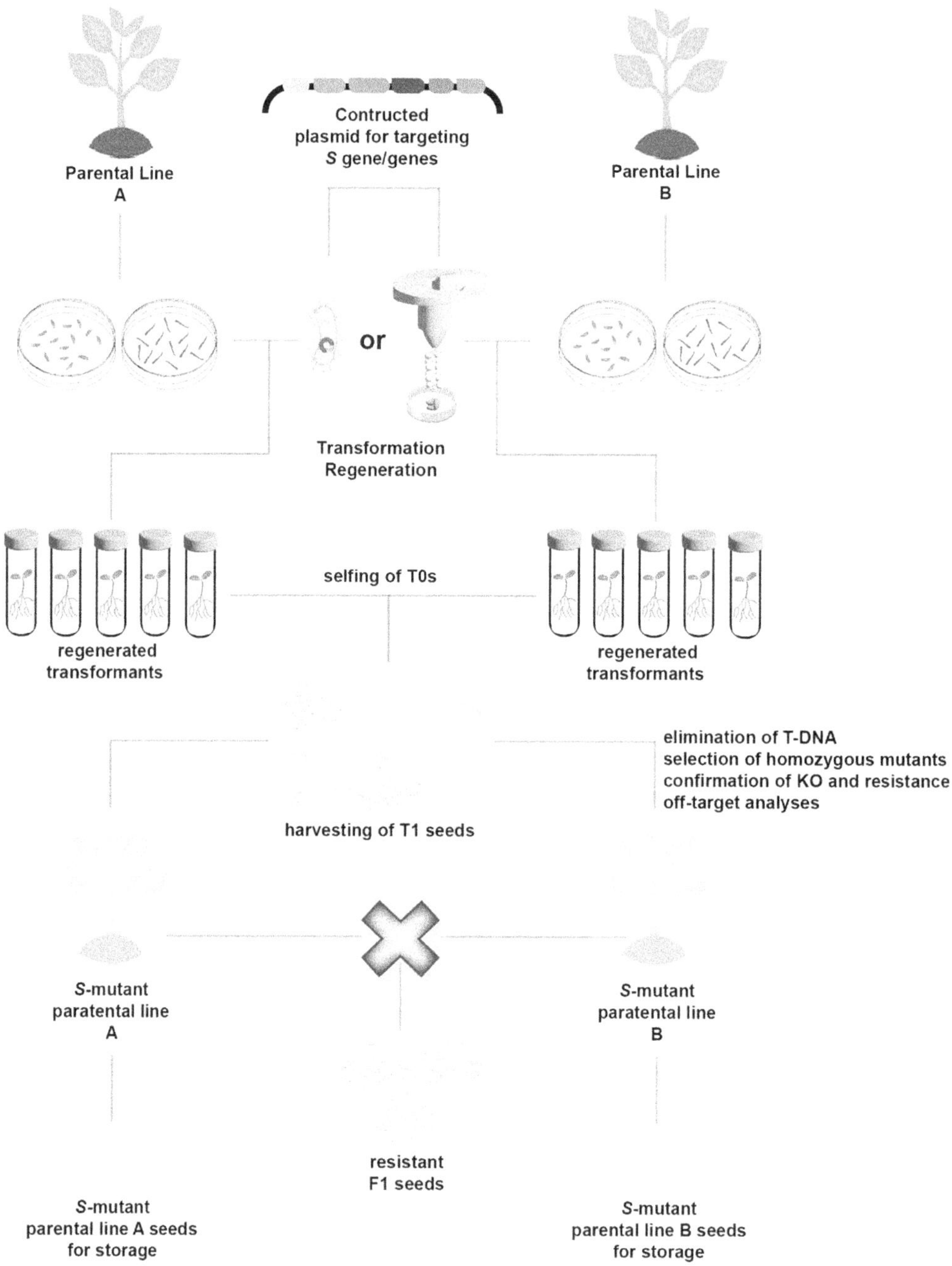

FIGURE 11.2 Route for developing resistant cultivars against plant pathogens using CRISPR/Cas9 by targeting *S* genes in the host. High-performing susceptible inbred lines (line A and line B) with desirable agronomic traits are selected as plant material, and explants are prepared for transformation. After generating the T0 generation, T1 seeds are harvested and screened to identify mutants. T-DNA-carrying plants can be eliminated in segregated T1 lines using PCR. Furthermore, off-target effects and mutation types are determined to prevent chimerism and off-target effects. Mutant T2 parental lines with the same *S* gene mutation are crossed to generate F1 plants. Meanwhile, seeds from T2 lines can also be harvested after the self-crossing for storage of parental mutant lines. The edited F1 plants are evaluated for any potential side effects, such as decreased quality or yield, by comparison with non-edited F1 plants. Finally, resistant F1 plants with the desired *S* gene knockout mutations are generated and can be commercially utilized following registration.

melting analysis (qPCR-HRM) has been used to detect mutants sensitively and rapidly, including heterozygous mutants, compared to other methods [109]. However, these methods only allow the detection of potential mutants and not the determination of mutation types. Therefore, sequencing approaches have been employed to obtain data for identifying the mutation types, using bioinformatic tools such as CRISPR-GE [25] or ICE (SYNTHEGO). Whole-genome sequencing (WGS) approaches are particularly valuable for analyzing on-target mutations and off-target effects of the CRISPR applications with high precision [110]. These advanced sequencing techniques provide comprehensive insights into the genetic changes introduced by CRISPR and allow for a thorough evaluation of the editing outcomes.

The biggest challenge in developing CRISPR-edited resistant plants against plant pathogens is the massive production process. While traditional crossing or mutation breeding methods typically take 8–10 years, the use of CRISPR has significantly reduced this time to 4–6 years for converting disease-susceptible elite varieties into resistant ones [111]. Even this reduced time is still considered long-term for generating new cultivars. To address this, various studies have focused on generating disease-resistant cultivars by targeting the *S*-genes of parental inbred lines. This approach offers advantages because the commercial progeny (F1) resulting from crossing these parental inbred lines already possess desirable traits such as high yield and quality. Resistant F1 plants can be generated by introducing mutations in the *S*-genes, while still retaining the favorable agronomic characteristics of the parental lines [78]. Furthermore, after the selection of mutants, the T-DNA can be eliminated through self-crossing in subsequent generations [112]. This allows for the harvesting of transgene-free F1 seeds that are suitable for commercial purposes and possess disease resistance conferred by *S*-gene mutations. In addition, homozygous *S*-mutant inbred lines can be propagated through selfing without any penalties, as they are already pure lines and do not segregate for other characteristics.

In conclusion, CRISPR/Cas9 technology has revolutionized the study of plant–pathogen interactions and has opened up new possibilities for developing disease-resistant cultivars. By harnessing the potential of CRISPR/Cas9 and continuously refining its applications, we can pave the way for resilient crop varieties and mitigate the impact of plant diseases on global food security. The advances in omic technologies and our understanding of plant–pathogen interactions provide opportunities to overcome challenges and develop durable disease-resistant cultivars. By utilizing cutting-edge technologies such as genome editing and gene stacking, breeders can introduce multiple resistance genes or edit specific target genes to enhance disease resistance in cultivars. These approaches promote sustainable and environmentally friendly agricultural practices by providing long-lasting resistance and reducing the dependence on chemical pesticides. Overall, addressing the challenges of developing disease-resistant cultivars requires a comprehensive approach involving continuous pathogen monitoring, optimization of delivery systems, rigorous safety and stability testing, collaboration among stakeholders, and advancements in scientific knowledge. By tackling these challenges, we can enhance crop resilience, minimize yield losses, and contribute to a more sustainable and secure food production system.

REFERENCES

[1] Flor, H. H. Inheritance of Reaction to Rust in Flax. *J Agric Res*, **1947**, *74*(1), 241–262, https://handle.nal.usda.gov/10113/IND43970158

[2] Flor, H. H. Current Status of the Gene-For-Gene Concept. *Annu Rev Phytopathol*, **1971**, *9* (1), 275–296. https://doi.org/10.1146/annurev.py.09.090171.001423.

[3] Wang, J. Y.; Doudna, J. A. CRISPR Technology: A Decade of Genome Editing Is Only the Beginning. *Science*, **2023**, 379 (6629). https://doi.org/10.1126/science.add8643

[4] Schiml, S.; Puchta, H. Revolutionizing Plant Biology: Multiple Ways of Genome Engineering by CRISPR/Cas. *Plant Methods*, **2016**, p 8. https://doi.org/10.1186/s13007-016-0103-0.

[5] Staskawicz, B. J. Genetics of Plant-Pathogen Interactions Specifying Plant Disease Resistance. *Plant Physiol*, **2001**, *125* (1), 73–76. https://doi.org/10.1104/PP.125.1.73.

[6] Wang, W.; Feng, B.; Zhou, J. M.; Tang, D. Plant Immune Signaling: Advancing on Two Frontiers. *J Integr Plant Biol*, **2020**, *62* (1), 2–24. https://doi.org/10.1111/JIPB.12898.
[7] Jones, J. D. G.; Dangl, J. L. The Plant Immune System. *Nature*, **2006**, *444* (7117), 323–329. https://doi.org/10.1038/nature05286.
[8] Tsuda, K.; Sato, M.; Stoddard, T.; Glazebrook, J.; Katagiri, F. Network Properties of Robust Immunity in Plants. *PLoS Genet*, **2009**, *5* (12), e1000772. https://doi.org/10.1371/JOURNAL.PGEN.1000772.
[9] Ahmed, A. A.; Mclellan, H.; Aguilar, G. B.; Hein, I.; Thordal-Christensen, H.; Birch, P. R. J. Engineering Barriers to Infection by Undermining Pathogen Effector Function or by Gaining Effector Recognition. In *Plant Pathogen Resistance Biotechnology*, **2016**, pp. 21–50. https://doi.org/10.1002/9781118867716.CH2.
[10] De Wit, P. J. G. M.; Joosten, M. H. A. J.; Thomma, B. H. P. J.; Stergiopoulos, I. Gene for Gene Models and Beyond: The Cladosporium fulvum-Tomato Pathosystem. *The Mycota*, **2009**, 135–156. https://doi.org/10.1007/978-3-540-87407-2_7.
[11] Paul, N. C.; Park, S. W.; Liu, H.; Choi, S.; Ma, J.; MacCready, J. S.; Chilvers, M. I.; Sang, H. Plant and Fungal Genome Editing to Enhance Plant Disease Resistance Using the CRISPR/Cas9 System. *Front Plant Sci*, **2021**, 12, 700925. https://doi.org/10.3389/FPLS.2021.700925/BIBTEX.
[12] Nejat, N.; Mantri, N. Plant Immune System: Crosstalk between Responses to Biotic and Abiotic Stresses the Missing Link in Understanding Plant Defence. *Curr Issues Mol Biol* **2017**, *23* (1), 1–16. https://doi.org/10.21775/CIMB.023.001.
[13] Ijaz, M.; Khan, F.; Zaki, H. E. M.; Khan, M. M.; Radwan, K. S. A.; Jiang, Y.; Qian, J.; Ahmed, T.; Shahid, M. S.; Luo, J.; et al. Recent Trends and Advancements in CRISPR-Based Tools for Enhancing Resistance against Plant Pathogens. *Plants* **2023**, *12* (9), 1911. https://doi.org/10.3390/PLANTS12091911.
[14] Zhao, Y.; Zhu, X.; Chen, X.; Zhou, J. M. From Plant Immunity to Crop Disease Resistance. *J Genet Genomics*, **2022**, *49* (8), 693–703. https://doi.org/10.1016/J.JGG.2022.06.003.
[15] Belhaj, K.; Chaparro-Garcia, A.; Kamoun, S.; Patron, N. J.; Nekrasov, V. Editing Plant Genomes with CRISPR/Cas9. *Curr Opin Biotechnol*, **2015**, *32*, 76–84. https://doi.org/10.1016/J.COPBIO.2014.11.007.
[16] Zhang, Y.; Iaffaldano, B.; Qi, Y. CRISPR Ribonucleoprotein-Mediated Genetic Engineering in Plants. *Plant Commun*, **2021**, *2* (2). https://doi.org/10.1016/J.XPLC.2021.100168.
[17] Niu, D.; Wei, H. J.; Lin, L.; George, H.; Wang, T.; Lee, I. H.; Zhao, H. Y.; Wang, Y.; Kan, Y.; Shrock, E.; et al. Inactivation of Porcine Endogenous Retrovirus in Pigs Using CRISPR-Cas9. *Science*, 2017, 357 (6357), 1303–1307. https://doi.org/10.1126/SCIENCE.AAN4187.
[18] Ihry, R. J.; Worringer, K. A.; Salick, M. R.; Frias, E.; Ho, D.; Theriault, K.; Kommineni, S.; Chen, J.; Sondey, M.; Ye, C.; et al. P53 Inhibits CRISPR-Cas9 Engineering in Human Pluripotent Stem Cells. *Nat Med*, **2018**, *24* (7), 939–946. https://doi.org/10.1038/s41591-018-0050-6.
[19] Enache, O. M.; Rendo, V.; Abdusamad, M.; Lam, D.; Davison, D.; Pal, S.; Currimjee, N.; Hess, J.; Pantel, S.; Nag, A.; et al. Cas9 Activates the P53 Pathway and Selects for P53-Inactivating Mutations. *Nat Genet*, **2020**, *52* (7), 662–668. https://doi.org/10.1038/s41588-020-0623-4.
[20] Schenke, D.; Cai, D. Applications of CRISPR/Cas to Improve Crop Disease Resistance: Beyond Inactivation of Susceptibility Factors. *iScience*, **2020**, *23* (9). https://doi.org/10.1016/J.ISCI.2020.101478.
[21] van Schie, C. C. N.; Takken, F. L. W. Susceptibility Genes 101: How to Be a Good Host. *Annu Rev Phytopathol*, **2014**, *52* (1), 551–581. https://doi.org/10.1146/annurev-phyto-102313-045854.
[22] Eckardt, N. A. Plant Disease Susceptibility Genes? *Plant Cell*, **2002**, *14* (9), 1983–1986. https://doi.org/10.1105/tpc.140910.
[23] Zaidi, S. S. E. A.; Mukhtar, M. S.; Mansoor, S. Genome Editing: Targeting Susceptibility Genes for Plant Disease Resistance. *Trends Biotechnol*, **2018**, *36* (9), 898–906. https://doi.org/10.1016/J.TIBTECH.2018.04.005.
[24] Concordet, J. P.; Haeussler, M. CRISPOR: Intuitive Guide Selection for CRISPR/Cas9 Genome Editing Experiments and Screens. *Nucleic Acids Res*, **2018**, *46* (W1), W242–W245. https://doi.org/10.1093/NAR/GKY354.
[25] Xie, X.; Ma, X.; Zhu, Q.; Zeng, D.; Li, G.; Liu, Y. G. CRISPR-GE: A Convenient Software Toolkit for CRISPR-Based Genome Editing. *Mol Plant*, **2017**, *10* (9), 1246–1249. https://doi.org/10.1016/J.MOLP.2017.06.004.
[26] Montague, T. G.; Cruz, J. M.; Gagnon, J. A.; Church, G. M.; Valen, E. CHOPCHOP: A CRISPR/Cas9 and TALEN Web Tool for Genome Editing. *Nucleic Acids Res*, **2014**, *42* (Web Server issue), W401. https://doi.org/10.1093/NAR/GKU410.
[27] Xu, Z.; Xu, X.; Gong, Q.; Li, Z.; Li, Y.; Wang, S.; Yang, Y.; Ma, W.; Liu, L.; Zhu, B.; et al. Engineering Broad-Spectrum Bacterial Blight Resistance by Simultaneously Disrupting Variable TALE-Binding Elements of Multiple Susceptibility Genes in Rice. *Mol Plant*, **2019**, *12* (11), 1434–1446. https://doi.org/10.1016/J.MOLP.2019.08.006.

[28] Zeng, X.; Luo, Y.; Vu, N. T. Q.; Shen, S.; Xia, K.; Zhang, M. CRISPR/Cas9-Mediated Mutation of *OsSWEET14* in Rice Cv. Zhonghua11 Confers Resistance to *Xanthomonas* oryzae Pv. oryzae without Yield Penalty. *BMC Plant Biol*, **2020**, *20* (1), 1–11. https://doi.org/10.1186/S12870-020-02524-Y/FIGURES/5.

[29] Peng, A.; Chen, S.; Lei, T.; Xu, L.; He, Y.; Wu, L.; Yao, L.; Zou, X. Engineering Canker-Resistant Plants through CRISPR/Cas9-Targeted Editing of the Susceptibility Gene *CsLOB1* Promoter in Citrus. *Plant Biotechnol J*, **2017**, *15* (12), 1509–1519. https://doi.org/10.1111/PBI.12733.

[30] Jia, H.; Zhang, Y.; Orbović, V.; Xu, J.; White, F. F.; Jones, J. B.; Wang, N. Genome Editing of the Disease Susceptibility Gene CsLOB1 in Citrus Confers Resistance to Citrus Canker. *Plant Biotechnol J*, **2017**, *15* (7), 817–823. https://doi.org/10.1111/PBI.12677.

[31] Perk, E. A.; Arruebarrena Di Palma, A.; Colman, S.; Mariani, O.; Cerrudo, I.; D'Ambrosio, J. M.; Robuschi, L.; Pombo, M. A.; Rosli, H. G.; Villareal, F.; et al. CRISPR/Cas9-Mediated Phospholipase C 2 Knock-out Tomato Plants are More Resistant to Botrytis Cinerea. *Planta*, **2023**, *257* (6), 117. https://doi.org/10.1007/S00425-023-04147-7.

[32] Ortigosa, A.; Gimenez-Ibanez, S.; Leonhardt, N.; Solano, R. Design of a Bacterial Speck Resistant Tomato by CRISPR/Cas9-mediated Editing of SlJAZ2. *Plant Biotechnol J*, **2019**, *17* (3), 665. https://doi.org/10.1111/PBI.13006.

[33] de Toledo Thomazella, D. P.; Seong, K.; Mackelprang, R.; Dahlbeck, D.; Geng, Y.; Gill, U. S.; Qi, T.; Pham, J.; Giuseppe, P.; Lee, C. Y.; et al. Loss of Function of a DMR6 Ortholog in Tomato Confers Broad-Spectrum Disease Resistance. *Proc Natl Acad Sci U S A*, **2021**, *118* (27). https://doi.org/10.1073/PNAS.2026152118/-/DCSUPPLEMENTAL.

[34] Falak, N.; Imran, Q. M.; Hussain, A.; Yun, B. W. Transcription Factors as the "Blitzkrieg" of Plant Defense: A Pragmatic View of Nitric Oxide's Role in Gene Regulation. *Int J Mol Sci*, **2021**, *22* (2), 1–23. https://doi.org/10.3390/IJMS22020522.

[35] Zhou, Y.; Xu, S.; Jiang, N.; Zhao, X.; Bai, Z.; Liu, J.; Yao, W.; Tang, Q.; Xiao, G.; Lv, C.; et al. Engineering of Rice Varieties with Enhanced Resistances to Both Blast and Bacterial Blight Diseases via CRISPR/Cas9. *Plant Biotechnol J*, **2022**, *20* (5), 876. https://doi.org/10.1111/PBI.13766.

[36] Wang, F.; Wang, C.; Liu, P.; Lei, C.; Hao, W.; Gao, Y.; Liu, Y. G.; Zhao, K. Enhanced Rice Blast Resistance by CRISPR/Cas9-Targeted Mutagenesis of the ERF Transcription Factor Gene OsERF922. *PLoS One*, **2016**, *11* (4), e0154027. https://doi.org/10.1371/JOURNAL.PONE.0154027.

[37] Wang, X.; Guo, R.; Tu, M.; Wang, D.; Guo, C.; Wan, R.; Li, Z.; Wang, X. Ectopic Expression of the Wild Grape WRKY Transcription Factor VqWRKY52 in Arabidopsis thaliana Enhances Resistance to the Biotrophic Pathogen Powdery Mildew But Not to the Necrotrophic Pathogen *Botrytis cinerea*. *Front Plant Sci*, **2017**, *8* (JANUARY). https://doi.org/10.3389/FPLS.2017.00097.

[38] Wang, X.; Tu, M.; Wang, D.; Liu, J.; Li, Y.; Li, Z.; Wang, Y.; Wang, X. CRISPR/Cas9-mediated Efficient Targeted Mutagenesis in Grape in the First Generation. *Plant Biotechnol J*, **2018**, *16* (4), 844. https://doi.org/10.1111/PBI.12832.

[39] Büschges, R.; Hollricher, K.; Panstruga, R.; Simons, G.; Wolter, M.; Frijters, A.; Van Daelen, R.; Van der Lee, T.; Diergaarde, P.; Groenendijk, J.; et al. The Barley Mlo Gene: A Novel Control Element of Plant Pathogen Resistance. *Cell*, **1997**, *88* (5), 695–705. https://doi.org/10.1016/S0092-8674(00)81912-1.

[40] Consonni, C.; Humphry, M. E.; Hartmann, H. A.; Livaja, M.; Durner, J.; Westphal, L.; Vogel, J.; Lipka, V.; Kemmerling, B.; Schulze-Lefert, P.; et al. Conserved Requirement for a Plant Host Cell Protein in Powdery Mildew Pathogenesis. *Nat Genet*, **2006**, *38* (6), 716–720. https://doi.org/10.1038/NG1806.

[41] Wang, Y.; Cheng, X.; Shan, Q.; Zhang, Y.; Liu, J.; Gao, C.; Qiu, J. L. Simultaneous Editing of Three Homoeoalleles in Hexaploid Bread Wheat Confers Heritable Resistance to Powdery Mildew. *Nat Biotechnol*, **2014**, *32* (9), 947–951. https://doi.org/10.1038/NBT.2969.

[42] Zheng, Z.; Nonomura, T.; Appiano, M.; Pavan, S.; Matsuda, Y.; Toyoda, H.; Wolters, A. M. A.; Visser, R. G. F.; Bai, Y. Loss of Function in Mlo Orthologs Reduces Susceptibility of Pepper and Tomato to Powdery Mildew Disease Caused by *Leveillula Taurica*. *PLoS One*, **2013**, *8* (7), e70723. https://doi.org/10.1371/JOURNAL.PONE.0070723.

[43] Humphry, M.; Reinstädler, A.; Ivanov, S.; Bisseling, T.; Panstruga, R. Durable Broad-Spectrum Powdery Mildew Resistance in Pea Er1 Plants Is Conferred by Natural Loss-of-Function Mutations in PsMLO1. *Mol Plant Pathol*, **2011**, *12* (9), 866–878. https://doi.org/10.1111/j.1364-3703.2011.00718.x.

[44] Berg, J. A.; Appiano, M.; Santillán Martínez, M.; Hermans, F. W. K.; Vriezen, W. H.; Visser, R. G. F.; Bai, Y.; Schouten, H. J. A Transposable Element Insertion in the Susceptibility Gene CsaMLO8 Results in Hypocotyl Resistance to Powdery Mildew in Cucumber. *BMC Plant Biol*, **2015**, *15* (1). https://doi.org/10.1186/s12870-015-0635-x.

[45] Berg, J. A.; Appiano, M.; Bijsterbosch, G.; Visser, R. G. F.; Schouten, H. J.; Bai, Y. Functional Characterization of Cucumber (Cucumis Sativus L.) Clade V MLO Genes. *BMC Plant Biol*, **2017**, *17* (1), 80. https://doi.org/10.1186/s12870-017-1029-z.
[46] Nie, J.; Wang, Y.; He, H.; Guo, C.; Zhu, W.; Pan, J.; Li, D.; Lian, H.; Pan, J.; Cai, R. Loss-of-Function Mutations in CsMLO1 Confer Durable Powdery Mildew Resistance in Cucumber (Cucumis Sativus L.). *Front Plant Sci*, **2015**, *6* (DEC), 1155. https://doi.org/10.3389/fpls.2015.01155.
[47] Kusch, S.; Pesch, L.; Panstruga, R. Comprehensive Phylogenetic Analysis Sheds Light on the Diversity and Origin of the MLO Family of Integral Membrane Proteins. *Genome Biol Evol*, **2016**, *8* (3), 878–895. https://doi.org/10.1093/gbe/evw036.
[48] Nekrasov, V.; Wang, C.; Win, J.; Lanz, C.; Weigel, D.; Kamoun, S. Rapid Generation of a Transgene-Free Powdery Mildew Resistant Tomato by Genome Deletion. *Sci Rep*, **2017**, *7* (1). https://doi.org/10.1038/s41598-017-00578-x.
[49] Xuebo, W.; Dandan, L.; Xiaolei, T.; Changchun, C.; Xinyao, Z.; Zhan, S.; Aiguo, Y.; Xiankui, F.; Dan, L. CRISPR/Cas9-Mediated Targeted Mutagenesis of Two Homoeoalleles in Tobacco Confers Resistance to Powdery Mildew. *Euphytica*, **2023**, *219* (6), 1–11. https://doi.org/10.1007/S10681-023-03196-Z/FIGURES/4.
[50] Shnaider, Y.; Elad, Y.; David, D. R.; Pashkovsky, E.; Leibman, D.; Kravchik, M.; Shtarkman-Cohen, M.; Gal-On, A.; Spiegelman, Z. Development of Powdery Mildew (*Podosphaera xanthii*) Resistance in Cucumber (*Cucumis sativus*) Using CRISPR/Cas9-Mediated Mutagenesis of *CsaMLO8*. *Phytopathology®*, **2022**, *113* (5), 786–790. https://doi.org/10.1094/PHYTO-06-22-0193-FI.
[51] Devoto, A.; Piffanelli, P.; Nilsson, I. M.; Wallin, E.; Panstruga, R.; Von Heijne, G.; Schulze-Lefert, P. Topology, Subcellular Localization, and Sequence Diversity of the Mlo Family in Plants. *J Biol Chem*, **1999**, *274* (49), 34993–35004. https://doi.org/10.1074/jbc.274.49.34993.
[52] Xu, X.; Liu, X.; Yan, Y.; Wang, W.; Gebretsadik, K.; Qi, X.; Xu, Q.; Chen, X. Comparative Proteomic Analysis of Cucumber Powdery Mildew Resistance between a Single-Segment Substitution Line and Its Recurrent Parent. *Hortic Res*, **2019**, *6* (1), 1–13. https://doi.org/10.1038/s41438-019-0198-3.
[53] Assaad, F. F.; Qiu, J. L.; Youngs, H.; Ehrhardt, D.; Zimmerli, L.; Kalde, M.; Wanner, G.; Peck, S. C.; Edwards, H.; Ramonell, K.; et al. The PEN1 Syntaxin Defines a Novel Cellular Compartment upon Fungal Attack and Is Required for the Timely Assembly of Papillae. *Mol Biol Cell*, **2004**, *15* (11), 5118–5129. https://doi.org/10.1091/MBC.E04-02-0140.
[54] Kwon, C.; Neu, C.; Pajonk, S.; Yun, H. S.; Lipka, U.; Humphry, M.; Bau, S.; Straus, M.; Kwaaitaal, M.; Rampelt, H.; et al. Co-Option of a Default Secretory Pathway for Plant Immune Responses. *Nature*, **2008**, *451* (7180), 835–840. https://doi.org/10.1038/NATURE06545.
[55] Böhlenius, H.; Mørch, S. M.; Godfrey, D.; Nielsen, M. E.; Thordal-Christensen, H. The Multivesicular Body-Localized GTPase ARFA1b/1c Is Important for Callose Deposition and ROR2 Syntaxin-Dependent Preinvasive Basal Defense in Barley. *Plant Cell*, **2010**, *22* (11), 3831–3844. https://doi.org/10.1105/TPC.110.078063.
[56] Humphry, M.; Consonni, C.; Panstruga, R. Mlo-Based Powdery Mildew Immunity: Silver Bullet or Simply Non-Host Resistance? *Mol Plant Pathol*, **2006**, *7* (6), 605–610. https://doi.org/10.1111/J.1364-3703.2006.00362.X.
[57] Micali, C.; Göllner, K.; Humphry, M.; Consonni, C.; Panstruga, R. The Powdery Mildew Disease of Arabidopsis: A Paradigm for the Interaction between Plants and Biotrophic Fungi. *Arabidopsis Book*, **2008**, *6*, e0115. https://doi.org/10.1199/TAB.0115.
[58] Kim, M. C.; Lee, S. H.; Kim, J. K.; Chun, H. J.; Choi, M. S.; Chung, W. S.; Moon, B. C.; Kang, C. H.; Park, C. Y.; Yoo, J. H.; et al. Mlo, a Modulator of Plant Defense and Cell Death, Is a Novel Calmodulin-Binding Protein. Isolation and Characterization of a Rice Mlo Homologue. *J Biol Chem*, **2002**, *277* (22), 19304–19314. https://doi.org/10.1074/jbc.M108478200.
[59] Wan, D. Y.; Guo, Y.; Cheng, Y.; Hu, Y.; Xiao, S.; Wang, Y.; Wen, Y. Q. CRISPR/Cas9-Mediated Mutagenesis of *VvMLO3* Results in Enhanced Resistance to Powdery Mildew in Grapevine (*Vitis vinifera*). *Hortic Res*, **2020**, *7* (1), 1–14. https://doi.org/10.1038/s41438-020-0339-8.
[60] Tek, M. I.; Calis, O.; Fidan, H.; Shah, M. D.; Celik, S.; Wani, S. H. CRISPR/Cas9 Based Mlo-Mediated Resistance against Podosphaera xanthii in Cucumber (*Cucumis sativus L.*). *Front Plant Sci*, **2022**, *13*, 1081506. https://doi.org/10.3389/FPLS.2022.1081506/BIBTEX.
[61] Jacobs, A. K.; Lipka, V.; Burton, R. A.; Panstruga, R.; Strizhov, N.; Schulze-Lefert, P.; Fincher, G. B. An Arabidopsis Callose Synthase, GSL5, Is Required for Wound and Papillary Callose Formation. *Plant Cell*, **2003**, *15* (11), 2503–2513. https://doi.org/10.1105/TPC.016097.
[62] Nishimura, M. T.; Stein, M.; Hou, B. H.; Vogel, J. P.; Edwards, H.; Somerville, S. C. Loss of a Callose Synthase Results in Salicylic Acid-Dependent Disease Resistance. *Science*, **2003**, *301* (5635), 969–972. https://doi.org/10.1126/SCIENCE.1086716.

[63] Ellinger, D.; Naumann, M.; Falter, C.; Zwikowics, C.; Jamrow, T.; Manisseri, C.; Somerville, S. C.; Voigt, C. A. Elevated Early Callose Deposition Results in Complete Penetration Resistance to Powdery Mildew in Arabidopsis. *Plant Physiol*, **2013**, *161* (3), 1433–1444. https://doi.org/10.1104/PP.112.211011.

[64] Eggert, D.; Naumann, M.; Reimer, R.; Voigt, C. A. Nanoscale Glucan Polymer Network Causes Pathogen Resistance. *Sci Rep*, **2014**, *4*. https://doi.org/10.1038/SREP04159.

[65] Santillán Martínez, M. I.; Bracuto, V.; Koseoglou, E.; Appiano, M.; Jacobsen, E.; Visser, R. G. F.; Wolters, A. M. A.; Bai, Y. CRISPR/Cas9-Targeted Mutagenesis of the Tomato Susceptibility Gene PMR4 for Resistance against Powdery Mildew. *BMC Plant Biol*, **2020**, *20* (1), 1–13. https://doi.org/10.1186/S12870-020-02497-Y/FIGURES/6.

[66] Zhang, Y.; Bai, Y.; Wu, G.; Zou, S.; Chen, Y.; Gao, C.; Tang, D. Simultaneous Modification of Three Homoeologs of TaEDR1 by Genome Editing Enhances Powdery Mildew Resistance in Wheat. *Plant J*, **2017**, *91* (4), 714–724. https://doi.org/10.1111/TPJ.13599.

[67] Cao, Y.; Zhou, H.; Zhou, X.; Li, F. Control of Plant Viruses by CRISPR/Cas System-Mediated Adaptive Immunity. *Front Microbiol*, **2020**, *11*, 593700. https://doi.org/10.3389/FMICB.2020.593700/BIBTEX.

[68] Wang, A.; Krishnaswamy, S. Eukaryotic Translation Initiation Factor 4E-Mediated Recessive Resistance to Plant Viruses and Its Utility in Crop Improvement. *Mol Plant Pathol*, **2012**, *13* (7), 795–803. https://doi.org/10.1111/J.1364-3703.2012.00791.X.

[69] Wittmann, S.; Chatel, H.; Fortin, M. G.; Laliberté, J. F. Interaction of the Viral Protein Genome Linked of Turnip Mosaic Potyvirus with the Translational Eukaryotic Initiation Factor (iso) 4E of *Arabidopsis thaliana* Using the Yeast Two-Hybrid System. *Virology*, **1997**, *234* (1), 84–92. https://doi.org/10.1006/VIRO.1997.8634.

[70] Léonard, S.; Plante, D.; Wittmann, S.; Daigneault, N.; Fortin, M. G.; Laliberté, J.-F. Complex Formation between Potyvirus VPg and Translation Eukaryotic Initiation Factor 4E Correlates with Virus Infectivity. *J Virol*, **2000**, *74* (17), 7730–7737. https://doi.org/10.1128/JVI.74.17.7730-7737.2000.

[71] Ruffel, S.; Dussault, M. H.; Palloix, A.; Moury, B.; Bendahmane, A.; Robaglia, C.; Caranta, C. A Natural Recessive Resistance Gene against Potato Virus Y in Pepper Corresponds to the Eukaryotic Initiation Factor 4E (ElF4E). *Plant J*, **2002**, *32* (6), 1067–1075. https://doi.org/10.1046/J.1365-313X.2002.01499.X.

[72] Ruffel, S.; Gallois, J. L.; Moury, B.; Robaglia, C.; Palloix, A.; Caranta, C. Simultaneous Mutations in Translation Initiation Factors ElF4E and ElF(Iso)4E are Required to Prevent Pepper Veinal Mottle Virus Infection of Pepper. *J Gen Virol*, **2006**, *87* (7), 2089–2098. https://doi.org/10.1099/VIR.0.81817-0.

[73] Nicaise, V.; German-Retana, S.; Sanjuán, R.; Dubrana, M. P.; Mazier, M.; Maisonneuve, B.; Candresse, T.; Caranta, C.; LeGall, O. The Eukaryotic Translation Initiation Factor 4E Controls Lettuce Susceptibility to the Potyvirus Lettuce Mosaic Virus. *Plant Physiol*, **2003**, *132* (3), 1272–1282. https://doi.org/10.1104/PP.102.017855.

[74] Kanyuka, K.; Druka, A.; Caldwell, D. G.; Tymon, A.; McCallum, N.; Waugh, R.; Adams, M. J. Evidence That the Recessive Bymovirus Resistance Locus rym4 in Barley Corresponds to the Eukaryotic Translation Initiation Factor 4E Gene. *Mol Plant Pathol*, **2005**, *6* (4), 449–458. https://doi.org/10.1111/J.1364-3703.2005.00294.X.

[75] Diaz-Pendon, J. A.; Truniger, V.; Nieto, C.; Garcia-Mas, J.; Bendahmane, A.; Aranda, M. A. Advances in Understanding Recessive Resistance to Plant Viruses. *Mol Plant Pathol*, **2004**, *5* (3), 223–233. https://doi.org/10.1111/J.1364-3703.2004.00223.X.

[76] Pyott, D. E.; Sheehan, E.; Molnar, A. Engineering of CRISPR/Cas9-Mediated Potyvirus Resistance in Transgene-Free Arabidopsis Plants. *Mol Plant Pathol*, **2016**, *17* (8), 1276–1288. https://doi.org/10.1111/MPP.12417.

[77] Chandrasekaran, J.; Brumin, M.; Wolf, D.; Leibman, D.; Klap, C.; Pearlsman, M.; Sherman, A.; Arazi, T.; Gal-On, A. Development of Broad Virus Resistance in Non-Transgenic Cucumber Using CRISPR/Cas9 Technology. *Mol Plant Pathol*, **2016**, *17* (7), 1140–1153. https://doi.org/10.1111/MPP.12375.

[78] Fidan, H.; Calis, O.; Ari, E.; Atasayar, A.; Sarikaya, P.; Tek, M. I.; Izmirli, A.; Oz, Y.; Firat, G. Knockout of ElF4E Using CRISPR/Cas9 for Large-Scale Production of Resistant Cucumber Cultivar against WMV, ZYMV, and PRSV. *Front Plant Sci*, **2023**, *14*, 1143813. https://doi.org/10.3389/FPLS.2023.1143813/BIBTEX.

[79] Macovei, A.; Sevilla, N. R.; Cantos, C.; Jonson, G. B.; Slamet-Loedin, I.; Čermák, T.; Voytas, D. F.; Choi, I. R.; Chadha-Mohanty, P. Novel Alleles of Rice EIF4G Generated by CRISPR/Cas9-Targeted Mutagenesis Confer Resistance to Rice Tungro Spherical Virus. *Plant Biotechnol J*, **2018**, *16* (11), 1918–1927. https://doi.org/10.1111/PBI.12927.

[80] Gomez, M. A.; Lin, Z. D.; Moll, T.; Chauhan, R. D.; Hayden, L.; Renninger, K.; Beyene, G.; Taylor, N. J.; Carrington, J. C.; Staskawicz, B. J.; et al. Simultaneous CRISPR/Cas9-Mediated Editing of Cassava EIF4E Isoforms NCBP-1 and NCBP-2 Reduces Cassava Brown Streak Disease Symptom Severity and Incidence. *Plant Biotechnol J*, **2019**, *17* (2), 421–434. https://doi.org/10.1111/PBI.12987.

[81] Nishikiori, M.; Mori, M.; Dohi, K.; Okamura, H.; Katoh, E.; Naito, S.; Meshi, T.; Ishikawa, M. A Host Small GTP-Binding Protein ARL8 Plays Crucial Roles in Tobamovirus RNA Replication. *PLoS Pathog*, 2011, *7* (12), e1002409. https://doi.org/10.1371/JOURNAL.PPAT.1002409.

[82] Salem, N.; Mansour, A.; Ciuffo, M.; Falk, B. W.; Turina, M. A New Tobamovirus Infecting Tomato Crops in Jordan. *Arch Virol*, **2016**, *161* (2), 503–506. https://doi.org/10.1007/S00705-015-2677-7.

[83] Luria, N.; Smith, E.; Reingold, V.; Bekelman, I.; Lapidot, M.; Levin, I.; Elad, N.; Tam, Y.; Sela, N.; Abu-Ras, A.; et al. A New Israeli Tobamovirus Isolate Infects Tomato Plants Harboring Tm-22 Resistance Genes. *PLoS One*, **2017**, *12* (1), e0170429. https://doi.org/10.1371/JOURNAL.PONE.0170429.

[84] Vossenberg, B. T. L. H. van de; Dawood, T.; Woźny, M.; Botermans, M. First Expansion of the Public Tomato Brown Rugose Fruit Virus (ToBRFV) Nextstrain Build; Inclusion of New Genomic and Epidemiological Data. *PhytoFrontiers™*, **2021**, *1* (4), 359–363. https://doi.org/10.1094/PHYTOFR-01-21-0005-A.

[85] Ishikawa, M.; Yoshida, T.; Matsuyama, M.; Kouzai, Y.; Kano, A.; Ishibashi, K. Tomato Brown Rugose Fruit Virus Resistance Generated by Quadruple Knockout of Homologs of *TOBAMOVIRUS MULTIPLICATION1* in Tomato. *Plant Physiol*, **2022**, *189* (2), 679–686. https://doi.org/10.1093/PLPHYS/KIAC103.

[86] Pramanik, D.; Shelake, R. M.; Park, J.; Kim, M. J.; Hwang, I.; Park, Y.; Kim, J. Y. CRISPR/Cas9-Mediated Generation of Pathogen-Resistant Tomato against Tomato Yellow Leaf Curl Virus and Powdery Mildew. *Int J Mol Sci*, **2021**, *22* (4), 1–18. https://doi.org/10.3390/IJMS22041878.

[87] Baltes, N. J.; Hummel, A. W.; Konecna, E.; Cegan, R.; Bruns, A. N.; Bisaro, D. M.; Voytas, D. F. Conferring Resistance to Geminiviruses with the CRISPR-Cas Prokaryotic Immune System. *Nat Plants*, **2015**, *1* (10). https://doi.org/10.1038/NPLANTS.2015.145.

[88] Ali, Z.; Abulfaraj, A.; Idris, A.; Ali, S.; Tashkandi, M.; Mahfouz, M. M. CRISPR/Cas9-Mediated Viral Interference in Plants. *Genome Biol*, **2015**, *16* (1), 1–11. https://doi.org/10.1186/S13059-015-0799-6/FIGURES/4.

[89] Ji, X.; Zhang, H.; Zhang, Y.; Wang, Y.; Gao, C. Establishing a CRISPR-Cas-like Immune System Conferring DNA Virus Resistance in Plants. *Nat Plants*, **2015**, *1* (10), 1–4. https://doi.org/10.1038/nplants.2015.144.

[90] Tripathi, J. N.; Ntui, V. O.; Ron, M.; Muiruri, S. K.; Britt, A.; Tripathi, L. CRISPR/Cas9 Editing of Endogenous *Banana Streak Virus* in the B Genome of *Musa* Spp. Overcomes a Major Challenge in Banana Breeding. *Commun Biol*, **2019**, *2* (1), 1–11. https://doi.org/10.1038/s42003-019-0288-7.

[91] Kis, A.; Hamar, É.; Tholt, G.; Bán, R.; Havelda, Z. Creating Highly Efficient Resistance against Wheat Dwarf Virus in Barley by Employing CRISPR/Cas9 System. *Plant Biotechnol J*, **2019**, *17* (6), 1004–1006. https://doi.org/10.1111/PBI.13077.

[92] Yin, K.; Han, T.; Xie, K.; Zhao, J.; Song, J.; Liu, Y. Engineer Complete Resistance to Cotton Leaf Curl Multan Virus by the CRISPR/Cas9 System in *Nicotiana Benthamiana*. *Phytopathol Res*, **2019**, *1* (1), 1–9. https://doi.org/10.1186/S42483-019-0017-7.

[93] Tashkandi, M.; Ali, Z.; Aljedaani, F.; Shami, A.; Mahfouz, M. M. Engineering Resistance against Tomato Yellow Leaf Curl Virus via the CRISPR/Cas9 System in Tomato. *Plant Signal Behav*, **2018**, *13* (10). https://doi.org/10.1080/15592324.2018.1525996.

[94] Aman, R.; Ali, Z.; Butt, H.; Mahas, A.; Aljedaani, F.; Khan, M. Z.; Ding, S.; Mahfouz, M. RNA Virus Interference via CRISPR/Cas13a System in Plants. *Genome Biol*, **2018**, *19* (1). https://doi.org/10.1186/S13059-017-1381-1.

[95] Zhan, X.; Zhang, F.; Zhong, Z.; Chen, R.; Wang, Y.; Chang, L.; Bock, R.; Nie, B.; Zhang, J. Generation of Virus-Resistant Potato Plants by RNA Genome Targeting. *Plant Biotechnol J*, **2019**, *17* (9), 1814–1822. https://doi.org/10.1111/PBI.13102.

[96] Zhang, T.; Zheng, Q.; Yi, X.; An, H.; Zhao, Y.; Ma, S.; Zhou, G. Establishing RNA Virus Resistance in Plants by Harnessing CRISPR Immune System. *Plant Biotechnol J*, **2018**, *16* (8), 1415. https://doi.org/10.1111/PBI.12881.

[97] De Almeida Engler, J.; Favery, B.; Engler, G.; Abad, P. Loss of Susceptibility as an Alternative for Nematode Resistance. *Curr Opin Biotechnol*, **2005**, *16* (2), 112–117. https://doi.org/10.1016/J.COPBIO.2005.01.009.

[98] Van Schie, C. C. N.; Takken, F. L. W. Susceptibility Genes 101: How to Be a Good Host. *Annu Rev Phytopathol*, **2014**, *52*, 551–581. https://doi.org/10.1146/ANNUREV-PHYTO-102313-045854.

[99] Dangl, J. L.; Horvath, D. M.; Staskawicz, B. J. Pivoting the Plant Immune System from Dissection to Deployment. *Science*, **2013**, *341* (6147), 746–751. https://doi.org/10.1126/SCIENCE.1236011.

[100] Li, S.; Lin, D.; Zhang, Y.; Deng, M.; Chen, Y.; Lv, B.; Li, B.; Lei, Y.; Wang, Y.; Zhao, L.; et al. Genome-Edited Powdery Mildew Resistance in Wheat without Growth Penalties. *Nature*, **2022**, *602* (7897), 455–460. https://doi.org/10.1038/s41586-022-04395-9.

[101] Sandhya, D.; Jogam, P.; Allini, V. R.; Abbagani, S.; Alok, A. The Present and Potential Future Methods for Delivering CRISPR/Cas9 Components in Plants. *J Genet Eng Biotechnol*, **2020**, *18* (1), 1–11. https://doi.org/10.1186/S43141-020-00036-8/FIGURES/3.

[102] Ghogare, R.; Ludwig, Y.; Bueno, G. M.; Slamet-Loedin, I. H.; Dhingra, A. Genome Editing Reagent Delivery in Plants. *Transgenic Res*, **2021**, *30* (4), 321–335. https://doi.org/10.1007/S11248-021-00239-W.

[103] Laforest, L. C.; Nadakuduti, S. S. Advances in Delivery Mechanisms of CRISPR Gene-Editing Reagents in Plants. *Front Genome Ed*, **2022**, *4*, 830178. https://doi.org/10.3389/FGEED.2022.830178/BIBTEX.

[104] González-Salitre, L.; Román-Gutiérrez, A.; Contreras-López, E.; Bautista-Ávila, M.; Rodríguez-Serrano, G.; González-Olivares, L.; Del, A.; De Hidalgo, E.; De, I.; De, C.; et al. Promising Use of Selenized Yeast to Develop New Enriched Food: Human Health Implications. *Food Rev Int*, **2023**, *39*(3), 1594–1611. https://doi.org/10.1080/87559129.2021.1934695.

[105] Liang, Z.; Chen, K.; Zhang, Y.; Liu, J.; Yin, K.; Qiu, J. L.; Gao, C. Genome Editing of Bread Wheat Using Biolistic Delivery of CRISPR/Cas9 in Vitro Transcripts or Ribonucleoproteins. *Nat Protoc*, **2018**, *13* (3), 413–430. https://doi.org/10.1038/NPROT.2017.145.

[106] Klimek-Chodacka, M.; Gieniec, M.; Baranski, R. Multiplex Site-Directed Gene Editing Using Polyethylene Glycol-Mediated Delivery of CRISPR GRNA:Cas9 Ribonucleoprotein (RNP) Complexes to Carrot Protoplasts. *Int J Mol Sci*, **2021**, *22* (19). https://doi.org/10.3390/IJMS221910740.

[107] Guo, J.; Li, K.; Jin, L.; Xu, R.; Miao, K.; Yang, F.; Qi, C.; Zhang, L.; Botella, J. R.; Wang, R.; et al. A Simple and Cost-Effective Method for Screening of CRISPR/Cas9-Induced Homozygous/Biallelic Mutants. *Plant Methods*, **2018**, *14* (1). https://doi.org/10.1186/S13007-018-0305-8.

[108] Wang, C.; Wang, K. Rapid Screening of CRISPR/Cas9-Induced Mutants Using the ACT-PCR Method. *Methods Mol Biol*, **2019**, 1917, 27–32. https://doi.org/10.1007/978-1-4939-8991-1_2.

[109] Li, R.; Ba, Y.; Song, Y.; Cui, J.; Zhang, X.; Zhang, D.; Yuan, Z.; Yang, L. Rapid and Sensitive Screening and Identification of CRISPR/Cas9 Edited Rice Plants Using Quantitative Real-Time PCR Coupled with High Resolution Melting Analysis. *Food Control*, **2020**, *112*, 107088. https://doi.org/10.1016/J.FOODCONT.2020.107088.

[110] Liu, G.; Qi, Y.; Zhang, T. Analysis of Off-Target Mutations in CRISPR-Edited Rice Plants Using Whole-Genome Sequencing. *Methods Mol Biol*, **2021**, *2238*, 145–172. https://doi.org/10.1007/978-1-0716-1068-8_10.

[111] Chen, K.; Wang, Y.; Zhang, R.; Zhang, H.; Gao, C. CRISPR/Cas Genome Editing and Precision Plant Breeding in Agriculture. *Ann Rev Plant Biol*, **2019**, *70*, 667–697. https://doi.org/10.1146/annurev-arplant-050718-100049.

[112] Gao, C. Genome Engineering for Crop Improvement and Future Agriculture. *Cell*, **2021**, *184* (6), 1621–1635. https://doi.org/10.1016/J.CELL.2021.01.005.

12 Advancement in Gene Editing for Crop Improvement Highlighting the Application of Pangenomes

Biswajit Pramanik, Sandip Debnath, and Anamika Das

12.1 INTRODUCTION

As the world's population is projected to reach 10 billion by 2050 (FAO, 2017), the most urgent challenge of our time is to sustainably feed this expanding population. The Green Revolution and the progress in plant breeding methodologies have played a pivotal role in attaining present-day crop yields that are capable of catering to the food requirements of a significant proportion of the global populace. In light of the developing climate change circumstances and the progressively decreasing accessibility of cultivable land, it appears that our capacity for food production is currently experiencing a state of stasis if not regression. According to Springmann et al. (2018), to sufficiently provide sustenance for a worldwide populace of 10 billion individuals, it is projected that agricultural output must experience a significant increase of 60%. Therefore, it is imperative to enhance agricultural sustainability and productivity through the utilization of state-of-the-art technological advancements in crop cultivation.

The development of robust crop cultivars capable of withstanding environmental perturbations represents a promising strategy for achieving worldwide food security. The development of such varieties can be achieved through the utilization of existing genetic diversity. Crop improvement is a crucial aspect of agricultural research, and various methods have been employed to achieve this goal. The four commonly used methods include cross-breeding, mutation breeding, transgenic breeding, and gene editing (Chen et al., 2019). The restricted genetic variability in superior germplasm results in the incorporation of traits that are already present within the parental genomes through cross-breeding. The process of mutation breeding, while possessing potential value, is a laborious and time-consuming endeavor. This is due to the requirement of identifying desirable traits in a vast population of mutagenized plants. The integration of genes from diverse organisms into crops, known as transgenic breeding, represents a noteworthy advancement in the field of plant breeding. The potential of foreign DNA integration into plant genomes has been limited to a few crops due to the random nature of the process and the stringent safety regulations surrounding genetically modified organisms (GMOs) (Das et al., 2023). Therefore, genome editing, due to its ability to make accurate and predictable genetic alterations, is becoming a preferred technique for precision breeding (Gao, 2021) (Figure 12.1a–d).

The notion of a pangenome encompasses the core genome, which is conserved across all individuals of a given species, and the accessory genome, which varies among individuals and reflects the species' overall genetic diversity. Tettelin et al. first introduced the idea of pangenomes in bacterial research in 2005. Pangenomes provide a comprehensive understanding of the genetic diversity present within a species, which can be achieved through the sequencing of individual genomes or

 DOI: 10.1201/9781003387060-12

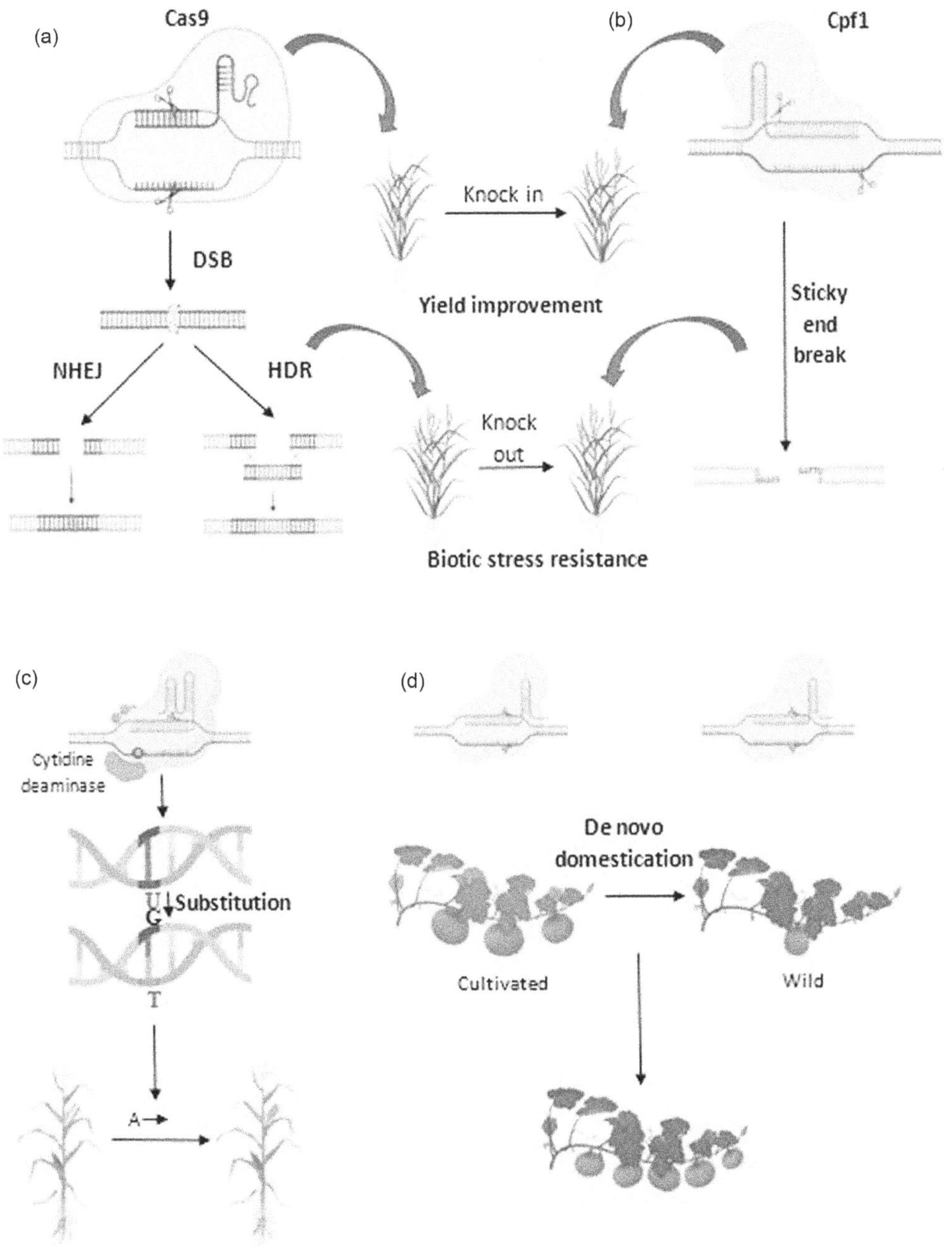

FIGURE 12.1 Schematic representation of several types of genome editing. (a) Mechanism of Cas9 mediated genome editing, (b) Mechanism of Cpf1 mediated gene editing, (c) Cytosine deaminase assisted base substitution for crop improvement, (d Multiplex editing using de novo domestication. (Created in Biorender; https://biorender.com) (Adapted from Razzaq et al. 2021.)

by analyzing gene variations within the species (see Figure 12.2). The utilization of pangenomes provides a precise depiction of intricate DNA polymorphisms present in a species, encompassing substantial insertions, deletions, duplications, translocations, presence-or-absence variations (PAVs), and copy number variations (CNVs). The aforementioned insights possess the potential

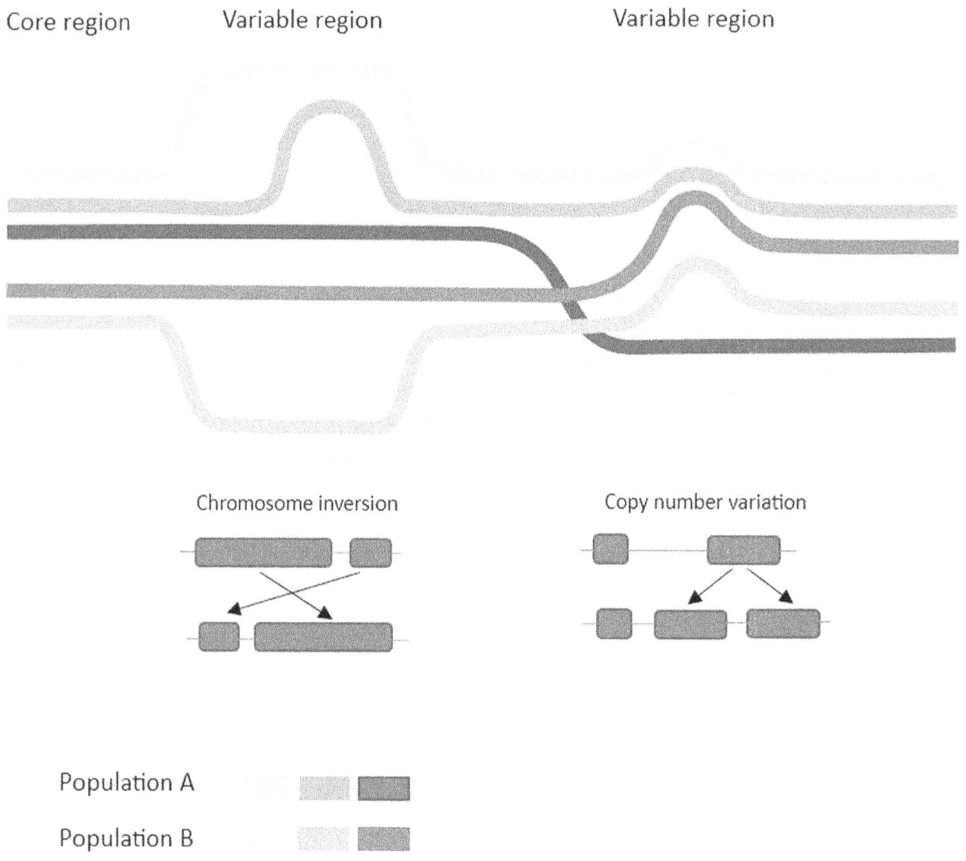

FIGURE 12.2 Schematic representation of an assembly of pangenome consisting of genomes from six individuals from two different populations showing the genetic variation observed in the variable regions due to chromosomal inversion or copy number variation. (Adapted from Tay Fernandez et al. 2022.)

to bring about a revolutionary transformation in the domain of crop enhancement and sustainable agriculture (Table 12.1).

12.2 TECHNIQUES FOR ASSEMBLY OF PANGENOME

The construction of pangenomes is typically achieved through one of three primary methodologies, each of which presents distinct advantages and disadvantages. The initial methodology involves de novo sequencing and comparison, which encompasses the sequencing, assembly, and comparative analysis of diverse genomes to identify both core and variable genes (refer to Figure 12.3). The employed methodology facilitates the accurate determination of the physical coordinates of individual genes and other genetic components, thereby affording comprehensive genetic cartography. Assembly and annotation errors may result in misinterpretation of variations. The application of this method is limited to a relatively small number of individuals due to the substantial cost and extensive sequencing coverage required, as reported by Bayer et al. (2020).

The iterative mapping and assembly approach was employed to construct the pangenome utilizing a single reference genome as depicted in Figure 12.3. The present methodology commences by aligning the sequencing data of complete genomes obtained from diverse individuals to a reference

TABLE 12.1
An Overview of the Pangenome Studies in Some Major Crops in Recent Times (in Chronological Order)

Species	Number of Accessions	Approaches of Assembly	Ploidy Level	Domestication Status	Traits Studied	Causes of Variations	Pangenome Size	References
Triticum aestivum	19	De novo	Hexaploid	Cultivated	Disease resistance, stress resistance	PAV	~11 Gb; 140,500 genes	Montenegro et al. (2017)
Brassica napus	53	Iterative	Tetraploid	Cultivated	Flowering time, disease resistance, acyl lipid and glucosinolate metabolism	PAV	1.1 Gb; 94,013 genes	Hurgobin et al. (2018)
Oryza sativa *O. glaberrima*	3010	Map-to-pan	Diploid	Cultivated	Flowering time, grain length and width, disease resistance	SNP, PAV	268 Mb, 12,465 genes	Wang et al. (2018)
O. sativa *O. rufipogon*	67	Iterative	Diploid	Cultivated Wild	Flowering time, stress tolerance, grain weight, tiller angle and plant height, hull color	SNP, PAV	430 Mb; 42,580 genes	Zhao et al. (2018)
Solanum lycopersicum *S. cheesmaniae* *S. galapagense*	725	De novo	Diploid	Cultivated Wild Wild	Disease resistance, fruit flavor	PAV	1179 Mb, 40,369 genes	Gao et al. (2019)
Helianthus annus	493	Iterative	Diploid	Cultivated	Downey mildew resistance	SNP	3.6 Gb; 61,205 genes	Hübner et al. (2019)
Sesamum indicum	5	De novo	Diploid	Cultivated	Oil content, yield	PAV	26,472 gene clusters	Yu et al. (2019)
Hordeum vulgare	20	De novo	Diploid	Cultivated	Yield	PAV inversions	40,176 orthologues	Jayakodi et al. (2020
Glycine max	26	De novo	Diploid	Cultivated	Yield attributing traits	PAV	1011.6 Mb; 57,492 gene clusters	Liu et al. (2020b)
B. napus	9	De novo	Tetraploid	Cultivated	Siliqua length, flowering time, seed weight	PAV	1033 Mb; 105,672 gene clusters	Song et al. (2020)
Cajanus cajan	89	Iterative	Diploid	Cultivated	Disease resistance	PAV	622 Mb	Zhao et al. (2020)
Zea mays	26	De novo	Diploid	Cultivated	Disease resistance, flowering time	TE-insertion; SNP, PAV	103,538 genes	Hufford et al. (2021)
Sorghum bicolor	177	Iterative	Diploid	Cultivated	Drought resistance	SNP, PAV	883.3 Mb, 35,719 genes	Ruperao et al. (2021)

Abbreviations: Gb, Giga base pairs; Mb, Mega base pairs; PAV, Presence or absence of variation; SNP, Single nucleotide polymorphism; TE, Transposable elements.

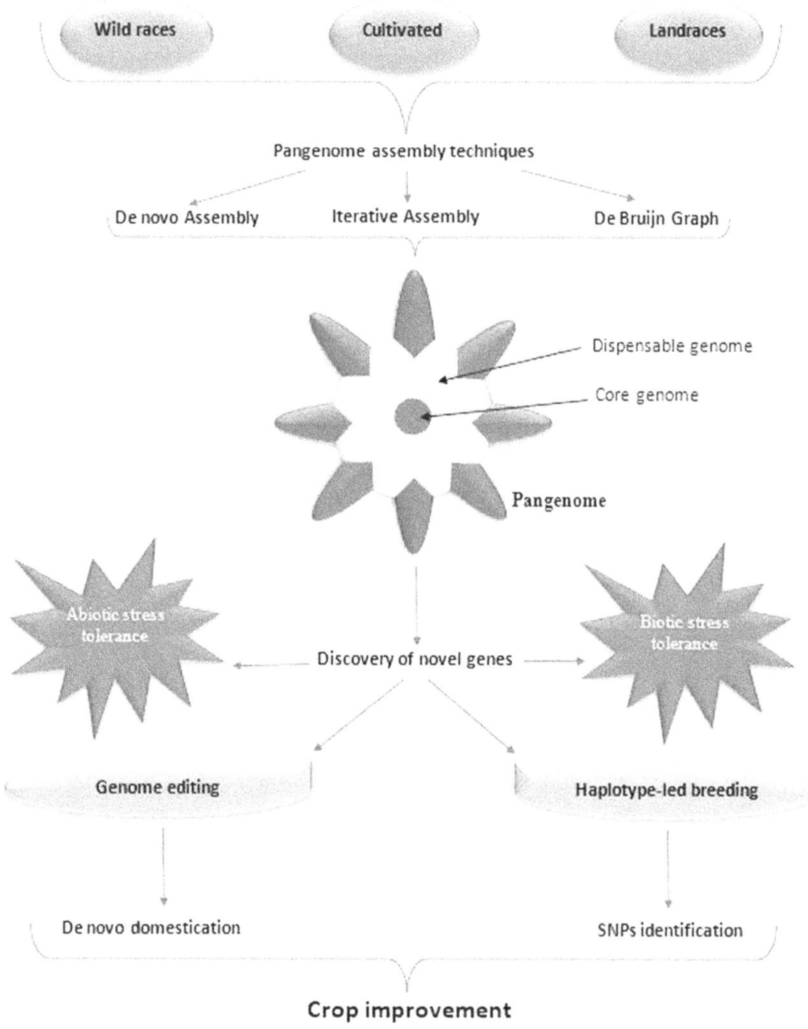

FIGURE 12.3 Various techniques of pangenome assembly for crop improvement through the introduction of novel genes. (Adapted from Razzaq et al. 2021.)

genome. After the alignment process, any reads containing non-aligning sequences were subjected to assembly and subsequently integrated into the reference genome, thereby generating the pangenome. Compared to de novo assembly, the present method is characterized by its cost-effectiveness, lower data requirements, and ability to facilitate the analysis of a larger population with only moderate sequencing coverage. Following the assembly process, gene presence or absence can be ascertained by realigning the sequencing data of each individual to the pangenome. The methodology predominantly detects the variations of presence and absence (PAVs) in genes and necessitates additional scrutiny to precisely position the non-reference contigs in a genome, as stated by Danilevicz et al. (2020). The present study demonstrates that the utilization of recurrent assembly through a diverse set of low-coverage individuals, in conjunction with de novo assembly for a limited number of representative individuals, facilitates the acquisition of genomic context and identification of PAVs within genes at the population level. This approach enables comprehensive diversification studies.

The third methodology utilizes graph-based approaches, encompassing authentic haplotype graphs and sequence and variation graphs (Figure 12.3). Pangenome graphs can be constructed through the utilization of de novo graph genomic assembly or complete genome assemblies. The graphical representation of genomic variations is presented as an interconnected network of pathways, rather than a singular sample sequence. The present study unifies shared sequence regions among individuals into a single path in the graph. In addition, structural variants are integrated as nodes at the loci where they initially occur. The present study provides a comprehensive depiction of genomic variation and sequence conservation through the depiction of variant information for variable regions as separate paths along the graph, as described by Zanini et al. (2022). The construction of graph-based pangenomes demands substantial computational resources. However, the quality of the resultant graphs is directly proportional to the quality of the data utilized for their construction. With the ongoing progress in sequencing techniques and data processing, especially with the advent of high-quality long-read data, it is anticipated that pangenome graphs will replace previous methods of constructing pangenomes (Tay Fernandez et al., 2022).

Irrespective of the mode of construction, pangenomes function as all-inclusive data repositories for trait association and guiding CRISPR-Cas design, thereby enabling accuracy in gene editing. With the growing demand for efficient genome editing and breeding, the significance of crop pangenome assemblies as a crucial resource in agricultural biotechnology is on the rise.

12.3 PANGENOME-MEDIATED ASSOCIATION ANALYSIS FOR GENE EDITING

The utilization of CRISPR-Cas technologies has emerged as a pivotal approach for gene expression regulation and functional genome analysis, thereby facilitating crop enhancement (Sar et al., 2022; Das et al., 2023). The effectiveness of gene editing in plant research is frequently dependent on the genotype and the particular loci of interest, as noted by Li et al. (2018b). The mutation rate facilitated by CRISPR-Cas technology can be influenced by various variables such as the GC content, accessibility of the target site due to chromatin structure state, and the secondary structure of single guide RNA (sgRNA) (Decaestecker et al., 2019).

Polyploid plants pose a significant challenge for gene editing due to their ability to modify multiple alleles or overcome allelic redundancy, which can obscure the phenotypic effects of gene editing interventions (Li et al., 2020). The phenotypic expression of plants can be influenced by various alleles in diverse manners. Therefore, the development of allele-specific CRISPR-sgRNA can be expedited by detecting and charting multiple variant alleles in a pangenome. The utilization of the *mlo* gene in *Hordeum vulgare* (barley) has been a well-established strategy for imparting horizontal resistance against powdery mildew in various crop species. However, owing to the pleiotropic effects of the gene, it can also exert a deleterious influence on the yield. The integration of diverse *mlo* alleles can lead to a wide spectrum of pleiotropic resistances against multiple pathogens, enhancing the overall robustness of the plants. In a recent study by Gruner et al. (2020), it was observed that wheat plants carrying a specific allele of the *mlo* gene exhibited an increased susceptibility to powdery mildew. In addition to barley, the utilization of allele-specific phenotypes has been implemented to enhance resistance against abiotic stress, herbicides, and disease, and to enhance overall yield in polyploid crops like wheat (*Triticum aestivum*) and camelina (*Camelina sativa*) (Morineau et al., 2017; Li et al., 2019).

Selective mutations of the three delta-12-desaturase genes (FAD2) at different loci in Camelina have been reported to result in an increase in oleic acid production and a decrease in polyunsaturated fatty acids (PUFAs) in the oil, as observed by Morineau et al. (2017). The CRISPR-Cas system has enabled the evaluation of the effects of minor genetic variations against a specific genetic background with high precision. The utilization of CRISPR-Cas enables the exploration of the impact of gene products produced per allele by knocking out or down-regulating specific variant alleles, as demonstrated by Schaart et al. (2021). The integration of pangenomic investigation with phenotypic information can facilitate the identification of variant alleles and the determination of CRISPR-Cas target sites, thereby advancing the development of more effective field varieties. The analysis of pangenomes can reveal structural variations, which can aid in the identification of new alleles for functional genomic investigations. In addition, pangenomes can offer detailed insights into the location and accessibility of the target allele within the genome.

The pangenomes possess a significant advantage in their capacity to depict the consequences of chromosomal inversions. These inversions can be specifically targeted for re-inversion through CRISPR-Cas mediated modifications, as illustrated in Figure 12.4. The impact of chromosomal inversions on crop breeding is significant due to their ability to impede the recombination of inverted regions. In this study, a pangenome analysis was conducted to examine chromosomal inversions in maize. A total of 66 inbred lines were utilized as reference genomes. Significantly, the study by Schwartz et al. (2020) revealed the presence of considerable insertion-deletion (In-Del) chromosomal rearrangements across all ten chromosomes. The most extensive pericentric inversion was observed on the second chromosome, measuring up to 75.5 Mb. The detection of essential structural variations and the consequent reversion of the genomic segment using CRISPR-Cas9

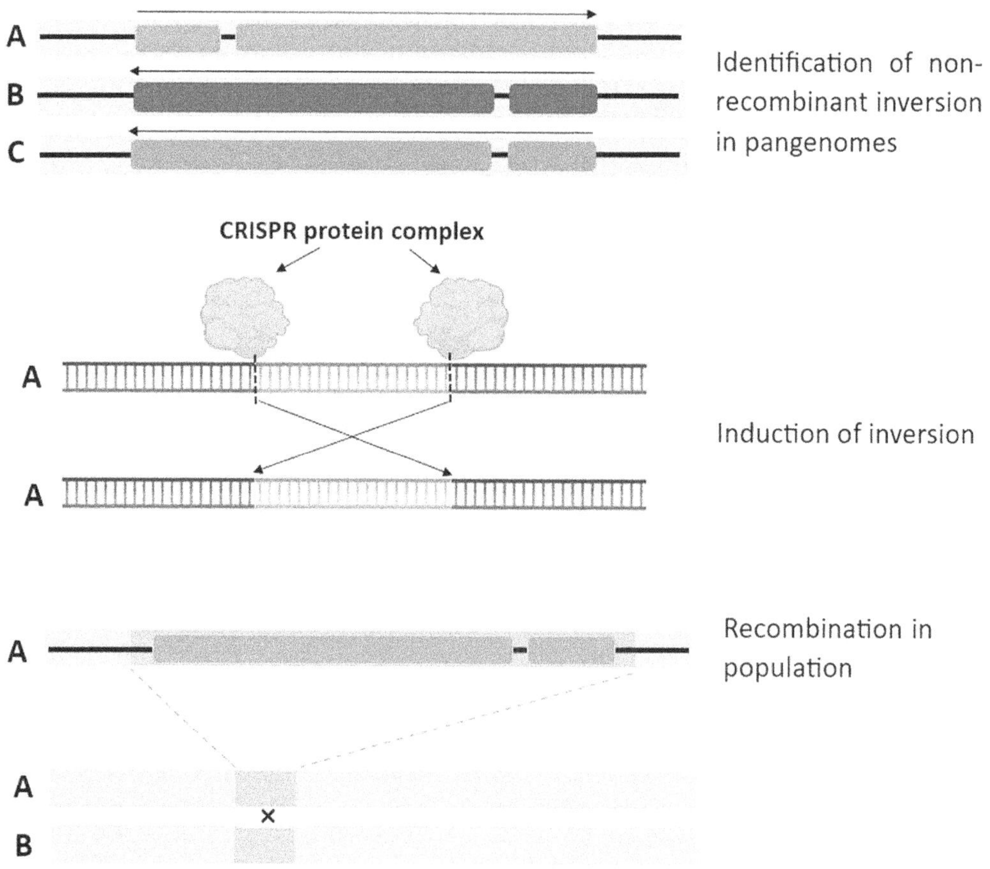

FIGURE 12.4 CRISPR-mediated inversion reversal to allow inverted genes to cross. When comparing individual A to individuals B and C, a non-recombinant inversion is found using a pangenome. The inverted area is then reversed using CRISPR-Cas proteins, making the locked genes accessible for recombination and causing double-stranded breaks at specified target locations. Individual A's previously inverted genes can subsequently be crossed with those of other people in the community. (Created in Biorender; https://biorender.com) (Adapted from Tay Fernandez et al. 2022.)

by the pangenome is imperative for accessing locus-specific genes for recombination with other inbred lines, as reported by Zaman et al. (2019). The present study highlights the significance of pangenomes in precisely detecting the rearranged chromosomal borders during inversional events in maize chromosomes. This, in turn, facilitates the CRISPR-Cas genome editing of crucial chromosomal regions within these boundaries.

12.4 PANGENOME-DIRECTED TARGETED MUTAGENESIS

The establishment of a correlation between genotypes present in a pangenome assembly is a crucial step in distinguishing domesticated species from their wild counterparts. The aforementioned data can facilitate the identification of agriculturally relevant genomic alterations, thereby establishing a foundation for precise mutagenesis. In this investigation, a pangenome of rice was analyzed, comprising 66 accessions. The study revealed certain features such as decreased height and premature flowering, which are suggestive of the Green Revolution. These characteristics played a pivotal role in the identification of the genes responsible for these traits (Huang et al., 2018). In this study, we performed a comprehensive analysis of structural variations in the pangenome of rice. Our findings reveal the presence of 129 conserved gene loci that are associated with common traits. Subsequently, a study utilizing the CRISPR-Cas knockout technique has identified 31 genes that are correlated with amplified yield, among which six genes have been previously reported, including the semi-dwarf gene sd1 (Huang et al., 2018).

A pangenome analysis was conducted on medical cannabis (*Cannabis sativa*) to identify 145 sgRNAs that target genes involved in the cannabinoid biosynthesis pathways. Pedigree-based pangenome analyses were employed to identify conserved genes linked to the phenotype of interest through directional selection. These findings highlight the potential of pangenome data in evaluating the influence of genetic variations on observable traits.

The comprehensive understanding of gene function within pangenomes can be enhanced by the incorporation of diverse omics data pertaining to the transcriptome, metabolome, and proteomes. A transcription-wide association study was conducted to identify quantitative trait loci (QTLs) associated with oil production regulation across eight different conditions, using an established Brassica napus pangenome. In the present study, a multi-omics-based dataset and a gene prioritization framework were employed to identify a cluster of genes associated with seed oil content. The findings were further validated through the use of CRISPR-Cas9 and T-DNA mutants, confirming their involvement in the regulation of oil production (Tang et al., 2021).

In a recent investigation conducted by Martina et al. (2021), 359 known QTLs associated with taste and fragrance were mapped onto the tomato pangenome. This approach enabled the identification of potential target regions for the improvement of tomato lines. The FLORAL4 gene, which has been exclusively identified in wild tomato relatives and was lost during the domestication process, was successfully mapped in the tomato pangenome through a search for promoter regions that are linked to aroma-associated QTLs, as reported by Martina et al. (2021). The present studies highlight the effectiveness of omics-based analysis of diverse pangenome regions in elucidating the conservation of QTL regions during domestication, identification of novel alleles and structural variations, and mapping of known QTLs that are advantageous for mutation studies.

Computational association studies that employ small variants, such as Genome-Wide Association Studies (GWAS), are frequently constrained to detecting an intra-haplotype cluster of linked co-inherited loci. The presence/absence variable regions of the genome have been shown to contain trait-associated Single Nucleotide Polymorphisms (SNPs) in rice, which may be overlooked by a single reference genome up to 41.6% of the time (Yao et al., 2015). The utilization of pangenomes as a haplotyping resource is advantageous due to its ability to encompass variations present in diverse populations, thus providing a comprehensive set of targets for CRISPR-Cas modification (Schatz et al., 2014). The utilization of CRISPR-Cas to disrupt the promoter region is a viable approach for inferring gene function when a haplotype associated with a trait is situated near or within a gene, as demonstrated by Huang et al. (2018).

In 2014, the first instance of multiplexed editing in plants utilizing the CRISPR-Cas toolkit was conducted. Subsequently, comprehensive editing was conducted in 2017 to generate rice mutant libraries that cover over one hundred thousand sites (Meng et al., 2017). After the aforementioned, a plethora of crops, such as *Zea mays* (maize), *Glycine max* (soybean), *Brassica napus*, and *Oryza sativa* (rice), have undergone multiplexed editing, as reported by Li et al. (2018a), Bai et al. (2020), and Liu et al. (2020a). Despite the potential of high-throughput mutagenesis studies, the functional validation of minor genetic variations within and among plant populations remains a challenge. In contrast, the CRISPR-Cas system has been utilized for saturation editing to investigate almost all SNPs across 13 exons that encode the functional domains of the BRCA1 gene, which is associated with breast cancer susceptibility. This approach has enabled precise functional validation in human subjects, as reported by Findlay et al. (2018).

The utilization of multiplexing techniques for targeted saturation editing of haplotypes associated with specific traits in pangenomic datasets is a promising strategy for the efficient and cost-effective identification of small variants in plants. The elucidation of biochemical mechanisms underlying diverse traits can be facilitated by comprehending the impact of minor variations and allelic recombination in pangenomes. The aforementioned observation can serve as a valuable reference for plant breeders in the development of tailored cultivars, ultimately leading to a thorough comprehension of the genetic variations that form the basis of agronomic characteristics (Bai et al., 2020).

12.5 OFF-TARGETS EFFECTS IN MULTIPLEXED GENOME EDITING

Multiplexed genome editing, a technique that involves the simultaneous editing of multiple loci, is frequently associated with off-target effects when utilizing CRISPR-Cas systems. The impact of certain factors is significantly evident during the assembly, expression, and processing of sgRNA arrays, as well as in the utilization of prevailing transformation methodologies for proficient delivery, as reported by Hashimoto et al. (2018). The utilization of CRISPR-Cas for editing purposes has shown great potential in enhancing editing efficiency. However, its application in high-throughput mutagenesis poses several technical challenges.

A significant challenge in the field of genome editing is the possibility of decreased specificity and efficacy of binding resulting from the simultaneous editing of multiple loci. This issue is further complicated by the unforeseen interactions between various CRISPR components that arise due to multiplexing, as reported by Zhang and Voigt (2018). Moreover, the design and synthesis throughput of single guide RNA (sgRNA) is subject to certain limitations, as reported by McCarty et al. (2020). As the number of targeted sites increases, the issue of off-target binding becomes more pronounced, thereby complicating the direct design process. The issue of cultivar-specific sgRNA design is of particular relevance, given the potential presence of variations in the target sequence or protospacer adjacent motif (PAM) site within a given population. In instances of this nature, the utilization of pangenomic references is imperative owing to their capacity to detect all conceivable off-target locations within the specified population.

The implementation of pangenomes and their associated revelations regarding an individual's genetic composition can augment the accuracy of genome editing methodologies. The PAM site plays a pivotal role in the functionality of Clustered Regularly Interspaced Short Palindromic Repeats (CRISPR) systems. The corresponding single-guide RNA (sgRNA) is engineered to complement specific genomic sequences, thereby facilitating the recruitment of the Cas protein to a specific genomic locus. The selection of the target sequence is crucial for the effectiveness of the system. Regions exhibiting high sequence similarity to the target site can lead to significant off-target effects (Zhang et al., 2015), which are typically undesired and have been observed in various plant species, including cotton (*Gossypium* sp.), grapevines (*Vitis vinifera*), and rice (Xu et al., 2006).

To minimize the potential impact of off-target modifications, it is imperative to possess comprehensive genetic data pertaining to the individual under investigation. The implementation of a computational methodology for genome editing must not solely rely on singular reference data but rather delve extensively into the wealth of genetic databases at hand. The incorporation of a wider spectrum of variant data in the analysis can potentially reduce the occurrence of off-target events, as suggested by Grohmann et al. in their 2019 publication. The vast availability of data enables researchers to devise customized single-guide RNAs (sgRNAs) that precisely target particular allelic regions and circumvent mismatches arising from sequence variation, as reported by Scheben and Edwards (2017). The functional traits of crop species can be significantly altered by focusing on specific variations in allele sequences, such as PAVs and SNPs revealed through pangenome analysis.

12.6 ENHANCED DISEASE RESISTANCE THROUGH PANGENOMICS AND GENOME ENGINEERING

The enhancement of disease resistance in plants has been a prominent focus in the field of plant breeding. Recent developments in pangenomics and genome editing have expanded our ability to fortify plants against a diverse range of diseases. The relevance of this phenomenon is further amplified in the context of climate change, as variable weather patterns may catalyze the expansion of extant and novel plant pathogens (Das et al., 2023; Razzaq et al., 2021).

The comprehensive approach of pangenomics has enabled the discovery of a broader spectrum of disease-resistance genes among various accessions of a given species. The present study employs a comprehensive approach to investigating genetic diversity, thereby uncovering novel and distinct genetic variations that may potentially confer resistance to diseases (Bayer et al., 2020; Danilevicz et al., 2020; Jayakodi et al., 2021). The study conducted by Hübner et al. (2019) revealed alterations in gene content and disease resistance due to hybridization through the unveiling of the pangenome of sunflowers. The pangenome of hexaploid bread wheat, comprising both cultivated varieties and wild relatives, has been found to reveal a greater reservoir of disease-resistance genes that are absent in the reference genome (Montenegro et al., 2017).

Concurrently, the CRISPR-Cas9 genome editing technique has been employed to investigate the operational mechanism of distinct disease-resistance genes and to generate innovative resistance characteristics (Decaestecker et al., 2019; Liu et al., 2020b). The CRISPR-TSKO technique has been employed in Arabidopsis to investigate the impact of mutating disease-resistance genes on plant physiology (Decaestecker et al., 2019). This technique facilitates accurate mutagenesis in particular tissues or organs. The application of the aforementioned tactics in food crops, such as rice and maize, has resulted in the identification of new resistance genes (Liu et al., 2020a; Huang et al., 2018).

Although these advancements exhibit promise, their implementation remains intricate. The utilization of CRISPR-Cas9 in polyploid species is a challenging task due to the presence of multiple genomes, which necessitates the editing of all gene copies to attain the intended phenotype (Schaart et al., 2021). Furthermore, the outcomes of gene editing may exhibit unpredictability owing to the interdependence of biological systems (Grohmann et al., 2019). The integration of pangenomics and genome editing presents significant potential for enhancing disease resistance, thereby contributing to the improvement of crop yield and global food security, despite the challenges associated with this approach (Khan et al., 2020; Gao, 2021).

12.7 FUTURE PROSPECTS OF PANGENOMES IN GENOME EDITING

Restoring agronomically advantageous genes that are abandoned from cultivated crop species but preserved in wild relatives is a worthwhile target for genome editing. These deleted genes in cultivars due to domestication and/or breeding may have the ability to perform useful agronomic tasks including enhancing nutrient uptake efficiency, enhancing disease resistance, or adapting to harsh conditions like heat and drought in wild relatives. For instance, such as the bread wheat Yr36 gene for rust resistance and rice and sorghum (*Sorghum bicolor*) disease resistance genes have been lost as a result of domestication selection. As evidenced by the identification of the TomLoxC promoter allele associated with the taste that was restored into presently cultivated tomato cultivars, wild intrusion is a viable method for reintroducing variable genes (Tay Fernandez et al., 2022). Wild introgression, however, has the potential to introduce harmful genes too, such as those that modify blooming time or diminish plant growth. Genome editing with CRISPR-Cas could make it possible to reintroduce some features without the associated deleterious allelic variation by multiplexing editing of SVs relevant to those traits. (Zanini et al., 2022). However, this needs an extensive analysis of the total gene pool would improve target specificity and guard against off-target consequences. A combination of several pangenomes from many species within a group of taxa to show the genetic makeup of that group above the species level constitutes a super-pangenome (Figure 12.5). By studying super-pangenomes, it is possible to incorporate markers associated with desirable traits in wild relatives into domesticated crops (Khan et al., 2020).

The reintroduction of agriculturally beneficial genes, which were previously discarded from domesticated crop species but conserved in their wild counterparts, represents a significant goal for

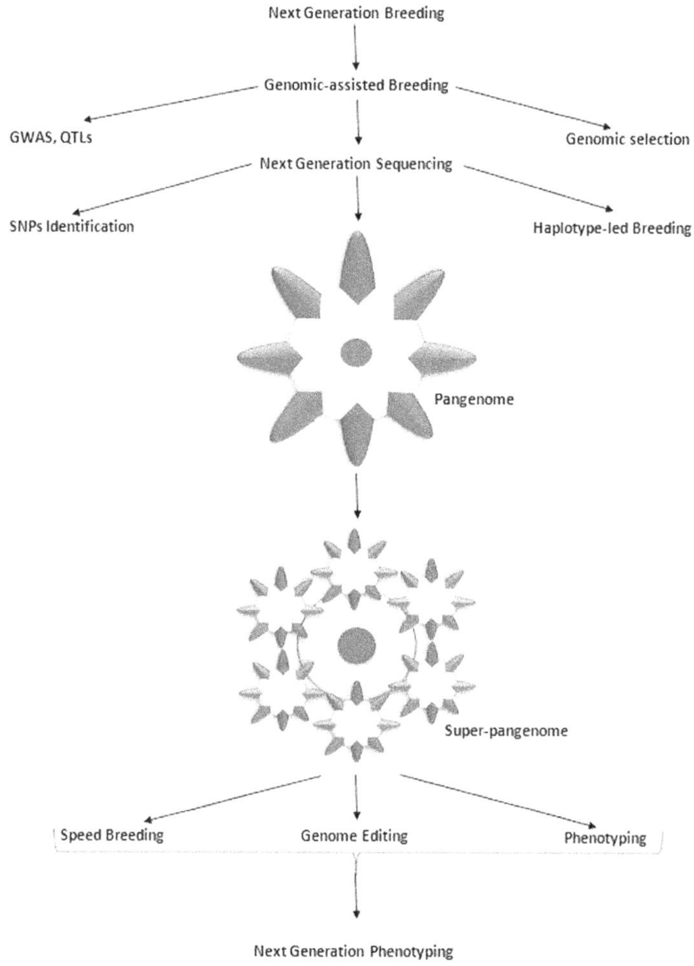

FIGURE 12.5 Diagrammatic representation of the future of pangenome. (Adapted from Razzaq et al. 2021.)

genome editing. The loss of certain genes in cultivars as a result of domestication or breeding practices may confer substantial agronomic advantages, such as improved efficiency of nutrient uptake, a heightened resistance to diseases, or the ability to adapt to harsh environmental conditions such as heat and drought. The phenomenon of lost genes has been observed in various crop species. For instance, bread wheat (*Triticum aestivum*) has lost the Yr36 gene, which was responsible for conferring rust resistance. Similarly, rice (Oryza sativa) and sorghum (*Sorghum bicolor*) have lost certain disease-resistance genes due to domestication selection.

The reintroduction of genes through wild introgression has been demonstrated to be a successful process for restoring desirable traits in cultivated tomato cultivars. For instance, Tay Fernandez et al. (2022) have shown that the TomLoxC promoter allele, which is associated with flavor, can be restored through this process. Nonetheless, the process of wild introgression poses a potential hazard of incorporating genes that may have deleterious effects, such as modifying the timing of flowering or hindering plant development. The utilization of CRISPR-Cas for genome editing presents a potential resolution for the reintroduction of advantageous traits while avoiding unfavorable allelic variations. This can be achieved through the multiplexing of the editing of structural variations (SVs) that are pertinent to said traits, as demonstrated by Zanini et al. (2022). The implementation of this approach necessitates a thorough examination of the complete genetic repertoire to guarantee precise targeting and avert unintended impacts on non-targeted genetic elements.

In this context, the notion of a super-pangenome emerges, denoting an amalgamation of multiple pangenomes from diverse species within a taxonomic assemblage, furnishing an illustration of the genetic composition of the assemblage that transcends the species level. The integration of markers linked to favorable traits from wild relatives into cultivated crops can be facilitated by super-pangenomes, as reported by Khan et al. (2020). The utilization of pangenomes to map PAVs can facilitate the restoration of advantageous traits that have been lost in crops. This can be achieved through the modification of specific alleles using CRISPR-Cas, while simultaneously circumventing the incorporation of deleterious alleles.

The utilization of super-pangenomes presents a promising avenue for the prospective domestication of untamed crop relatives, thereby introducing novel food sources and augmenting the existing crop varieties. The identification of potential domestication genes associated with polymorphic adenine variants (PAVs) and structural variants (SVs) could facilitate targeted CRISPR-Cas modifications in a particular species. The utilization of multiplexed CRISPR-Cas has demonstrated its efficacy in the manipulation of domestication genes in wild tomato relatives, resulting in the generation of novel lines that exhibit larger, more productive, and nutrient-rich fruits, as well as heightened resistance to biotic and abiotic stress when compared to their cultivated counterparts (Scheben et al., 2017).

The utilization of CRISPR-Cas presents a promising avenue for domesticating various wild crop relatives. For instance, it is possible to create a novel cereal crop with high genetic diversity by domesticating teosinte (*Zea mays* ssp. parviglumis). In addition, pennycress (*Thlaspi arvense*) can be utilized to develop an oilseed crop with cold tolerance, while weeping grass (*Microlaena stipoides*) can be engineered to produce a cereal crop with abiotic stress tolerance. Furthermore, contemporary varieties have the potential to exchange traits through transfer mechanisms. In this study, Tay Fernandez et al. (2022) propose the transfer of disease-resistance genes from rust-resistant wheat types to other Poaceae species, such as barley or sorghum. In addition, the authors suggest the alteration of disease-resistance genes in Brassica species using CRISPR-Cas technology, to transfer these genes to other Brassicaceae species. The process of domesticating wild plants or improving existing crops through the utilization of potential domestication genes in related species or genera is a crucial step toward ensuring food security in the future. The augmentation of genetic diversity in agricultural systems is expected to result in the development of cultivars that are better adapted to confront the exigencies of future environmental conditions.

The utilization of super-pangenomes, in conjunction with CRISPR-Cas editing, has the potential to enable the transmission of advantageous environmental adaptations among multiple plant genera.

The swift progress in pangenome assembly suggests that pangenomes encompassing various eukaryotic genera or families may imminently become attainable. The investigation of favorable characteristics in the aforementioned genera and/or families that are associated with agricultural crop species can be utilized to develop multiplexed CRISPR-Cas systems. These systems can be employed to integrate such traits into various crop variants, as reported by Li et al. (2022). This study highlights potential adaptations that may confer improved resilience to climate change. These adaptations are based on the identification of SNPs that are associated with tolerance to precipitation and temperature variability. In addition, the study suggests that enhanced photosynthetic efficiency in C3 plants, such as rice, may also contribute to improved resilience. Finally, the study draws inspiration from phosphorus-saving strategies employed by non-mycotrophic Proteaceae plants in South-West Australia, which may lead to improved nutrient use efficiency. To ensure the sustainability of food production in the face of global challenges such as climate change and emerging diseases, the utilization of multiplexed CRISPR-Cas systems is imperative. These systems will play a pivotal role in the development of novel crop varieties possessing the desired traits.

12.8 CHALLENGES AND CONSIDERATIONS

The primary limitations associated with utilizing pangenomes for directing CRISPR-Cas editing in crops pertain to the magnitude and quantity of concurrent targets. In general, window widths or base editors span from 4–5 base pairs (bp) to 50–150 bp, thereby enabling the induction of point mutations within the target. Various CRISPR techniques have been employed in plant systems to achieve larger cassette insertions, including non-homologous end joining (NHEJ). This technique facilitates the insertion and deletion of a few kilobases in size, as reported by Son and Park (2022). The efficacy of editing is observed to decrease considerably with an increase in the magnitude of the edit, thereby impeding the editing of SVs that are detected through pangenomics.

To surmount these constraints, plausible tactics encompass the fabrication of innovative Cas proteins that necessitate diverse or adaptable PAM sequence sites. The utilization of these techniques would enhance the capacity to target a wider spectrum of genomic loci, especially when employed in combination with multiplexed CRISPR-Cas systems. SVs exert a substantial influence on crop evolution and diversity. Therefore, forthcoming investigations ought to concentrate on investigating approaches to expand the catalytic range for CRISPR-Cas editing. The implementation of these technological advancements will enable us to optimize the utilization of the advancements achieved in the field of pangenomics.

The pangenomic approach poses certain challenges, wherein the iterative and de novo assembly of pangenomes may not effectively determine the chromosomal positioning of variable regions that are not present in the primary reference. Theoretically, CRISPR can still be employed to target the extra contigs in individuals where they exist. However, the comprehension of the regulatory mechanisms that impact their expression may be restricted due to the dearth of information concerning their genomic context. The utilization of various graph techniques can potentially surmount this limitation; however, it is contingent upon the acquisition of deep sequencing or long reads to precisely capture SVs (Tay Fernandez et al., 2022). Therefore, the progress in sequencing methodologies that enable the economical identification of SVs is imperative to augment the effectiveness of pangenomic information for CRISPR-Cas editing.

12.9 CONCLUSIONS

Genome editing technology holds significant promise for addressing the mounting global food demand due to population growth and anticipated climate change impacts. By facilitating the creation of crops with superior yields, enhanced nutritional value, pest resistance, and tolerance to abiotic stress, genome editing can play a pivotal role in promoting food security across developed and developing nations. It has the potential to accelerate and enhance the efficacy of agricultural development programs. Several key challenges must be addressed to establish an effective genome editing process for crop improvement. These include the assembly of high-quality pangenome references, systematic identification of potential editing sites through functional genomics, improved delivery methods for genome editing systems, and minimizing off-target editing. Pangenomes serve as a rich resource for identifying critical genetic variations beneficial for agronomy. They have been instrumental in facilitating high-throughput mutagenesis methods, linking significant SVs to traits, identifying and targeting genomic regions of interest in CRISPR experiments, and augmenting our comprehension of QTL regions. Pangenomic studies have enhanced our understanding of functional traits that were either lost or previously unidentified in crop species. In addition, they have enabled researchers to identify beneficial traits in wild crop species that could be incorporated into existing or novel crops. CRISPR-Cas technology's potential to create highly precise CRISPR target sequences has been significantly boosted by the knowledge gleaned from pangenomic research. This allows for targeted alterations in genome content and gene expression. Although the application of CRISPR-Cas systems derived from pangenomic studies is currently limited by the size of the catalytic windows required for modifications and the number of SVs and PAVs associated with specific phenotypes, advancements in pangenome construction technology continue to progress. Improvements in the precision, cost, and accessibility of genomic annotation and sequencing are occurring in tandem with these developments. Future crop gene editing research will increasingly rely on pangenomes to bolster global food security and resilience to climate change.

REFERENCES

Bai, M.; Yuan, J.; Kuang, H.; Gong, P.; Li, S.; Zhang, Z.; Liu, B.; Sun, J.; Yang, M.; Yang, L.; et al. Generation of a multiplex mutagenesis population via pooled CRISPR-Cas9 in soya bean. *Plant Biotechnol. J.* **2020**, 18, 721–731. https://doi.org/10.1111/pbi.13239

Bayer, P. E.; Golicz, A. A.; Scheben, A.; Batley, J.; Edwards, D. Plant pan-genomes are the new reference. *Nat. Plants* **2020**, 6, 914–920. https://doi.org/10.1038/s41477-020-0733-0

Chen, K.; Wang, Y.; Zhang, R.; Zhang, H.; Gao, C. CRISPR/Cas genome editing and precision plant breeding in agriculture. *Annu. Rev. Plant Biol.* **2019**, 70, 667–697. https://doi.org/10.1146/annurev-arplant-050718-100049

Danilevicz, M. F.; Tay Fernandez, C.G.; Marsh, J. I.; Bayer, P. E.; Edwards, D. Plant pangenomics: Approaches, applications and advancements. *Curr. Opin. Plant Biol.* **2020**, 54, 18–25. https://doi.org/10.1016/j.pbi.2019.12.005

Das, A.; Mahanta, M.; Pramanik, B.; Purkayastha, S. Genetically modified crops and crop species adapted to global warming in dry regions. In: A Naorem, D Machiwal (eds.) *Enhancing Resilience of Dryland Agriculture under Changing Climate*, Springer, Singapore, **2023**, pp. 385–409. https://doi.org/10.1007/978-981-19-9159-2_19

Decaestecker, W.; Buono, R.A.; Pfeiffer, M.L.; Vangheluwe, N.; Jourquin, J.; Karimi, M.; Van Isterdael, G.; Beeckman, T.; Nowack, M.K.; Jacobs, T.B. CRISPR-TSKO: A technique for efficient mutagenesis in specific cell types, tissues, or organs in Arabidopsis. *Plant Cell* **2019**, 31, 2868–2887. https://doi.org/10.1105/tpc.19.00454

FAO. *The Future of Food and Agriculture: Trends and Challenges.* FAO, United Nations, 2017.

Findlay, G.M.; Daza, R.M.; Martin, B.; Zhang, M.D.; Leith, A.P.; Gasperini, M.; Janizek, J.D.; Huang, X.; Starita, L.M.; Shendure, J. Accurate classification of BRCA1 variants with saturation genome editing. *Nature* **2018**, 562, 217–222. https://doi.org/10.1038/s41586-018-0461-z

Gao, C. Genome engineering for crop improvement and future agriculture. *Cell* **2021**, 184(6), 1621–1635. https://doi.org/10.1016/j.cell.2021.01.005

Gao, L.; Gonda, I.; Sun, H.; Ma, Q.; Bao, K.; Tieman, D.M.; Burzynski-Chang, E.A.; Fish, T.L.; Stromberg, K.A.; Sacks, G.L.; Thannhauser, T.W. The tomato pan-genome uncovers new genes and a rare allele regulating fruit flavor. *Nat. Genet.* **2019**. 51(6), 1044–1051. https://doi.org/10.1038/s41588-019-0410-2

Grohmann, L.; Keilwagen, J.; Duensing, N.; Dagand, E.; Hartung, F.; Wilhelm, R.; Bendiek, J.; Sprink, T. Detection and Identification of genome editing in plants: Challenges and opportunities. *Front. Plant Sci.* **2019**, 10, 236. https://doi.org/10.3389/fpls.2019.00236

Gruner, K.; Esser, T.; Acevedo-Garcia, J.; Freh, M.; Habig, M.; Strugala, R.; Stukenbrock, E.; Schaffrath, U.; Panstruga, R. Evidence for allele-specific levels of enhanced susceptibility of wheat mlo mutants to the hemibiotrophic fungal pathogen *Magnaporthe oryzae* pv. Triticum. *Genes*. **2020**, 11, 517. https://doi.org/10.3390/genes11050517

Hashimoto, R.; Ueta, R.; Abe, C.; Osakabe, Y.; Osakabe, K. Efficient multiplex genome editing induces precise, and self-ligated type mutations in tomato plants. *Front. Plant Sci.* **2018**, 9, 916. https://doi.org/10.3389/fpls.2018.00916

Huang, J.; Li, J.; Zhou, J.; Wang, L.; Yang, S.; Hurst, L.D.; Li, W.-H.; Tian, D. Identifying a large number of high-yield genes in rice by pedigree analysis, whole-genome sequencing, and CRISPR-Cas9 gene knockout. *Proc. Natl. Acad. Sci. USA* **2018**, 115, E7559. https://doi.org/10.1073/pnas.1806110115

Hübner, S.; Bercovich, N.; Todesco, M.; Mandel, J.R.; Odenheimer, J.; Ziegler, E.; Lee, J.S.; Baute, G.J.; Owens, G.L.; Grassa, C.J.; Ebert, D.P. Sunflower pan-genome analysis shows that hybridization altered gene content and disease resistance. *Nat. Plants.* **2019**, 5(1), 54–62. https://doi.org/10.1038/s41477-018-0329-0

Hufford, M.B.; Seetharam, A.S.; Woodhouse, M.R.; Chougule, K.M.; Ou, S.; Liu, J.; Ricci, W.A.; Guo, T.; Olson, A.; Qiu, Y.; Della Coletta, R. De novo assembly, annotation, and comparative analysis of 26 diverse maize genomes. *Science* **2021**, *373*(6555), 655–662. https://doi.org/10.1126/science.abg5289

Hurgobin, B.; Golicz, A.A.; Bayer, P.E.; Chan, C.K.K.; Tirnaz, S.; Dolatabadian, A.; Schiessl, S.V.; Samans, B.; Montenegro, J.D.; Parkin, I.A.; Pires, J.C. Homoeologous exchange is a major cause of gene presence/absence variation in the amphidiploid Brassica napus. *Plant Biotechnol. J.* **2018**, 16(7), 1265–1274. https://doi.org/10.1111/pbi.12867

Jayakodi, M.; Schreiber, M.; Stein, N.; Mascher, M. Building pan-genome infrastructures for crop plants and their use in association genetics. *DNA Res.* **2021**, 28(1), dsaa030. https://doi.org/10.1093/dnares/dsaa030

Khan, A. W.; Garg, V.; Roorkiwal, M.; Golicz, A. A.; Edwards, D.; Varshney, R. K. Super-pangenome by integrating the wild side of a species for accelerated crop improvement. *Trends Plant. Sci.* **2020**, 25, 148–158. https://doi.org/10.1016/j.tplants.2019.10.012

Li, C.; Hao, M.; Wang, W.; Wang, H.; Chen, F.; Chu, W.; Zhang, B.; Mei, D.; Cheng, H.; Hu, Q. An efficient CRISPR/Cas9 platform for rapidly generating simultaneous mutagenesis of multiple gene homoeologs in allotetraploid oilseed rape. *Front. Plant Sci.* **2018a**, 9, 442. https://doi.org/10.3389/fpls.2018.00442

Li, F.; Wen, W.; Liu, J.; Zhang, Y.; Cao, S.; He, Z.; Rasheed, A.; Jin, H.; Zhang, C.; Yan, J.; et al. Genetic architecture of grain yield in bread wheat based on genome-wide association studies. *BMC Plant Biol.* **2019**, 19, 168. https://doi.org/10.1186/s12870-019-1781-3

Li, Q.; Sapkota, M.; van der Knaap, E. Perspectives of CRISPR/Cas-mediated cis-engineering in horticulture: Unlocking the neglected potential for crop improvement. *Hortic. Res.* **2020**, 7, 36. https://doi.org/10.1038/s41438-020-0258-8

Li, W.; Liu, J.; Zhang, H.; Liu, Z.; Wang, Y.; Xing, L.; He, Q.; Du, H. Plant pan-genomics: Recent advances, new challenges, and roads ahead. *J. Genet. Genomics* **2022**, 49(9), 833–846. https://doi.org/10.1016/j.jgg.2022.06.004

Li, X.; Wang, Y.; Chen, S.; Tian, H.; Fu, D.; Zhu, B.; Luo, Y.; Zhu, H. Lycopene is enriched in tomato fruit by CRISPR/Cas9-mediated multiplex genome editing. *Front. Plant Sci.* **2018b**, 9, 559. https://doi.org/10.3389/fpls.2018.00559

Liu, H.-J.; Jian, L.; Xu, J.; Zhang, Q.; Zhang, M.; Jin, M.; Peng, Y.; Yan, J.; Han, B.; Liu, J.; et al. High-throughput CRISPR/Cas9 mutagenesis streamlines trait gene identification in maize[OPEN]. *Plant Cell* **2020a**, 32, 1397–1413. https://doi.org/10.1105/tpc.19.00934

Liu, Y.; Du, H.; Li, P.; Shen, Y.; Peng, H.; Liu, S.; Zhou, G.A.; Zhang, H.; Liu, Z.; Shi, M.; Huang, X. Pan-genome of wild and cultivated soybeans. *Cell* **2020b**, 182(1), 162–176. https://doi.org/10.1016/j.cell.2020.05.023

Martina, M.; Tikunov, Y.; Portis, E.; Bovy, A.G. The genetic basis of tomato aroma. *Genes* **2021**, 12, 226. https://doi.org/10.3390/genes12020226

McCarty, N.S.; Graham, A.E.; Studená, L.; Ledesma-Amaro, R. Multiplexed CRISPR technologies for gene editing and transcriptional regulation. *Nat. Commun.* 2020, 11, 1281. https://doi.org/10.1038/s41467-020-15053-x

Meng, X.; Yu, H.; Zhang, Y.; Zhuang, F.; Song, X.; Gao, S.; Gao, C.; Li, J. Construction of a genome-wide mutant library in rice using CRISPR/Cas9. *Mol. Plant* **2017**, 10, 1238–1241. https://dx. doi.org/10.1016/j.molp.2017.06.006

Montenegro, J.D.; Golicz, A.A.; Bayer, P.E.; Hurgobin, B.; Lee, H.; Chan, C.K.K.; Visendi, P.; Lai, K.; Doležel, J.; Batley, J.; Edwards, D. The pangenome of hexaploid bread wheat. *Plant J.* **2017**, 90(5), 1007–1013. https://doi.org/10.1111/tpj.13515

Morineau, C.; Bellec, Y.; Tellier, F.; Gissot, L.; Kelemen, Z.; Nogué, F.; Faure, J.-D. Selective gene dosage by CRISPR-Cas9 genome editing in hexaploid *Camelina sativa*. *Plant Biotechnol. J.* **2017**, 15, 729–739. https://doi.org/10.1111/pbi.12671

Razzaq, A.; Kaur, P.; Akhter, N.; Wani, S.H.; Saleem, F. Next-generation breeding strategies for climate-ready crops. *Front. Plant Sci.* **2021**, 12, 620420. https://doi.org/10.3389/fpls.2021.620420

Ruperao, P.; Thirunavukkarasu, N.; Gandham, P.; Selvanayagam, S.; Govindaraj, M.; Nebie, B.; Manyasa, E.; Gupta, R.; Das, R.R.; Odeny, D.A.; Gandhi, H. Sorghum pan-genome explores the functional utility for genomic-assisted breeding to accelerate the genetic gain. *Front. Plant Sci.* **2021**, 963, 666342. https://doi.org/10.3389/fpls.2021.666342

Sar, P.; Pramanik, B.; Debnath, S. CRISPR/Cas: A new horizon in crop improvement. In: Debnath, S.; Manjula, I.K.; Shah, K.R.; Bhattacharjee, S. (eds.) *Futuristic Trends in Biotechnology*. IIP International Publishers, **2022**, pp. 9–27. https://rsquarel.org/assets/docupload/rsl2023D884B54EF9ADCCF.pdf

Schaart, J.G.; van deWiel, C.C.M.; Smulders, M.J.M. Genome editing of polyploid crops: Prospects, achievements and bottlenecks. *Transgenic Res.* **2021**, 30, 337–351. https://doi.org/10.1007/s11248-021-00251-0

Schatz, M.C.; Maron, L.G.; Stein, J.C.; Wences, A.H.; Gurtowski, J.; Biggers, E.; Lee, H.; Kramer, M.; Antoniou, E.; Ghiban, E.; et al. Whole genome *de novo* assemblies of three divergent strains of rice, *Oryza sativa*, document novel gene space of *aus* and *indica*. *Genome Biol.* **2014**, 15, 506. https://doi.org/10.1186/s13059-014-0506-z

Scheben, A.; Edwards, D. Genome editors take on crops. *Science* **2017**, 355, 1122. https://doi.org/10.1126/science.aal4680

Scheben, A.; Wolter, F.; Batley, J.; Puchta, H.; Edwards, D. Towards CRISPR/Cas crops-bringing together genomics and genome editing. *New Phytol.* **2017**, 216, 682–698. https://doi.org/10.1111/nph.14702.

Schwartz, C.; Lenderts, B.; Feigenbutz, L.; Barone, P.; Llaca, V.; Fengler, K.; Svitashev, S. CRISPR-Cas9-mediated 75.5-Mb inversion in maize. *Nat. Plants* **2020**, 6, 1427–1431. https://doi.org/10.1038/s41477-020-00817-6

Son, S.; Park, S.R. Challenges facing CRISPR/Cas9-based genome editing in plants. *Front. Plant Sci.* **2022,** 13, 902413. https://doi.org/10.3389/fpls.2022.902413

Song, J.M.; Guan, Z.; Hu, J.; Guo, C.; Yang, Z.; Wang, S.; Liu, D.; Wang, B.; Lu, S.; Zhou, R.; Xie, W.Z. Eight high-quality genomes reveal pan-genome architecture and ecotype differentiation of *Brassica napus*. *Nat. Plants* **2020**, 6(1), 34–45. https://doi.org/10.1038/s41477-019-0577-7

Springmann, M.; Clark, M.; Mason-D'Croz, D.; Wiebe, K.; Bodirsky, B.L.; Lassaletta, L.; de Vries, W.; Vermeulen, S.J.; Herrero, M.; Carlson, K.M.; Jonel, M. Options for keeping the food system within environmental limits. *Nature* **2018**, 562(7728), 519–525. https://doi.org/10.1038/s41586-018-0594-0

Tang, S.; Zhao, H.; Lu, S.; Yu, L.; Zhang, G.; Zhang, Y.; Yang, Q.-Y.; Zhou, Y.; Wang, X.; Ma, W.; et al. Genome- and transcriptome-wide association studies provide insights into the genetic basis of natural variation of seed oil content in *Brassica napus*. *Mol. Plant.* **2021**, 14, 470–487. https://doi.org/10.1016/j.molp.2020.12.003

Tay Fernandez, C.G.; Nestor, B.J.; Danilevicz, M.F.; Marsh, J.I.; Petereit, J.; Bayer, P.E.; ... & Edwards, D. Expanding gene-editing potential in crop improvement with pangenomes. *Intr. J. Mol. Sci.* **2022**, 23(4), 2276. https://doi.org/10.3390/ijms23042276

Tettelin, H.; Masignani, V.; Cieslewicz, M.J.; Donati, C.; Medini, D.; Ward, N.L.; Angiuoli, S.V.; Crabtree, J.; Jones, A.L.; Durkin, A.S.; DeBoy, R.T. Genome analysis of multiple pathogenic isolates of *Streptococcus agalactiae*: Implications for the microbial "pan-genome". *Proc. Natl. Acad.* Sci. *USA*, **2005,** 102, 13950–13955. https://doi.org/10.1073/pnas.0506758102

Wang, W.; Mauleon, R.; Hu, Z.; Chebotarov, D.; Tai, S.; Wu, Z.; Li, M.; Zheng, T.; Fuentes, R.R.; Zhang, F.; Mansueto, L. Genomic variation in 3,010 diverse accessions of Asian cultivated rice. *Nature.* **2018**, 557(7703), 43–49. https://doi.org/10.1038/s41586-018-0063-9

Xu, K.; Xu, X.; Fukao, T.; Canlas, P.; Maghirang-Rodriguez, R.; Heuer, S.; Ismail, A.M.; Bailey-Serres, J.; Ronald, P.C.; Mackill, D.J. Sub1A is an ethylene-response-factor-like gene that confers submergence tolerance to rice. *Nature.* **2006**, 442, 705–708. https://doi.org/10.1038/nature04920

Yao, W.; Li, G.; Zhao, H.; Wang, G.; Lian, X.; Xie, W. Exploring the rice dispensable genome using a metagenome-like assembly strategy. *Genome Biol.* **2015**, 16, 187. https://doi.org/10.1186/s13059-015-0757-3

Yu, J.; Golicz, A.A.; Lu, K.; Dossa, K.; Zhang, Y.; Chen, J.; Wang, L.; You, J.; Fan, D.; Edwards, D.; Zhang, X. Insight into the evolution and functional characteristics of the pan-genome assembly from sesame landraces and modern cultivars. *Plant Biotechnol J.* **2019**, 17(5), 881–892. https://doi.org/10.1111/pbi.13022

Zaman, Q.U.; Li, C.; Cheng, H.; Hu, Q. Genome editing opens a new era of genetic improvement in polyploid crops. *Crop J.* **2019**, 7, 141–150. https://doi.org/10.1016/j.cj.2018.07.004

Zanini, S.F.; Bayer, P.E.; Wells, R.; Snowdon, R.J.; Batley, J.; Varshney, R.K.; Nguyen, H.T.; Edwards, D.; Golicz, A.A. Pangenomics in crop improvement-from coding structural variations to finding regulatory variants with pangenome graphs. *The Plant Genome*, **2022**, 15(1), e20177. https://doi.org/10.1002/tpg2.20177

Zhang, S.; Voigt, C.A. Engineered dCas9 with reduced toxicity in bacteria: Implications for genetic circuit design. *Nucleic Acids Res.* **2018**, 46, 11115–11125. https://doi.org/10.1093/nar/gky884

Zhang, X.-H.; Tee, L.Y.; Wang, X.-G.; Huang, Q.-S.; Yang, S.-H. Off-target effects in CRISPR/Cas9-mediated genome engineering. *Mol. Ther. Nucleic Acids* **2015**, 4, e264. https://doi.org/10.1038/mtna.2015.37

Zhao, J.; Bayer, P.E.; Ruperao, P.; Saxena, R.K.; Khan, A.W.; Golicz, A.A.; Nguyen, H.T.; Batley, J.; Edwards, D.; Varshney, R.K. Trait associations in the pangenome of pigeon pea (*Cajanus cajan*). *Plant Biotechnol. J.* **2020**, 18(9), 1946–1954. https://doi.org/10.1111/pbi.13354

Zhao, Q.; Feng, Q.; Lu, H.; Li, Y.; Wang, A.; Tian, Q.; Zhan, Q.; Lu, Y.; Zhang, L.; Huang, T.; Wang, Y. Pangenome analysis highlights the extent of genomic variation in cultivated and wild rice. *Nat. Genet.* **2018**, 50(2), 278–284. https://doi.org/10.1038/s41588-018-0041-z

13 Characterization of CRISPR-Associated Protein for Epigenetic Manipulation in Plants

Ayyadurai Pavithra, Chinnasamy Sashtika, Arumugam Vijaya Anand, Senthil Kalaiselvi, Santhanu Krishnapriya, and Natchiappan Senthilkumar

LIST OF ABBREVIATIONS

ABA	Abscisic Acid
APX2	Ascorbate Peroxidase2
AtHAC1	Arabidopsis Histone Acetyl Transferase1
AtHAT1	Arabidopsis Histone Acetyltransferase1
BBM	BABY BOOM ()
CMT	Chromo Methylase
CMT3	Chromomethylase3
CMV	Cucumber Mosaic Virus
CS	Chromo Shadow
dCas9	Dead Cas9
DNMTs	DNA Methyltransferases
DRM	Domains Reorganized and Rearrange Methyltransferase
EZH2	Zeste Homolog 2
FT	Flowering Locus T
FWA	Flowering Wageningen
gRNA	Guide RNA
H3K27	Histone H3 lysine 27
H3K4	Histone H3 Lysine 4
H3K4me3	H3K4 Trimethylation
HATs	Histone Acetyltransferases
HDACs	Histone eacetylases
HMTs	Histone Methyltransferases
HPLC	High-Performance Liquid Chromatography
HS	Heat Stress
JMJ	Jumanji
KRAB	Kruppel Associated Box
LC-MS	Liquid Chromatography-Mass Spectrometry
ncRNAs	Non-coding RNAs
NLS	Nuclear Localization Signals

DOI: 10.1201/9781003387060-13

PDS	Phytoene Desaturase
RNP	Ribonucleoprotein
sgRNA	Single-Guide RNA
TEs	Transposable Elements
TMV	Tobacco Mosaic Virus
UFGT	UDP-Glucose Flavonoids Glycosyltransferases
ZF	Zinc Finger

13.1 INTRODUCTION

Knowing about the plant's physiology is a wrap of the whole chemical and physical processes associated with life that occurs in plants. It is essential to have a comprehensive knowledge of the types of responses by plants to their environment so that methods to manipulate them become easy. Sessile agricultural plants are vulnerable to a variety of abiotic stresses, which significantly reduce yields by an average of more than 50% and seriously affect world harvests. In the past, stress-tolerant crop genotypes have been created using plant breeding technological advances and genetic engineering techniques (Mushtaq et al., 2018). The creation of instruments for the precise modification of the epigenetics of plants remains a huge task for epigenomics. Hopefully, the CRISPR gene editing tools would make it possible to transit from correlation-based to causal-based discoveries, which is the requirement for drawing conclusions based on mechanistic principles.

The current dCas9 technique acts through RNA-DNA interaction, enabling better flexibility and modularity for tool techniques to manipulate the epigenetics of plant physiology. The benefits of transient genome editing and diminished off-target effects, direct delivery of the CRISPR/Cas9 system as an RNP (ribonucleoprotein) complex composed of the Cas9 protein and sgRNA (single-guide RNA) has become a potent and popular technique for editing genomes (Zhang et al., 2021; López-Calleja et al., 2019). The epigenome is a heritable layer of data that isn't stored in the genomic DNA sequence but rather in chemical modifications that are tagged to histones or superficially to the DNA. Together with transcription factors, these chemical alterations control the spatiotemporal regulators of gene expression.

Controlled site-specific modification of epigenetic information is necessary for dissecting epigenome function. To function as targeted transcription factors or epigenetic modifiers, customized DNA-binding platforms in conjunction with effector domains can be used (Brocken et al., 2018). Recently, "second generation" chimeric dCas9 system improvements that aim to improve targeting efficiency and modifier capability have been tried in the plants and have shown encouraging outcomes. The use of dCas9 as a cutting-edge and adaptable tool for fundamental studies on chromatin structure, transcription control, and epigenetic landscapes plays a vital role in the systematic and desirable modification of plant physiology. In this chapter, we outline the benefits and drawbacks of tools and technologies created to affect epigenetic tags that could be used to examine their direct impact on nuclear epigenomics, chromatin structure, and transcriptional levels in turn predicting their additional role in the plant cell fate and development.

13.2 CRISPR REGULATIONS IN EPIGENETICS

CRISPR-Cas9 is an excellent technique for accurate methylation editing and gene control because the CRISPR-Cas9 protein with a direction RNA may target DNA sequences regardless of the methylation of the target site's state. It has been demonstrated that dCas9 fusion proteins used in targeted methylation editing techniques are very successful at regulating genes without changing the DNA sequence. Sequence-specific methylation and de-methylation technologies could be very helpful clinically since abnormal regulation of DNA methylation in promoters of tumor suppressor genes or proto-oncogenes may direct the onset and development of several forms of cancer (Sung and Yim, 2020).

13.2.1 CRISPR-Induced Methylation

13.2.1.1 DNA Methylation on Cytosines of CpG Dinucleotides

In mammalian cells, methylation of DNA on the cytosines of CpG dinucleotides is a well-known mechanism for controlling epigenetic regulation. But in plants, it is unclear how epigenetic alterations affect the effectiveness of double-stranded DNA breaks caused by CRISPR/Cas9 and the subsequent DNA repair process. When next-generation sequencing to examine the impact of the methylation of cytosine on the results of CRISPR/Cas9-induced mutations at various Cas9 target locations in *Nicotiana benthamiana* leaf cells, high promoter methylation levels, but not methylation of gene body, were observed to reduce the frequency of Cas9-mediated mutations.

DNA methylation had a target-specific impact on the proportion of insertions to deletions and possibly the kind of Cas9 cleavage. In addition, an excess of deletion events at Cas9-induced DNA breaks were controlled by a single 5′-terminal nucleotide. DNA methylation can affect the activity of Cas9 and following DNA repair in an indirect manner, most likely by altering the structure of local chromatin. In addition to the well-known blunt-end double-stranded DNA breaks caused by Cas9, evidence was presented for staggered DNA cuts caused by Cas9 in the plant cells. Both sorts of cuts might trigger a novel, as of yet unidentified, mechanism that directs microhomology-mediated DNA repair (Přibylová et al., 2022).

13.2.1.2 DNA Methylation in the Non-CG Context

DNA methylation in the non-CG is common in plants. Despite being co-ordinatedly controlled by the domains reorganized and rearranged methyltransferase (DRM) and chromo methylase (CMT) proteins in Arabidopsis thaliana, non-CG methylation in key crops has not yet been thoroughly investigated since genetic materials are difficult to come by. Single- and multiple-knockout mutants were produced for each of the nine DNA methyltransferases (DNMTs) in rice (*Oryza sativa*) using the incredibly effective multiplex CRISPR-Cas9 genome editing technology and it is evaluated their whole-genome methylation level at single-nucleotide resolution. Unexpectedly, the combined loss of DRM2, chromomethylase3 (CMT3), and CMT3 activities only slightly decreased non-CG methylation in rice as opposed to fully erasing it in Arabidopsis. In the triple mutant rice Os-dcc, areas that were extensively methylated in non-CG contexts showed high GC contents. Furthermore, the Os-ddccc quintuple mutant and the Os-ddcc quadruple mutant both showed that OsCMT3b regulates non-CG methylation in the absence of other significant methyltransferases. According to the above-mentioned findings, OsCMT3b is sub-functionalized to make room for a distinctive group of non-CG-methylated sites at very GC-rich areas of the rice genome (Hu et al., 2021).

13.2.1.3 Effects of CRISPR/Cas9 Guided Mutagenesis on Epigenetics

Precision breeding and plant genome editing have been accelerated by the overall benefits of CRISPR/Cas9 in a variety of plant species. Beyond off-target nucleotide changes, unintended repercussions are still not fully understood. The promoters of inherently hypermethylated and hypomethylated genes from Arabidopsis were investigated for changes in DNA methylation. Through the use of floral dip transformation and Agrobacterium, transgenic plants were created. From segregated T2 plants, homozygous altered lines were chosen to utilize an *in vitro* digestion assay. To find modifications in DNA methylation at the targeted loci, bisulfite sequencing comparisons were conducted between the paired groups of edited and unedited groups of plants. CRISPR/Cas9-guided mutagenesis had no unexpected effects on morphology or epigenetics. Wild-type, transgenic empty vector, and transgenic altered plants all had comparable phenotypes. According to epigenetic profiles, transgenic empty vector, wild-type, and transgenic edited plants, all had promoter regions flanking their target sequences that were identically methylated. The type of mutation has no impact on the epigenetic state. In the modified plants, we also examined off-target mutagenesis consequences. Sequencing was done on possible off-target locations with up to 4-bp mismatches between the targets. In candidate sites, no off-target mutations have been discovered (Li et al., 2018; Gallego-Bartolomé et al., 2018; Parrilla-Doblas et al., 2019; Oberkofler and Bäurle, 2022; Lee et al., 2020).

13.2.2 CRISPR-Induced Demethylation

An essential epigenetic change involved in transposable element silencing and gene control is DNA methylation. Numerous aspects of plant development are controlled by CRISPR-induced DNA demethylation, which is the enzymatic elimination of methylated cytosine. The base excision repair pathway, which is started by the Repressor of the Silencing 1/Demeter family of bifunctional DNA glycosylases, is involved in active DNA demethylation in *Arabidopsis*. The regulation of the development of various activities in plants (development of seed and stomatal, fruit ripening, etc.) are only a few examples of the varied developmental processes in which CRISPR-induced active DNA demethylation plays a role in different plant species (Li et al., 2018).

13.2.2.1 Targeted DNA Demethylation for FWA Up-regulation, and a Heritable Late-Flowering Phenotype

The development of persistent epialleles as a result of heritable changes in DNA methylation. DNA demethylation in the Flowering Wageningen (FWA) gene, which results in a heritable late-flowering phenotype and the up-regulation of FWA, is a well-known model of a steady epiallele in plants. It is caused by the defeat of DNA cytosine methylation (5mC) in the promoter of the FWA gene. Here, we show that a fusion of the human demethylase TET1cd's catalytic domain with a synthetic zinc finger (ZF) made to target the FWA promoter can lead to highly effective targeted demethylation, heritable late-flowering phenotype, and FWA up-regulation. Making use of the TET1cd and a modified SunTag system, CRISPR/dCas9-based targeted demethylation system was created. When referred to the FWA or CACTA1 loci, the SunTag-TET1cd system can target demethylation and trigger the expression of the gene, which is comparable to the ZF-TET1cd fusions (Gallego-Bartolomé et al., 2018).

13.2.2.2 DNA Glycolases in Active DNA Demethylation Mechanism

The active DNA demethylation mechanism found in plants uses DNA glycosylases to remove 5-meC and start the base excision repair pathway to replace it with unmodified C. This mechanism is specific to plants. DNA methylation patterns in plants are balanced yet flexible thanks to a complicated mechanism called base excision repair-mediated DNA demethylation that involves many proteins and extra regulatory elements that prevent the build-up of potentially damaging intermediates and coordinate DNA methylation as well as demethylation. One of the main goals of active DNA demethylation is to prevent methylation from spreading to neighboring genes. This process combats extreme methylation at transposable elements (TEs), primarily in euchromatic areas. Additionally, it plays a role in the transcriptional activation of TEs and TE-derived sequences in the companion cells of both female and male gametophytes. This strengthens transposon silencing in gametes and also aids in gene imprinting in the endosperm. Additional physiological processes that are affected by plant 5-meC DNA glycosylases include fruit ripening, development and germination of seed, and plant responses to a wide range of both the biotic and abiotic environmental stressors (Parrilla-Doblas et al., 2019).

13.2.3 CRISPR-Induced Histone Modification

13.2.3.1 Jumonji (JMJ) Histone H3 Lysine 4 (H3K4) Demethylase Domain Modification Using a Heat-inducible dCas9

Histone alterations are essential for integrating environmental cues into the regulation of gene expression. The critical need for strategies that enable locus-specific modulation of histone modifications (preferably in an inducible manner), is brought on by the fact that genetic and pharmaceutical interference frequently results in pleiotropic consequences. The plant's modified epigenome uses a heat-inducible dCas9 system that allows targeting a Jumanji (JMJ) histone H3 lysine 4 (H3K4) demethylase domains to a locus of interest in *Arabidopsis thaliana*. Ascorbate peroxidase2 (APX2) served as a model locus because it exhibits transcriptional memory following heat stress (HS), which is correlated with H3K4 hyper-methylation. The dCas9-JMJ targets APX2 in an HS-dependent way, and when dCas9-JMJ binds to the locus, the HS-induced over-accumulation of H3K4 trimethylation (H3K4me3) reduces. As a result, the APX2 locus has less HS-mediated transcriptional memory. The dCas9-JMJ fusion protein may function in part independent of the activity of demethylase because the impact was not entirely conditional on the enzymatic activity of the rubber domain (Oberkofler and Bäurle, 2022).

13.2.4 CRISPR-Induced Non-Coding DNA Regulators

Non-coding RNAs (ncRNAs), which have little to no ability to code for proteins but are nevertheless functional, make up a sizable and substantial percentage of the transcriptomes of eukaryotic organisms (Figure 13.1). Numerous ncRNAs, including both small RNAs and long ncRNAs (lncRNAs), regulate the expression of gene at both the transcriptional and post-transcriptional stages, playing crucial regulatory roles in nearly all biological processes (Fal et al., 2021; Yu et al., 2019). Recently,

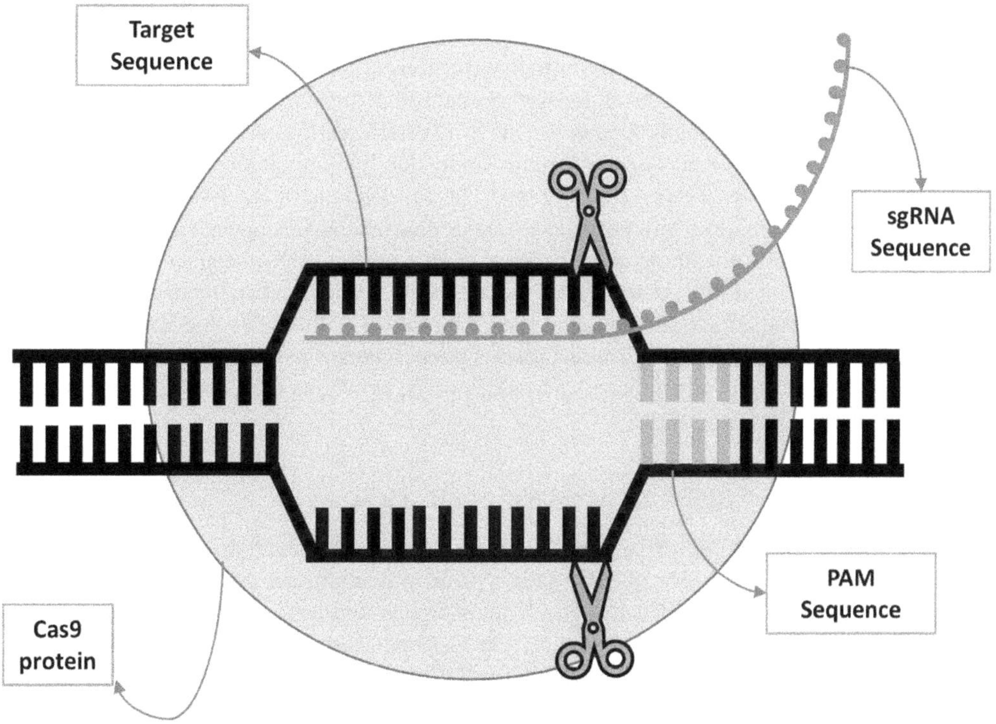

FIGURE 13.1 CRISPR/CAS9 System.

ncRNAs have become versatile master regulators of biological processes. MicroRNAs (miRNAs) are short, self-complementary ncRNAs that are produced naturally and range in size from 18 to 24 nucleotides. Plant miRNAs serve important roles in gene regulatory networks and a variety of biological processes in addition to their direct engagement in developmental processes. Instead, lncRNAs are a sizable and varied class of transcribed ncRNAs that are longer than 200 nucleotides. Different RNA polymerases are used to transcribe plant lncRNAs, which have a variety of structural characteristics. Plant lncRNAs have a significant role in controlling gene expression in different biological processes. In the past 10 years, the CRISPR-Cas9 technology has revolutionized genome editing. The CRISPR loci are translated into ncRNA, which eventually combines with Cas9 to produce a functional complex that is further directed to cleave corresponding invader DNA. Important crops like wheat, maize, and rice as well as model plants like Arabidopsis and tobacco have all benefited from the convenience of the CRISPR-Cas system. All of these studies, however, concentrate on genes that code for proteins. There is little available data on non-coding gene targeting. Until now, the CRISPR-Cas system has only been utilized to modify miRNA and lncRNAs in vertebrate systems; however, it is still largely unknown in plant species (Basak and Nithin, 2015).

13.3 MODIFIED FORM OF CRISPR/Cas SYSTEM; THE CRISPR/dCas SYSTEM

The Cas proteins can be classified into two main classes with three broad subtypes each. Type-I, type-III and type-IV fall under class-I and type-II, type-V and type-VI fall under class II. Each of the types is associated with a group of proteins that have a specific target, on which they act (Table 13.1). Out of all the types, type-II, type-V and type-VI from class-II have been well-founded in recent times (Bhardwaj et al., 2022). The reason behind this is the presence of single protein effector modules in class II which makes them less complicated as compared to class-I which has multi-subunit effector complexes (Koonin et al., 2017). The CRISPR/Cas system undergoes both natural as well as artificially induced modifications. This system saves and sustains the lives of nearly 95% of Archaea-bacteria and 48% of Eubacteria by executing a defense mechanism against the phage attack. At the time when there is a new anti-CRISPR mechanism adopted by the phage, the bacteria are under pressure to modify themselves for an efficient counterattack. This co-evolution of the phage and bacterial defense mechanism contributes to the natural modification of the CRISPR/Cas system (Bondy-Denomy et al., 2013). Different types of Cas proteins with their unique structural and functional units are studied individually in an attempt to modify them artificially in controlled laboratory experimentations. The development of catalytically inactive forms of Cas proteins which are often termed as dead-Cas or deactivated-Cas opened new possibilities to modify the gene expression without actually changing the original DNA sequence. This CRISPR/dCas system, a deactivated form of the CRISPR/Cas system takes a part an important role in delivering the transcriptional effector to the target gene. The delivery of this effector can activate or halt a particular gene expression. When there is transcriptional repression, the system is called CRISPR interference (CRISPRi), and when there is a transcriptional activation the CRISPR system is called CRISPRa (Kazi and Biswas, 2021).

13.3.1 Structure of CRISPR/dCas System and Its Domains

The dCas9 system is made up of several significant domains that support its ability to regulate gene expression (Figure 13.2). The dCas9 system's primary domains are outlined as; Cas9 nuclease domain, RNA-binding domains, nuclear localization signal, and transcriptional effector domains (Activation domains and Repression domains). The Cas9 nuclease domain is responsible for DNA binding and cleavage in the wild-type Cas9 protein. They are inactivated by induced mutations that abolish their nuclease activity in the dCas9 system. They are still able to link together to target DNA sequences but fail to cleave DNA sequences (Jinek et al., 2012). RNA-binding domains help to form the dCas9-sgRNA complex which guides dCas9 to the target DNA sequence. The sgRNA-binding

TABLE 13.1
Comparative Study between CRISPR/Cas9 and dCas9

Characters	CRISPR/Cas	CRISPR/dCas
Abbreviation	Clustered Regularly Interspaced Short Palindromic Repeats with related protein 9.	CRISPR-associated catalytically deactivated Cas9 (dead Cas9)
Functionality	Double-strand breaks in DNA can be created by CRISPR Cas proteins like Cas9 and Cas12a, which can then be used for gene editing and DNA alterations.	CRISPR dCas proteins, in contrast, lack nuclease activity and are unable to directly alter DNA sequences.
Mechanism of action	Methylation/demethylation, histone modification and Non-coding RNA.	Activation and Repression by effector molecules.
DNA cleavage ability	DNA double-strand breaks can result from CRISPR Cas proteins' capacity to cleave DNA at specified target locations. This cleavage starts DNA repair processes that can be used for gene editing.	On the other hand, CRISPR dCas proteins do not break DNA and do not activate DNA repair pathways.
Genetic modifications	At the targeted DNA site, CRISPR Cas proteins have the ability to deliver a variety of genetic alterations, including insertions, deletions, and replacements. Genes can be disrupted, mutations can be fixed, and new genetic sequences can be inserted using these alterations.	CRISPR dCas proteins, in contrast, do not alter the DNA sequence directly. They do not cause genetic changes, but they can control gene expression by interfering with transcriptional mechanisms, such as by inhibiting or activating gene expression.
Targeting range	Wide-ranging CRISPR Cas proteins can be used to alter genes in a variety of species and genetic backgrounds. The gRNA sequence largely determines how precisely their target is recognized.	Comparatively, the CRISPR dCas proteins' targeting range is also reliant on the gRNA, but their effects are mainly limited to controlling the expression of a specific gene at the target location as opposed to causing DNA changes.
Precision	Accurate gene editing is possible thanks to the targeted DNA cleavage that CRISPR Cas proteins may introduce. However, Cas proteins have the potential to accidentally cleave undesired DNA sequences, leading to off-target consequences.	Because they cannot break DNA, CRISPR dCas proteins do not have off-target effects associated with DNA cleavage.
Protein engineering	The efficiency, specificity, and off-target effects of DNA cleavage are frequently optimized while creating Cas proteins.	Engineering efforts for dCas proteins are typically focused on enhancing their capacity for gene regulation, such as boosting activation or repression efficiency or generating additional functionality, such as protein recruitment or epigenetic changes.
Ethical considerations	CRISPR Cas proteins create ethical questions when it comes to germline editing or making heritable changes in embryos because of their capacity to manipulate DNA directly. There are ongoing ethical discussions surrounding the use of CRISPR Cas proteins in these situations.	Since they do not change the DNA sequence, CRISPR dCas proteins, which are largely utilized for gene regulation, do not have the same ethical ramifications.
Research applications	Creating gene knockouts, inducing specific mutations, or inserting desired genetic sequences are just a few of the gene editing uses for CRISPR Cas proteins.	The primary uses of CRISPR dCas proteins are in gene regulation, including regulating biological processes, investigating gene function, and regulating gene expression levels.

domain and the tracrRNA-binding domain are the two RNA-binding domains present in the dCas9 system (Jinek et al., 2013). The dCas9 protein is often given nuclear localization signals (NLS) to ensure that it is transported effectively into the target cells' nuclei. The NLS makes it easier for dCas9 to become nuclearly localized, where it interacts with the target DNA and performs its gene

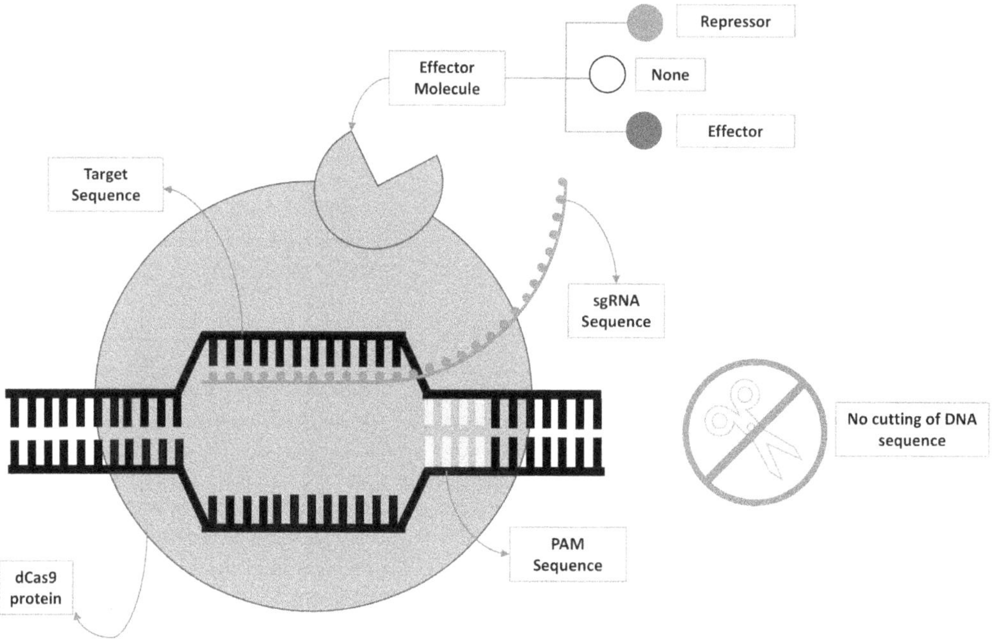

FIGURE 13.2 CRISPR/dCAS9 System.

regulatory actions (Mali et al., 2013). To achieve gene regulation, dCas9 is often fused with transcriptional effector domains that can modulate gene expression. Commonly used effector domains include the activator domain; to increase the expression and repressor domains to bring down the gene expression.

13.3.2 Mechanism Adapted by CRISPR/dCas System

The CRISPR/dCas system varies from its original wild-typer CRISPR/Cas system by lacking the ability to cleave the double-stranded DNA sequence. However, they do not lose the ability to bind to specific DNA sequences guided by a guide RNA (gRNA) or sgRNA (Gilbert et al., 2013). The dCas9-sgRNA partnership can be designed to target particular gene enhancers, promoters, or other regulatory regions to precisely and adaptably modify gene expression. Transcriptional activation and repression are the two primary ways by which the dCas9 system functions. An outline of these mechanisms is given below:

13.3.2.1 Transcriptional Activation by CRISPR/dCas System

A sgRNA molecule requires the dCas9 protein to particular genomic sites where the target DNA sequence and the sgRNA are meant to complement each other. The binding occurs due to the base-pairing that forms a dCas9-sgRNA complex at the target region. The dCas9 protein is then linked to transcriptional activation domains like VP64, p65, or VPR where the genes are activated multi-fold. Most often these activation domains collaborate with co-activators and the transcriptional machinery for better gene expression (Chavez et al., 2015).

13.3.2.2 Transcriptional Repression by CRISPR/dCas System

Similar to transcriptional activation, the dCas9 protein is guided to specific genomic regions by a sgRNA molecule that forms a dCas9-sgRNA complex at the target site. The dCas9 protein attaches to transcriptional repressor domains like the KRAB (Kruppel associated box) domain of Kox1 or SRDX, where these repressor domains enhance the transcriptional repression or chromatin-modifying enzymes to halt the expression of the target genes (Thakore et al., 2015). Other transcriptional repressor domains that are most commonly exploited are the CS (Chromo Shadow) domain of HP1α, the WRPW domain of Hes1, the SID4X domain and the enhancer of zeste homolog 2 (EZH2) (O'Geen et al., 2019).

13.3.3 Various Types of CRISPR/dCas System Enhancing Epigenetic Approaches in the Manipulation of Plant Physiology

The essential ability of the dCas system to modulate gene expression by adding effectors acts as the key mechanism for the exploitation of the CRISPR/dCas system characterization. They can also manifest the plant physiology in other ways including, methylation, histone modification, and non-coding RNA manipulation. In specific genomic loci in plants, DNA methylation can occur where CRISPR/dCas9 can be employed to add or delete DNA methylation tags. By combining dCas9 with DNMTs or DNA demethylases, this can be accomplished. While the dCas9-DNA demethylase fusion can erase DNA methylation marks, the dCas9-DNMT fusion can attract DNMTs to specific locations and cause DNA methylation (Ghoshal and Gardiner, 2021). Histone modifications, which are essential for controlling gene expression, can be altered using CRISPR/dCas9 technology. To introduce certain histone acetylation or methylation marks, respectively, into target genomic areas, dCas9 can be coupled with histone acetyltransferases (HATs) or histone methyltransferases (HMTs). On the other hand, to add or repress particular histone modifications, dCas9 can be coupled with histone deacetylases (HDACs) or histone demethylases (Puchta, 2016). In the modulation of ncRNAs, CRISPR/dCas9 has been used to control the expression or activity of ncRNAs for the regulation of plants' genes. To activate or repress the transcriptional activity of ncRNA genes, for example, dCas9 can be guided to their regulatory regions. Targeting the precursors or binding sites of particular ncRNAs, including miRNAs or lncRNAs, dCas9 can also be utilized to obstruct their processing or functionality (Deng et al., 2022). A few of the latest CRISPR/dCas system modifications that alter the epigenetics of plant physiology are discussed one after the other in this section.

13.3.3.1 CRISPRa: CRISPR/dCas9-VP64 and dCas9-TV in Grape Plants for Gene Activation in Normal as well as Stress-induced Studies

The transcriptional activation of endogenous genes in grapes was examined using the two transcriptional activators TV and VP64. These domains are responsible for the activation of the grape cell gene for UDP-glucose flavonoids glycosyltransferases (UFGT). It was found that the UFGT gene was transcriptionally activated in grape cells by utilizing both the dCas9-VP64 and dCas9-TV platforms. The dCas9-VP64 system was approximately 1.6–5.6 times more effective in UFGT activation than the dCas9-TV system, which was approximately 5.7–7.2 times more effective. In both cases, gene expression increased noticeably. The same study with dCas9-TV technology showed stimulated the cold-responsive transcription factor gene CBF4 in grape plants. In transgenic plants, the expression of CBF4 was enhanced 3.7–42.3 times, where CBF4-activated plants showed less electrolyte loss after exposure to cold than wild-type plants (Ren et al., 2022).

13.3.3.2 CRISPRa: CRISPR/dCas9-VP64 and MS2-p65-HSF1 in Maize Plants to Study Transcriptional Activation for Induction of Parthenogenesis

The AP2/ERF family transcription factor BABY BOOM (BBM) is an essential regulator of plant cell totipotency. The BBM gene's expression was first discovered in *Brassica napus*, a species commonly known as Rapeseed. The BBM gene has been related to play a significant role as a gene marker in several developmental and signaling pathways in plants that involve apogamy or parthenogenesis, cell division, regeneration, plant transformation, and embryogenesis (Jha and Kumar, 2018). A study has taken advantage of the use of *in vivo* CRISPRa targeting maize BABY BOOM2 (ZmBBM2) which acts as a fertilization checkpoint, providing a way to come up with parthenogenesis.

The dCas9-VP64 and MS2-p65-HSF1 effectors from the CRISPR/dCas9 system were utilized for targeting these genes in an attempt to induce high activation potential. The conclusion was drawn after investigating the immature embryo development before fertilization, where 3.9% of kernels developed double embryo sacs in ovaries. The indication of an embryo developed as a maternal clone confirmed transcriptionally activated parthenogenesis in maize plants (Qi et al., 2023).

13.3.3.3 CRISPRi: CRISPR/dCas9-SRDX in Rice Plants to Study Gene Silencing of Fungal Pathogens

Worldwide rice production is under threat from two filamentous fungal pathogens well known as *Magnaportheoryzae and Ustilaginoidea virens*. The CRISPRi system uses the nuclease activity of dead Cas9 (dCas9) to silence the promoters of target genes *MoATG3*, *MoATG7*, and *UvPal1* in these fungal pathogens. MoATG3 (Magnaportheoryzae ATG3) genes are found to code for proteins that are engaged in autophagy, a cellular process in which cellular components are degraded and recycled (Yin et al., 2019). MoATG7 (Magnaportheoryzae ATG7) is essential for the conjugation of the protein ATG8 to phosphatidylethanolamine (PE), which is the primary step in the process of autophagosomes. The UvPal1 (Ustilagoviolacea Pal1) gene is implicated in the control of melanin manufacturing, which contributes to understanding the dark pigmentation in certain fungi. The very effective CRISPRi system in *U. virens* makes it a valuable tool for quickly identifying gene function in pathogenic fungal species. The study showed the targeted genes were >100-fold suppressed, and the intended phenotypic characteristics were observed inCRISPRi strains (Zhang et al., 2023).

13.3.3.4 CRISPRi: CRISPR/dCas9-3×SRDX in Wheat Plants for Gene Repression in Leaf Bleaching Phenotype Studies

To figure out the observable leaf bleaching phenotype in the common wheat plant (*Triticum aestivum*) the phytoene desaturase gene (PDS) gene was targeted. To conduct the knock-out or knock-down of the targeted gene, the synthetic transcriptional repressor dCas9-3×SRDX (dCas9-SRDX) was constructed by the fusion of 3×SRDX to C-terminus of dCas9. According to reports, PDS is crucial for carotenoid biosynthesis and only exists in one copy in common wheat (Kumagai et al., 1995; Cong et al., 2010). Data showed that the CRISPRi system could effectively suppress the target gene's transcription level in stable transgenic seedlings, despite certain differences between vegetative and reproductive organs. The TaPDS target gene's real-time quantitative PCR analysis persuaded that the TRD (3 SRDX) could effectively repress the target gene's expression. They can also pass on their activating or repressive effects to the following generation. The study also suggested that there might be a threshold for the onset of leaf bleaching in plants when PDS was reduced. Where the predicted bleaching phenotype will not occur when the PDS expression level is above this threshold (Zhou et al., 2022).

13.3.4 Histone Modification

13.3.4.1 CRISPR/dCasHAT System Enhancing the Drought Tolerance Capacity

The consequences of drought episodes which reduce plant growth and yield can be minimized via the use of CRISPR/dCas tools. Many studies give evidence that the histone modification by the CRISPR/dCas system can improve plants' drought tolerance capacity. In plants, water stress is naturally detected and responded to using many signaling pathways. After sensing these signals, the plants adapt themself physiologically and developmentally by activating certain plant hormones. A vital positive regulator for the drought stress response is the abscisic acid (ABA) and its responsive element binding protein1/ABRE with its binding factor (AREB1/ABF2). It has been identified that Arabidopsis histone acetyltransferase1 (AtHAT1) enhances the activation of gene expression by putting chromatin in a relaxed state. The catalytic core of the Arabidopsis histone acetyl transferase1 gene (AtHAC1, AT1G79000) was linked to the N-terminal region of dCas9 and cloned to form an altered variant of the plant binary vector pGreenII. The results of the study with qRT-PCR tests showed that the control plants' gene expression levels were lower than those of the AREB1-regulated gene. The plants produced here had greater survival rates following drought stress, as well as increased content of chlorophyll and faster stomatal aperture under a severe deficiency of water. AREB1 expression levels significantly increased in dCas9HAT-sgA2 plants and dCas9HAT-sgA1 plants, by 2-2.6-fold. By this, we conclude that the positive regulation of AREB1 by CRISPRa dCas9HAT constitutes a promising biotechnological technique for enhancing resistance to drought and its tolerance (Roca Paixão et al., 2019).

13.3.4.2 Histone modification: MS2-CRISPR/dCas9 System Altering the Flowering Time

The main mechanism by which histone modification boosts gene expression is the addition or removal of acetyl group on histone. A study with Arabidopsis thaliana targeting the genes responsible for the floral regulator flowering locus T (FT). It has been demonstrated that specific target genes in Arabidopsis thaliana plants can have their transcriptional activity and epigenetic status modified using the MS2-CRISPR/dCas9 system. The dCas9-p300 acetyltransferase increases the acetylation of histone H3 lysine 27 (H3K27) near the promoter and enhancer vicinity. Histone H3K27 acetylation improves DNA accessibility by increasing nucleosome space and thus reducing DNA-histone binding affinity. Thus dCas9-p300 chimeric protein increases the acetylation of H3K27 (histone H3 lysine in the promoter site to enhance gene expression (Jogam et al., 2022; Lee et al., 2019).

13.3.4.3 Histone Modification: CRISPR/p300-dCas9 System Regulating the Production of Secondary Metabolites

In a study, the secondary metabolic gene from *Aspergillus niger* was targeted and this histone modification by the p300-dCas9 system was explained. Here they first built a p300-dCas9 system and demonstrated that an EGFP fluorescent reporter could be activated. Second, the transcription of the secondary metabolic gene *breF* was induced by specifically localized histone acetylase p300 on the ATG close by region. Third, the native polyketide synthase (PKS) gene *fuml* was targeted by p300-dCas9, increasing the production of the substance fumonisin B2, which was identified by high-performance liquid chromatography (HPLC) and liquid chromatography-mass spectrometry (LC-MS). The transcription of *breF* was then inhibited by the endogenous histone acetylase GcnE-dCas9 and the HDACs HosA-dCas9 and RpdA-dCas9. Finally, the histone deacetylase HosA stimulated the expression of the pigment gene *fwnA* and accelerated the manufacture of melanin by targeting HosA-dCa9 fusion to this gene was demonstrated. The results of the study demonstrate that systems for histone epigenetic modification led by CRISPR/dCas9 efficiently reprogrammed the target gene expression in *Aspergillus niger* (Li et al., 2021).

13.3.5 Non-Coding RNA: CRISPR/CasRx System Conferring Resistance to Plant RNA Viruses

There are numerous RNA-targeted against virus immunity in plants, such as RNA decay, RNA silencing, and RNA quality management systems. Some of these systems have been used to prevent and control plant RNA viruses, including efficient RNA silencing-based anti-viral engineering (Li and Wang, 2019). It was also shown that Cas13a can be used to successfully prevent plant RNA virus infection. The most often used host susceptibility genes modified by CRISPR/Cas9 systems to protect plant RNA viruses are the eukaryotic translation initiation factors eIF4E, eIF4G, and its isoforms eIF(iso)4E and eIF(iso)4G (Zhang et al., 2018; Cao et al., 2020). Researchers altered the *Nicotiana benthamiana* plants' CasRx-mediated RNA interference system. We tested the CRISPR/CasRx system on two plant RNA viruses from distinct families, specifically cucumber mosaic virus (CMV) and tobacco mosaic virus (TMV), to investigate the CRISPR/CasRx technology's potential for universal use. The systemic condition of TMV-GFP or the local expression of CMV-GFP, as well as the accumulation of GFP protein, were dramatically attenuated by the expression of the CasRx duplexed with the sgRNA targeting GFP caused by the PEBV promoter in the TRV2 vector.

It should be noted that the CMV CP was changed to GFP in this vector, preventing the virus from spreading to systemic leaves and these findings imply that by concentrating on the conserved area of the viral genomic RNA, the CasRx-mediated anti-viral mechanism can effectively prevent the invasion of a variety of viruses (Cao et al., 2021).

13.4 CHALLENGES FACED BY CRISPR

CRISPR/Cas9 is a promising strategy, because of its recent discovery and application. Despite the quick advancements in basic research and clinical trials, CRISPR faces numerous fundamental claims like editing effectiveness, relative delivery difficulty, off-target consequences, polymorphism, delivery method, ethics, immunogenicity, etc., (Liu et al., 2021). CRISPR/Cas9 can treat a wide range of complicated genetic illnesses that currently go untreated. Recent advancements in sequence-specific nuclease technology have resulted in a significant acceleration of medicinal genome editing methods that target disease-causing genes or mutant genes. The therapeutic potential of genome editing has not yet been fully investigated, and there are still many problems that raise the possibility of more mutations. Here, the primary obstacles that CRISPR are briefly discussed (Rasul et al., 2022).

Off-target effects happen when the gene editing technique unintentionally changes parts of the genome that weren't meant to be altered. The key off-target repercussions affecting the dCas system include biased or unbiased off-target detection, prime editing, adenine or cytosine base editors, dCas9, Cas9 paired nickase, RNP delivery, and truncated gRNAs. To improve the specificity of CRISPR-based gene editing, studies are ongoing in the field of reducing off-target consequences (Naeem et al., 2020). CRISPRi is a recently created method for selective gene silencing. It has excellent potential for use in identifying gene regulatory components and analyzing gene function. However, ideal conditions for effective sgRNA design for CRISPRi are not yet fully established. This is the reason that the effectiveness of gene manipulation turns out to be the main difficulty (Radzisheuskaya et al., 2016). It is essential to use efficient delivery techniques to make sure the gene editing tool is delivered to the correct area of the plant system. The most challenging obstacle to future CRISPR/Cas9 *in vivo* usage is transport means and proper delivery. The reported CRISPR/Cas9 cargos and delivery methods presently in use include physical delivery techniques, viral delivery techniques, and non-viral delivery techniques. Development in these tools is necessary for successful gene manipulation using CRISPR/Cas9 (Lino et al., 2018). When it comes to large-scale gene disruption investigations, such as single-gene and multiple-gene knock-out, knock-down, and knock-up libraries, as well as their related screening assays, CRISPR-Cas9 technology is applied. Multiple gene regulation at once can be harder to manage. HERE, Using CRISPR-dCas9 to coordinate the expression or suppression of numerous genes calls for careful planning and optimization (Deyell et al., 2019). Agricultural biotechnology and breeding could be revolutionized by CRISPR/Cas genome editing. Regarding consumer and producer choices, food ethics, and government, CRISPR/Cas raises new issues. CRISPR/Cas is a viable alternative to traditional breeding due to its accuracy, simplicity, and low cost. Nature-identical GMOs, on the other hand, blur the border between nature and technology, making modifications impossible to identify. that is the reason why the current regulatory approaches need to be reconsidered and raise new challenges based on ethics (Zhang et al., 2020).

13.5 SCOPE

CRISPR is considered the biotechnology innovation of the century. Technologies for modifying the genome have advanced rapidly. The benefits of CRISPR-Cas9 in the embryos of humans have sparked controversy and multidisciplinary arguments, even though the editing of the genome in species, such as plants, animals as well as microbes are widely accepted techniques. This presents numerous difficulties for both scientists and those who must control the application of these methods (Lima and Martínez, 2021).

Creating the greatest Cas9 endonuclease proteins and single-regulated RNA would help handle diseases of multiple origins (Tiruneh et al., 2021). This intends to give an overview of CRISPR-Cas9 technology, as well as a look at some of the method's present and future uses, prospects, and mechanisms of action in treating various diseases.

13.6 CONCLUSION

The chapter throws light upon the different strategies adopted by the CRISPR gene editing tool to modify the epigenetics of plant physiology. The various mechanisms of epigenetic modification in plants using both CRISPR/Cas as well as CRISPR/dCas system have been briefly reviewed. The major pathways including; methylation/demethylation, histone modification and non-coding RNA's role in epigenetic manipulation have been studied with the CRISPR/Cas system. Recently, the most researched form of the CRISPR/Cas system namely CRISPR/dCas is analyzed in the later part of this chapter. The details about the major domains of the dCas proteins such as the Cas9 nuclease domain, RNA-binding domains, nuclear localization signal, and transcriptional effector domains have been structured. The action of the gene expression where they are activated and repressed is demonstrated with CRISPRa and CRISPRi mechanisms respectively. The role of various types of dCas modifications altering the epigenetics of a variety of plant physiology is illustrated with numerous studies to support our chapter. Our chapter ends with the possibilities of the practical challenges with the current research methodologies and their potential outcome for the same. This chapter promises that the loopholes of this research can 1 day be corrected, which takes the epigenetics of plant physiology rise to the next level.

ACKNOWLEDGMENT

All the authors are thankful to their respective universities and institutions for their valuable support.

DATA AVAILABILITY

The authors confirm that the data supporting the findings of this study are available within the book/chapter.

REFERENCES

Basak, J., & Nithin, C. (2015). Targeting non-coding RNAs in plants with the CRISPR-Cas technology is a challenge yet worth accepting. *Frontiers in Plant Science*, *6*, 1001. DOI: 10.3389/fpls.2015.01001

Bhardwaj, P., Kant, R., Behera, S. P., Dwivedi, G. R., & Singh, R. (2022). Next-generation iagnostic with CRISPR/Cas: beyond nucleic acid detection. *International Journal of Molecular Sciences*, *23*(11), 6052. DOI: 10.3390/ijms23116052

Bondy-Denomy, J., Pawluk, A., Maxwell, K. L., & Davidson, A. R. (2013). Bacteriophage genes that inactivate the CRISPR/Cas bacterial immune system. *Nature*, *493*(7432), 429–432. DOI: 10.1038/nature11723

Brocken, D. J., Tark-Dame, M., & Dame, R. T. (2018). dCas9: a versatile tool for epigenome editing. *Current Issues in Molecular Biology*, *26*(1), 15–32. DOI: 10.21775/cimb.026.015

Cao, Y., Zhou, H., Zhou, X., & Li, F. (2020). Control of plant viruses by CRISPR/Cas system-mediated adaptive immunity. *Frontiers in Microbiology*, *11*, 593700. DOI: 10.3389/fmicb.2020.593700

Cao, Y., Zhou, H., Zhou, X., & Li, F. (2021). Conferring resistance to plant RNA viruses with the CRISPR/CasRx system. *Virologica Sinica*, *36*(4), 814–817.. DOI: 10.1007/s12250-020-00338-8

Chavez, A., Scheiman, J., Vora, S., Pruitt, B. W., Tuttle, M., PR Iyer, E., Lin, S., Guzman, C. D., Wiegand, D. J., & Ter-Ovanesyan, D. (2015). Highly efficient Cas9-mediated transcriptional programming. *Nature Methods*, *12*(4), 326–328. DOI: 10.1038/nmeth.3312

Cong, L., Wang, C., Li, Z., Chen, L., Yang, G., Wang, Y., & He, G. (2010). cDNA cloning and expression analysis of wheat (*Triticum aestivum* L.) phytoene and ζ-carotene desaturase genes. *Molecular Biology Reports*, *37*, 3351–3361. DOI: 10.1007/s11033-009-9922-7

Deng, F., Zeng, F., Shen, Q., Abbas, A., Cheng, J., Jiang, W., Chen, G., Shah, A. N., Holford, P., Tanveer, M., & Zhang, D. (2022). Molecular evolution and functional modification of plant miRNAs with CRISPR. *Trends in Plant Science*, *27*(9), 890–907. DOI: 10.1016/j.tplants.2022.01.009

Deyell, M., Ameta, S., & Nghe, P. (2019). Large scale control and programming of gene expression using CRISPR. In *Seminars in Cell & Developmental Biology*, Vol. *96*, pp. 124–132. DOI: 10.1016/j.semcdb.2019.05.013

Fal, K., Tomkova, D., Vachon, G., Chaboute, M. E., Berr, A., & Carles, C. C. (2021). Chromatin manipulation and editing: challenges, new technologies and their use in plants. *International Journal of Molecular Sciences*, *22*(2), 512. DOI: 10.3390/ijms22020512

Gallego-Bartolomé, J., Gardiner, J., Liu, W., Papikian, A., Ghoshal, B., Kuo, H. Y., Zhao, J. M. C., Segal, D. J., & Jacobsen, S. E. (2018). Targeted DNA demethylation of the Arabidopsis genome using the human TET1 catalytic domain. *Proceedings of the National Academy of Sciences of the United States of America*, *115*(9), E2125–E2134. DOI: 10.1073/pnas.1716945115

Ghoshal, B., & Gardiner, J. (2021). CRISPR-dCas9-based targeted manipulation of DNA methylation in plants. *CRISPR-Cas Methods*, 2, 57–71. DOI: 10.1007/978-1-0716-1657-4_5

Gilbert, L. A., Larson, M. H., Morsut, L., Liu, Z., Brar, G. A., Torres, S. E., Stern-Ginossar, N., Brandman, O., Whitehead, E. H., Doudna, J. A., & Lim, W. A. (2013). CRISPR-mediated modular RNA-guided regulation of transcription in eukaryotes. *Cell*, *154*(2), 442–451. DOI: 10.1016/j.cell.2013.06.044

Hu, D., Yu, Y., Wang, C., Long, Y., Liu, Y., Feng, L., Lu, D., Liu, B., Jai, J., Xia, R., & Du, J. (2021). Multiplex CRISPR-Cas9 editing of DNA methyltransferases in rice uncovers a class of non-CG methylation specific for GC-rich regions. *The Plant Cell*, *33*(9), 2950–2964. DOI: 10.1093/plcell/koab162

Jha, P., & Kumar, V. (2018). BABY BOOM (BBM): a candidate transcription factor gene in plant biotechnology. *Biotechnology Letters*, *40I*(10–11), 1467–1475. DOI: 10.1007/s10529-018-2613-5

Jinek, M., Chylinski, K., Fonfara, I., Hauer, M., Doudna, J. A., & Charpentier, E. (2012). A programmable dual-RNA-guided DNA endonuclease in adaptive bacterial immunity. *Science*, *337*(6096), 816–821. DOI: 10.1126/science.1225829

Jinek, M., East, A., Cheng, A., Lin, S., Ma, E., & Doudna, J. (2013). RNA-programmed genome editing in human cells. *Elife*, *2*, e00471. DOI: 10.7554/eLife.00471

Jogam, P., Sandhya, D., Alok, A., Peddaboina, V., Allini, V. R., & Zhang, B. (2022). A review on CRISPR/Cas-based epigenetic regulation in plants. *International Journal of Biological Macromolecules*. *219*, 1261–1271. DOI: 10.1016/j.ijbiomac.2022.08.182

Kazi, T. A., & Biswas, S. R. (2021). CRISPR/dCas system as the modulator of gene expression. *Progress in Molecular Biology and Translational Science*, *178*, 99–122. DOI: 10.1016/bs.pmbts.2020.12.002

Koonin, E. V., Makarova, K. S., & Zhang, F. (2017). Diversity, classification and evolution of CRISPR-Cas systems. *Current Opinion in Microbiology*, *37*, 67–78. DOI: 10.1016/j.mib.2017.05.008

Kumagai, M. H., Donson, J., Della-Cioppa, G., Harvey, D., Hanley, K., & Grill, L. (1995). Cytoplasmic inhibition of carotenoid biosynthesis with virus-derived RNA. *Proceedings of the National Academy of Sciences*, *92*(5), 1679–1683. DOI: 10.1073/pnas.92.5.1679

Lee, J. E., Neumann, M., Duro, D. I., & Schmid, M. (2019). CRISPR-based tools for targeted transcriptional and epigenetic regulation in plants. *PLoS One*, *14*(9), e0222778. DOI: 10.1371/journal.pone.0222778

Lee, J. H., Mazarei, M., Pfotenhauer, A. C., Dorrough, A. B., Poindexter, M. R., Hewezi, T., Lenaghan, S. C., Graham, D. E., & Stewart Jr, C. N. (2020). Epigenetic footprints of CRISPR/Cas9-mediated genome editing in plants. *Frontiers in Plant Science*, *10*, 1720. DOI: 10.3389/fpls.2019.01720

Li, F., & Wang, A. (2019). RNA-targeted antiviral immunity: more than just RNA silencing. *Trends in Microbiology*, *27*(9), 792–805. DOI: 10.1016/j.tim.2019.05.007

Li, X., Huang, L., Pan, L., Wang, B., & Pan, L. (2021). CRISPR/dCas9-mediated epigenetic modification reveals differential regulation of histone acetylation on Aspergillus niger secondary metabolite. *Microbiological Research*, *245*, 126694. DOI: 10.1016/j.micres.2020.126694

Li, Y., Kumar, S., & Qian, W. (2018). Active DNA demethylation: mechanism and role in plant development. *Plant Cell Reports*, *37*, 77–85. DOI: 10.1007/s00299-017-2215-z

Lima, N. S., & Martínez, A. G. (2021). Biotechnological challenges: the scope of genome editing. *JBRA Assisted Reproduction*, *25*(1), 150. DOI: 10.5935/1518-0557.20200038

Lino, C. A., Harper, J. C., Carney, J. P., & Timlin, J. A. (2018). Delivering CRISPR: a review of the challenges and approaches. *Drug Delivery*, *25*(1), 1234–1257. DOI: 10.1080/10717544.2018.1474964

Liu, W., Li, L., Jiang, J., Wu, M., & Lin, P. (2021). Applications and challenges of CRISPR-Cas gene-editing to disease treatment in clinics. *Precision Clinical Medicine*, *4*(3), 179–191. DOI: 10.1093/pcmedi/pbab014

López-Calleja, A. C., Vizuet-de-Rueda, J. C., & Alvarez-Venegas, R. (2019). Targeted epigenome editing of plant defense genes via CRISPR activation (CRISPRa). In *Epigenetics in Plants of Agronomic Importance: Fundamentals and Applications: Transcriptional Regulation and Chromatin Remodelling in Plants*, pp. 267–289. DOI: 10.1007/978-3-030-14760-0_10

Mali, P., Yang, L., Esvelt, K. M., Aach, J., Guell, M., DiCarlo, J. E., Norville, J. E., & Church, G. M. (2013). RNA-guided human genome engineering via Cas9. *Science*, *339*(6121), 823–826. DOI: 10.1126/science.1232033

Mushtaq, M., Bhat, J. A., Mir, Z. A., Sakina, A., Ali, S., Singh, A. K., Tyagi, A., Salgotra, R. K., Dar, A. A., & Bhat, R. (2018). CRISPR/Cas approach: a new way of looking at plant-abiotic interactions. *Journal of Plant Physiology*, *224*, 156–162. DOI: 10.1016/j.jplph.2018.04.001

Naeem, M., Majeed, S., Hoque, M. Z., & Ahmad, I. (2020). Latest developed strategies to minimize the off-target effects in CRISPR-Cas-mediated genome editing. *Cells*, *9*(7), 1608. DOI: 10.3390/cells9071608

Oberkofler, V., & Bäurle, I. (2022). Inducible epigenome editing probes for the role of histone H3K4 methylation in Arabidopsis heat stress memory. *Plant Physiology*, *189*(2), 703–714. DOI: 10.1093/plphys/kiac113

O'Geen, H., Bates, S. L., Carter, S. S., Nisson, K. A., Halmai, J., Fink, K. D., Rhie, S.K., Farmham, P.J., & Segal, D. J. (2019). Ezh2-dCas9 and KRAB-dCas9 enable engineering of epigenetic memory in a context-dependent manner. *Epigenetics Chromatin*, *12*(1), 26. DOI: 10.1186/s13072-019-0275-8.

Parrilla-Doblas, J. T., Roldán-Arjona, T., Ariza, R. R., & Córdoba-Cañero, D. (2019). Active DNA demethylation in plants. *International Journal of Molecular Sciences*, *20*(19), 4683. DOI: 10.3390/ijms20194683

Přibylová, A., Fischer, L., Pyott, D. E., Bassett, A., & Molnar, A. (2022). DNA methylation can alter CRISPR/Cas9 editing frequency and DNA repair outcome in a target-specific manner. *New Phytologist*, *235*, 2285–2299. DOI: 10.1111/nph.18212

Puchta, H. (2016). Using CRISPR/Cas in three dimensions: towards synthetic plant genomes, transcriptomes and epigenomes. *The Plant Journal*, *87*(1), 5–15. DOI: 10.1111/tpj.13100

Qi, X., Gao, H., Lv, R., Mao, W., Zhu, J., Liu, C., Mao, L., & Xie, C. (2023). CRISPR/dCas-mediated gene activation toolkit development and its application for parthenogenesis induction in maize. *Plant Communications*, *4*(2), 100449. DOI: 10.1016/j.xplc.2022.100449

Radzisheuskaya, A., Shlyueva, D., Müller, I., & Helin, K. (2016). Optimizing sgRNA position markedly improves the efficiency of CRISPR/dCas9-mediated transcriptional repression. *Nucleic Acids Research*, *44*(18), e141–e141. DOI: 10.1093/nar/gkw583

Rasul, M. F., Hussen, B. M., Salihi, A., Ismael, B. S., Jalal, P. J., Zanichelli, A., & Taheri, M. (2022). Strategies to overcome the main challenges of the use of CRISPR/Cas9 as a replacement for cancer therapy. *Molecular Cancer*, *21*(1), 64. DOI: 10.1186/s12943-021-01487-4

Ren, C., Li, H., Liu, Y., Li, S., & Liang, Z. (2022). Highly efficient activation of endogenous gene in grape using CRISPR/dCas9-based transcriptional activators. *Horticulture Research*, *9*, uhab037 DOI: 10.1093/hr/uhab037

Roca Paixão, J. F., Gillet, F. X., Ribeiro, T. P., Bournaud, C., Lourenço-Tessutti, I. T., Noriega, D. D., Melo, B. P. D., de Almedia-Engler, J., & Grossi-de-Sa, M. F. (2019). Improved drought stress tolerance in Arabidopsis by CRISPR/dCas9 fusion with a histone acetyl transferase. *Scientific Reports*, *9*(1), 8080. DOI: 10.1038/s41598-019-44571-y

Sung, C. K., & Yim, H. (2020). CRISPR-mediated promoter de/methylation technologies for gene regulation. *Archives of Pharmacal Research*, *43*(7), 705–713. DOI: 10.1007/s12272-020-01257-8

Thakore, P. I., D'ippolito, A. M., Song, L., Safi, A., Shivakumar, N. K., Kabadi, A. M., Reddy, T. E., Crawford, G. E., & Gersbach, C. A. (2015). Highly specific epigenome editing by CRISPR-Cas9 repressors for silencing of distal regulatory elements. *Nature Methods*, *12*(12), 1143–1149. DOI: 10.1038/nmeth.3630

Tiruneh, G., Medhin, M., Chekol Abebe, E., Sisay, T., Berhane, N., Bekele, T., & Asmamaw Dejenie, T. (2021). Current applications and future perspectives of CRISPR-Cas9 for the treatment of lung cancer. *Biologics: Targets and Therapy*, *15*, 199–204. DOI: 10.2147/BTT.S310312

Yin, Z., Chen, C., Yang, J., Feng, W., Liu, X., Zuo, R., Wang, J., Yang, J., Zhong, K., Gao, C., & Zhang, H. (2019). Histone acetyltransferase MoHat1 acetylates autophagy-related proteins MoAtg3 and MoAtg9 to orchestrate functional appressorium formation and pathogenicity in *Magnaporthe oryzae*. *Autophagy*, *15*(7), 1234–1257. DOI: 10.1080/15548627.2019.1580104

Yu, Y., Zhang, Y., Chen, X., & Chen, Y. (2019). Plant noncoding RNAs: hidden players in development and stress responses. *Annual Review of Cell and Developmental Biology*, *35*, 407–431. DOI: 10.1146/annurev-cellbio-100818-125218

Zhang, C., Konermann, S., Brideau, N. J., Lotfy, P., Wu, X., Novick, S. J., Strutzenberg, T., Griffin, P. R., Hsu, P. D., & Lyumkis, D. (2018). Structural basis for the RNA-guided ribonuclease activity of CRISPR-Cas13d. *Cell*, *175*(1), 212–223e17. DOI: 10.1016/j.cell.2018.09.001

Zhang, D., Hussain, A., Manghwar, H., Xie, K., Xie, S., Zhao, S., Larkin, R. M., Qing, P., Jin, S., & Ding, F. (2020). Genome editing with the CRISPR-Cas system: an art, ethics and global regulatory perspective. *Plant Biotechnology Journal*, *18*(8), 1651–1669. DOI: 10.1111/pbi.13383

Zhang, S., Shen, J., Li, D., & Cheng, Y. (2021). Strategies in the delivery of Cas9 ribonucleoprotein for CRISPR/Cas9 genome editing. *Theranostics*, *11*(2), 614–648. DOI: 10.7150/thno.47007

Zhang, Y. M., Zheng, L., & Xie, K. (2023). CRISPR/dCas9-mediated gene silencing in two plant fungal pathogens. *Msphere*, *8*(1), e00594–22. DOI: 10.1128/msphere.00594-22

Zhou, H., Xu, L., Li, F., & Li, Y. (2022). Transcriptional regulation by CRISPR/dCas9 in common wheat. *Gene*, *807*, 145919. DOI: 10.1016/j.gene.2021.145919

14 Advancement in Bioinformatics Tools in the Era of Genome Editing-Based Functional Genomics

Karma Landup Bhutia, Sarita Kumari, Kumari Anjani, Bhavna Borah, Anuradha, Vinay Kumar Sharma, Anima Kisku, Rajalingam Amutha Sudhan, Mahtab Ahmad, Bharti Lap, Nangsol Dolma Bhutia, Ponnuchamy Mugudeshwari, and Akbar Hossain

14.1 INTRODUCTION

The rediscovery of Mendelian genetics has made a paradigm shift in the ways of analyzing organismal genetics which is now advanced to the applications of gene & genomes, molecular markers, and organism's sequence-based trait dissection. Similarly, the development and advancement of nucleotide and amino acid sequencing technologies have not only revolutionized the ways of looking for insight into the genetic basis of different phenotypic traits but also accelerated the rate of crop breeding approaches with more efficiency and accuracy (Varshney et al. 2021). The efficiency of advanced techniques of sequencing aided by bioinformatics tools can be measured by looking at more than 1,000 plants' draft genomes representing 788 species that are already available in the public domain (Sun et al. 2021). The efficient utilization of bioinformatics tools is visible when we look at not only draft genomes but also high-quality reference genomes of several crop plants like rice (Zhou et al. 2020). Likewise, the development of pan-genomes and super-pangenomes is underway in many crop species (Khan et al. 2020). This rapid generation of biological data and storage in the public domain has now shifted the task more to retrieval and understanding of complex biological data using computer analysis (Kathiresan et al. 2017). The bioinformatics tools and algorithms need to be updated with an increase in data volume. Bioinformatics is a multifaceted field of science that combines biology, computer science, and statistics for analyzing and interpreting biological data, especially in the sphere of genetics and genomics. In the past few decades, the development of high-efficiency techniques such as DNA sequencing, genome sequencing, and microarrays has led to an exponential increase in the amount of data that needs to be stored and analyzed. It has a number of applications in the field of biology and medicine including genomics, proteomics, pharmaco-genomics, and personalized medicine. Bioinformatics has also helped to get a better understanding of topics such as genetic variation, genetic disorders, and the evolution of species. But, to unleash the full potential of bioinformatics in the field of science and technology certain computer-based software and algorithms are used which are known as the bioinformatics tools. These tools are used to process, analyze, interpret, and visualize biological data. These tools can be used to process large amounts of data, identify patterns and relationships, and make predictions about biological molecules and systems.

DOI: 10.1201/9781003387060-14

14.2 BIOLOGICAL DATABASES

Biological databases are of three types *i.e.,* public repositories (large scale), community, and project-specific databases. Public repositories (large scale) were developed for long-term data storage and maintained by government agencies, and national or international projects. Some of the large-scale public repositories are GenBank (Benson et al. 2012), INSDC (Cochrane et al. 2016), DDBJ (Mashima et al. 2016), EMBL (Aken et al. 2016) for genomic sequences, PDB (Berman et al. 2000) for protein structures and UniProt (Boutet et al. 2007) for protein information, Gene Expression Omnibus (GEO) (Edgar et al. 2002) and Array Express (Parkinson et al. 2005) for microarray data. Community-specific databases provide information for a specific research community like plant databases *i.e.* Gramene (Tello-Ruiz et al. 2016), PlantGDB (Dong et al. 2004), PLEXdb (Dash et al. 2012), Planteome (Cooper et al. 2016), ENSEMBL Plants (Bolser et al. 2016), Gene Expression Atlas (Petryszak et al. 2016) and species-specific databases like RAPdb (Ohyanagi et al. 2006), Beijing Genomics Institute -Rice Information System (He and Wang 2007), Rice SNP-Seek Database (Alexandrov et al. 2015; Mansueto et al. 2016), Community-specific database also include a specific type of database for protein modification (Tchieu et al. 2003) and metabolism (Zhang and Horvath 2005). The third category includes short-lived, smaller-scale databases developed for the management of project data. These databases are mostly maintained only during the funding period of projects.

14.3 BIOINFORMATICS TOOLS FOR DNA SEQUENCING AND ALIGNMENT

Sanger sequencing and Maxam and Gilbert sequencing were available as the first-generation sequencing technologies for sequencing both smaller and larger genomes. Since the discovery of these initial techniques, numerous techniques of sequencing have evolved during the last 10–15 years which has led to ever increase in data output, read lengths, and greater applications more particularly the increase in sequencing efficiencies and accuracy. The general flow chart of genome/exome sequencing is represented in Figure 14.1.

The period of second-generation sequencing technologies provided a reduction in sequencing time and increased the sequencing depth and read length with a significant reduction in the overall cost. Simultaneously, sequencing of short reads (~600 bp) using short-read sequencing technologies was frequently utilized in genomic research due to its low-cost feature and applicability in a wide range of biostatistical analyses (Heather and Chain 2016). As the genome sizes of any organism (particularly eukaryotes) are large, the short sequence reads were difficult to utilize in the reconstruction of the larger or original molecules due to the presence of homopolymers which has led to the sequencing of longer reads. Long-read sequencing (~10 kb) facilitated the efficient de novo assembly and also helped in the identification of structural variants and transcript isoform or novel isoform. For example in rice, third-generation sequencing with long reads was used in the construction of 105 accessions pan-genome which has resulted in identifications of 604 Mb novel sequences as compared to the reference genome (Zhang et al. 2022). Many tools have been developed for long-read sequencing and more than 350 long-read analysis tools are available and are frequently utilized in Nanopore and SMART sequencing platforms (Amarasinghe et al. 2020). Based

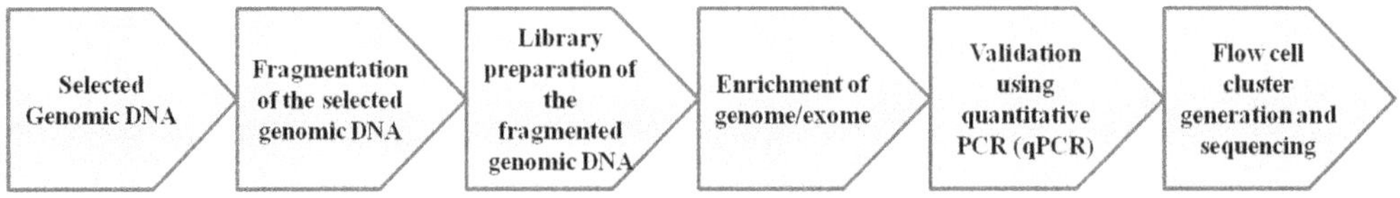

FIGURE 14.1 Workflow of genome or exome sequencing.

on various parameters and applications, anyone can choose the appropriate long-read tool from a publically available database named "long-read-tools.org" (Amarasinghe et al. 2021).

The analysis of sequencing data includes advanced analysis such as raw read quality control, sequence alignment, variant calling, genome assembly, and genome annotation. Sequence analysis utilizes a variety of bioinformatics tools that have been created (Table 14.1). To assure the quality of any subsequent analysis, it is crucial to examine the original sequence data. It can provide a comprehensive summary of read counts, read lengths, coverage reads, contaminating sequences, and the degree of sequence duplication. Through a quality assessment procedure, adaptor sequences and low-quality sequences are isolated from whole genome sequencing data in the first stage. FastQC is a well-known bioinformatics tool for determining sequencing read quality control (Andrews 2010). Fastp is a tool that has more recently been used for filtering, base correction, and quality control of sequencing reads and it is two to five times faster than earlier approaches (Chen et al. 2018). The second stage is read/sequence alignment, which aligns the sequences with the reference genome. In the absence of a reference genome, the overlapping areas are aligned to create contigs using the de novo genome assembly approach. The most significant and critical phase in the entire workflow is this one. Using a number of tools and algorithms, the sequence reads are accurately and swiftly aligned to the relevant locations of the reference genome. Sequence alignment has been made possible using a number of tools, some of which are Novoalign (http://novocraft.com/), MOSAIK (Lee et al. 2014), CUSHAW3 (Liu et al. 2014), Bowtie2 (Langmead and Salzberg 2012) and BWA (Li and Durbin 2009). The mapping tool that can now align reads generated by all the main sequencing technologies is called MOSAIK.

A variation call is made in the third stage. Compared to the reference sequence, the output sequences may differ, and these differences are referred to as variants. Variant calling techniques are used to find SNPs, INDELs, presence/absence variations (PAVs), copy number variations, and haplotype blocks. SAM tools (Li et al. 2009), Torrent Variant Caller (TVC) (Life Technologies, Rockville, MD), SNPSVM (O'Fallon et al. 2013), varScan (Koboldt et al. 2013), DeepVariant

TABLE 14.1
Biointormatics Tools Used in Next-Generation DNA Sequencing and Analysis

Purpose	Bioinformatics Tool	Link
Quality check	FastQC	https://www.bioinformatics.babraham.ac.uk/projects/fastqc
	fastp	https://github.com/OpenGene/fastp
Sequence alignment	Bowtie2	http://bowtie-bio.sourceforge.net/bowtie2/index.shtml
	BWA	http://bio-bwa.sourceforge.net/
	SOAP3-dp	http://soap.genomics.org.cn/
	CUSHAW3	http://cushaw3.sourceforge.net/homepage.htm#latest
	MAQ	http://maq.sourceforge.net/
	Novoalign	http://www.novocraft.com/products/novoalign/
	MOSAIK	https://github.com/wanpinglee/MOSAIK
Variant calling	GATK	https://software.broadinstitute.org/gatk/
	Freebayes	https://github.com/ekg/freebayes
	DeepVariant	https://github.com/google/deepvariant
	Platypus	http://www.well.ox.ac.uk/platypus
	VarScan	http://dkoboldt.github.io/varscan/
	ToTem	https://totem.software/
	Appreci8	https://hub.docker.com/r/wwuimi/appreci8/
Data visualization	IGV	https://igv.org/
	VISTA	https://genome.lbl.gov/vista/index.shtml

(Poplin et al. 2018), and Freebayes (Garrison and Marth 2012) are some of the tools used for variant calling. The variant calling procedure has been streamlined by a number of automated operations. To provide an end-to-end solution, these workflows combine multiple aligners and variant calling tools with additional upstream and downstream technologies. Tools like ToTem and Appreci8 (Tom et al. 2018; Sandmann et al. 2018) are pipelines for calling variants that are entirely automated. Because of its automatic pipeline optimization and effective analysis management, ToTem is a solution that is growing in popularity. Appreci8 uses eight independent tools to complete the same work, which results in accurate variant calling. The last phase is data visualization. Depending on the experiments and the goals of the research, many tools are available for data visualization. Integrated genome viewer is one of the most widely used options for reference genome visualization tools (if they are available) (Thorvaldsdottir et al. 2013). A visualization tool called VISTA can be used to compare the differences between two genetic sequences. Desktop solutions for a variety of genomic analysis needs, such as transcriptomics, variant calling, epigenomics, metagenomics, and comparative genomics, are available to help biologists who are unfamiliar with or have limited experience using the perl/python languages. These workflow systems include Qiagen CLC Genomics Workbench, geWorkbench, Partek Genomics Suite, JMP Genomics, and DNA Baser-NextGen Sequence Workbench.

14.4 BIOINFORMATICS TOOLS FOR SEQUENCE ANALYSIS

Recent advances in instrumentation and sequencing technologies produce a huge volume of data. A large amount of data can't be analyzed or solved by humans thus there is a need to use computation-based analysis, tools, and software packages for biological data. The National Institute of Health defines bioinformatics as the discipline related to genetics, and genomics that involves computer technology for collecting, storing, analyzing, and disseminating biological data and information such as DNA, RNA, and amino acid sequences or annotations of those sequences. Biological data include DNA, RNA, and protein sequences involved in fundamental biological functions. Traditionally, the information was obtained from textbooks, research articles, and scientific publications. In recent years, genome experimental data and sequences have been available in online open-source and biological databases. Several databases (~1685) publicly available biological databases are mentioned in Nucleic Acid Research Molecular Biology Database Collection (Rigden et al. 2016) and crop-specific databases. Bioinformatics tools also contributed to gene editing using CRISPR-cas9 technology as it helped in identifying matched spacer sequences across different genomes and finding variants of Cas nuclease (Cas10, Cas12, Cas12a, etc.) (Naeem and Alkhnbashi 2023).

14.4.1 Sequence Alignment and Gene Identification

Traditionally sequencing approaches include library construction, cloning, sequencing, and electrophoresis. In the modern era, numerous improvements in sequencing technology i.e., next-generation sequencing (NGS) technologies developed large volumes of high throughput data. Bioinformatics tools are used for managing, processing, and analyzing these sequences. Software packages including Arachne, Phred/Phrap/Consed, and GAP4 were mainly used for the sequence assembly (Gibbs and Weinstock 2003). TIGR is an open-source package used for comparative genome assembly (Pop et al. 2004). In the NGS, the BioNano platform uses the alignment tool -RefAligner, ONT platform uses PoreSeq and Nanocorr for de novo sequence analysis (Wee et al. 2019). MHAP is mainly used to identify overlaps among the long reads (Berlin et al. 2015). PBJelly program uses a scaffolding approach for closing gaps in genome assembling (English et al. 2012). FALCON is a hierarchical haplotype genome assembly tool. For genome-based sequence alignment rHATis used for long-read alignment. Some of the well-known computer programs for protein-coding gene identification are Genscan, GRAIL, Genie, GeneMarkHMM, and Glimmer (Mignone et al. 2003).

Structural annotation of the genome identifies genes and transposons in genomic sequences. Genomic tools like SynBrowse and VISTA are used for accurate gene identification (Pan et al. 2005; Frazer et al. 2004). GoMapMania is a web-accessible open-source resource for gene functional annotations in several plant species (Ramšak et al. 2014). DAVID (Database for Annotation, Visualization and Integrated Discovery) is a functional annotation tool. It helps to identify the biological meaning of genes (Sherman et al. 2022). Another important aspect of genome analysis is the identification of repetitive sequences. Repeat Masker can be used for the identification of repetitive sequences in the genome (Chen 2004). Novel repeats in the genome can be identified by Generic Repeat Finder. It is a highly sensitive tool that finds out terminal direct repeats, terminal inverted repeats, and interspaced repeats (Shi and Liang 2019).

14.4.2 Sequence Comparison

Comparison of different sequences allows for a better interpretation of the function, structure and evolution of genomes and genes. Methods in sequence comparison are grouped into pairwise, profile-profile, and sequence profile comparison. FASTA and BLAST (Basic Local Alignment Search Tool) are the most popular in pairwise comparison. PSI-BLAST is a popular sequence profile alignment tool. HMMER, SAM, and Meta-MEME are more accurate than PSI-BLAST but they perform slowly. Besides, several sequence-based protein family databases were built that integrate domain information from multiple protein domains i.e. InterPro (Blum et al. 2021). The phylogenetic tree shows the evolutionary relationships among proteins and it can be used in comparative genomics and gene function prediction. ClustalW is a tool used for aligning multiple nucleotides and related protein sequences (Larkin et al. 2007). TM- Aligner is a multiple sequence alignment tool used for transmembrane protein sequence alignment (Bhat et al. 2017). As of late, a tool called NGenomeSyn has been created for the publication-ready visualization of syntenic relationships of the entire genome or small region as well as genomic characteristics (such as repeats, structural variants, and genes) across several genomes (He et al. 2023). Other popular tools used for the identification and comparison of whole genome alignment between two or more genomes include MCScanX, SyRl, and GENESPACE (Goel et al. 2019; Lovell et al. 2022).

14.4.3 Transcriptome Analysis

Transcriptome analysis explains how changes in transcript affect the growth and development of an individual with respect to the environment. DNA Microarrays are powerful techniques used for detecting the transcriptional profile of genes. Gene Traffic, GeneSpring, Affymetrix's Gene Chop Operating Software, Cluster, CaARRAY, and BASE are the notable commercial software available for performing a variety of analyses on microarray data sets (Saal et al. 2002). The Gene Expression Omnibus (GEO) is a public repository for expression data maintained by the National Centre for Biotechnological Information (NCBI). GEO has several basic analysis tools such as GDS and GEO_2R. The GDS tools provide clustering and differential expression of curated genes in GEO and GEO2R is a tool that provides differential expression testing (Barrett et al. 2012). The large array of data analysis uses a variety of tools (Table 14.2) for deriving meaningful information. Shiny GEO is another microarray analysis tool that performs on many microarray datasets (Dumas et al. 2016). Recently Amaral et al. (2018) introduced BART (Bioinformatics Array Research Tool), an R-Shiny web application that downloads microarray data and processes for analysis across different microarray platforms automatically. Similarly, an online web tool that specifically analyzes the protein-based microarray data is PAWER. Gene expression studies help in the identification of the functional roles of the genes and further contribute to the deciphering of genomc function. The large amount of data generated by these experiments can be critically analyzed to gain information about the functional and regulatory roles of the genes. Some of the tools used in microarray-based gene expression studies are listed in Table 14.3

TABLE 14.2
Some of the Important Tools Used for Gene Expression data

SL. No.	Name	Link	Function
	ASDB	http://cbcg.nersc.gov/asdb	A resource for probing protein functions with small molecules
	Body Map	http://bodymap.ims.u-tokyo.ac.jp	Identification of genes, collection of gene expression information for the human and mouse genomes
	GeneCodis	http://genecodis.cnb.csic.es. http://genecodis.cbi.pku.edu.cn.	Modular enrichment analysis
	ChipCodis	http://chipcodis.cnb. csic.es	Regulatory systems in yeast using chip-on-chip data using a similar approach.
	SENT	http://sent.cnb.csic.es	Text mining tools
	Moara	http://moara.cnb.csic.es	Text mining tools
	AlignACE	http://arep.med.harvard.edu/ mrnadata/mrnasoft.html	Identification of common motifs
	GibbsDNA	https://www.familytreedna.com/ groups/gibbs/	Identification of common motifs

TABLE 14.3
Tools and Database Related to Microarray Technology

SL. No.	Name	Link	Function
1	Engene	http://engene.cnb.csic.es	Microarray data analysis, from preprocessing to clustering and classification.
2	MARQ	http://marq.cnb.csic.es	Compilation of microarray gene expression signature database of model organisms
3	CIMminer	http://discover.nci.nih.gov/ cimminer/	Representation of high dimensional data sets such as gene expression profiles.
4	SpliceMiner	http://www.tigerteamconsulting. com/ SpliceMiner/intro.jsp/	Gene's splice variants display in a non-redundant manner
5	AffyProbeMiner	http://gauss.dbb.georgetown. edu/liblab/affyprobeminer/	re-defining chip definition files (CDFs) for Affymetrix chips

14.4.4 Gene Ontology Tools

Gene Ontology (GO) describes molecular functions, biological processes, and cellular components. It will help in annotating genes and splitting them into different functional groups. GOnet is an open-source tool that facilitates the biological interpretation of omics data (Pomaznoy et al. 2018). Other some of web-based applications are Gorilla (Eden et al. 2009), NaviGO (Wei et al. 2017), DAVID (Huang et al. 2009), and AmiGO (Carbon et al. 2009).

14.5 TOOLS FOR ANALYZING SEQUENCE HOMOLOGY AND SIMILARITY

A significant change in biological research has been brought about by the development of bioinformatics tools and the interaction of the biological and computer sciences (Akhtar et al. 2016). Understanding evolutionary links and determining functional annotations are both important aspects of modern biology and both depend on the homology sequence search. Sequence homology describes the evolutionary relationship between sequences which suggests that the sequences are related and that they come from a single ancestor sequence. Due to their shared evolutionary past, such homologous sequences are predicted to have comparable properties in terms of their structures or activities (Pearson 2013). Whereas, the degree of resemblance between two sequences is referred to as sequence similarity; and several approaches, including sequence alignment techniques like BLAST (Basic Local Alignment Search Tool) or FASTA (Fast Alignment Search Tool), are used to quantify it. In a sequence alignment, extra spaces (typically indicated by a '-' symbol) are added to the two given sequences at the proper points to enhance sequence similarity. The word's level or degree of similarity, alignment with optimal similarity, alignment's positional identity percentage, and the likelihood of alignment may all be used when referring to sequence similarities. Any similarity, including sequence similarity, should simply be called similarity. On the basis of shared sequence motifs, patterns, or conserved residues, these algorithms compare the sequences and pinpoint areas of similarity. They should be backed up by appropriate statistical analysis.

Phylogenetic analysis is mostly used for sequence alignment, alignment optimization, concatenation of alignments, selection of the best-fit evolutionary models and partitioning schemes, phylogeny reconstruction, and finally visualization and annotation of the phylogram. If various programs are used for unique operations, this can be time-consuming, especially as they typically have different input file format requirements that can even call for manual file editing. As a result, a wide range of evolutionary biologists are depending more and more on software programs that serve many purposes and support workflows. As multigene or genomic data sets are quickly being replaced by single-gene data sets as the tool of choice for phylogenetic reconstruction, automated gene extraction from genomic data, batch manipulation in some of the above steps, such as alignment, and automated concatenation are becoming increasingly important (Rivera-Rivera and Montoya-Burgos 2016). However, when annotating and comparing molecular sequence data, multiple sequence alignments (MSAs) are commonly used to uncover medically relevant substitutions (Higgs and Wood 2008), deduce the history of species (Misof et al. 2014), detect lineage and site-specific changes in the evolutionary processes(Jayaswal et al. 2014) as well as generate new enzymes (Wilding et al. 2017). In addition to the fully specified nucleotides or amino acids, MSAs may also contain ambiguous characters (nucleotides or amino acids that have not been fully specified). Additionally, alignment gaps are frequently inserted between some of the nucleotides or amino acids in the sequences to increase the homology of residues from different sequences (Wong et al. 2020).

14.5.1 Tools for Sequence Analyses

With the rapidly expanding sequence databanks, sequence similarity searching is an essential component of molecular biology and bioinformatics and is frequently used to infer the structures and functions of unannotated sequences, gain insight into relationships between sequences, and build phylogenetic trees. The most commonly used tool for nucleotide or protein sequence similarity search is BLAST (Basic Local Alignment Search Tool) on the web page of the NCBI which allows users to find related sequences and infer functional and evolutionary linkages by comparing a query sequence to a database of sequences (Altschul et al. 1990). A version of BLAST known as BLAT (BLAST-Like Alignment Tool) identifies similar sequences within closely related species or the same species (Bhagwat et al. 2012). Another tool routinely used for sequence similarity search is EMBOSS (European Molecular Biology Open Software Suite) which was created as free and open-source software to meet the demands of the molecular biology and bioinformatics user community. It is a comprehensive suite of sequence analysis tools that covers a wide range of tasks, including sequence alignment, motif searching, primer design, and protein domain analysis (Rice et al. 2000). MEME Suite: The MEME Suite provides a collection of tools for motif discovery and analysis. It includes MEME, a widely used tool for de novo motif discovery, as well as other tools for motif comparison, visualization, and enrichment analysis (Bailey et al. 2009). GROMACS:

GROMACS is a powerful software package for molecular dynamics simulations of biomolecules. It is commonly used for studying the behavior and interactions of proteins, nucleic acids, and lipids at an atomic level (Van Der Spoel et al. 2012). Geneious: Geneious is a comprehensive bioinformatics software platform that integrates multiple tools and functionalities for sequence analysis, including sequence alignment, assembly, phylogenetics, primer design, and gene expression analysis (Kearse et al. 2012). MUSCLE (Multiple Sequence Comparison by Log-Expectation): It is a widely used MSA tool that employs a progressive algorithm combined with a refinement phase. It is known for its accuracy and scalability (Edgar 2004). MAFFT (Multiple Alignment using Fast Fourier Transform): It is a versatile MSA tool that offers various algorithms and strategies for different alignment scenarios. It is known for its speed and accuracy (Katoh and Standley 2013). Clustal Omega: Clustal Omega is a popular MSA tool that employs a progressive alignment algorithm. It offers high scalability and can handle large-scale sequence alignments efficiently (Sievers and Higgins 2018). T-Coffee (Tree-based Consistency Objective Function for Alignment Evaluation): It is a comprehensive MSA tool that integrates information from multiple sources, including sequence similarity, structural information, and domain information (Notredame et al. 2000). ProbCons: ProbCons is a probabilistic consistency-based MSA tool that utilizes a hidden Markov model (HMM) framework. It considers both sequence similarity and evolutionary information to generate alignments (Pei et al. 2009). MSAProbs: MSAProbs is a probabilistic multiple sequence alignment tool that utilizes a HMM framework. It employs an iterative algorithm to refine the alignment by considering both local and global alignment consistency (Liu et al. 2010). Kalign: it is a progressive multiple sequence alignment tool that incorporates an accurate scoring system based on a combination of sequence identity and secondary structure prediction. It performs well in aligning structurally conserved regions and can handle large-scale alignments efficiently (Lassmann 2020). PASTA: PASTA (Practical Alignment using SATé and TrAnsitivity) is an iterative MSA tool that combines the accuracy of SATé with the speed of fast tree estimation methods. It uses transitivity-based consistency transformations to refine the alignment iteratively (Mirarab et al. 2020). PROMALS3D: PROMALS3D is a multiple sequence alignment tool that integrates sequence, structure, and evolutionary information. It employs a HMM approach combined with a 3D structural modeling strategy to generate accurate alignments for distantly related sequences (Pei and Grishin 2014). InterProScan: The tool InterProScan uses a number of databases and algorithms to forecast protein function and detect sequence homology. It searches for conserved domains, motifs, and other sequence traits to derive functional annotations (Jones et al. 2014). Some of the major sequence alignment tools are listed in Table 14.4

14.6 BIOINFORMATICS TOOLS FOR STRUCTURAL AND FUNCTIONAL PREDICTION OF RNA AND PROTEINS

The RNA is the linking entity between genes and proteins. Apart from its function as a messenger molecule, RNA is also involved in regulatory functions. A large number of different types of RNA molecules are present in a cell which plays a crucial role in genome function. Comparative studies can be effectively used to identify putative RNAs and their functions. Table 14.5 lists a few alignment-based tools employed for RNA structure studies.

In the context of structural and functional prediction of proteins, bioinformatics tools play a crucial role. Proteins are complex biomolecules with a specific three-dimensional structure and a unique function. Determining the structure and function of proteins is essential for understanding their role in biological processes and identifying potential targets for drug development for various diseases and infections. Structural prediction tools use various computational methods to predict the 3D- structure of proteins based on their amino acid sequence. These methods include homology modeling, ab initio prediction, molecular dynamics simulations, etc. Predicting the structure of proteins enables researchers and scientists to understand how they interact with other molecules,

TABLE 14.4
Alignment Tools Commonly Used in Functional Genomics

S. No.	Name	Link	Function
	BLAST	http://blast.ncbi.nlm.nih.gov/Blast.cgi	Sequence search and alignment tool based on a significant statistical calculation
	GeneOrder 2.0	http://binf.gmu.edu:8080/GeneOrder2.0/	Alignment of small GenBank genome sequences (up to 0.25Mb).
	CoreGenes	http://binf.gmu.edu:8080/CoreGenes1.0/	Analysis of two to five genomes simultaneously, the discovery of orthologs and putative orthologs.
	ClustalW	http://www.ebi.ac.uk/Tools/msa/clustalw2/	Multiple Sequence Alignment search tool
	PROBCONS	http://probcons.stanford.edu/index.html	Multiple alignments of protein sequences.
	T-COFFEE	http://tcoffee.vital-it.ch/cgibin/Tcoffee/tcoffee_cgi/index.cgi	Multiple Sequence Alignment search tool, accurate for sequences with less than 30% identity
	PipeAlign	http://bips.u-strasbg.fr/PipeAlign/	Protein family analysis
	HOMSTRAD	http://www-cryst.bioc.cam.ac.uk/~homstrad	Alignment database of homologous proteins

TABLE 14.5
Tools Used for RNA Structure Prediction

SL. No.	Name	Link	Function
	MirAlign	http://bioinfo.au.tsinghua.edu.cn/miralign/	Identification of new miRNA based on available known miRNA in the databases.
	Mirscan	http://genes.mit.edu/mirscan/	Identification of microRNA-like hairpin sequences from an arbitrary set of candidates.
	ERPIN	http://rna.igmors.upsud.fr/download/	Easy RNA Profile Identification: a software program for RNA motif identification.
	RNAsoft	http://www.rnasoft.ca/	Computational Prediction and design of RNA/DNA structure
	mfold software	http://mobyle.pasteur.fr/cgi-bin/portal.py?#forms::mfold	Web server for nucleic acid folding and hybridization prediction.
	RNAz	http://rna.tbi.univie.ac.at/cgi-bin/RNAz.cgi	Prediction of structurally conserved and thermodynamically stable RNA secondary structure in multiple sequence alignment
	SFold	https://sfold.wadsworth.org/	RNA folding
	Alifold	http://rna.tbi.univie.ac.at/cgibin/RNAalifold.cgi	Prediction of consensus secondary structures for the set of aligned RNA and DNA sequences.
	Evofold	http://users.soe.ucsc.edu/~jsp/EvoFold/	A comparative method for identifying functional RNA structures in multiple sequence alignments.

and identify potential drug targets. Furthermore, functional prediction tools analyze the amino acid sequence of a protein to predict its biological function. These methods include sequence alignment, protein domain analysis, and phylogenetic analysis. Overall, bioinformatics tools are essential for predicting the structure and function of proteins. They provide a powerful set of computational methods and algorithms for processing and analyzing large amounts of biological data, which can lead to new insights and discoveries in the field of biomedicine.

14.6.1 Bioinformatics Tools Used for Predicting the 3D Structure of Proteins

There are several bioinformatics tools used for predicting the 3D structure of proteins out of which some of the most commonly used methods are homology modeling, Ab initio prediction, molecular dynamics and fold recognition. With the development of advanced computer software and technology, some other bioinformatics tools were also developed for predicting the structure of proteins like deep learning, single-cell sequencing, network analysis, AlphaFold, MMseqs 2, PconsFold, etc. Homology Modeling: by using the known structure of a similar protein as a guide, homology modeling predicts the structure of a protein. This approach is based on the idea that proteins with comparable sequences would probably have related structures. MODELLER is one of the most popular software programs for homology modeling (Šali and Blundell 1993). Ab initio prediction: the structure of the protein is predicted based on first principles such as energy calculations of physics in the Ab initio method. Rosetta is one of the most widely used software tools for Ab initio predictions (Rohl et al. 2004). Molecular Dynamics (MD): a computational model that uses physics-based force fields to predict the behavior of atoms and molecules. MD simulations can be used to predict the movement and behavior of a protein, which can provide insights into its structure and function (Karplus and McCammon 2002). Fold Recognition: it involves predicting the structure of a protein based on identifying its fold in a database of known protein structures. One of the most widely used software tools for fold recognition is Phyre2 (Kelley and Sternberg 2009). Deep Learning: it is a subfield of machine learning that involves training artificial neural networks to recognize patterns in data. In bioinformatics, deep learning has been used for a variety of applications, including protein structure prediction and drug discovery (Senior et al. 2020). Single-Cell Sequencing: it is a technique that allows analyzing the gene expression profiles of individual cells. In bioinformatics, single-cell sequencing has been used for understanding cell heterogeneity and identifying new cell types (Zhang et al. 2017). Network Analysis: it involves the study of interactions between biological molecules, such as proteins or genes. In bioinformatics, network analysis has been used for identifying new drug targets and understanding the mechanisms of disease (Zhang and Horvath 2005). Alpha Fold: it is a deep learning-based method for protein structure prediction developed by the Deep-Mind team. In November 2020, they published a paper in Nature describing how AlphaFold had achieved a significant breakthrough in predicting the 3D structure of proteins with high accuracy (Jumper et al. 2021). MMseqs2: it is a fast and scalable sequence clustering method developed by researchers at the Technical University of Munich. It has been used for protein structure prediction and metagenomics analysis (Steinegger and Söding 2017).

Similarly, SWISS-MODEL is a widely used protein structure prediction tool that uses homology modeling to predict protein structures based on the known structures of related proteins (Waterhouse et al. 2018). I-TASSER: A popular protein structure prediction tool that uses iterative threading assembly refinement to predict protein structures from amino acid sequences (Zhang 2008). Rosetta: A widely used protein structure prediction tool that uses a combination of Monte Carlo simulations and energy minimization to predict protein structures (Leaver-Fay et al. 2011). RaptorX: A protein structure prediction server that combines template-based and template-free methods for improved accuracy (Källberg et al. 2012). QUARK: a fully automated protein structure prediction server that uses both ab initio and template-based modeling methods (Xu and Zhang 2012). HHpred: a protein structure prediction tool that uses profile hidden Markov models and pairwise comparison methods for template-based modeling (Söding et al. 2005). These are just a few examples of bioinformatics tools discovered in the past few decades and years for predicting protein structures. The field of bioinformatics is constantly evolving, and new methods are being developed and refined all the time. The tools and databases containing this information are very elaborate (Table 14.6) and can help study protein function.

TABLE 14.6
Protein Structure Prediction Tools and Databases

SL. No.	Name	Link	Function
1	ASTRAL	https://astral.berkeley.edu/	Database and tools useful for analyzing protein structures and their sequences
2	Protein Data Bank (PDB)	https://www.rcsb.org/	Global repository for 3D structural data of proteins, DNA, RNA and their complexes with small molecules.
3	MMDB	https://www.ncbi.nlm.nih.gov/Structure/MMDB/mmdb.shtml	NCBI Structure Database of the 3D structure of biomolecules
4	Cn3-D	https://www.ncbi.nlm.nih.gov/Structure/CN3D/cn3d.shtml	Viewer of structural sequence alignment for MMDB database.
5	SWISS PDB Viewer	https://spdbv.unil.ch/	Network for analysis of several proteins, usual tools for visualization of protein structure.
6	Chemscape Chime, Rasmol and protein explorer	http://www.openrasmol.org/	Usual tools for visualization of protein structure, reading molecular structure files from PDB
7	Mage and Kinemages	https://www.umass.edu/microbio/rasmol/mage.htm	Tool for animated protein structure visualization
8	STRING	http://string.embl.de	physical interactions and functional associations prediction of protein
9	YASPIN	http://www.ibi.vu.nl/programs/yaspinwww/	Protein-protein interaction and site prediction.
10	J Pred	http://www.compbio.dundee.ac.uk/www-jpred/	Protein secondary structure prediction
11	VAST	https://structure.ncbi.nlm.nih.gov/Structure/VAST/vast.shtml	NCBI tool, identification of similar proteins with 3D structure
12	DALI	http://ekhidna2.biocenter.helsinki.fi/dali/	Protein structure alignment tool used for comparison of protein structure in 3D
13	Ligplot	https://www.ebi.ac.uk/thornton-srv/software/LigPlus/	Prediction of interaction between protein and ligand, representation of hydrogen and hydrophobic contacts
14	Conserved Domain Database	https://www.ncbi.nlm.nih.gov/Structure/cdd/cdd.shtml	The database contains sequence alignment and profiles, showing protein domain conserved during the molecular evolution course.
15	CDART: (Conserved Domain Architecture Retrieval Tool)	https://www.ncbi.nlm.nih.gov/Structure/lexington/docs/cdart_about.html	Searching proteins having similar domain architectures.
16	FAT-PTM database	https://Bioinformatics.cse.unr.edu/fat-ptm/	Functional Analysis Tools for Post-Translational Modifications

14.6.2 Protein Structures Discovered Using Bioinformatics Tools

Bioinformatics tools have played a critical role in discovering new protein structures, particularly in recent years. Protein structures that have been discovered using bioinformatics tools in recent years, for example, (1) Cas13: in 2017, a team of scientists from the Broad Institute used bioinformatics tools to discover the structure of the RNA-targeting protein Cas13. This protein is a key component of the CRISPR-Cas13 system, which can be used for gene editing (Smargon et al. 2017) (2) Tau protein: in 2018, a team of researchers from the University of Cambridge used bioinformatics tools to identify the structure of the tau protein, which is implicated in Alzheimer's disease. The researchers used cryo-electron microscopy aided with bioinformatics tools to determine the structure of tau fibrils, which are associated with the disease (Falcon et al. 2018) (3) G protein-coupled receptors (GPCRs): in 2019, a team of scientists from the University of Michigan used bioinformatics tools to develop a new computational method for predicting the structures of GPCRs. GPCRs are a large family of proteins that play a critical role in cell signaling and are targets for many drugs (Zhang et al. 2019) (4) SARS-CoV-2 spike protein: in 2020, a team of scientists from the University of Texas used bioinformatics tools to determine the structure of the spike protein on the surface of the SARS-CoV-2 virus, which causes COVID-19. This structure has been crucial for the development of vaccines and treatments for the disease (Wrapp et al. 2020). These are a few examples of the many protein structures that have been discovered using bioinformatics tools in recent years. Bioinformatics continues to play a critical role in advancing our understanding of the structure and function of proteins, which has important implications for drug discovery and the development of new treatments for diseases.

14.7 BIOINFORMATICS TOOLS FOR GENOME EDITING

The three major genome editing tools are widely being used for genetic manipulations are; zinc finger nucleases (ZFNs), transcription activator-like effector nucleases (TALENs), and clustered regularly interspaced short palindromic repeat (CRISPR)–Cas-associated nucleases (Li et al.2020). Bioinformatics plays a crucial role in designing the construct required for the synthesis of an array of helical proteins with variable repeat against triplet nucleotide sets in target DNA for ZFNs and variable di repeats (VDR) for TALEN that is linked with FOK-I nuclease for induction of cleavage in the target DNA. Various bioinformatics tools are being used for the customization of zinc finger nuclease-based genome editing tools out of which a few popular tools are; Zinc Finger tools, ZiFiT (Zinc Finger Targeter), ZFNGenome, etc. Zinc Finger tools are user-friendly tools for the customization of DNA binding motifs and prediction of zinc finger protein binding sites in the genome (Mandell and Barbas 2006). Similarly, ZiFiT is a web-based tool for the identification of potential target sites of zinc finger proteins (Sander et al. 2010) and ZFNGenome is a GBrowse-based tool for the prediction of genome binding region of ZFN with visualization. TALEN are more efficient genome editing tools as a repeat variable di-repeat (RVD) binds to a unique base of DNA. Bioinformatics tools TAL Effector-Nucleotide Targeter (TALE-NT) (Doyle et al. 2012), E-TALEN (Heigwer et al. 2013), Mojo Hand (Neff et al. 2013) and CHOPCHOP (Montague et al. 2014) are the web-based tools play role in customization of RVD, design of effector against a target, target recognition, predict probable binding sites in the genome and design an efficient construct for genetic manipulation in the target host. CRISPR/Cas tools have become the most prevalent for genetic manipulation in the field of life sciences in the last decades (Moon et al. 2019). The RNA/DNA-based target recognition and ease of customization make the CRISPR/Cas tools more popular than protein/DNA-based ZFN and TALEN (Mohanta et al. 2017).

CRISPR/Cas is highly efficient for the target recognition, insertion, deletion, and modification of genes of interest that can be achieved under in vivo and in vitro conditions. However, CRISPR/Cas tools have evolved as highly efficient genome editing tools but the rate and efficiency of target modification depend upon how meticulously the guide RNA has been designed. In the last decades, an array of bioinformatics tools have been developed for designing high throughput CRISPR/Cas tools to enhance efficiency and reduce the off-target effect. Bioinformatics tools that play a vital role in the customization of guide RNA are CRISPR-P, CRISPR-PLANT, CRISPR/Cas9 Designer, Multiple targets CRISPR, sgRNA Designer, CRISPRitz, and CRISPR-ERA. The identification of potential off-target regions associated with the CRISPR/Cas system within the genome can be predicted with bioinformatics tools are; GT-Scan, CCTop, Off-Spotter, and efficiency of target cleavage with the online tools; Cas-OFFinder and CasOT. The types of bioinformatics tools for various genome editing tools with their features are mentioned in Table 14.7.

TABLE 14.7
Bioinformatics Tools for Identification and Designing of Genome Editing System for the Specific Region

Bioinformatics Tools	Features	References
Zinc Finger tools	Customization of DNA binding domain	Mandell and Barbas (2006)
ZiFiT	Identify the potential off-target region	Sander et al. (2010)
ZFNGenome	Predicting zinc finger nuclease target sites in the genome	Reyon et al. (2011)
TALE-NT	Tool for designing target effector motif and prediction of target binding region	Doyle et al. (2012)
E-TALEN	TALEN for genome engineering	Heigwer et al. (2013)
Mojo Hand	Design TAL and TALEN and predict the genomic region	Neff et al. (2013)
CHOPCHOP	Tool for selecting an optimal target for CRISPR/Cas and TALEN system	Montague et al. (2014)
Cas-OFFinder	Search for off-target sites for Cas	Bae et al. (2014)
CasOT	Search for Cas9/guide RNA off-target effect	Xiao et al. (2014)
CRISPR-P	Tool for designing guide RNA for plant system	Lee et al. (2014)
CRISPR-PLANT	Tools for designing specific guide RNA for eight plant species	Xiao et al. (2014)
GT-Scan	System for prediction of the optimal number of target sites within the reference genome	O'Brien and Bailey (2014)
CCTop	Search for specific target sites based on the off-target effect	Stemmer et al. (2015)
CRISPR Direct	Predict cleavage sites, Tm and GC percentage of the target region with PAM positions	Naito et al. (2015)
CRISPR/Cas9 Designer	Customization of guide RNA for Knockout function	Park et al. (2015)
Multiple Target CRISPR	Design highly specific guide RNA	Prykhozhij et al. (2015)
sgRNA Designer	Design specific guide RNA for the gene of interest	Brazelton et al. (2015)
Off-Spotter	Identify guide RNA and match genomic region with PAM sequences	Pliatsika and Rigoutsos (2015)
CRISPRitz	A suite of software for designing the CRISPR/Cas-based experiments	Cancellieri et al. (2020)
CRISPR-ERA	Different search options with single guide RNA design	Liu and Wang (2021)

Most of the CRISPR/Cas system is based on knockout and knock-in approaches that are the aftermath of a specific double-stranded break followed by endogenous DNA repair systems; non-homologous end joining (NHEJ) and homology-directed repair (HDR). Thus, several web-designed tools are needed for the customization of the CRISPR/Cas-based editing system with minimal off-target effect. Moreover, this CRISPR/Cas system has been also designed to edit the genome without a double-stranded break called a base editor. CRISPR/Cas-based base editor tools edit and modify target base using linked enzymes; cytidine deaminase, adenine base editor, adenine deaminase with deactivated Cas9 (dCas9) and Cas9 nickase (nCas9) (Bikard et al. 2013; Ran et al. 2013; Zong et al. 2017; Billon et al. 2017). Moreover, the bioinformatics tools against the CRISPR/Cas-based base editor system that can predict the probable off-target and other deleterious effects on the genome are scarce in number. A recent study reported the development of tools with multiple functions for the CRISPR-based base editing system are; Base-Designer and Base-Analyser (Hwang et al. 2018).

14.8 CONCLUSION

Bioinformatics is the application of a computation system to retrieve, analyze, store, and visualization of biological data. The advancement of storage systems and computational capacity makes it liable for the interpretation of large genomic datasets. It is an indispensable technology in the field of biotechnology and molecular biology and is recognized as an omics tool for mapping, sequencing, and interpreting the sequencing dataset and displaying them as a database. Similarly, significant change in biological research has been brought about by the development of bioinformatics tools and the interaction of the biological and computer sciences. These tools give users the ability to gather, manage, evaluate, and go over raw data that has been gathered by employing cutting-edge methods (NGS) and reliable studies that are published in the public domain. The value and distinctive qualities of these devices need to be constantly enhanced. Even for seasoned researchers, accessing and organizing such vast amounts of data can be time-consuming and challenging. This opens up a wide range of research opportunities for researchers and other people who require advanced computer skills. The amount of genetic data that may currently be accessed through public databases has greatly increased as a result of developments in advanced bioinformatics tools.

REFERENCES

Aken BL, Ayling S, Barrell D, Clarke L, Curwen V, Fairley S, Fernandez Banet J, Billis K, GarcíaGirón C, Hourlier T, Howe K. 2016. The Ensembl gene annotation system. *Database*. 2016: baw093.

Akhtar MM, Micolucci L, Islam MS, Olivieri F, Procopio AD. 2016. Bioinformatic tools for microRNA dissection. *Nucleic Acids Research*. 44(1):24–44.

Alexandrov N, Tai S, Wang W, Mansueto L, Palis K, Fuentes RR, Ulat VJ, Chebotarov D, Zhang G, Li Z, Mauleon R. 2015. SNP-Seek database of SNPs derived from 3000 rice genomes. *Nucleic Acids Research*. 43(D1):D1023–7.

Altschul SF, Gish W, Miller W, Myers EW, Lipman DJ. 1990. Basic local alignment search tool. *Journal of Molecular Biology*. 215(3):403–10.

Amaral ML, Erikson GA, Shokhirev MN. 2018. BART: Bioinformatics array research tool. *BMC Bioinformatics*. 19(1):1–6.

Amarasinghe SL, Ritchie ME, Gouil Q, 2021. Long-read-tools. org: An interactive catalogue of analysis methods for long-read sequencing data. *GigaScience*. 10(2): 1–7.

Amarasinghe SL, Su S, Dong X, Zappia L, Ritchie ME, Gouil Q. 2020. Opportunities and challenges in long-read sequencing data analysis. *Genome Biology*. 21(1): 1–16.

Andrews S. 2010. FastQC: A quality control tool for high throughput sequence data. https://www.bioinformatics.babraham.ac.uk/projects/fastqc

Bae S, Park J, Kim JS. 2014. Cas-OFFinder: A fast and versatile algorithm that searches for potential off-target sites of Cas9 RNA-guided endonucleases. *Bioinformatics*. 30(10):1473–5.

Bailey TL, Boden M, Buske FA, Frith M, Grant CE, Clementi L, Ren J, Li WW, Noble WS. 2009. MEME SUITE: Tools for motif discovery and searching. *Nucleic Acids Research*. 37(suppl_2):W202–8.

Barrett T, Wilhite SE, Ledoux P, Evangelista C, Kim IF, Tomashevsky M, Marshall KA, Phillippy KH, Sherman PM, Holko M, Yefanov A. 2012. NCBI GEO: Archive for functional genomics data sets-update. *Nucleic Acids Research*. 41(D1):D991–5.

Benson DA, Karsch-Mizrachi I, Clark K, Lipman DJ, Ostell J, Sayers EW. 2012. GenBank. *Nucleic Acids Res*. 40:D48–53

Berlin K, Koren S, Chin CS, Drake JP, Landolin JM, Phillippy AM. 2015. Assembling large genomes with single-molecule sequencing and locality-sensitive hashing. *Nature Biotechnology*. 33(6):623–30.

Berman HM, Westbrook J, Feng Z, Gilliland G, Bhat TN, Weissig H, Shindyalo/v IN, Bourne PE. 2000. The protein data bank. *Nucleic Acids Research*. 28(1):235–42.

Bhagwat M, Young L, Robison RR. 2012. Using BLAT to find sequence similarity in closely related genomes. *Current Protocols in Bioinformatics*. 37(1):10–8.

Bhat B, Ganai NA, Andrabi SM, Shah RA, Singh A. 2017. TM-Aligner: Multiple sequence alignment tool for transmembrane proteins with reduced time and improved accuracy. *Scientific Reports*. 7(1):12543.

Bikard D, Jiang W, Samai P, Hochschild A, Zhang F, Marraffini LA. 2013. Programmable repression and activation of bacterial gene expression using an engineered CRISPR-Cas system. *Nucleic Acids Research*. 41(15):7429–37.

Billon P, Bryant EE, Joseph SA, Nambiar TS, Hayward SB, Rothstein R, Ciccia A. 2017. CRISPR-mediated base editing enables efficient disruption of eukaryotic genes through induction of STOP codons. *Molecular Cell*. 67(6):1068–79.

Blum M, Chang HY, Chuguransky S, Grego T, Kandasaamy S, Mitchell A, Nuka G, Paysan-Lafosse T, Qureshi M, Raj S, Richardson L. 2021. The InterPro protein families and domains database: 20 years on. *Nucleic Acids Research*. 49(D1):D344–54.

Bolser D, Staines DM, Pritchard E, Kersey P. 2016. Ensembl plants: Integrating tools for visualizing, mining, and analyzing plant genomics data. In: Edwards, D. (ed) *Plant Bioinformatics. Methods in Molecular Biology*, Vol 1374. Humana Press, New York. https://doi.org/10.1007/978-1-4939-3167-5_6

Boutet E, Lieberherr D, Tognolli M, Schneider M, Bairoch A. 2007. UniProtKB/Swiss-Prot. In: Edwards, D. (ed) *Plant Bioinformatics. Methods in Molecular Biology*™, Vol 406. Humana Press. https://doi.org/10.1007/978-1-59745-535-0_4

Brazelton Jr VA, Zarecor S, Wright DA, Wang Y, Liu J, Chen K, Yang B, Lawrence-Dill CJ. 2015. A quick guide to CRISPR sgRNA design tools. *GM Crops & Food*. 6(4):266–76.

Cancellieri S, Canver MC, Bombieri N, Giugno R, Pinello L. 2020. CRISPRitz: Rapid, high-throughput and variant-aware in silico off-target site identification for CRISPR genome editing. *Bioinformatics*. 36(7):2001–8.

Carbon S, Ireland A, Mungall CJ, Shu S, Marshall B, Lewis S, AmiGO Hub, Web Presence Working Group. 2009. AmiGO: Online access to ontology and annotation data. *Bioinformatics*. 25(2):288–9.

Chen N. 2004. Using repeat masker to identify repetitive elements in genomic sequences. *Current Protocols in Bioinformatics*. 5(1):4–10.

Chen S, Zhou Y, Chen Y, Gu J. 2018. fastp: An ultra-fast all-in-one FASTQ preprocessor. *Bioinformatics*, 34(17): 884–890.

Cochrane G, Karsch-Mizrachi I, Takagi T, Sequence Database Collaboration IN. 2016. The international nucleotide sequence database collaboration. *Nucleic Acids Research*. 44(D1):D48–50.

Cooper L, Meier A, Elser J, Preece J, Xu X, Kitchen R, Qu B, Zhang E, Todorovic S, Jaiswal P, Laporte MA. 2016. The Planteome project. In P. Jaiswal (Ed.), *Proceedings of the Joint International Conference on Biological Ontology and BioCreative* (pp. 2016). Corvallis, OR.

Dash S, Van Hemert J, Hong L, Wise RP, Dickerson JA. 2012. PLEXdb: Gene expression resources for plants and plant pathogens. *Nucleic Acids Research*. 40(D1):D1194–201.

Dong Q, Schlueter SD, Brendel V. 2004. PlantGDB, plant genome database and analysis tools. *Nucleic Acids Research*. 32(suppl_1):D354–9.

Doyle EL, Booher NJ, Standage DS, Voytas DF, Brendel VP, VanDyk JK, Bogdanove AJ. 2012. TAL Effector-Nucleotide Targeter (TALE-NT) 2.0: Tools for TAL effector design and target prediction. *Nucleic Acids Research*. 40(W1):W117–22.

Dumas J, Gargano MA, Dancik GM. 2016. shinyGEO: A web-based application for analyzing gene expression omnibus datasets. *Bioinformatics*. 32(23):3679–81.

Eden E, Navon R, Steinfeld I, Lipson D, Yakhini Z. 2009. GOrilla: A tool for discovery and visualization of enriched GO terms in ranked gene lists. *BMC Bioinformatics*. 10(1):1–7.

Edgar R, Domrachev M, Lash AE. 2002. Gene Expression Omnibus: NCBI gene expression and hybridization array data repository. *Nucleic Acids Research*. 30(1):207–10.

Edgar RC. 2004. MUSCLE: Multiple sequence alignment with high accuracy and high throughput. *Nucleic Acids Research*. 32(5):1792–7.

English AC, Richards S, Han Y, Wang M, Vee V, Qu J, Qin X, Muzny DM, Reid JG, Worley KC, Gibbs RA. 2012. Mind the gap: Upgrading genomes with Pacific Biosciences RS long-read sequencing technology. *PLoS One*. 7(11):e47768.

Falcon B, Zhang W, Murzin AG, Murshudov G, Garringer HJ, Vidal R, Crowther RA, Ghetti B, Scheres SH, Goedert M. 2018. Structures of filaments from Pick's disease reveal a novel tau protein fold. *Nature*. 561(7721):137–40.

Frazer KA, Pachter L, Poliakov A, Rubin EM, Dubchak I. 2004. VISTA: Computational tools for comparative genomics. *Nucleic Acids Research*. 32(suppl_2):W273–9.

Garrison E, Marth G. 2012. Haplotype-based variant detection from short-read sequencing. arXiv, 1207: 3907.

Gibbs RA, Weinstock GM. 2003. Evolving methods for the assembly of large genomes. In *Cold Spring Harbor Symposia on Quantitative Biology*. 68: 189–94.

Goel M, Sun H, Jiao WB, Schneeberger K. 2019. SyRI: Finding genomic rearrangements and local sequence differences from whole-genome assemblies. *Genome Biology*. 20(1):1–3.

He W, Yang J, Jing Y, Xu L, Yu K, Fang X. 2023. NGenomeSyn: An easy-to-use and flexible tool for publication-ready visualization of syntenic relationships across multiple genomes. *Bioinformatics*. 39(3):btad121.

He X, Wang J. 2007. BGI-RIS V2. In: Edwards, D. (ed) *Plant Bioinformatics. Methods in Molecular Biology™*, vol 406. Humana Press. https://doi.org/10.1007/978-1-59745-535-0_13

Heather JM, Chain B. 2016. The sequence of sequencers: The history of sequencing DNA. *Genomics*. 107(1): 1–8

Heigwer F, Kerr G, Walther N, Glaeser K, Pelz O, Breinig M, Boutros M. 2013. E-TALEN: A web tool to design TALENs for genome engineering. *Nucleic Acids Research.* 41(20):e190.

Higgs DR, Wood WG. 2008. Genetic complexity in sickle cell disease. *Proceedings of the National Academy of Sciences of the United States of America.* 105(33):11595–6.

Huang DW, Sherman BT, Lempicki RA. 2009. Systematic and integrative analysis of large gene lists using DAVID bioinformatics resources. *Nature Protocols*. 4(1):44–57.

Jayaswal V, Wong TK, Robinson J, Poladian L, Jermiin LS. 2014. Mixture models of nucleotide sequence evolution that account for heterogeneity in the substitution process across sites and across lineages. *Systematic Biology*. 63(5):726–42.

Jones P, Binns D, Chang HY, Fraser M, Li W, McAnulla C, McWilliam H, Maslen J, Mitchell A, Nuka G, Pesseat S. 2014. InterProScan 5: Genome-scale protein function classification. *Bioinformatics*. 30(9):1236–40.

Jumper J, Evans R, Pritzel A, Green T, Figurnov M, Ronneberger O, Tunyasuvunakool K, Bates R, Žídek A, Potapenko A, Bridgland A. 2021. Highly accurate protein structure prediction with AlphaFold. *Nature*. 596(7873):583–9.

Källberg M, Wang H, Wang S, Peng J, Wang Z, Lu H, Xu J. 2012. Template-based protein structure modeling using the RaptorX web server. *Nature Protocols*. 7(8):1511–22.

Karplus M, McCammon JA. 2002. Molecular dynamics simulations of biomolecules. *Nature Structural Biology*. 9(9):646–52.

Kathiresan N, Temanni R, Almabrazi H, Syed N, Jithesh PV, Al-Ali R. 2017. Accelerating next generation sequencing data analysis with system level optimizations. *Scientific Reports*. 7(1): 1–11.

Katoh K, Standley DM. 2013. MAFFT multiple sequence alignment software version 7: Improvements in performance and usability. *Molecular Biology and Evolution*. 30(4):772–80.

Kearse M, Moir R, Wilson A, Stones-Havas S, Cheung M, Sturrock S, Buxton S, Cooper A, Markowitz S, Duran C, Thierer T. 2012. Geneious basic: An integrated and extendable desktop software platform for the organization and analysis of sequence data. *Bioinformatics*. 28(12):1647–9.

Kelley LA, Sternberg MJ. 2009. Protein structure prediction on the Web: A case study using the Phyre server. *Nature Protocols*. 4(3):363–71.

Khan AW, Garg V, Roorkiwal M, Golicz AA, Edwards D, Varshney RK. 2020. Super-pangenome by integrating the wild side of a species for accelerated crop improvement. *Trends in Plant Science*. 25(2): 148–58.

Koboldt DC, Larson DE, Wilson RK. 2013. Using VarScan 2 for germline variant calling and somatic mutation detection. *Current Protocols in Bioinformatics*. 44(1): 15–4

Langmead B, Salzberg SL. 2012. Fast gapped-read alignment with Bowtie 2. *Nature Methods*. 9(4): 357–359.

Larkin MA, Blackshields G, Brown NP, Chenna R, McGettigan PA, McWilliam H, Valentin F, Wallace IM, Wilm A, Lopez R, Thompson JD, Gibson TJ, Higgins DG. 2007. Clustal W and Clustal X version 2.0. *Bioinformatics*. 23(21):2947–8.

Lassmann, T. 2020. Kalign 3: Multiple sequence alignment of large datasets. *Bioinformatics*. 36(6):1928–9.

Leaver-Fay A, Tyka M, Lewis SM, Lange OF, Thompson J, Jacak R, Kaufman KW, Renfrew PD, Smith CA, Sheffler W, Davis IW. 2011. ROSETTA3: An object-oriented software suite for the simulation and design of macromolecules. *Methods in Enzymology*, 487: pp. 545–574.

Lee WP, Stromberg MP, Ward A, Stewart C, Garrison EP, Marth GT. 2014. MOSAIK: A hash-based algorithm for accurate next-generation sequencing short-read mapping. *PLoS One*. 9(3): e90581

Li H, Durbin R. 2009. Fast and accurate short read alignment with Burrows-Wheeler transform. *Bioinformatics*. 25(14): 1754–60.

Li H, Handsaker B, Wysoker A, Fennell T, Ruan J, Homer N, Marth G, Abecasis G, Durbin R. 2009. The sequence alignment/map format and SAMtools. *Bioinformatics*. 25(16): 2078–79.

Li H, Yang Y, Hong W, Huang M, Wu M, Zhao X. 2020. Applications of genome editing technology in the targeted therapy of human diseases: Mechanisms, advances and prospects. *Signal Transduction and Targeted Therapy*. 5(1):1.

Liu H, Wang X. 2021. CRISPR-ERA: A webserver for guide RNA design of gene editing and regulation. *Computational Methods in Synthetic Biology*. 2189:65–9.

Liu Y, Popp B, Schmidt B. 2014. CUSHAW3: Sensitive and accurate base-space and color-space short-read alignment with hybrid seeding. *PLoS One*. 9(1): e86869

Liu Y, Schmidt B, Maskell DL. 2010. MSAProbs: Multiple sequence alignment based on pair hidden Markov models and partition function posterior probabilities. *Bioinformatics*. 26(16):1958–64.

Lovell JT, Sreedasyam A, Schranz ME, Wilson M, Carlson JW, Harkess A, Emms D, Goodstein DM, Schmutz J. 2022. GENESPACE tracks regions of interest and gene copy number variation across multiple genomes. *Elife*. 11:e78526.

Mandell JG, Barbas CF. 2006. Zinc finger tools: Custom DNA-binding domains for transcription factors and nucleases. *Nucleic Acids Research*. 34(suppl_2):W516–23.

Mansueto L, Fuentes RR, Chebotarov D, Borja FN, Detras J, Abriol-Santos JM, Palis K, Poliakov A, Dubchak I, Solovyev V, Hamilton RS. 2016. SNP-Seek II: A resource for allele mining and analysis of big genomic data in *Oryza sativa*. *Current Plant Biology*. 7:16–25.

Mashima J, Kodama Y, Kosuge T, Fujisawa T, Katayama T, Nagasaki H, Okuda Y, Kaminuma E, Ogasawara O, Okubo K, Nakamura Y. 2016. DNA data bank of Japan (DDBJ) progress report. *Nucleic Acids Research*. 44(D1):D51–7.

Mignone F, Grillo G, Liuni S, Pesole G. 2003. Computational identification of protein coding potential of conserved sequence tags through cross-species evolutionary analysis. *Nucleic Acids Research*. 31(15):4639–45.

Mirarab, B., Nguyen, N., Warnow, T. 2020. PASTA: Ultra-large multiple sequence alignment in the julia programming language. *Bioinformatics*, 6(8):2549–2551.

Misof B, Liu S, Meusemann K, Peters RS, Donath A, Mayer C, Frandsen PB, Ware J, Flouri T, Beutel RG, Niehuis O. 2014. Phylogenomics resolves the timing and pattern of insect evolution. *Science*. 346(6210):763–7.

Mohanta TK, Bashir T, Hashem A, Abd_Allah EF, Bae H. 2017. Genome editing tools in plants. *Genes*. 8(12):399.

Montague TG, Cruz JM, Gagnon JA, Church GM, Valen E. 2014. CHOPCHOP: A CRISPR/Cas9 and TALEN web tool for genome editing. *Nucleic Acids Research*. 42(W1):W401–7.

Moon SB, Kim DY, Ko JH, Kim YS. 2019. Recent advances in the CRISPR genome editing tool set. *Experimental & Molecular Medicine*. 51(11):1–11.

Naeem M, Alkhnbashi OS. 2023. Current bioinformatics tools to optimize CRISPR/Cas9 experiments to reduce off-target effects. *International Journal of Molecular Sciences*. 24(7):6261.

Naito Y, Hino K, Bono H, Ui-Tei K. 2015. CRISPRdirect: Software for designing CRISPR/Cas guide RNA with reduced off-target sites. *Bioinformatics*. 31(7):1120–3.

Neff KL, Argue DP, Ma AC, Lee HB, Clark KJ, Ekker SC. 2013. Mojo Hand, a TALEN design tool for genome editing applications. *BMC Bioinformatics*. 14(1):1–7.

Notredame C, Higgins DG, Heringa J. 2000. T-Coffee: A novel method for fast and accurate multiple sequence alignment. *Journal of Molecular Biology*. 302(1):205–17.

O'Brien A, Bailey TL. 2014. GT-Scan: Identifying unique genomic targets. *Bioinformatics*. 30(18):2673–5.

O'Fallon BD, Wooderchak-Donahue W, Crockett DK. 2013. A support vector machine for identification of single-nucleotide polymorphisms from next-generation sequencing data. *Bioinformatics*, 29(11): 1361–1366

Ohyanagi H, Tanaka T, Sakai H, Shigemoto Y, Yamaguchi K, Habara T, Fujii Y, Antonio BA, Nagamura Y, Imanishi T, Ikeo K. 2006. The Rice Annotation Project Database (RAP-DB): Hub for *Oryza sativa* ssp. *japonica* genome information. *Nucleic Acids Research*. 34(suppl_1):D741–4.

Pan X, Stein L, Brendel V. 2005. SynBrowse: A synteny browser for comparative sequence analysis. *Bioinformatics*. 21(17):3461–8.

Park J, Bae S, Kim JS. 2015. Cas-Designer: A web-based tool for choice of CRISPR-Cas9 target sites. *Bioinformatics*. 31(24):4014–6.

Parkinson H, Sarkans U, Shojatalab M, Abeygunawardena N, Contrino S, Coulson R, Farne A, Garcia Lara G, Holloway E, Kapushesky M, Lilja P. 2005. ArrayExpress-a public repository for microarray gene expression data at the EBI. *Nucleic Acids Research*. 33(suppl_1):D553–5.

Pearson WR. 2013. An introduction to sequence similarity ("homology") searching. *Current Protocols in Bioinformatics*. 42(1):3–1.

Pei J, Grishin NV. 2014. PROMALS3D: Multiple protein sequence alignment enhanced with evolutionary and three-dimensional structural information. In: Russell, D. (ed) *Multiple Sequence Alignment Methods. Methods in Molecular Biology*, vol 1079. Humana Press, Totowa, NJ. https://doi.org/10.1007/978-1-62703-646-7_17

Pei, I., Sadreyev, L. D. and Grishin, N. V. 2009. PCMA: Fast and accurate multiple sequence alignment based on probabilistic consistency. *Bioinformatics*, 25(9):1260–1266.

Petryszak R, Keays M, Tang YA, Fonseca NA, Barrera E, Burdett T, Füllgrabe A, Fuentes AM, Jupp S, Koskinen S, Mannion O. 2016. Expression Atlas update-an integrated database of gene and protein expression in humans, animals and plants. *Nucleic Acids Research*. 44(D1):D746–52.

Pliatsika V, Rigoutsos I. 2015. "Off-Spotter": Very fast and exhaustive enumeration of genomic lookalikes for designing CRISPR/Cas guide RNAs. *Biology Direct.* 10:1–10.

Pomaznoy M, Ha B, Peters B. 2018. GOnet: A tool for interactive gene ontology analysis. *BMC Bioinformatics.* 19(1):1–8.

Pop M, Phillippy A, Delcher AL, Salzberg SL. 2004. Comparative genome assembly. *Briefings in Bioinformatics.* 5(3):237–48.

Poplin R, Chang PC, Alexander D, Schwartz S, Colthurst T, Ku A, Newburger D, Dijamco J, Nguyen N, Afshar PT, Gross SS. 2018. A universal SNP and small-indel variant caller using deep neural networks. *Nature Biotechnology*, 36(10): 983–987

Prykhozhij SV, Rajan V, Gaston D, Berman JN. 2015. CRISPR multitargeter: A web tool to find common and unique CRISPR single guide RNA targets in a set of similar sequences. *PLoS One.* 10(3):e0119372.

Ramšak Ž, Baebler Š, Rotter A, Korbar M, Mozetič I, Usadel B, Gruden K. 2014. GoMapMan: Integration, consolidation and visualization of plant gene annotations within the MapMan ontology. *Nucleic Acids Research.* 42(D1):D1167–75.

Ran FA, Hsu PD, Lin CY, Gootenberg JS, Konermann S, Trevino AE, Scott DA, Inoue A, Matoba S, Zhang Y, Zhang F. 2013. Double nicking by RNA-guided CRISPR Cas9 for enhanced genome editing specificity. Cell. 154(6):1380–9.

Reyon D, Kirkpatrick JR, Sander JD, Zhang F, Voytas DF, Joung JK, Dobbs D, Coffman CR. 2011. ZFNGenome: A comprehensive resource for locating zinc finger nuclease target sites in model organisms. *BMC Genomics.* 12:1–9.

Rice P, Longden I, Bleasby A. 2000. EMBOSS: The European molecular biology open software suite. *Trends in Genetics.* 16(6):276–7.

Rigden DJ, Fernández-Suárez XM, Galperin MY. 2016. The 2016 database issue of nucleic acids research and an updated molecular biology database collection. *Nucleic Acids Research.* 44(D1):D1–6.

Rivera-Rivera CJ, Montoya-Burgos JI. 2016. LS³: A method for improving phylogenomic inferences when evolutionary rates are heterogeneous among taxa. *Molecular Biology and Evolution.* 33(6):1625–34.

Rohl CA, Strauss CE, Misura KM, Baker D. 2004. Protein structure prediction using Rosetta. *Methods in Enzymology.* 383: 66–93.

Saal LH, Troein C, Vallon-Christersson J, Gruvberger S, Borg Å, Peterson C. 2002. BioArray Software Environment (BASE): A platform for comprehensive management and analysis of microarray data. *Genome Biology.* 3:1–6.

Šali A, Blundell TL. 1993. Comparative protein modelling by satisfaction of spatial restraints. *Journal of Molecular Biology.* 234(3):779–815.

Sander JD, Maeder ML, Reyon D, Voytas DF, Joung JK, Dobbs D. 2010. ZiFiT (Zinc Finger Targeter): An updated zinc finger engineering tool. *Nucleic Acids Research.* 38(suppl_2):W462–8.

Sandmann S, Karimi M, de Graaf AO, Rohde C, Gollner S, Varghese J, Ernsting J, Walldin G, van der Reijden BA, Müller-Tidow C, Malcovati L. 2018. appreci8: A pipeline for precise variant calling integrating 8 tools. *Bioinformatics*, 34(24): 4205–12.

Senior AW, Evans R, Jumper J, Kirkpatrick J, Sifre L, Green T, Qin C, Žídek A, Nelson AW, Bridgland A, Penedones H. 2020. Improved protein structure prediction using potentials from deep learning. *Nature.* 577(7792):706–10.

Sherman BT, Hao M, Qiu J, Jiao X, Baseler MW, Lane HC, Imamichi T, Chang W. 2022. DAVID: A web server for functional enrichment analysis and functional annotation of gene lists (2021 update). *Nucleic Acids Research.* 50(W1):W216–21.

Shi J, Liang C. 2019. Generic repeat finder: A high-sensitivity tool for genome-wide de novo repeat detection. *Plant Physiology.* 180(4):1803–15.

Sievers F, Higgins DG. 2018. Clustal Omega for making accurate alignments of many protein sequences. *Protein Science.* 27(1):135–45.

Smargon AA, Cox DB, Pyzocha NK, Zheng K, Slaymaker IM, Gootenberg JS, Abudayyeh OA, Essletzbichler P, Shmakov S, Makarova KS, Koonin EV. 2017. Cas13b is a type VI-B CRISPR-associated RNA-guided RNase differentially regulated by accessory proteins Csx27 and Csx28. *Molecular Cell.* 65(4):618–30.

Söding J, Biegert A, Lupas AN. 2005. The HHpred interactive server for protein homology detection and structure prediction. *Nucleic Acids Research.* 33(suppl_2):W244–8.

Steinegger M, Söding J. 2017. MMseqs2 enables sensitive protein sequence searching for the analysis of massive data sets. *Nature Biotechnology.* 35(11):1026–8.

Stemmer M, Thumberger T, del Sol Keyer M, Wittbrodt J, Mateo JL. 2015. CCTop: An intuitive, flexible and reliable CRISPR/Cas9 target prediction tool. *PLoS One.* 10(4):e0124633.

Sun Y, Shang L, Zhu QH, Fan L, Guo L. 2021. Twenty years of plant genome sequencing: Achievements and challenges. *Trends in Plant Science*, 27(4): 391–401.

Tello-Ruiz MK, Stein J, Wei S, Preece J, Olson A, Naithani S, Amarasinghe V, Dharmawardhana P, Jiao Y, Mulvaney J, Kumari S. 2016. Gramene 2016: Comparative plant genomics and pathway resources. *Nucleic Acids Research*. 44(D1):D1133–40.

Thorvaldsdottir H, Robinson JT, Mesirov JP. 2013. Integrative Genomics Viewer (IGV): High-performance genomics data visualization and exploration. *Briefings in Bioinformatics*, 14(2): 178–92.

Tom N, Tom O, Malcikova J, Pavlova S, Kubesova B, Rausch T, Kolarik M, Benes V, Bystry V, Pospisilova S. 2018. ToTem: A tool for variant calling pipeline optimization. *BMC Bioinformatics*, 19(1): 1–9.

Van der Spoel D, van Maaren PJ, Caleman C. 2012. GROMACS molecule & liquid database. *Bioinformatics*. 28(5):752–3.

Varshney RK, Bohra A, Yu J, Graner A, Zhang Q, Sorrells ME. 2021. Designing future crops: Genomics-assisted breeding comes of age. *Trends in Plant Science*, 26(6): 631–49.

Waterhouse A, Bertoni M, Bienert S, Studer G, Tauriello G, Gumienny R, Heer FT, de Beer TA, Rempfer C, Bordoli L, Lepore R. 2018. SWISS-MODEL: Homology modelling of protein structures and complexes. *Nucleic Acids Research*. 46(W1):W296–303.

Wee Y, Bhyan SB, Liu Y, Lu J, Li X, Zhao M. 2019. The bioinformatics tools for the genome assembly and analysis based on third-generation sequencing. *Briefings in Functional Genomics*. 18(1):1–2.

Wei Q, Khan IK, Ding Z, Yerneni S, Kihara D. 2017. NaviGO: Interactive tool for visualization and functional similarity and coherence analysis with gene ontology. *BMC Bioinformatics*. 18:1–3.

Wilding M, Peat TS, Kalyaanamoorthy S, Newman J, Scott C, Jermiin LS. 2017. Reverse engineering: Transaminase biocatalyst development using ancestral sequence reconstruction. *Green Chemistry*. 19(22):5375–80.

Wong TK, Kalyaanamoorthy S, Meusemann K, Yeates DK, Misof B, Jermiin LS. 2020. A minimum reporting standard for multiple sequence alignments. *NAR Genomics and Bioinformatics*. 2(2):lqaa024.

Wrapp D, Wang N, Corbett KS, Goldsmith JA, Hsieh CL, Abiona O, Graham BS, McLellan JS. 2020. Cryo-EM structure of the 2019-nCoV spike in the prefusion conformation. *Science*. 367(6483):1260–3.

Xiao A, Cheng Z, Kong L, Zhu Z, Lin S, Gao G, Zhang B. 2014. CasOT: A genome-wide Cas9/gRNA off-target searching tool. *Bioinformatics*. 30(8):1180–2.

Xu D, Zhang Y. 2012. Ab initio protein structure assembly using continuous structure fragments and optimized knowledge-based force field. *Proteins: Structure, Function, and Bioinformatics*. 80(7):1715–35.

Zhang B, Horvath S. 2005. A general framework for weighted gene co-expression network analysis. *Statistical Applications in Genetics and Molecular Biology*. 4(1). https://doi.org/10.2202/1544-6115.1128

Zhang F, Xue H, Dong X, Li M, Zheng X, Li Z, Xu J, Wang W, Wei C. 2022. Long-read sequencing of 111 rice genomes reveals significantly larger pan-genomes. *Genome Research*. 32(5): 853–63.

Zhang Y, Sun B, Feng D, Hu H, Chu M, Qu Q, Tarrasch JT, Li S, Sun Kobilka T, Kobilka BK, Skiniotis G. 2017. Cryo-EM structure of the activated GLP-1 receptor in complex with a G protein. *Nature*. 546(7657):248–53.

Zhang Y. 2008. I-TASSER server for protein 3D structure prediction. *BMC Bioinformatics*. 9:1–8.

Zhou Y, Chebotarov D, Kudrna D, Llaca V, Lee S, Rajasekar S, Mohammed N, Al-Bader N, Sobel-Sorenson C, Parakkal P, Arbelaez LJ. 2020. A platinum standard pan-genome resource that represents the population structure of Asian rice. *Scientific Data*, 7(1): 1–11.

Zong Y, Wang Y, Li C, Zhang R, Chen K, Ran Y, Qiu JL, Wang D, Gao C. 2017. Precise base editing in rice, wheat and maize with a Cas9-cytidine deaminase fusion. *Nature Biotechnology*. 35(5):438–40.

15 Gene Editing Using CRISPR/Cas9 System
Methods and Applications

Gopika, Boro Arthi, Arumugam Vijaya Anand, Natchiappan Senthilkumar, Senthil Kalaiselvi, and Santhanu Krishnapriya

LIST OF ABBREVIATIONS

BE	Base editing
CAS	CRISPR-associated genes
CBE	Cytosine base editors
CRISPR	Clustered regularly interspaced short palindromic repeats-associated nuclease
DSB	double-strand breaks
gRNA	guide RNA
NASEM	National Academics of Science, Engineering and Medicine
NBT	New breeding techniques
PAM	Protospacer adjacent motif
PASTE	Programmable addition via site-specific targeting elements
SNV	Single nucleotide variants
TALEN	Transcription activator-like effectors nucleases
ZFN	Zinc finger nucleases

15.1 INTRODUCTION

Gene editing is one of the revolutionary steps forward in the field of biological science developed in the past decade, the development of this technique has empowered target manipulation of the genes by site-specific modification by deletion, substitution, and addition of the gene sequence (Figure 15.1) (Dasgupta et al., 2021). Editing of a gene using genetic engineering technique is of great interest for its array in multiple scientific domains like the development of plants that are drought resistant and others, modification of the pluripotent stem cells, and generation of genetically modified organisms (Loureiro and da Silva, 2019). In the last few decades, revolutions have been found in the technologies related to gene editing in genetically modified organisms, this technology is been considered one of the important technologies and is one of the rational production systems in the agricultural field. One of the gene editing techniques is 'new breeding techniques' (NBT) has emerged as a technique in which the DNA molecule is inserted modified, replaced, and deleted in the predetermined location in the genome of the organism (Kang et al., 2022). This technology is also one of the emerging therapeutic procedures by modifying the genome in the eukaryotes by targeting specific nuclease that expedites the proper sequence in the genes by precise strategies such as engineering and delivery of gene editing nuclease by different systems that include zinc finger nuclease (ZFN), transcription activator-like effector-nuclease (TALEN), clustered regularly interspaced short palindromic

 DOI: 10.1201/9781003387060-15

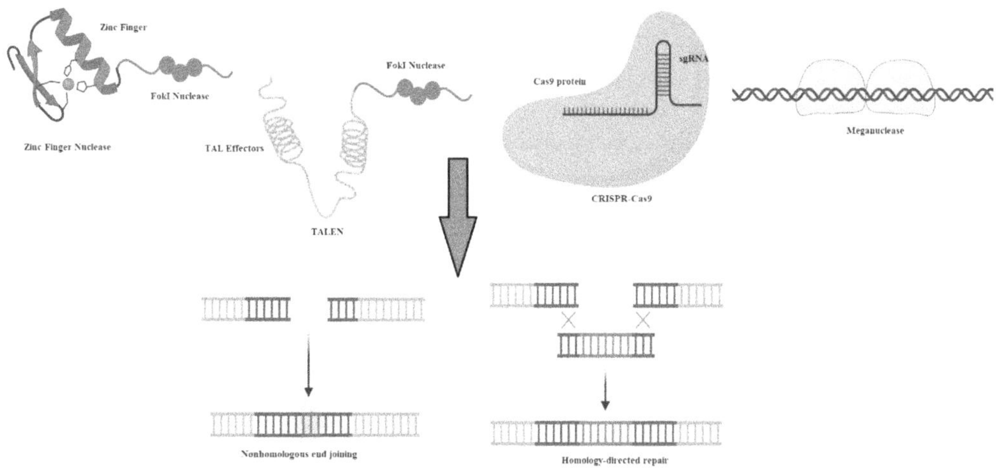

FIGURE 15.1 Genetic engineering of a DNA sequence. Engineering of the genome sequence is carried out by various system that replaces a particular sequence of the genome by using the components and then replaces the sequence with a target sequence.

repeats-associated nuclease (CRISPR) (Shim et al., 2017). These technologies provide opportunities to face challenges that are faced by humanity. The recent development of these technologies raises an expectation to control infectious diseases in livestock (Petersen et al., 2022).

The expansion of these technologies has driven the evolution of the biomedical world toward the progression in fundamental research like epigenetics, development of drugs, disease modeling, and diagnosis, and also in the treatment of various human diseases by the usage of therapeutic gene editing (Zhou et al., 2020). The improvement of these techniques in the field of research in human has increased the international interest to hold sway over these technologies by some of the organizations like the U.S National Academics of Science, Engineering and Medicine (NASEM) which are formed for the generation of reports along with other bodies that belong to professional societies and international commissions that call for neutralizing actions which also includes halting of hereditary gene editing, anticipatory public engagement, and pathways for responsible clinical translation (Waltz et al., 2021). The technique of gene editing is the potential to cure treat and prevent a wide array of acquired and inherited diseases, in which the genome can be modified in precise locations. The clinical trial of these techniques has advanced in multiple diseases till now (Ates et al., 2020).

15.2 GENE EDITING TECHNOLOGY

15.2.1 Zinc Finger Nuclease (ZFN)

ZFNs are reagent that cleaves DNA and is adopted as one of the gene targeting tools. In recent times, it has proven to be one of the versatile and efficient gene targeting tools that have separate domains for DNA binding and domains for cleavage of DNA. It is a type of synthetic protein that was first observed by Chandrasegaran in which he observed that the *FokI* a natural type IIS restriction enzyme has two separate domains for activities of binding and cleavage. He also observed that the domain of cleavage does not have any sequence specificity (Carroll, 2011). The nucleases in this system are generated by fusing the zinc-finger DNA binding domain to any non-specific nuclease domain (Table 15.1); the designed ZFN can introduce double-strand breaks (DSB) in a target sequence and increase the gene targeting to 100-fold by stimulating DNA repair pathways in the targeted cells. The repair of the DSB by ZFN can be done by two DNA repair pathways which are nonhomologous end joining and homology-directed repair (Table 15.1) (Cathomen and Joung, 2008). In studies carried out in model organisms, the gene targeting agent ZFN is found to be effective in introducing mutation for genetic study and can be encouraged for their use in gene therapy. The first animal model that ZFN is using successfully is *Drosophila* melanogaster (Porteus and Carroll, 2005).

15.2.2 Transcription Activators like Effector-Nuclease (TALEN)

TALEN is one of the gene editing tools in place of ZFN as an alternative. It is an effective gene editing tool that can be easily and rapidly constructed and can be used to target any DNA sequence due to its flexible nature. It also uses the non-specific *FokI* domain as the domain for introducing targeted DSBs in the sequence as dimers but the DNA binding domain of TALEN (Table 15.1) contains tandem repeats that comprise 33–35 amino acids that recognize a single nucleotide (Sun and Zhao, 2013). This technology has been reported to induce pluripotent stem cells, plants, nematodes, and zebrafish but in rodent embryos, this technology is less effective a combined expression of Exouclease 1 and TALEN is found to have disrupted the albino (Tyr) gene in the zygotes of rat producing a knockout rat (Mashimo, 2014). TALEN is a cheaper, safer, and an efficient gene editing technology in targeting a specific region in the genome of the organisms. The TALE protein of TALEN contains three domains amino-terminal domain with a transporting signal, a DNA binding tool composed of repeats of 34 amino acids and a carboxyl-terminal domain with nuclear localization and transcription activation domain. This system has an advantage over ZFN as the DNA binding domain recognizes only one nucleotide in place of 3 nucleotides like ZFN (Khalil, 2020).

15.2.3 Base Editing (BE)

Base editing is a new genome editing technology. It is a technology in which the point mutations are installed in the cellular DNA or RNA without any introduction of DSBs using the components of the CRISPR system and other protein enzymes (Rees and Liu, 2018). This technology is the most advanced genome editing technique in which Single nucleotide variants (SNV) (Table 15.1) are introduced into a living cell's DNA or RNA (Porto et al., 2020). It is a novel technique that

TABLE 15.1
Different Genome Editing Technology

Genome Editing System	Components	Function
Zinc Finger Nuclease (ZFN)	Zinc-finger DNA binding domain and Non- specific nuclease domain	Insertion of DBSs and then repair by nonhomologous end joining and homology-directed repair
Transcription activator-like effector-nuclease (TALEN)	TALE and Non-specific *FokI* nuclease domain	Insertion DSBs and then repair by nonhomologous end joining and homology-directed repair
Base editing (BE)	The inactive domain of CRISPR-Cas9 and Adenosine or Cytosine deaminase domain	Introduction of single nucleotide variants
Programmable addition via site-specific targeting elements (PASTE)	Cas9-nickase, Reverse transcriptase and Serine integrase	Integration of DNA sequence at targeted sites
Homing endonuclease	Nuclease enzyme	Insertion of DSBs and then repair by nonhomologous end joining and homology-directed repair

enables the substitution of nucleotides in a manner without disrupting the gene, a base editor is a component in this technology which is a fusion of the inactive catalytic domain of CRISPR-Cas9 and an adenosine or cytosine deaminase domain (Table 15.1) that can convert one base to another. This technique can revert the change in a single base and minimize the deletion or duplication. The two base editors cytosine and adenine have emerged as precise tools for genome modification recently (Mishra et al., 2020). There are two types of base editors, DNA base editors induce mutations in the DNA and RNA base editors convert one nucleotide to another in the RNA sequence. Cytosine base editors (CBE) are the first to be developed which enable a transition from Cytosine to Guanine. It contains a Cas9 nickase, fused to cytidine deaminase and uracil glycosylase inhibitor (Molla et al., 2021).

15.2.4 Paste

Programmable addition via site-specific targeting elements also known as PASTE is a new approach that includes site-specific integrase with programmable CRISPR-based editing which makes the possibility of integrating large DNA sequences at defined target regions. This technique employs Cas9-nickase attached with reverse transcriptase and serine integrase (Table 15.1) for a protein complex that has the ability of 5%–50% precise insertions (Mahmood and Mansoor, 2023). This strategy is reported by Yarnall et al. It has emerged to be a two-step phenomenon that includes targeted insertion and integration into the desired template. It is proven to have higher integration efficiency (Awan et al., 2023). PASTE can edit similarly to DNA repair pathways such as homologous recombination repair and nonhomologous end joining (Table 15.1) in the non-dividing cells and *in vivo* and with capabilities of multiplex gene insertion without dependence on the pathways of DNA repairing (Yarnall et al., 2023).

15.2.5 Homing Endonuclease

Also known as meganuclease belongs to a nuclease family and is named due to the presence of conserved amino acids within the enzyme sequence that interacts with DNA. They make sequence-specific interactions with DNA (Gaj et al., 2016). These are DNA-cutting enzymes that are encoded by the homologous endonuclease genes which are found to be encoded in archaeal introns and are embedded in self-splicing elements for group I, II for RNA and inteins for encoded primary proteins sequence. These genes encode the endonuclease (Table 15.1) protein that promotes their and other genetic sequence (Hafez and Hausner, 2012). These are specific endonucleases that contain recognition sequence and cause DSBs (Table 15.1) which activates the recombination repair system in the cells that uses homologous chromosomes to introduce homologous endonuclease genes for repair (Windbichler et al., 2011).

15.3 CRISPR

CRISPR is one of the indispensable tools in research in the field of the biological world, is a short palindromic repeat element that was first initially noticed in the *Escherichia coli* (E coli) genome by Dr Nakata and group. These tandem repeats are separated by spacers which are non-repeating DNA sequences. CRISPR is present in 40% of the sequenced bacterial organisms and 90% of sequenced organisms belong to Achaea and are present adjacent to conserved genes called CRISPR-associated genes (CAS) (Adli, 2018). It was mentioned for the first time in the year 1987 by a scientist who was studying the activity of the IAP genes, and its intriguing function was brought to light after 20 years of its finding. It is one of the crucial elements in the defense system of bacterial organisms, by sequencing the spacer of CRISPR, its sequence was revealed to be homologous to sequences of bacteriophage and plasmid which led to hypothesized that CRISPR contains a defense mechanism of the bacterial system against external objects (Loureiro and da Silva, 2019).

With the development of this technology, the ability to edit genomes in both mammalian and other cell types is enhanced greatly. It is one of the most favored approaches, and with the rapid evolution of this technology, it is becoming a challenge for the user of this technology to keep up with the latest developments in this field to design and implement CRISPR-based experiments (Graham and Root, 2015). It is an acquired immune system in Achaea and bacteria that modifies the invading nucleotides and the system by which it operates including three phases that are adaptation or spacer integration, expression, and interference (Moon et al., 2019). In the CRISPR-Cas system, there are two proteins Cas1 and Cas2 that are present majorly in the modifying system that is required for inserting the spacers in the CRISPR cassettes and the complex of these two proteins represent the module of information processing in the CRISPR-Cas system. There are three major types of CRISPR–Cas system as Cas3 in type I system, Cas9 in type II, and Cas10 in type III system. Cas1 protein is the most conserved protein present in the CRISPR-Cas system and is also found to evolve slower than the other Cas proteins (Makarova et al., 2015). CRISPR along with the related Cas genes are found to be distributed widely in prokaryotes, in the *Lactobacillus brevis* strains a diversity of CRISPR-Cas was found. The CRISPR-Cas system found in this strain may have a role as active immunity against foreign invasive DNA, plasmids, and other mobile genetic elements (Panahi et al., 2022; Pei et al., 2021).

15.4 CRISPR/CAS

The CRISPR-Cas system is one of the favorable genome editing tools in the molecular biology field and is found to achieve many achievements in correcting pathogenic mutations, genes for immunotherapy in cancer, solving problems related to organ transplantation, and others (Xu and Li, 2020). This is an immune system that belongs to the archaea and bacteria and is encoded by operons that have diverse architecture and a high rate of evolution. This system mediates immunity based on three stages that are adaptation, expression, and interference. During the stage of adaptation short pieces of the viral or plasmid DNA will be integrated into the loci of CRISPR, which are approximately 30 base pair long viral-derived resistance-conferring spacer in the leader of the loci of CRISPR and the event of insertion, duplication of repeats is carried out which leads to the creation of newer space repeat unit. In the expression stage, the primary transcript of the CRISPR locus is generated and is processed to small cr-RNA which is catalyzed by the endoribonucleases. In the final interference stage, the foreign DNA or RNA are targeted and cleaved and crRNA guides the complex of Cas proteins to the target sequence which matches the spaces further cleavage will be carried out by a nuclease enzyme like *E coli* in which the cleavage is catalyzed by HD endonuclease domain in the Cas3 protein (Makarova et al., 2011). The CRISPR-Cas system consists of two classes Class I and Class II, Class I is a pol subunit effector complex and Class II is a mono-protein effector module and is again divided into six types each class containing three types (Zhou et al., 2021). Due to fast evolution and variability classifying this system is very intimidating and due to frequent recombination and the absence of a universal Cas gene a single classifying criterion cannot be practiced (Koonin et al., 2017). The Class I of the system consists of the diversified and common type I and type III and a comparatively rare type IV, the effector modules of this type are multiple Cas protein complexes and the backbone of these complexes consists of paralogous repeat-associated mysterious proteins and contain RNA recognition motif and C terminal glycine-rich loop of diagnostic sequence with large and small subunits. In comparison to that Class II has a simple organization with an effector domain that consists of single, large, multi-domain, and protein of multiple functions. This class contains the most abundant type II and rare types, type V and type VI (Makarova et al., 2018).

15.5 CRISPR/CAS9

The CRISPR Cas9 system is a defense mechanism that has been a powerful RNA-guided DNA targeting platform for processes such as editing of the genome, transcriptional perturbation, epigenetic modulation, and genome imaging which uses to manipulate the gene sequence by guide RNA (gRNA) which are short stretch RNA and their functions in development and progression of the disease, correction of mutation that causes disease, activation of the tumor suppressor and inactivation of the oncogenes (Jiang and Doudna, 2017). CRISPR-Cas9 is a powerful genome editing tool and has been explored in various fields for its high precision and efficiency. CRISPR-Cas9 a type II CRISPR/Cas system can cleave endogenous genomic sites in both human and mouse cells and then repair by either nonhomologous end joining or homology-directed repair (Figure 15.2) (Wang et al., 2022). Among all the CRISPR systems CRISPR-Cas9 is the routinely used gene editing system to successfully edit the genes in mammalian cells (Liu et al., 2017a). This system is a promising gene editing technology, to make it more effective various computational tools are been developed to assist in the development of gRNA that have specific cleavage efficiency and the ability for avoiding the non-target sequence (Konstantakos et al., 2022). The Cas9 of this system introduces DSBs in the target sequences which is complementary to a spacer sequence of 20 nucleotides in the gRNA bond to the enzyme, another requirement is a protospacer adjacent motif (PAM) present adjacent to the target sequence (Riesenberg et al., 2022). SpCas9 is the most widely used CRISPR protein in genome editing. This is the first CRISPR system that was investigated and was found to have high gene modification efficiency but also found to contain side effects like generating off-targets. FrCas9 a CRISPR-Cas9 system derived from *Faecalibaculum rodentium* is found to have the same efficiency and as effective as SpCas9 containing simple PAM and 21–22 bp gRNA.

The following CRISPR protein possesses double editing windows and was able to target the TATA box in the promoter region in eukaryotes in the TATA box-related diseases which establishes this protein to be safer for gene editing applications in different fields (Cui et al., 2022). The cleavage of foreign DNA in this technology is carried out by the components of the CRISPR-Cas9 system which are Cas9 and sgRNA, Cas9 is the DNA endonuclease that can be derived from various bacterial species from which this protein can be isolated like *Brecibacillus laterosporus*, *Staphylococcus aureus*, *Staphylococcus pyogenes*, and *Streptococcus thermophilus*. This protein is found to have

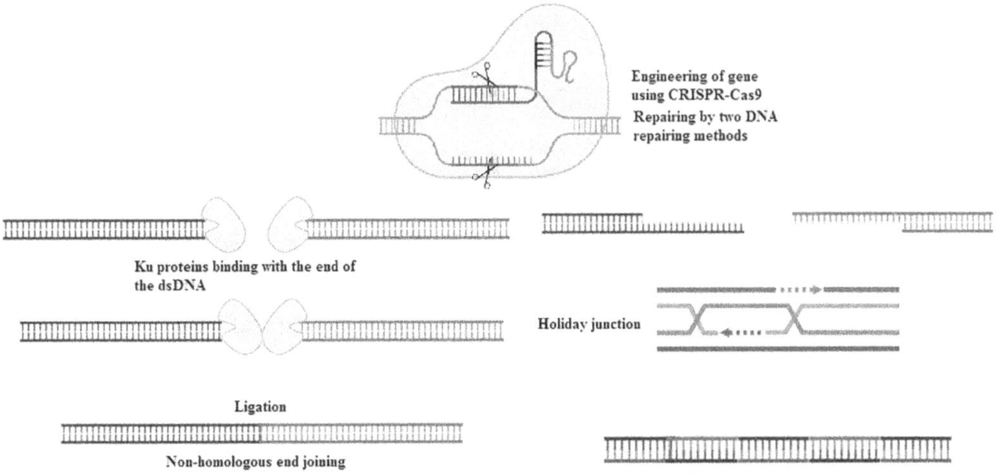

FIGURE 15.2 Genome engineering by CRISPR/Cas9 system. The process of genome engineering by the CRISPR/Cas9 system includes two component of the system one is the Cas9 proteins and other is sgRNA which binds at a sequence in the genome and then carries out the process of modification which is further repaired by nonhomologous end joining and homology-directed repair.

two domains HNH domain and a RucV-like domain. The former domain cuts the complementary strand of crRNA and the RucV-like domain cleaves the opposite strand of double-stranded DNA, sgRNA a synthetic RNA of 100 nucleotides identifies the target sequence and is found to contain a PAM sequence. It contains a loop structure in the 3′ end that attaches to the target sequence and forms a complex with Cas9 that produces double-stranded breaks in dsDNA (Liu et al., 2017b). There are mainly three strategies for engineering genes using CRISPR-Cas9. The first strategy is a straightforward method in which a plasmid-based CRISPR system is used and is the simplest and also avoids multiple time transfections and exhibits more stability than other strategies. The second strategy is carried out to deliver the complex of mRNA of Cas9 and sgRNA and the third is delivery of the complex of Cas9 proteins and sgRNA into the targeted cells (Liu et al., 2017a).

15.6 DELIVERY OF CRISPR-CAS9

The delivery of the CRISPR-Cas9 system can be carried out by both physical and non-physical or viral vector approaches. In physical approaches, the delivery of this system is approached by processes like electroporation, nanoparticles, or by microinjection (Table 15.2) in the targeted cells. Electroporation (Table 15.2) is the most used method to introduce proteins into mammalian cells. It is the most suitable method for delivery of the CRISPR system into the targeted cells for all types of CRISPR-Cas systems. In genome editing for organogenesis, this method is mostly used to introduce the plasmid-based CRISPR system into fish and other organisms for the generation of fish fins, embryonic cells axolotl regenerations, development of chicks, and the brain of mice. It is also used in the transferring of the system into cells like cancer cells, T cells, and stem cells for treatment (Liu et al., 2017b).

Nanocarriers are an ideal delivery system for introducing the CRISPR system into the targets. These include cationic lipid-based nanoparticles (LNP), lipid complexes, DNA nanoparticles (Table 15.2), etc. which are been used in *in vitro*. Lipid-based nanoparticles in the classic and the most extensively studied delivery system among all, plasmids or any mRNA delivered by these nanoparticles protect them from degradation by nuclease enzymes. DNA nanostructures are the recent emerging delivery system that has an advantage over others due to strong loading capacities, biocompatibility and biodegradability, these structures can be targeted to specific membrane receptors or receptor-mediated pathways. In some studies carried out the complex of Cas9/sgRNA/DNA nanomolecule is found to have a genomic editing efficiency of about 28% (Li et al., 2021). Another type of nano molecules that are used as carriers are cationic polymer nanoparticles (Table 15.2) which are found to have more chemical diversity and diversity in their structural designs and functional potential than lipid-based carriers. One example of a suitable polymer carrier is chitosan which forms a red fluorescent protein and delivers Cas9 protein with glutamate residues. Another group of carriers are gold nanoparticles (Table 15.2) which are a new type of delivery vehicle for CRISPR-Cas9 protein (Duan et al., 2021).

The most common delivery system for the CRISPR system is the microinjection which requires a micromanipulator and an inverted microscope for successful delivery of the CRISPR-Cas9

TABLE 15.2
Types of Delivery System for CRISPR-Cas System

Types of the Delivery System	Types
Physical method	Microinjection, Electroporations
Viral method	Adeno-associated virus, Adenovirus, and Lentivirus
Nonviral method	Lipid-based nanoparticles, lipid complexes, DNA nanoparticles, Gold nanoparticles, and polymer-based nanoparticles

(Xu et al., 2020). It is considered to be the gold standard for introducing CRISPR into the cells, in this method plasmid encoding both the Cas9 and sgRNA or the mRNA encoding the Cas9 and sgRNA are delivered to the target cells. This method is found to be the best method for *in vitro* and *ex vivo* works (Lino et al., 2018).

The viral approach to the delivery of the CRISPR system consists of various methods which include adeno-associated viruses, lentivirus vectors, and adenovirus vectors. Adeno-associated virus vectors (Table 15.2) are the common vectors for delivery; they contain some unique properties which has grasped the interest of using these vectors for delivery in *vivo* applications. But for the CRISPR system, the adeno-associated virus is useful in only *in vitro* applications for genome editing. These vectors are used in the knocking of genes by homology-directed repair pathways. Lentiviral vectors (Table 15.2) are another vectors that are derived from the single-stranded spherical RNA virus which possesses merits like mild immunogenicity, stable expression, high packaging capacity, and high efficiency for infection in both dividing and non-dividing cells (Cheng et al., 2021). These delivery vectors are used for the delivery of CRISPR-Cas9, these vectors have a generous capacity in cloning than the adeno-associated viral vectors and are also less laborious. Adenovirus vectors (Table 15.2) are the most laborious but are widely used for delivery of genes for clinical trials, these vectors can be used to transduce both dividing and non-dividing cells and the transfected sequence, does not integrate into the host genome. The effective delivery by the viral vectors for the CRISPR system needs to be more edited than other nonviral methods (Yip, 2020).

15.7 APPLICATION OF CRISPR-CAS9

15.7.1 Genetically Modified Organisms

CRISPR-Cas9 is a versatile tool which is been used to produce both knockin and knockout models of animals and also cell lines (Çiçek et al., 2019). The modification of the genome in the animal embryos by the CRISPR-Cas9 system includes three steps which include isolation of zygotes, delivery of the sgRNA and Cas9 into a zygote and then transfer of the modified embryo to the pseudo-pregnant mothers. This method is rapid, convenient, and efficient for a one-step process for producing transgenic organisms. Using this technology model chimeric mice were produced by editing the genome of embryonic stem cells to study human limb malformation by altering the genes by inducing deletion, inversion, and duplication (Jin and Li, 2016). Editing of the genome in the livestock in biotechnology produced livestock with modifications that increase milk production, meat production, and disease-free livestock. The use of CRISPR-Cas9 in livestock improves the reproductive traits and also makes disease resistant. The insertion of the gene NRAMP1 in the genome of the cattle resisted bovine tuberculosis in cattle by *Mycobacterium bovis* (Jabbar et al., 2021).

In plants, the major application of CRISPR-Cas9 is to explicate the function of single or multiple genes by editing the gene sequence and by transcriptional regulation (Liu et al., 2017a). For plant species like maize, soybean, wheat, rice, tomato, lettuce, Arabidopsis, petunia, grapevine, and tobacco, the plasmids are isolated and are transfected with the CRISPR-Cas9 system to edit the genome. CRISPR-Cas9 use in the modification of the genes in plants is found to be effective in various pathways (El-Mounadi et al., 2020). Knockout of the functional genes in the plant disrupts the enzyme production by translation disruption and has been used in plants like rice, tobacco, sorghum and Arabidopsis. High-efficient multiplex genome editing depends on the expression of Cas9/gRNA and also on the capacity of the gRNA cassette. The promoter for the expression of gRNA is much more restricted for species than the expression promoter for Cas9 for which promoters like snoRNA, U3 and U6 in different plants. The off-target activity of Cas9 is one of the drawbacks of this technology, but high throughput analysis and proper design of highly specific gRNA can reduce the risk of off-targets. Crop breeding can be of great value by site specifically integrating genes in the genome of the plants which is successfully carried out by the CRISPR system in plants such as rice which conferred resistance to bispyribac sodium a herbicide by substituting two nucleotides in acetolactate synthase (Ding et al., 2016). The highly flexible recognition of DNA by the CRISPR system, and its ability to modulate the expression of genes from its promoter or enhancer is key for researchers in gene editing. The CRISPR-Cas9 system can be adapted for regulating transcription by forming a complex of specific effector domains that binds with the carboxyl terminus of a catalytically inactive protein Cas9 by substituting two amino acids in the RuvC and NHN domains of the protein to form dCas9 which can bind to DNA in RNA-directed manner and can regulate expression of the target gene (Gaj et al., 2016).

15.7.2 Application in Agriculture

The field of agriculture has been using advanced genome editing to get the desired traits and increase in quantity the yield of crops. CRISPR-Cas9 is one of them in the field of genome engineering that can reproduce plants with the desired characteristics. Editing of the genes in the plants by CRISPR is used as cargo that integrates into the genome of the plants. The CRISPR-Cas editing system is introduced in the plant cells that are recipients by ways of traditional methods such as particle bombardment or *Agrobacterium*-mediated transformation. CRISPR-Cas9 is been used to improve the quality, the yield of the plants along with an increase in tolerance to both biotic and abiotic stress. The most promising approach in genetic engineering is eliminating the unwanted traits which can be achieved by suppressing or negatively regulating their expression. The simplest way carried out by CRISPR-Cas9 is knocking out unwanted traits (Rao and Wang, 2021). An increase in the nutritional value of the food is the main factor along with an increase in the total yield from plants. By use of the CRISPR-Cas9 technique in rice the nutritional value is been preserved. Many genes in the rice plant are been edited by knocking out using CRISPR-Cas9 which resulted in the loss of function and controlled contents such as amylase, amylopectin, and mineral content like cadmium and aroma (Mehta et al., 2020). Several factors cause stress in the plant during its growth, harvesting, and others which limit their survival and production in large yield. Modification carried out by this CRISPR system could help in adapting plants to certain conditions. The precise use of the system in the editing of the locus of ARDOS8 an ethylene response negative regulator has generated a variant that has an improvement in yield than the wild type in maize (Eş et al., 2019).

15.7.3 Application in Therapeutics

CRISPR-Cas9 is found to have applications in many fields, one of them being its role in therapy. In recent years this system has been used in the field of therapeutic by correcting the disease-causing gene mutation or engineering the cells in immunotherapy in cancer (Liu et al., 2017b). In the delivery of these components for therapeutic, once delivered starts manipulating the genetic material by various ways for mutation correction, and insertion of desired genes (Mout et al., 2017). The development of the CRISPR system has revolutionized the field of gene therapy, CRISPR-Cas9 can efficiently induce loss of function and gain of function in cancer, in both *in vitro* and *in vivo* and can correct various genetic aberrations that drive the pathogenesis of cancer and their development. Targeting the genes that are related to the survival of cancer cells and their proliferation with CRISPR-Cas9, can decrease the growth of the tumors (Yi and Li, 2016). Non-sense mutation in the gene Rpe65 which causes Leber congenital amaurosis, a childhood blindness is corrected therapeutically using CRISPR-Cas9 using adeno-associated virus vectors in the rd12 mice model (Jo et al., 2019). This system is been used in the treatment of cancer by using processes such as immunotherapy, cancer genome and epigenome manipulation, and inactivation of infections by viruses. In cardiovascular disease, the CRISPR system targets the gene PCSK9 gene which lowers levels of LDLC, and thereby it may be one of the therapeutic targets in cardiovascular disease. Mutation in the MYBPC3 that causes hypertrophic cardiomyopathy is been corrected by CRISPR-Cas9 in the embryos of humans. CRISPR-Cas9 can edit the gene Crygc in a homology-directed repair manner and improve cataracts in the eyes (Mollanoori and Teimourian, 2018). The therapeutic approach in *ex vivo* genome editing is carried out by editing the genome of cells in *in vitro* and then introducing it to the patients. This approach of therapy is in direct contrast to *in vivo* editing of genome which is delivered by CRISPR-Cas9 or other gene editing systems. This approach can be a promising method for delivering the genes to the targeted organs, but many steps are required for this strategy including a collection of cells, isolation expansion, editing, selection, and finally transplantation (Li et al., 2020). Application of CRISPR-Cas9 in Alzheimer's disease has shown correcting the mutation in the genes associated with the diseases like APP, PSEN1, and PSEN2 and has approached the development of a model of Alzheimer's disease and also shown hope for an effective approach in therapeutic (Barman et al., 2020). Duchenne muscular dystrophy an X-linked disease is characterized by progressive muscle weakness with a short life span, CRISPR-Cas9 technology is used to study in the mdx mice model to correct the Dmd gene mutation with Cas9 and sgRNA *in vivo* for cure (Jacinto et al., 2020).

15.8 CONCLUSION

Genomic engineering is the revolutionary change in the biological world that has been carried out using different technologies to modify the genome in organisms. Many technologies have been adopted to get good results in the fields of agriculture, food, medicine, and others. CRISPR is a new technology whose primary response is immunity in bacteria and archaea but is been used in the engineering of the genome due to its ability to insert genomic sequences. It can be one of the most powerful technologies in the future for achieving better prospects in many fields, but still, there are some outcomes that unmatch the prospect, which can be overcome by further research and study.

ACKNOWLEDGMENT

All the authors are thankful to their respective universities and institutions for their valuable support.

DATA AVAILABILITY

The authors confirm that the data supporting the findings of this study are available within the book/chapter.

REFERENCES

Adli, M., (2018). The CRISPR tool kit for genome editing and beyond. *Nat Commun, 9(1)*, 1911. DOI: 10.1038/s41467-018-04252-2

Ates, I., Rathbone, T., Stuart, C., Bridges, P.H., & Cottle, R.N., (2020). Delivery approaches for herapeutic genome editing and challenges. *Genes, 11(10)*, 1113. DOI: 10.3390/genes11101113

Awan, M.J., Mahmood, M.A., Naqvi, R.Z., & Mansoor, S., (2023). PASTE: A high-throughput method for large DNA insertions. *Trends Plant Sci, 28(5)*, 509–11. DOI: 10.1016/j.tplants.2023.02.013

Barman, N.C., Khan, N.M., Islam, M., Nain, Z., Roy, R.K., Haque, A., & Barman, S.K., (2020). CRISPR-Cas9: A promising genome editing therapeutic tool for Alzheimer's disease-A narrative review. *Neurol Ther, 9(2)* 419–34. DOI: 10.1007/s40120-020-00218-z

Carroll, D., (2011). Genome engineering with zinc-finger nucleases. *Genetics, 188(4)*, 773–82. DOI: 10.1534/genetics.111.131433

Cathomen, T., & Joung, J.K., (2008). Zinc-finger nucleases: The next generation emerges. *Mol Ther, 16(7)*, 1200–7. DOI: 10.1038/mt.2008.114

Çiçek, Y.A., Luther, D.C., Kretzmann, J.A., & Rotello, V.M., (2019). Advances in CRISPR/Cas9 technology for in vivo translation. *Biological and Pharmaceutical Bulletin, 42(3)*, 304–311. doi: 10.1248/bpb.b18-00811.

Cheng, H., Zhang, F., & Ding, Y., (2021). CRISPR/Cas9 delivery system engineering for genome editing in therapeutic applications. *Pharmaceutics*, *13(10)*, 1649. DOI: 10.3390/pharmaceutics13101649

Cui, Z., Tian, R., Huang, Z., Jin, Z., Li, L., Liu, J., Huang, Z., Xie, H., Liu, D., Mo, H., Zhou, R., Lang, B., Meng, B., Weng, H., & Hu, Z., (2022). FrCas9 is a CRISPR/Cas9 system with high editing efficiency and fidelity. *Nat Commun*, *13(1)*, 1425. DOI: 10.1038/s41467-022-29089-8

Dasgupta, I., Flotte, T.R., & Keeler, A.M., (2021). CRISPR/Cas-dependent and nuclease-free *in vivo* therapeutic gene editing. *Hum Gene Ther*, *32(5–6)*, 275–93. DOI: 10.1089/hum.2021.013

Ding, Y., Li, H., Chen, L.L., & Xie, K., (2016). Recent advances in genome editing using CRISPR/Cas9. *Front Plant Sci*, *7*, 703. DOI: 10.3389/fpls.2016.00703

Duan, L., Ouyang, K., Xu, X., Xu, L., Wen, C., Zhou, X., Qin, Z., Xu, Z., Sun, W., & Liang, Y., (2021). Nanoparticle delivery of CRISPR/Cas9 for genome editing. *Front Genet*, *12*, 673286. DOI: 10.3389/fgene.2021.673286

El-Mounadi, K., Morales-Floriano, M.L., & Garcia-Ruiz, H., (2020). Principles, applications, and biosafety of plant genome editing using CRISPR-Cas9. *Front Plant Sci*, *11*, 56. DOI: 10.3389/fpls.2020.00056

Eş, I., Gavahian, M., Marti-Quijal, F.J., Lorenzo, J.M., Khaneghah, A.M., Tsatsanis, C., Kampranis, S.C., & Barba, F.J., (2019). The application of the CRISPR-Cas9 genome editing machinery in food and agricultural science: Current status, future perspectives, and associated challenges. *Biotechnol Adv*, *37(3)*, 410–21. DOI: 10.1016/j.biotechadv.2019.02.006

Gaj, T., Sirk, S.J., Shui, S.L., & Liu, J., (2016). Genome-editing technologies: Principles and applications. *Cold Spring Harb Perspect Biol*, *8(12)*, a023754. DOI: 10.1101/cshperspect.a023754

Graham, D.B., & Root, D.E., (2015). Resources for the design of CRISPR gene editing experiments. *Genome Biol*, *16*, 260. DOI: 10.1186/s13059-015-0823-x

Hafez, M., & Hausner, G., (2012). Homing endonucleases: DNA scissors on a mission. *Genome*, *55(8)*, 553–69. DOI:10.1139/G2012-049

Jabbar, A., Zulfiqar, F., Mahnoor, M., Mushtaq, N., Zaman, M.H., Din, A.S., Khan, M.A., & Ahmad, H.I., (2021). Advances and perspectives in the application of CRISPR-Cas9 in livestock. *Mol Biotechnol*, *63(9)*, 757–67. DOI: 10.1007/s12033-021-00347-2

Jacinto, F.V., Link, W., & Ferreira, B.I., (2020). CRISPR/Cas9-mediated genome editing: From basic research to translational medicine. *J Cell Mol Med*. *24(7)*, 3766–78. DOI: 10.1111/jcmm.14916

Jiang, F., & Doudna, J.A., (2017). CRISPR-Cas9 structures and mechanisms. *Ann Rev Biophys*, *46*, 505–29. DOI: 10.1146/annurev-biophys-062215-010822

Jin, L.F., & Li, J.S., (2016). Generation of genetically modified mice using CRISPR/Cas9 and haploid embryonic stem cell systems. *Dongwuxue Yanjiu*, *37(4)*, 205–13. DOI: 10.13918/j.issn.2095-8137.2016.4.205

Jo, D.H., Song, D.W., Cho, C.S., Kim, U.G., Lee, K.J., Lee, K., Park, S.W., Kim, D., Kim, J.H., Kim, J.S., & Kim, S., (2019) CRISPR-Cas9-mediated therapeutic editing of *Rpe65* ameliorates the disease phenotypes in a mouse model of Leber congenital amaurosis. *Sci Adv*, *5(10)*, eaax1210. DOI: 10.1126/sciadv.aax1210

Kang, Y., Deng, H., Pray, C., & Hu, R., (2022). Managers' attitudes toward gene-editing technology and companies' R&D investment in gene-editing: The case of Chinese seed companies. *GM Crops Food*, *13(1)*, 309–26. DOI: 10.1080/21645698.2022.2140567

Khalil, A.M., (2020). The genome editing revolution. *J Genet Eng Biotechnol*, *18(1)*, 1–6. DOI: 10.1186/s43141-020-00078-y

Konstantakos, V., Nentidis, A., Krithara, A., & Paliouras, G., (2022). CRISPR-Cas9 gRNA efficiency prediction: An overview of predictive tools and the role of deep learning. *Nucleic Acids Res*, *50(7)*, 3616–637. DOI: 10.1093/nar/gkac192

Koonin, E.V., Makarova, K.S., & Zhang, F., (2017). Diversity, classification and evolution of CRISPR-Cas systems. *Curr Opin Microbiol*, *37*, 67–78. DOI: 10.1016/j.mib.2017.05.008

Li, Y., Glass, Z., Huang, M., Chen, Z.Y., & Xu, Q., (2020). Ex vivo cell-based CRISPR/Cas9 genome editing for therapeutic applications. *Biomaterials*, *234*, 119711. DOI: 10.1016/j.biomaterials.2019.119711

Li, H., Yu, H., Du, S., & Li, Q. (2021). CRISPR/Cas9 mediated high efficiency knockout of myosin essential light chain gene in the Pacific Oyster (Crassostrea Gigas). *Marine Biotechnology*, *23*, 215–224. DOI: 10.1007/s10126-020-10016-1

Lino, C.A., Harper, J.C., Carney, J.P., & Timlin, J.A., (2018). Delivering CRISPR: A review of the challenges and approaches. *Drug Delivery*, *25(1)*, 1234–57. DOI: 10.1080/10717544.2018.1474964

Liu, C., Zhang, L., Liu, H., & Cheng, K., (2017a). Delivery strategies of the CRISPR-Cas9 gene-editing system for therapeutic applications. *J Control Release*, *266*, 17–26. DOI: 10.1016/j.jconrel.2017.09.012

Liu, X., Wu, S., Xu, J., Sui, C., & Wei, J., (2017b). Application of CRISPR/Cas9 in plant biology. *Acta Pharm Sin B*, *7(3)*, 292–302. DOI: 10.1016/j.apsb.2017.01.002

Loureiro, A., & da Silva, G.J., (2019). CRISPR-Cas: Converting a bacterial efence mechanism into a state-of-the-art genetic manipulation tool. *Antibiotics*, *8(1)*, 18. DOI: 10.3390/antibiotics8010018

Mahmood, M.A., & Mansoor, S., (2023). PASTE: The way forward for large DNA insertions. *CRISPR J*, *6(1)*, 2–4. DOI: 10.1089/crispr.2023.0001

Makarova, K.S., Haft, D.H., Barrangou, R., Brouns, S.J., Charpentier, E., Horvath, P., Moineau, S., Mojica, F.J., Wol, Y.I., Yakunin, A.F., van der Oost, J., & Koonin, E.V., (2011). Evolution and classification of the CRISPR-Cas systems. *Nat Rev Microbiol*, *9(6)*, 467–77. DOI: 10.1038/nrmicro2577

Makarova, K.S., Wolf, Y.I., Alkhnbashi, O.S., Costa, F., Shah, S.A., Saunders, S.J., ... & Koonin, E.V. (2015). An updated evolutionary classification of CRISPR–Cas systems. *Nature Reviews Microbiology*, *13(11)*, 722–736. doi: 10.1038/nrmicro3569

Makarova, K.S., Wolf, Y.I., & Koonin, E.V., (2018). Classification and nomenclature of CRISPR-Cas systems: Where from here? *CRISPR J*, *1(5)*, 325–36. DOI: 10.1089/crispr.2018.0033

Mashimo, T., (2014). Gene targeting technologies in rats: Zinc finger nucleases, transcription activator-like effector nucleases, and clustered regularly interspaced short palindromic repeats. Dev *Growth Differ*, *56(1)*, 46–52. DOI: 10.1111/dgd.12110

Mehta, S., Lal, S.K., Sahu, K.P., Venkatapuram, A.K., Kumar, M., Sheri, V., Varakumar, P., Vishwakarma, C., Yadav, R., Jameel, M.R., & Ali, M., (2020). CRISPR/Cas9-edited rice: A new frontier for sustainable agriculture. In *New Frontiers in Stress Management for Durable Agriculture*, pp. 427–58. DOI: 10.1007/978-981-15-1322-0_23

Mishra, R., Joshi, R. K., & Zhao, K. (2020). Base editing in crops: current advances, limitations and future implications. *Plant Biotechnology Journal*, *18(1)*, 20–31. doi: 10.1111/pbi.13225

Molla, K.A., Sretenovic, S., Bansal, K.C., & Qi, Y., (2021). Precise plant genome editing using base editors and prime editors. *Nature Plants*, *7(9)*, 1166–87. DOI: 10.1038/s41477-021-00991-1

Mollanoori, H., & Teimourian, S., (2018). Therapeutic applications of CRISPR/Cas9 system in gene therapy. *Biotechnol Lett*, *40(6)*, 907–14. DOI: 10.1007/s10529-018-2555-y

Moon, S.B., Kim, D.Y., Ko, J.H., & Kim, Y.S., (2019). Recent advances in the CRISPR genome editing tool set. *Exp Mol Med*, *51(11)*, 1–11. DOI: 10.1038/s12276-019-0339-7

Mout, R., Ray, M., Lee, Y.W., Scaletti, F., & Rotello, V.M., (2017). In vivo delivery of CRISPR/Cas9 for therapeutic gene editing: progress and challenges. *Bioconjug Chem*. *28(4)*, 880–4. DOI: 10.1021/acs.bioconjchem.7b00057

Panahi, B., Majidi, M., & Hejazi, M.A., (2022). Genome mining approach reveals the occurrence and diversity pattern of clustered regularly interspaced short palindromic repeats/CRISPR-associated systems in *Lactobacillus brevis* strains. *Front Microbiol*, *13*, 911706. DOI: 10.3389/fmicb.2022.911706

Pei, Z., Sadiq, F. A., Han, X., Zhao, J., Zhang, H., Ross, R. P., ... & Chen, W. (2021). Comprehensive scanning of prophages in Lactobacillus: distribution, diversity, antibiotic resistance genes, and linkages with CRISPR-Cas systems. *Msystems*, *6(3)*, e01211–20. doi: 10.1128/mSystems.01211-20

Petersen, G.E.L., Buntjer, J.B., Hely, F.S., Byrne, T.J., & Doeschl-Wilson, A., (2022). Modeling suggests gene editing combined with vaccination could eliminate a persistent disease in livestock. *Proc Natl Acad Sci U S A*, *119(9)*, e2107224119. DOI: 10.1073/pnas.2107224119

Porteus, M.H., & Carroll, D., (2005). Gene targeting using zinc finger nucleases. *Nat Biotechnol*, *23(8)*, 967–73. DOI: 10.1038/nbt1125

Porto, E. M., Komor, A. C., Slaymaker, I. M., & Yeo, G. W. (2020). Base editing: advances and therapeutic opportunities. *Nature Reviews Drug Discovery*, *19(12)*, 839–859. https://doi.org/10.1038%2Fs41573-020-0084-6

Rao, M.J., & Wang, L., (2021). CRISPR/Cas9 technology for improving agronomic traits and future prospective in agriculture. *Planta*, *254(4)*, 1–6. DOI: 10.1007/s00425-021-03716-y

Rees, H.A., & Liu, D.R., (2018). Base editing: Precision chemistry on the genome and transcriptome of living cells. *Nat Rev Genet*, *19(12)*, 770–88. DOI: 10.1038/s41576-018-0059-1

Riesenberg, S., Helmbrecht, N., Kanis, P., Maricic, T., & Pääbo, S., (2022). Improved gRNA secondary structures allow editing of target sites resistant to CRISPR-Cas9 cleavage. *Nat Commun*, *13(1)*, 489. DOI: 10.1038/s41467-022-28137-7

Shim, G., Kim, D., Park, G.T., Jin, H., Suh, S.K., & Oh, Y.K., (2017). Therapeutic gene editing: Delivery and regulatory perspectives. *Acta Pharmacol Sin*, *38(6)*, 738–53. DOI: 10.1038/aps.2017.2

Sun, N., & Zhao, H., (2013). Transcription activator-like effector nucleases (TALENs): A highly efficient and versatile tool for genome editing. *Biotechnol Bioeng*, *110(7)*, 1811–21. DOI: 10.1002/bit.24890

Waltz, M., Juengst, E.T., Edwards, T., Henderson, G.E., Kuczynski, K.J., Conley, J.M., Della-Penna, P., & Cadigan, R.J., (2021). The view from the Benches: Scientists' perspectives on the uses and governance of human gene-editing research. *CRISPR J*, *4(4)*, 609–15. DOI: 10.1089/crispr.2021.0038

Wang, S.W., Gao, C., Zheng, Y.M., Yi, L., Lu, J.C., Huang, X.Y., Cai, J.B., Zhang, P.F., Cui, Y.H., Ke, & A.W., (2022). Current applications and future perspective of CRISPR/Cas9 gene editing in cancer. *Mol Cancer, 21(1)*, 57. DOI: 10.1186/s12943-022-01518-8

Windbichler, N., Menichelli, M., Papathanos, P.A., Thyme, S.B., Li, H., Ulge, U.Y., Hovde, B.T., Baker, D., Monnat, R.J., Burt, A., & Crisanti, A., (2011). A synthetic homing endonuclease-based gene drive system in the human malaria mosquito. *Nature, 473(7346)*, 212–5. DOI: 10.1038/nature09937

Xu, S., Pham, T.P., & Neupane, S., (2020). Delivery methods for CRISPR/Cas9 gene editing in crustaceans. *Mar Life Sci Technol, 2(1)*, 1–5. DOI: 10.1007/s42995-019-00011-4

Xu, Y., & Li, Z., (2020). CRISPR-Cas systems: Overview, innovations and applications in human disease research and gene therapy. *Comput Struct Biotechnol J, 18*, 2401–2415. DOI: 10.1016/j.csbj.2020.08.031

Yarnall, M.T.N., Ioannidi, E.I., Schmitt-Ulms, C., Krajeski, R.N., Lim, J., Villiger, L., Zhou, W., Jiang, K., Garushyants, S.K., Roberts, N., Zhang, L., Vakulskas, C.A., Walker, J.A. 2nd, Kadina, A.P., Zepeda, A.E., Holden, K., Ma, H., Xie, J., Gao, G., Foquet, L., Bial, G., Donnelly, S.K., Miyata, Y., Radiloff, D.R., Henderson, J.M., Ujita, A., Abudayyeh, O.O., & Gootenberg, J.S., (2023). Drag-and-.drop genome insertion of large sequences without double-strand DNA cleavage using CRISPR-directed integrases. *Nat Biotechnol, 41(4)*, 500–12. DOI: 10.1038/s41587-022-01527-4

Yi, L., & Li, J., (2016). CRISPR-Cas9 therapeutics in cancer: Promising strategies and present challenges. *Biochim Biophys Acta Rev Cancer, 1866(2)*, 197–207. DOI: 10.1016/j.bbcan.2016.09.002

Yip, B.H., (2020). Recent advances in CRISPR/Cas9 delivery strategies. *Biomolecules, 10(6)*, 839. DOI: 10.3390/biom10060839

Zhou H, Su J, Hu X, Zhou C, Li H, Chen Z, Xiao Q, Wang BO, Wu W, Sun Y, Zhou Y., (2020). Glia-to-neuron conversion by CRISPR-CasRx alleviates symptoms of neurological disease in mice. *Cell, 181(3)*, 590–603.

16 CRISPR/Cas13 for the Control of Plant Viruses

Joana A. Ribeiro, Patrick Materatski, Carla M. R. Varanda, Maria Doroteia Campos, Mariana Patanita, André Albuquerque, Nicolás Garrido, Tomás Monteiro, Filipa Santos, and Maria do Rosário Félix

LIST OF ABBREVIATIONS

ABE	adenine base editor
C2c2	class 2 candidate 2
C2c6	class 2 candidate 6
Cas	CRISPR-associated proteins
CBE	cytosine base editor
CRISPR	clustered regularly interspaced short palindromic repeats
crRNAs	CRISPR RNAs
DNA	deoxyribonucleic acid
DR	direct repeat
DSB	double-strand break
dsRNA	double-stranded RNA
HDR	homology-directed repair
HEPN	higher eukaryotes and prokaryotes nucleotide-binding domains
Lba	*Lachnospiraceae bacterium*
Lbu	*Leptotrichia buccalis*
Lsh	*Leptotrichia sharii*
Lwa	*Leptotrichia wadei*
Lse	*Listeria seeligeri*
nt	nucleotide
NTD	N-terminal domain
NUC	nuclease
PFS	protospacer flanking site
REC	recognition
RNA	ribonucleic acid
sgRNA	single guide RNA
ssRNA	single-stranded RNA
TMV	*Tobacco mosaic virus*
TuMV	*Turnip mosaic virus*
tracrRNAs	transactivating CRISPR RNAs

DOI: 10.1201/9781003387060-16

16.1 INTRODUCTION

Viruses are among the most important pathogens in both animals and plants. These agents have RNA or DNA genomes, single or double-stranded, encoding some proteins responsible for their replication and transmission. Viruses can only replicate inside living cells of a host organism and their host range is usually relatively narrow [1].

Plant viruses have major implications on plant pathology and the study of virology, which dates back to early studies using *Tobacco mosaic virus* (TMV) which helped to understand the virus concept, uncovering chemical and physical characteristics of viruses in general [2]. Plant viruses can rapidly replicate and spread throughout a crop, being very difficult to monitor and control and, therefore, causing destructive diseases in many agricultural systems. These diseases can significantly reduce crop quality and yield, resulting in tremendous economic impacts all over the world and threatening food security and provision [3].

Unlike what happens with other plant disease-causing pathogens, there are no efficient chemical products that can eradicate a virus within a plant without disturbing host cells and the environment. Consequently, preventive sanitary measures, such as the use of viral-resistant plants, are usually the only options. Resistant plants were conventionally generated through a very time-consuming classical breeding process. However, nowadays, virus-resistant plants can be generated through molecular plant breeding, preventing and controlling viral diseases [4,5]. These molecular approaches can be based on genomic selection, molecular marker-assisted breeding, gene silencing, and pathogen-derived resistance, however, many setbacks have hampered their utility in agriculture. The major drawback is the fast adaptation and emergence of new viruses for which these techniques are not efficient enough. In addition, gene knockout in plants to prevent viral replication can compromise other desirable characteristics [3,6].

Over the past decade, a breakthrough has revolutionized plant breeding. The study of clustered regularly interspaced short palindromic repeats (CRISPR) and CRISPR-associated (Cas) protein systems allowed the development of a new technology that has opened up new horizons for plant breeding and improvement. The function of CRISPR/Cas is originally linked to the adaptative immune system present in prokaryotes to specifically target viruses. These systems may be used as gene editing tools and applied for the prevention and control of plant viruses in the field [4,7]. The first CRISPR/Cas systems studied were very useful for DNA targeting, however, more recently, the CRISPR/Cas13 was identified, which can specifically cleave single-stranded RNA (ssRNA) in eukaryotes. This has placed CRISPR/Cas as a promising tool for the development of immunity against a wide range of RNA viruses, which are the most abundant class of viruses in plants. In addition, among the existing DNA plant viruses, many contain an RNA intermediate [8].

The present chapter aims to bring together the latest information on CRISPR/Cas systems, including their origin, components and classification, and diverse applications, namely, to control plant viruses. Considering the unique characteristics of CRISPR/Cas13 systems, such as their robustness, preciseness, and versatility, this review will focus on CRISPR/Cas13 systems to control plant viruses. We also discuss the limitations and future challenges of CRISPR/Cas13 in the development of virus-resistant plants for future precision breeding and sustainable agriculture.

16.2 OVERVIEW OF CRISPR/CAS TECHNOLOGY

16.2.1 Origin of CRISPR/Cas Technology

For many years scientific efforts have been made to find new technologies able to modify eukaryotic genomes [9]. New solutions are often found in prokaryotes, which provide innovative and nature-based solutions for gene editing, such as reporter genes (lacZ) [10], strong inducible gene expression (tetracycline system) [11], and effective conditional mutagenesis (cre/loxP system) [12].

CRISPR/Cas is one of the most modern examples of genetic engineering tools that were found and developed from prokaryotes. Over the past decades, key findings and progress in prokaryote research allowed the launch of this technique [9]. Atsuo Nakata and his research team, in 1987, first reported DNA repeats with dyad symmetry, in Gram-negative bacteria *Escherichia coli* K12, which would become known as CRISPR [13]. The first studies with insights on CRISPR functionality were reported in archaea and published in 1993 and 1995 [14,15]. In 2006, CRISPR/Cas systems were suggested as bacterial defense mechanisms, due to the discovery of spacer sequences, which were homologous to DNA sequences from bacteriophages or plasmids [16]. One year later, this technology was experimentally demonstrated to take part in acquired immunity against bacteriophages, since it was confirmed that by acquiring spacers that match the viral genome, a sensitive bacterial strain can develop resistance to infection [17]. The following studies on CRISPR/Cas systems allowed comprehensive insights into its structure, components, and functions, leading to reports demonstrating functional CRISPR/Cas systems as competent genome editing tools in 2012 and 2013 [18,19].

Since 2013, when it was first applied in plants, CRISPR has been used for genome editing in a wide range of crops, many of which have high-value agricultural traits [20,21]. The most recent CRISPR/Cas technologies are particularly important because they can change nucleotides precisely, which can have major impacts on agriculture. In addition, this technology can go beyond editing specific *loci* for crop improvement, being capable of promoting gene regulation and protein engineering. Therefore, CRISPR/Cas technologies have already shown high potential for fundamental biological research and have raised the prospect for multiple new applications [22].

16.2.2 Components and Classification of CRISPR/Cas Systems

The CRISPR/Cas units, adaptative immune systems present in archaea and bacteria, offer protection against foreign DNA or, sometimes, foreign RNA by specifically recognizing sequences of the invader. CRISPR/Cas *loci* consist of a CRISPR array, which is composed of short direct repeats separated by spacers (short variable DNA sequences), bordered by different *cas* genes [23] (Figure 16.1).

CRISPR/Cas immunity comprises three basic steps: adaptation, expression, and interference (Figure 16.1). These systems work by memorizing phage infections, and the first step is adaptation or acquisition, which consists of the incorporation of protospacers, which are fragments of foreign DNA from invading organisms into the CRISPR array. After incorporation, the spacers allow a specific defense against following invasions [23]. Protospacer acquisition and insertion into the CRISPR array is mediated by Cas1 and Cas2, a complex of Cas proteins that together with additional proteins, regulate this process, being able to measure and cut out a piece of exactly the right size to insert a new spacer. All spacers are always flanked by repeats on each side, as the system incorporates a new repeat in the process [24].

CRISPR/Cas systems store these spacers in the DNA as a way of remembering the infection but do not use the DNA to directly recognize subsequent infections. Instead, the CRISPR array is transcribed as a precursor transcript (pre-crRNA) which is processed and matured into CRISPR RNAs (crRNAs). These crRNAs are used to find new invading viruses and can be degraded and recycled without destroying the original memory [23]. Pre-crRNAs can be bound to a single multidomain protein or a multisubunit effector complex. An endonuclease subunit of the multisubunit effector complex or an alternative mechanism involving bacterial RNase III and other RNA species called transactivating CRISPR RNA (tracrRNA), are responsible for the processing and maturation of the pre-crRNA into crRNAs [25].

The last phase in CRISPR immunity, interference, allows the cleavage of the invading phage's DNA upon infection. During this stage, mature crRNAs, aided by Cas proteins, the so-called effector complex or surveillance complex, recognize and cleave the cognate DNA or RNA. Once the phage's DNA or RNA is cleaved and its replication is incapacitated, the infection is over [23].

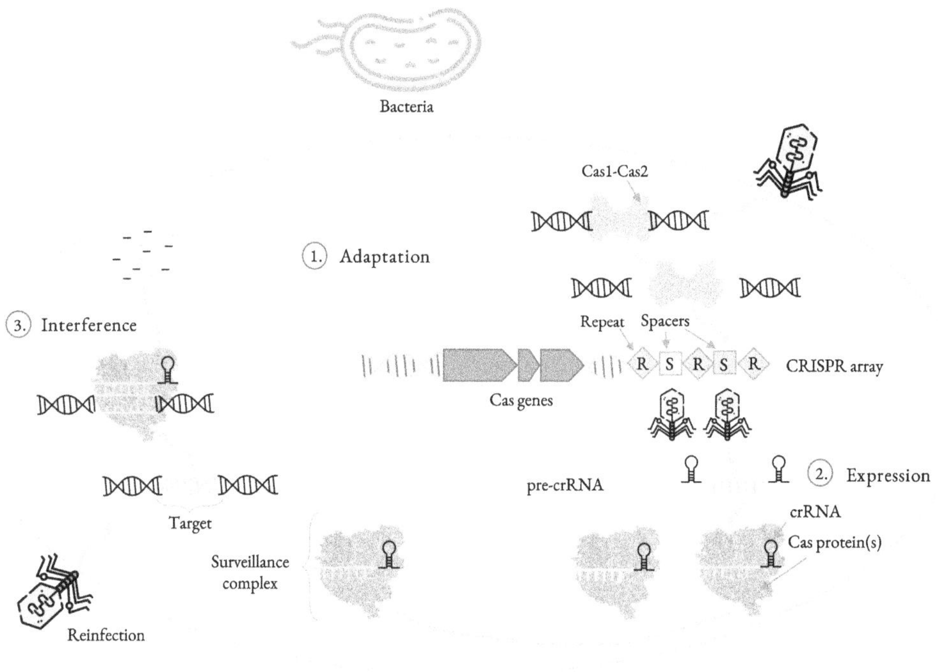

FIGURE 16.1 CRISPR/Cas systems components and immunity steps.

As a result of their evolutionary arms race with pathogens, the Cas protein sequences and the genomic architecture of CRISPR/Cas *loci* display a remarkable diversity typical of antiviral defense mechanisms [23,26,27]. However, this variability poses a major challenge for achieving a coherent annotation and a rather simple classification of CRISPR/Cas systems, which would clarify their origins and evolution and keep track of new variants. Nonetheless, for further progress in CRISPR research, a coherent classification scheme is essential [28,29].

CRISPR/Cas systems classification approaches have created and adopted a combined, semi-formal method, based on signature genes and distinctive gene architectures, allowing the assignment of these systems to types and subtypes. Therefore, signature genes are specific for each type and subtype of CRISPR/Cas systems. In addition, sequence similarity between multiple Cas proteins, the phylogeny of Cas1 (the best conserved Cas protein), the organization of the *loci* and the structure of the CRISPR themselves is of major importance for this classification [23,28,29].

As mentioned above, the classification of CRISPR/Cas systems is complicated because of their diversity and constant evolution. Not even Cas1 can be considered as a universal Cas protein since it fails to adequately represent the relationships between all CRISPR/Cas systems and cannot be used as a phylogenetic marker. Therefore, the application of the multiple criteria previously mentioned results in the classification scheme presently used for CRISPR/Cas systems, which separates them into two different classes, according to the design principles of the effector complex. These complexes can have several Cas proteins, having a multi-subunit design, which is the case of Class 1 systems. On the other hand, Class 2 systems have a single, large multidomain protein [23,28] (Figure 16.2).

Class 1 systems are classified into three types – I, III, and IV. Types II, V, and VI belong to Class 2 CRISPR/Cas systems. In each type, the design of the effector complex gene is different, having unique signature proteins, which allows its classification into several subtypes, encompassing subtler differences in *locus* organization and encoding subtype-specific Cas proteins [23,26,29,30] (Figure 16.2).

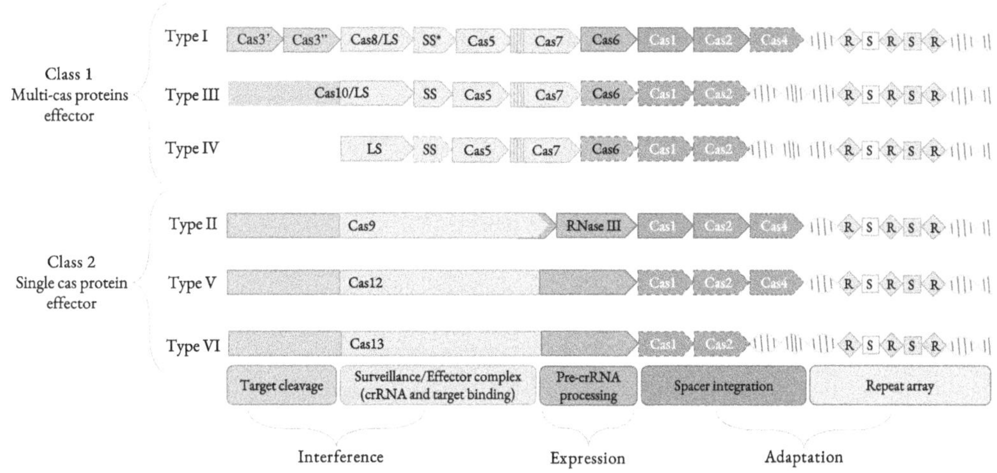

FIGURE 16.2 Class 1 and Class 2 CRISPR/Cas systems: key functions and gene organization. Genes are shown as blocks and the same color represents genes with a homologous function. Dashed outlines indicate unessential genes. An asterisk indicates the putative small subunit (SS) that is thought to be attached to the large subunit in many type I subtypes.

Accordingly, one of the main aspects that differentiate Class 1 from Class 2 systems is the pre-crRNA processing. In Class 1 systems, a complex of several Cas proteins – Cascade (CRISPR-associated complex for antiviral defense) – catalyzes the maturation of pre-crRNAs to crRNAs. The Cascade complex is responsible for binding the pre-crRNA and recruiting a supplementary Cas protein (nuclease), directly responsible for the processing (usually Cas 6, but it can also be Cas5) [31–33].

In Class 2 systems, for type II, considered as a prototype, an RNAse III enzyme catalyzes the processing of pre-crRNAs, aided by additional RNA species, the tracrRNA. On the other hand, in types V and VI, different nuclease activity, belonging to the same large effector protein, is observed, but this process is not completely understood. However, for subtype, V-B systems tracrRNAs have been identified, though RNAse III is not present, and the cleavage enzyme is still unknown [25,34,35]. DNA editing is possible using types II and V Cas proteins, whereas RNA editing can be done by applying type VI Cas proteins [4].

The interference stage is also different for each class of CRISPR/Cas systems. In Class 1 systems, the effector complex (crRNA and Cas proteins) recognizes the protospacer sequence in the target and recruits Cas3. The helicase domain of Cas3 unwinds the target dsDNA, and the nuclease domain cleaves the foreign DNA. In type III systems, the nuclease involved is a part of the effector complex. Hence, no helicase is involved but DNA cleavage first requires cleavage of RNA transcripts by a distinct CRISPR-associated RNase [36,37]. In Class 2 systems, cleavage is accomplished by the nuclease domain(s) of the large effector protein [25,35].

16.2.3 CRISPR/Cas Systems for Crop Improvement

CRISPR/Cas systems are highly specific and robust, allowing precise genome editing which can introduce beneficial traits to enhance agricultural sustainability. Therefore, this technology has altered plant molecular biology exceeding expectations. Furthermore, the variety of emerging technologies based on CRISPR/Cas systems has expanded the range of fundamental research and plant biology [22].

Genome editing using CRISPR/Cas has benefits compared to the basic strategies that use sequence-specific nucleases (meganucleases [38], zinc-finger nucleases [39], or transcription activator-like nucleases [40]) to induce DNA double-strand break (DSB) at a target site and homology-directed repair (HDR). These basic strategies have shown to be effective for plant genome editing, however, their creation requires complex protein engineering. On the other hand, CRISPR/Cas systems can be easily engineered to introduce DSBs at any chosen target location with reduced costs [41].

Precise genome editing in plants based on CRISPR/Cas systems, such as deaminase-mediated base editing or reverse transcriptase-mediated prime editing technologies, are alternative genome editing technologies in which DSB is not involved and a donor DNA is not needed, being more efficient than HDR in plants [22]. In addition, new technologies based on CRISPR/Cas9, such as cytosine base editor (CBE) [42], adenine base editor (ABE) [43], dual base editing [44], and CBE-based precise DNA deletion [45], were also developed for precise genome editing in plants.

CRISPR/Cas technologies can induce precise nucleotide changes, which can have a high impact on agriculture. Nevertheless, CRISPR/Cas potential goes much further than simply editing specific *loci* for crop improvement, as the development of new plant biotechnologies based on these systems has shown the capacity for gene regulation and protein engineering [22].

Regarding CRISPR/Cas upgrade applications for crop improvement, a great number of studies, mostly using CRISPR/Cas9, have demonstrated the successful improvement of several crop characteristics such as yield [46–48], quality [49–51], disease resistance [52,53] and herbicide resistance [54,55]. In addition, many applications in breeding technologies have emerged, targeting reproduction-related genes using CRISPR/Cas systems, such as haploid induction, generating male sterile lines, fixation of hybrid vigor, and manipulating self-incompatibility [22].

Concerning disease resistance, the role of CRISPR/Cas systems against plant viruses is highlighted in the present chapter. Over the past few years, researchers have been able to use CRISPR/Cas system-mediated gene editing to create resistance against pathogens, specifically against viruses, which are known to infect many economically important crops, being a great threat to food security worldwide. The first studies on CRISPR/Cas systems against plant viruses used CRISPR/Cas9 that targeted DNA viruses, but then the later use of CRISPR/Cas13 allowed to target RNA viruses. Generally, there are two main strategies to control plant viruses using CRISPR/Cas technologies: (1) targeting viral genome (DNA or RNA) to inhibit replication and infection or (2) manipulating host susceptibility factors essential for viral infection [4,56]

Targeting of the viral genome to protect plants against viruses using CRISPR/Cas technology was first studied and designed to target DNA viruses. Upon entry of a DNA virus into the plant cell, the single guide RNA (sgRNA) fused to crRNA and tracrRNA from CRISPR/Cas9, in this case, is complementary to a sequence from the DNA target. The Cas9/sgRNA first binds and then cleaves the DNA target [56]. Specifically, the CRISPR/Cas9-mediated editing tool was successfully used as a defense mechanism against DNA plant viruses such as members of the family *Geminiviridae* [57–59] and *Cauliflower mosaic virus* [60]. These studies were mostly done in model plants, which is the case of *Nicotiana benthamiana* and *Arabidopsis thaliana*, but more recent studies have attempted CRISPR/Cas-mediated resistance in crops. Accordingly, a CRISPR/Cas9 machinery was engineered in tomato plants to target *Yellow leaf curl virus* [61] and in barley against *Wheat dwarf virus* [62].

Nevertheless, RNA strands make up most plant viruses' genomes and even DNA plant viruses exhibit an RNA intermediate at some point in their life cycle, highlighting the importance of effectors with RNA specificity as systems of choice to target viral genomes and protect plants. For RNA viruses or RNA intermediates of pathogens with DNA genomes, studies were developed with both Cas9 and Cas13 proteins guided by a sgRNA or a crRNA, respectively. These systems have proven to be successful in targeting RNA viruses, being able to cleave their genome, and preventing further infection [56]. The first description of CRISPR/Cas-based plant immunity against an RNA virus targeted *Cucumber mosaic virus* and *Tomato mosaic virus* using FnCas9 variant. Reduced virus accumulation was observed in transgenic tobacco plants, as well as in *Arabidopsis* plants [63]. Applications of RNA virus interference by CRISPR/Cas13 in plants have also been described in recent literature, showing promising results. For instance, LshCas13a successfully interfered against *Turnip mosaic virus* in both *N. bentamiana* and *A. thaliana* [64]. In addition, CRISPR/Cas13 systems allow the targeting and degradation of viral RNA genomes, conferring resistance to an RNA virus in monocot plants. Namely, this was observed against *Rice black-streaked dwarf virus, Rice stripe mosaic virus,* in transgenic rice plants harboring the CRISPR/Cas13a system [65]; against multiple *Potato virus Y* strains in transgenic potato plants [66]; and against *Grapevine leafroll-associated virus 3* in grapevine [67]. These reports are examples of a great number of studies that are available and being developed worldwide on this revolutionizing technology for direct targeting of RNA viruses.

Moreover, CRISPR/Cas system-mediated resistance to plant viruses can also target host factors. This can be done by gene knockout of the susceptibility (S) genes, editing of the promoter regions, insertion of resistance genes by HDR, or mimicking polymorphisms by targeted nucleotide modification [4,56].

The selection of desirable traits commonly leads to a loss of genetic diversity and increased vulnerability to biotic and abiotic stresses [68]. Therefore, some studies point out the domestication of wild species or the use of semi-domesticated crops as an appealing way to help meet the continuously growing demand for food and nutrition, a consequence of a growing world population. Since the traditional process of wild species domestication is lengthy, because it involves many *loci* but just a few of them have key roles for the desired outcome, CRISPR/Cas is the perfect technology to accelerate the process, having the ability for precise genome editing [69]. Several pioneering and foundation studies on ways to accelerate this process have already been conducted, namely on *Solanum pimpinellifolium*, a putative ancestor of tomato [70,71], and *Oryza glaberrima*, the African rice [72]. Nevertheless, this process still includes several bottlenecks, and further studies are necessary to provide basic knowledge on the genetics of wild species and domestication genes.

Concerning plant biotechnology employing CRISPR/Cas systems, many studies have been carried out, specifically on CRISPR/Cas delivery in plants, gene regulation, multiplex genome editing, mutagenesis, and directed evolution. The application of this technology in plants requires a robust and universal delivery system. Biolistic bombardment and *Agrobacterium*-mediated delivery have been used for decades, but have some limitations. Biolistic bombardment can deliver genetic material beyond the rigid cell walls, using mechanical force however, efficiency is not very high and genome sequences can be damaged. In *Agrobacterium*-mediated delivery, although the integration of foreign DNA is inevitable, *Agrobacterium* can efficiently infect a large range of plants. Moreover, both methods require lengthy tissue culture procedures. *De novo* meristem induction, virus-assisted gene editing, and gene editing with haploid inducers, are delivery systems developed to undermine the limitations of traditional delivery systems. These new tools allow genome manipulation with no need for exogenous DNA, which has advantages over traditional breeding since target mutations are reduced and public concerns toward transgenic lines cease to be a problem [22].

16.3 CRISPR/Cas13 SYSTEMS

16.3.1 Discovery, Classification, and Structure of CRISPR/Cas13 Systems

CRISPR/Cas13 systems belong to Class 2 type V systems, with one multifunctional Cas13 effector protein, containing two higher eukaryotes and prokaryotes nucleotide-binding domains (HEPN) responsible for RNase activity. Similar to the previously mentioned for CRISPR/Cas9, the Cas13 associated with crRNA forms the effector complex, which in this case is an RNA-guided complex that targets and cleaves ssRNA. The nuclease domain(s) of the large effector protein is responsible for processing the pre-crRNA processing and cleavage of the ssRNA [8,25,35].

Applying data analysis and bioinformatics approaches, Shmakov and coworkers analyzed the whole microbial genome sequences from the National Centre for Biotechnology Information, based on the incidence of Cas1 (most conserved Cas protein gene) aiming to identify the unclassified candidate Class 2 CRISPR *loci*. As a result, a new Class 2 effector type was predicted, the C2c2 (Class 2 candidate2) or VI-A, using Cas1 as the seed [34]. This first presumed type VI effector, C2c2 or VI-A, demonstrated unique properties compared to any other Cas protein, therefore, it is now designated Cas13a and was assigned to a novel type (Class 2, type VI) [30,56,73]. The hypothesis that Cas13a presents an association between HEPN domains and RNase, acting as an RNA-guided RNase and being able to target RNA, was experimentally confirmed when Abudayayyeh and coworkers (2016), showed that type VI Cas13a effector possessed a ssRNA-targeting capability in RNA bacteriophage MS2 [74], facilitating interference and pre-crRNA processing [75].

Therefore, Cas13a was the first type VI ribonuclease identified (with an average size of 1250 amino acids) that can efficiently target and degrade ssRNA, but not double-stranded RNA (dsRNA). This system was characterized in *Leptotrichia sharii* (Lsh) [8,76], although it has many orthologs such as *Listeria seeligeri* (Lse), *Leptotrichia wadei* (Lwa) [77], *Leptotrichia buccalis* (Lbu) [75] and *Lachnospiraceae bacterium* (Lba) [78]. The Cas13a *locus* is composed of an adaptation module (Cas1 and Cas2), two HEPN domains, and a CRISPR array [73,74] (Figure 16.3). Thus, the crRNA-Cas13a complex is bilobed consisting of a nuclease lobe (NUC lobe) and a crRNA recognition lobe (REC lobe) [73]. NUC lobe contains HEPN domains, HEPN1 and HEPN2, with a linker domain in between, located on the outer surface and responsible for the cleavage of the target RNA outside the binding region. However, when this happens, the catalytic site of HEPN is exposed and available to all RNAs in a solution, which might result in some unspecific cleavage [56,79]. HEPN1 domain has the HEPN1 I subdomain and the HEPN1 II subdomain with a Helical-2 domain in between [79]. The REC lobe consists of an N-terminal domain (NTD) and a Helical-1 domain that catalyzes the maturation of the crRNA [79]. In these systems, the CRISPR array generally

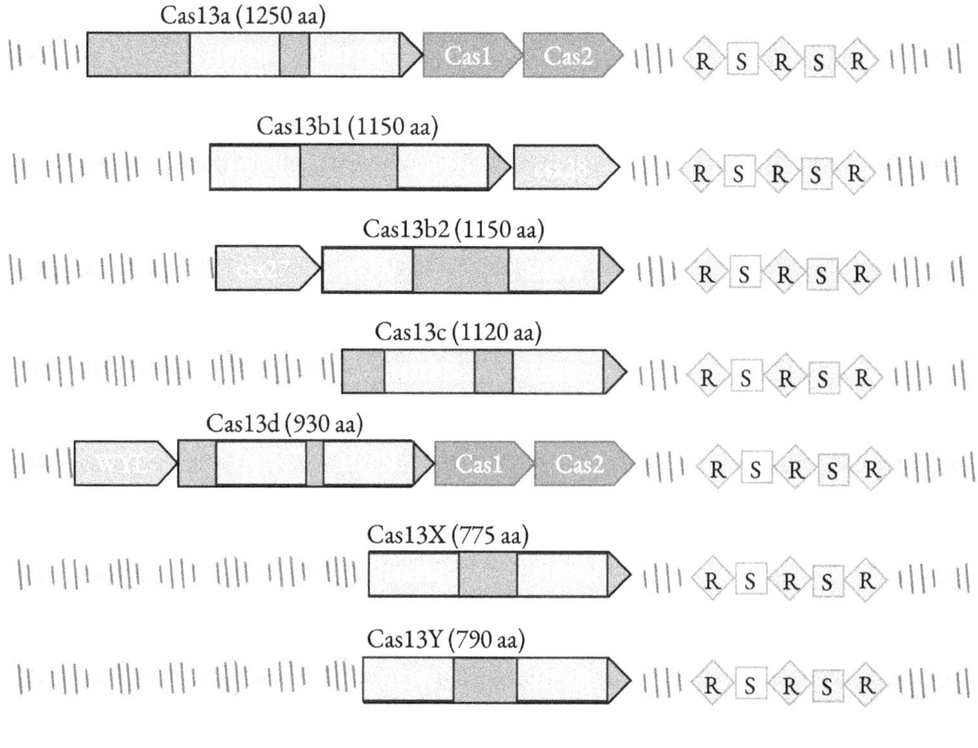

FIGURE 16.3 Constitutions of the different CRISPR-Cas13 subtypes. Genes are shown as arrows.

comprises a 5′ 28 nucleotides (nt) direct repeat (DR), which is typical of each ortholog, and a 28–30 nt spacer sequence (complementary to the RNA target). Some orthologs, such as LshCas13a, have a single base protospacer flanking site (PFS) preference, comprising A, U, or C nt bases (non-G) at the 3′ end of the spacer sequence (guide sequence) [8,74], whereas LwaCas13a and LbuCas13a do not show PFS preference. LwaCas13a system did not require a PFS motif, which improves the flexibility of Cas13a and makes it the variant of choice [77,80].

Nevertheless, not all CRISPR/Cas systems are autonomous, because Cas1 protein may be absent. In this case, they depend on adaptation modules (Cas1 and Cas2) of other CRISPR/Cas systems of the genome. Accordingly, the detection of these nonautonomous systems in the previous analysis, based on Cas1, was not possible [23,34,73]. However, following studies investigating the CRISPR/Cas *loci*, using the CRISPR repeat arrays as seed, were able to identify new Class 2 subtypes lacking the adaptation module. This research allowed the identification of two additional Class 2 type VI effector subtypes, containing HEPN domains: VI-B or C2c6 (Cas13b) and VI-C or C2c7 (Cas13c) [29,30,81,82].

The different subtypes that exist in type VI systems emphasize the diverse locations of HEPN domains of Cas13 and other additional features of the *locus* architecture. Also, the sequence similarity of the catalytic motif of the HEPN domain between the three groups (VI-A, VI-B, and VI-C) is extremely low which justifies their separation. Regarding subtype VI-B, although it also presents two HEPN domains and targets ssRNA, it encodes additional proteins that contain predicted transmembrane domains, which are significantly different from VI-A subtype, Csx27 and Csx28 (Figure 16.3). Csx27 can repress RNA targeting and Csx28 can enhance RNA cleavage. According to phylogenetic analysis, VI-B systems suffered an evolutionary divergence that resulted in VI-B1 and VI-B2 variants with distinct architectures of these associated predicted membrane proteins [56,73,82].

Cas13b was discovered using bioinformatic methods in species of Gram-negative bacteria, namely *Porphyromonas* sp. (PguCas13b) and *Prevotella* sp. (PspCas13b). However, there are other identified Cas13b orthologs such as *Bergeyella zoohelcum* (BzCas13b). These systems have the two HEPN domains positioned at N and C protein terminals and devoid of Cas1 and Cas2 proteins but include, as mentioned before, two additional proteins, Csx27 and Csx28 [82]. Cas13b has an average of 1150 amino acids and its DR on the 3′ end of crRNA (spacer length 30 nt) is in contrast to the 5′ DR present in Cas13a, and needs a double-sided PFS. BzCas13b and PguCas13b prefer 5′ PFS of A, U, or G (non-C) and 3′ PFS of NAN or NNA, maximizing their targeting ability. PspCas13b is an exception because it has no PFS requirement [8,80]. RNA-targeting in eukaryotes using PspCas13b has shown to have constantly greater efficiency than LwaCas13a, not only because it lacks the need for PFS, but also because it has demonstrated a lack of RNA collateral damage. Accordingly, PspCas13b is preferred for targeted RNA cleavage [73,83].

Cas13c was also first identified using a bioinformatic approach in *Fusobacterium* and *Clostirdium* but its functional characterization is much less complete than other Cas13 types. This protein has a similar *locus* and structure of the crRNA as Cas13a, with 1120 amino acids of length, however, the adaptation module (Cas1 and Cas2) is also absent, similar to Cas13b (Figure 16.3). This protein presents a DR on the 5′ end of crRNA with a spacer size of 28–30 nt. There is a lack of research available on this type of Cas13, mostly because it is less efficient at RNA targeting and interference in comparison to other Cas13 subtypes. The few studies that exist mostly employ the Cas13c ortholog *Fusobacterium perfoetens* (FpeCas13c) [8,73,80].

After upgrades on bioinformatic processes and with access to a higher number of datasets on genomics and metagenomics, it was possible to reveal a new Class 2 type VI effector protein, the Cas13d, subtype VI-D. This effector protein was identified mainly in *Eubacterium* and *Ruminococcus* [73,84]. The Cas13d from *Ruminococcus flavefaciens* XPD3002 (CasRx/RfxCas13d) is one of the most characterized variants [8]. Cas13d is smaller than Cas13a, Cas13b, and Cas13c effector proteins, with about 930 amino acids. REC lobe has NTD and Helical-1 domains and NUC lobe has domains HEPN-1, HEPN-2, and Helical-2. In addition, this system comprises a WYL domain with accessory proteins, one of which might positively modulate RNase activity, either targeted or collateral (Figure 16.3). The crRNA of Cas13d has a 30 nt 5′ DR followed by a variable spacer that can range from 14 to 26 nt. This CRISPR/Cas13 subtype is known for its versatility, since it has no PFS constraints, and employs rigorous sequence-specific RNA cleavage, which is promising for enhanced RNA interference compared to other systems [8,73,80].

Recently, other Cas13 protein variants were identified, Cas13X and Cas13Y, having the smallest size of 775 and 790 amino acids, respectively. Little is known about the structure of these proteins, they have two HEPN domains, located in the N- and C-terminus of the proteins, similar to the Cas13b subtype (Figure 16.3). Among Cas13X and Cas13Y, Cas13X1 exhibited the highest knockdown efficiency and showed no PFS bias [80,85].

BOX 1- CRISPR/CAS13 SYSTEMS IN A NUTSHELL

- Cas13 protein is the signature gene for type VI CRISPR systems. Based on the phylogeny of Cas13, features, and functional characterization, type VI CRISPR systems are classified into six subtypes: VI-A (effector protein is Cas13a/C2c2), VI-B (effector proteins are Cas13b1/C2c6 and Cas13b2), VI-C (effector protein is Cas13c/C2c7), VI-D (effector protein is Cas13d), VI-X (effector protein is Cas13X) and VI-Y (effector protein is Cas13Y) [8,29,30,73,74].
- The identification of these subtypes was first accomplished by using data mining and bioinformatic approaches, either by using Cas 1 as a seed (for the identification of Cas13a) or by using the CRISPR array as a seed (for the following Cas13 subtypes) [23,29,34,73].
- Although the effector protein sizes and primary sequence differ among Cas13 subtypes, they all share a common feature, which is the presence of two HEPN domains that provide RNase activity. An RNA-guided RNA targeting complex is formed with a crRNA to recognize and cleave ssRNA targets [8,30,34,73]. Therefore, all Cas13 proteins have two enzymatically distinct RNase activities, including processing pre-crRNA and degradation of target RNA [8,30,75].
- Cas13 effectors have a bilobed structure with NUC and REC lobes however, the sequence of nucleotide bases and the organization of the domains can be very different [79]. The NUC lobe contains HEPN domains, for RNA cleavage, which have different locations and are uniquely spaced based on the subtype of Cas13 protein [30,34,73]. In Cas13a, Cas13c, and Cas13d the HEPN domains are located at the centre and C-terminus. On the other hand, in Cas13b, Cas13X, and Cas13Y, HEPN domains are present at the N-terminus and C-terminus of the proteins. These domains can cleave the target RNA but also show nonspecific collateral cleavage that results in the degradation of the RNA near the Cas13 system. Nonetheless, some CRISPR/Cas13 variants showed a lack of collateral RNA damage, such as PspCas13b [73,83]. REC lobe has an NTD and a Helical 1 domain functional for pre-processing and interaction with sgRNA [79]. The length of the crRNA sequence varies from 24 to 30 nt.

Table 16.1 shows a schematic summary of CRISPR/Cas13 classification, according to the effector protein and structural composition.

16.3.2 Molecular Mechanism and Application of CRISPR/Cas13 Systems against Plant Viruses

As well as in other CRISPR/Cas systems the molecular mechanisms of adaptative immunity of CRISPR/Cas13 encompasses three steps: adaptation, expression, and interference. The adaptation

TABLE 16.1
CRISPR/Cas13 Classification

Type of Cas13	Length (aa)	Orthologs	Structural Composition	Functional Region	Reference
Cas13a or C2c2 (VI-A)	1,250	LshCas13a LseCas13a LwaCas13a LbuCas13a LbaCas13a	HEPN domains at center and C-terminus; 5′ end DR; 3′ non-G PFS preference (Lwa and LbuCas13a with no preference).	SsRNA (28–30nt spacer sequence)	[74,75]
Cas13b or C2c6 (VI-B)	1,150	PguCas13b PspCas13b BzCas13b	HEPN domains at N and C-terminus; 5′ non-C PFS preference; 3′ PFS NAN/NNA (except PspCas13b); 3′ end DR.	ssRNA (30nt spacer sequence)	[82,86]
Cas13c (VI-C)	1,120	FpeCas13c	HEPN domains at centre and C-terminus; 5′ end DR; No PFS preference.	ssRNA (28–30nt spacer sequence)	[30,87]
Cas13d (VI-D)	930	RfxCas13d	HEPN domains at centre and C-terminus; 5′ end DR; No PFS preference.	ssRNA (23–30nt spacer sequence)	[80,88]
Cas13X (VI-X)	775		HEPN domains at N and C-terminus; No PFS preference.	ssRNA	[80]
Cas13Y (VI-Y)	790		HEPN domains at N and C-terminus; No PFS preference.	ssRNA	[80]

Source: Adapted from Kavuri et al. [8].

step consists of acquiring new spacers; however, this mechanism is poorly understood in type VI systems compared to others. This is mainly because some CRISPR/Cas13 systems subtypes lack Cas1 and Cas2, the known adaptation modules, so they might require adaptation factors of other systems existing in the genome [34,89]. In the expression phase, the production of the crRNA and the effector nuclease takes place (surveillance complex). Transcription of the pre-crRNA into crRNA is accomplished by the REC lobe of Cas13, in contrast to other CRISPR/Cas systems that require either a tracrRNA or an endogenous RNase [75,89]. A hairpin flanked by a spacer sequence is formed by the DR region and the spacer within a crRNA is set for a precise RNA target. This complex will mediate recognition and binding of the CRISPR/Cas13 to the target RNA which will be surrounded by the nuclease core, and the catalytic nuclease is activated to efficiently cleave the target RNA [73,89]. Unlike, for instance, Cas9 effectors, Cas13 proteins do not need the existence of a protospacer adjacent motif (PAM) to recognize target RNA; however, some Cas13 subtypes show a preference for specific nucleotides flanking the 3′ region of the protospacer, the PFS [77]. In addition, collateral cleavage can also happen upon target recognition, which means not only do they target RNA but also can cleave indiscriminate bystander RNA. This is because the catalytic region and the crRNA: target RNA binding site are located on opposite sides, which makes it reachable from the outside of the complex. Nevertheless, this collateral cleavage activity of Cas13 is still controversial in eukaryotes, since it seems to be absent in some organisms [90].

The first CRISPR/Cas13 system to be studied for RNA-targeting used the effector protein ortholog LshCas13a. Immunity against *Turnip mosaic virus* (TuMV) in *N. benthamiana* and *A. thaliana* was developed and demonstrated that this system mediates specific RNA virus targeting in plants, although with moderate efficiency. LshCas13a interference was tested by transforming a plant codon-optimized LshCas13a into *N. benthamiana* leaves for transient and transgenic expression, using four different crRNAs (GFP1, GFP2, the *helper component proteinase silencing suppressor,* and *coat protein* sequences) to target the GFP-tagged TuMV genome. The observation of the GFP signal under UV light allowed the measurement of viral incidence. Collected data showed a 50% reduction in the transcript levels for GFP2 and the *helper component proteinase silencing suppressor*. Another similar study in *A. thaliana* proved the heritable immunity against TuMV up to T2 generation [64,91]. In addition, LshCas13a has become a promising Cas13 ortholog to accomplish immunity against a wide range of RNA viruses in different plant species, including against *Potato virus Y* in *Solanum tuberosum* [66] and against *Southern rice black-streaked dwarf virus* and *Rice stripe mosaic virus* in either *N. benthamiana* and *Oryza sativa* [65].

In addition to the well-characterized LshCas13a, other protein effectors have been studied to identify better variants against plant viruses, that would allow higher interference efficiency. For instance, LwaCas13a has been reported to mediate a stronger RNA-targeting activity than LshCas13a however, it does require a stabilizer fusion, such as msfGFP for an efficient interference activity [77]. In addition, other Cas13 subtypes were also exploited for developing resistance to plant viruses, such as BzCas13b, PspCas13b, and RfxCas13d [92].

Studies have shown that LshCas13a, LwaCas13a, BzCas13b, PspCas13b and RfxCas13d can be used against TuMV, TMV and *Potato virus X* in *N. benthamiana* [8,92] and LshCas13a, LwaCas13a, PspCas13b and RfxCas13d can also be efficient against *Cucumber mosaic virus* and *Sweet potato chlorotic stunt virus-RNase3* in *N. benthamiana* and *Ipomoea batatas* [8,93]. The virus RNA interference data suggested that all other variants (LwaCas13a, PspCas13b and RfxCas13d) were more efficient than LshCas13a [92]. Furthermore, RfxCas13d was identified as the most effective Cas13 ortholog for RNA targeting in *N. benthamiana* in transient and stable assays and showed to be a highly specific RNA targeting system that lacks collateral activities *in planta*. Multiplexed virus interference was also achieved using this ortholog, by targeting two different RNA viruses simultaneously [92].

16.3.3 Potential Limitations and Future Directions of CRISPR/Cas13 Systems

Despite all of the biotechnical and agricultural applications of CRISPR/Cas13 systems, there are still limitations regarding this technology. Despite the many studies that have arisen, it is a very recent technology that needs further investigation in different plant species and against diverse plant viruses. The newly identified subtypes of CRISPR/Cas13 systems (Cas13X and Cas13Y) are not fully characterized and their mechanism is not completely understood [8,73]. Therefore, additional information is needed on these effector proteins to comprehend their potential for RNA targeting and generating immunity against plant viruses.

In addition, the collateral RNase activity present in these systems may lead to the degradation of non-target RNAs, which was observed *in vitro* and bacterial cells [90]. However, in plant cells, no collateral activity was observed and even though there are Cas13 variants that showed to lack this collateral activity, it is important to ensure that this will not be a problem in future studies on CRISPR/Cas13 immunity in plants against RNA viruses [73].

Another important concern would be that these systems rely on a spacer in crRNA that is specific for a target ssRNA, so cleavage sites and cleavage patterns for a precise target transcript cannot be changed, otherwise, the system would not be successful for RNA interference [74,82]. Given the rapid adaptability of plant viruses, mutations in their genome can occur or new viruses may emerge, which will demand adjusting these systems to ensure plant immunity.

CRISPR/Cas13 RNA targeting systems are very promising RNA technologies with multiple advantages over the ones that previously existed, such as their robustness, specificity, easy design, and affordable price. Cas13 structural and functional variants that have a single effector protein allow not only studies on viral RNA interference, which have been the main focus but also Cas13-mediated knockdown of endogenous mRNA and targeting of non-coding RNAs (long non-coding RNA, microRNA, and circular RNA) to understand their role in plants [8,73].

16.4 CONCLUSION

Plant viruses can cause destructive diseases in several agricultural systems which highlights the need for new methods of control and monitoring, in view of a more sustainable agriculture, able to prevent extreme economic impacts worldwide and guaranteeing both food safety and food security. The development of virus-resistant plants is of utmost importance nowadays, since plant viruses can easily spread throughout crops and there are no chemical products available for their control in the field, making prevention essential to avoid catastrophic losses.

CRISPR/Cas systems have been studied over the last decades. These are adaptative immune systems that prokaryotes present as a result of their defense against viruses, and that have been found by scientists to have potential as gene editing tools. CRISPR/Cas immunity involves adaptation, expression, and interference stages and a variety of different systems has already been identified. According to multiple criteria, such as signature genes, the phylogeny of Cas1 (the best conserved Cas protein), the organization of the *loci,* and the structure of the CRISPR themselves, CRISPR/Cas systems are classified within Class 1 and Class 2, with different types and subtypes. Several applications for this technology have been found and some are still being studied, either for crop improvement (yield, quality, disease resistance, herbicide resistance, breeding technologies, or domestication of wild species) or plant biotechnology (delivery systems, gene regulation, multiplex editing and mutagenesis and directed evolution).

In this chapter, we focus on the use of CRISPR/Cas systems against plant viruses. The role of CRISPR/Cas13 was highlighted, mostly because of recently described systems that can target and cleave RNA virus genomes or transcripts, which is extremely important since most plant viruses are RNA viruses or have RNA intermediates. Cas13 is the signature gene for CRISPR/Cas13, type VI CRISPR systems, which can be classified into six subtypes (VI-A, effector protein Cas13a; VI-B, effector protein Cas13b; VI-C, effector protein Cas13c; VI-D, effector protein Cas13d; VI-X, effector protein Cas13X; and VI-Y, effector protein Cas13Y). Each subtype can have different variants and according to the available information, RfxCas13d was the most effective and promising effector protein variant. Although there are some limitations to the use of these systems, further analysis and more information are needed. Prospects indicate that these systems hold promise as technologies for creating virus-resistant plants to be used in the field.

REFERENCES

1. Wang, M.B.; Masuta, C.; Smith, N.A.; Shimura, H. RNA silencing and plant viral diseases. *Mol. Plant-Microbe Interact.* 2012, *25*, 1275–1285, doi:10.1094/MPMI-04-12-0093-CR.
2. Zaitlin, M.; Palukaitis, P. Advances in understantding plant ciruses and virus diseases. *Annu. Rev. Phytopathol.* 2000, *38*, 117–143, doi:10.1146/annurev.phyto.38.1.117.
3. Varanda, C.M.R.; Félix, M.D.R.; Campos, M.D.; Patanita, M.; Materatski, P. Plant viruses: From targets to tools for crispr. *Viruses* 2021, *13*, 1–19, doi:10.3390/v13010141.
4. Cao, Y.; Zhou, H.; Zhou, X.; Li, F. Control of plant viruses by CRISPR/Cas system-mediated adaptive immunity. *Front. Microbiol.* 2020, *11*, 1–9, doi:10.3389/fmicb.2020.593700.
5. Gómez, P.; Rodríguez-Hernández, A.M.; Moury, B.; Aranda, M. Genetic resistance for the sustainable control of plant virus diseases: Breeding, mechanisms and durability. *Eur. J. Plant Pathol.* 2009, *125*, 1–22, doi:10.1007/s10658-009-9468-5.
6. Galvez, L.C.; Banerjee, J.; Pinar, H.; Mitra, A. Engineered plant virus resistance. *Plant Sci.* 2014, *228*, 11–25, doi:10.1016/j.plantsci.2014.07.006.
7. Wang, T.; Zhang, H.; Zhu, H. CRISPR technology is revolutionizing the improvement of tomato and other fruit crops. *Hortic. Res.* 2019, *6*, doi:10.1038/s41438-019-0159-x.
8. Kavuri, N.R.; Ramasamy, M.; Qi, Y.; Mandadi, K. Applications of CRISPR/Cas13-based RNA editing in plants. *Cells* 2022, *11*, 1–11, doi:10.3390/cells11172665.
9. Mojica, F.J.M.; Montoliu, L. On the origin of CRISPR-Cas technology: From prokaryotes to mammals. *Trends Microbiol.* 2016, *24*, 811–820, doi:10.1016/j.tim.2016.06.005.

10. Gossler, A.; Joyner, A.L.; Rossant, J.; Skarnes, W.C. Mouse embryonc stem cells and reporter constructs. *Science* 1988, *244*, 463–466, doi:10.1126/science.2497519.
11. Gossen, M.; Bujard, H. Tight control of gene expression in mammalian cells by tetracycline-responsive promoters. *Proc. Natl. Acad. Sci. U. S. A.* 1992, *89*, 5547–5551, doi:10.1073/pnas.89.12.5547.
12. Gu, H.; Marth, J.D.; Orban, P.C.; Mossmann, H.; Rajewsky, K. Deletion of a DNA polymerase β gene segment in T cells using cell type-specific gene targeting. *Science.* 1994, *265*, 103–106, doi:10.1126/science.8016642.
13. Ishino, Y.; Shinagawa, H.; Makino, K.; Amemura, M.; Nakatura, A. Nucleotide sequence of the iap gene, responsible for alkaline phosphatase isoenzyme conversion in Escherichia coli, and identification of the gene product. *J. Bacteriol.* 1987, *169*, 5429–5433, doi:10.1128/jb.169.12.5429-5433.1987.
14. Mojica, F.J.M.; Juez, G.; Rodriguez-Valera, F. Transcription at different salinities of *Haloferax mediterranei* sequences adjacent to partially modified *Pst*I sites. *Mol. Microbiol.* 1993, *9*, 613–621, doi:10.1111/j.1365-2958.1993.tb01721.x.
15. Mojica, F.J.M.; Ferrer, C.; Juez, G.; Rodríguez-Valera, F. Long stretches of short tandem repeats are present in the largest replicons of the *Archaea Haloferax mediterranei* and *Haloferax volcanii* and could be involved in replicon partitioning. *Mol. Microbiol.* 1995, *17*, 85–93, doi:10.1111/j.1365-2958.1995.mmi_17010085.x.
16. Mojica, F.J.M.; Díez-Villaseñor, C.; García-Martínez, J.; Soria, E. Intervening sequences of regularly spaced prokaryotic repeats derive from foreign genetic elements. *J. Mol. Evol.* 2005, *60*, 174–182, doi:10.1007/s00239-004-0046-3.
17. Barrangou, R.; Fremaux, C.; Deveau, H.; Richardss, M.; Boyaval, P.; Moineau, S.; Romero, D.A.; Horvath, P.; Richards, M.; Boyaval, P.; et al. CRISPR provides against viruses in prokaryotes. *Science.* 2007, *315*, 1709–1712.
18. Jinek, M.; Chylinski, K.; Fonfara, I.; Hauer, M.; Doudna, J.A.; Charpentier, E. A programmable dual-RNA - guided. *Science.* 2012, *337*, 816–822.
19. Cong, L.; Ran, F.A.; Cox, D.; Lin, S.; Barretto, R.; Habib, N.; Hsu, P.D.; Wu, X.; Jiang, W.; Marraffini, L.A.; et al. Multiplex genome engineering using CRISPR/Cas systems. *Science.* 2013, *339*, 816–819.
20. Shan, Q.; Wang, Y.; Li, J.; Zhang, Y.; Chen, K.; Liang, Z.; Zhang, K.; Liu, J.; Xi, J.J.; Qiu, J.-L.; et al. Targeted genome modification of crop plants using a CRISPR-Cas system. *Nat. Biotechnol.* 2013, *31*, 8–10, doi:10.1038/nbt.2650.
21. Nekrasov, V.; Staskawicz, B.; Weigel, D.; Jones, J.D.G.; Kamoun, S. Targeted mutagenesis in the model plant *Nicotiana benthamiana* using Cas9 RNA-guided endonuclease. *Nat. Biotechnol.* 2013, *31*, 691–693, doi:10.1038/nbt.2655.
22. Zhu, H.; Li, C.; Gao, C. Applications of CRISPR-Cas in agriculture and plant biotechnology. *Nat. Rev. Mol. Cell Biol.* 2020, *21*, 661–677, doi:10.1038/s41580-020-00288-9.
23. Makarova, K.S.; Wolf, Y.I.; Alkhnbashi, O.S.; Costa, F.; Shah, S.A.; Saunders, S.J.; Barrangou, R.; Brouns, S.J.J.; Charpentier, E.; Haft, D.H.; et al. An updated evolutionary classification. *Nat. Rev. Microbiol.* 2015, *13*, 722–736, doi:10.1038/nrmicro3569.
24. Nuñez, J.K.; Kranzusch, P.J.; Noeske, J.; Wright, A. V; Davies, C.W.; Doudna, J.A. Cas1 - Cas2 complex formation mediates spacer acquisition during CRISPR - Cas adaptive immunity. *Nat. Struct. Mol. Biol.* 2014, *21*, doi:10.1038/nsmb.2820.
25. Deltcheva, E.; Chylinski, K.; Sharma, C.M.; Gonzales, K.; Chao, Y.; Pirzada, Z.A.; Charpentier, E.; Eckert, M.R. CRISPR RNA maturation by trans-encoded small RNA and host factor RNase III. *Nature* 2011, *471*, 602–607, doi:10.1038/nature09886.
26. Makarova, K.S.; Haft, D.H.; Barrangou, R.; Brouns, S.J.J.; Charpentier, E.; Horvath, P.; Moineau, S.; Mojica, F.J.M.; Wolf, Y.I.; Yakunin, A.F.; et al. Evolution and classification of the CRISPR-Cas systems. *Nat. Rev. Microbiol.* 2011, *9*, 467–477, doi:10.1038/nrmicro2577.
27. Stern, A.; Sorek, R. The phage-host arms race: Shaping the evolution of microbes. *BioEssays* 2011, *33*, 43–51, doi:10.1002/bies.201000071.
28. Makarova, K.S.; Wolf, Y.I.; Koonin, E. V. Classification and nomenclature of CRISPR-Cas systems: Where from here? *Cris. J.* 2018, *1*, 325–336, doi:10.1089/crispr.2018.0033.
29. Koonin, E. V.; Makarova, K.S.; Zhang, F. Diversity, classification and evolution of CRISPR-Cas systems. *Curr. Opin. Microbiol.* 2017, *37*, 67–78, doi:10.1016/j.mib.2017.05.008.
30. Shmakov, S.; Smargon, A.; Scott, D.; Cox, D.; Pyzocha, N.; Yan, W.; Abudayyeh, O.O.; Gootenberg, J.S.; Makarova, K.S.; Wolf, Y.I.; et al. Diversity and evolution of class 2 CRISPR-Cas systems. *Nat. Rev. Microbiol.* 2017, *15*, 169–182, doi:10.1038/nrmicro.2016.184.

31. Spilman, M.; Cocozaki, A.; Hale, C.; Shao, Y.; Ramia, N.; Terns, R.; Terns, M.; Li, H.; Stagg, S. Structure of an RNA silencing complex of the CRISPR-Cas immune system. *Mol. Cell* 2013, *52*, 146–152, doi:10.1016/j.molcel.2013.09.008.
32. Rouillon, C.; Zhou, M.; Zhang, J.; Politis, A.; Beilsten-Edmands, V.; Cannone, G.; Graham, S.; Robinson, C. V.; Spagnolo, L.; White, M.F. Structure of the CRISPR interference complex CSM reveals key similarities with cascade. *Mol. Cell* 2013, *52*, 124–134, doi:10.1016/j.molcel.2013.08.020.
33. Wiedenheft, B.; Lander, G.C.; Zhou, K.; Jore, M.M.; Brouns, S.J.J.; Van Der Oost, J.; Doudna, J.A.; Nogales, E. Structures of the RNA-guided surveillance complex from a bacterial immune system. *Nature* 2011, *477*, 486–489, doi:10.1038/nature10402.
34. Shmakov, S.; Abudayyeh, O.O.; Makarova, K.S.; Wolf, Y.I.; Gootenberg, J.S.; Semenova, E.; Minakhin, L.; Joung, J.; Konermann, S.; Severinov, K.; et al. Discovery and functional characterization of diverse Class 2 CRISPR-Cas systems. *Mol. Cell* 2015, *60*, 385–397, doi:10.1016/j.molcel.2015.10.008.
35. Liu, L.; Chen, P.; Wang, M.; Li, X.; Wang, J.; Yin, M.; Wang, Y. C2c1-sgRNA complex structure reveals RNA-guided DNA cleavage mechanism. *Mol. Cell* 2017, *65*, 310–322, doi:10.1016/j.molcel.2016.11.040.
36. Gong, B.; Shin, M.; Sun, J.; Jung, C.H.; Bolt, E.L.; Van Der Oost, J.; Kim, J.S. Molecular insights into DNA interference by CRISPR-associated nuclease-helicase Cas3. *Proc. Natl. Acad. Sci. U. S. A.* 2014, *111*, 16359–16364, doi:10.1073/pnas.1410806111.
37. Redding, S.; Sternberg, S.H.; Marshall, M.; Gibb, B.; Bhat, P.; Guegler, C.K.; Wiedenheft, B.; Doudna, J.A.; Greene, E.C. Surveillance and processing of foreign DNA by the Escherichia coli CRISPR-Cas system. *Cell* 2015, *163*, 854–865, doi:10.1016/j.cell.2015.10.003.
38. Wolter, F.; Puchta, H. The CRISPR/Cas revolution reaches the RNA world: Cas13, a new Swiss Army knife for plant biologists. *Plant J.* 2018, *94*, 767–775, doi:10.1111/tpj.13899.
39. Wright, D.A.; Townsend, J.A.; Winfrey, R.J.; Irwin, P.A.; Rajagopal, J.; Lonosky, P.M.; Hall, B.D.; Jondle, M.D.; Voytas, D.F. High-frequency homologous recombination in plants mediated by zinc-finger nucleases. *Plant J.* 2005, *44*, 693–705, doi:10.1111/j.1365-313X.2005.02551.x.
40. Christian, M.; Cermak, T.; Doyle, E.L.; Schmidt, C.; Zhang, F.; Hummel, A.; Bogdanove, A.J.; Voytas, D.F. Targeting DNA double-strand breaks with TAL effector nucleases. *Genetics* 2010, *186*, 756–761, doi:10.1534/genetics.110.120717.
41. Mali, P.; Yang, L.; Esvelt, K.M.; Aaach, J.; Guell, M.; DiCarlo, J.E.; Norville, J.E.; Church, G.M. RNA-guided human genome engineering via Cas9. *Science* 2013, *339*, 823–826, doi:10.1126/science.1232033.
42. Komor, A.C.; Kim, Y.B.; Packer, M.S.; Zuris, J.A.; Liu, D.R. Programmable editing of a target base in genomic DNA without double-stranded DNA cleavage. *Nature* 2016, *533*, 420–424, doi:10.1038/nature17946.
43. Gaudelli, N.M.; Komor, A.C.; Rees, H.A.; Packer, M.S.; Badran, A.H.; Bryson, D.I.; Liu, D.R. Programmable base editing of T to G C in genomic DNA without DNA cleavage. *Nature* 2017, *551*, 464–471, doi:10.1038/nature24644.
44. Li, C.; Zhang, R.; Meng, X.; Chen, S.; Zong, Y.; Lu, C.; Qiu, J.L.; Chen, Y.H.; Li, J.; Gao, C. Targeted, random mutagenesis of plant genes with dual cytosine and adenine base editors. *Nat. Biotechnol.* 2020, *38*, 875–882, doi:10.1038/s41587-019-0393-7.
45. Wang, S.; Zong, Y.; Lin, Q.; Zhang, H.; Chai, Z.; Zhang, D.; Chen, K.; Qiu, J.L.; Gao, C. Precise, predictable multi-nucleotide deletions in rice and wheat using APOBEC-Cas9. *Nat. Biotechnol.* 2020, *38*, 1460–1465, doi:10.1038/s41587-020-0566-4.
46. Wang, C.; Wang, G.; Gao, Y.; Lu, G.; Habben, J.E.; Mao, G.; Chen, G.; Wang, J.; Yang, F.; Zhao, X.; et al. A cytokinin-activation enzyme-like gene improves grain yield under various field conditions in rice. *Plant Mol. Biol.* 2020, *102*, 373–388, doi:10.1007/s11103-019-00952-5.
47. Zhang, Z.; Hua, L.; Gupta, A.; Tricoli, D.; Edwards, K.J.; Yang, B.; Li, W. Development of an *Agrobacterium*-delivered CRISPR/Cas9 system for wheat genome editing. *Plant Biotechnol. J.* 2019, *17*, 1623–1635, doi:10.1111/pbi.13088.
48. Yuste-Lisbona, F.J.; Fernández-Lozano, A.; Pineda, B.; Bretones, S.; Ortíz-Atienza, A.; García-Sogo, B.; Müller, N.A.; Angosto, T.; Capel, J.; Moreno, V.; et al. ENO regulates tomato fruit size through the floral meristem development network. *Proc. Natl. Acad. Sci. U. S. A.* 2020, *117*, 8187–8195, doi:10.1073/pnas.1913688117.
49. Dong, O.X.; Yu, S.; Jain, R.; Zhang, N.; Duong, P.Q.; Butler, C.; Li, Y.; Lipzen, A.; Martin, J.A.; Barry, K.W.; et al. Marker-free carotenoid-enriched rice generated through targeted gene insertion using CRISPR-Cas9. *Nat. Commun.* 2020, *11*, 1–10, doi:10.1038/s41467-020-14981-y.

50. Gao, H.; Gadlage, M.J.; Lafitte, H.R.; Lenderts, B.; Yang, M.; Schroder, M.; Farrell, J.; Snopek, K.; Peterson, D.; Feigenbutz, L.; et al. Superior field performance of waxy corn engineered using CRISPR-Cas9. *Nat. Biotechnol.* 2020, *38*, 579–581, doi:10.1038/s41587-020-0444-0.
51. Do, P.T.; Nguyen, C.X.; Bui, H.T.; Tran, L.T.N.; Stacey, G.; Gillman, J.D.; Zhang, Z.J.; Stacey, M.G. Demonstration of highly efficient dual gRNA CRISPR/Cas9 editing of the homeologous GmFAD2-1A and GmFAD2-1B genes to yield a high oleic, low linoleic and α-linolenic acid phenotype in soybean. *BMC Plant Biol.* 2019, *19*, 1–14, doi:10.1186/s12870-019-1906-8.
52. Peng, A.; Chen, S.; Lei, T.; Xu, L.; He, Y.; Wu, L.; Yao, L.; Zou, X. Engineering canker-resistant plants through CRISPR/Cas9-targeted editing of the susceptibility gene CsLOB1 promoter in citrus. *Int. J. Lab. Hematol.* 2017, *15*, 1509–1519, doi:10.1111/pbi.12733.
53. Xu, Z.; Xu, X.; Gong, Q.; Li, Z.; Li, Y.; Wang, S.; Yang, Y.; Ma, W.; Liu, L.; Zhu, B.; et al. Engineering broad-spectrum bacterial blight resistance by simultaneously disrupting variable TALE-binding elements of multiple susceptibility genes in rice. *Mol. Plant* 2019, *12*, 1434–1446, doi:10.1016/j.molp.2019.08.006.
54. Liu, L.; Kuang, Y.; Yan, F.; Li, S.; Ren, B.; Gosavi, G.; Spetz, C.; Li, X.; Wang, X.; Zhou, X.; et al. Developing a novel artificial rice germplasm for dinitroaniline herbicide resistance by base editing of OsTubA2. *Plant Biotechnol. J.* 2021, *19*, 5–7, doi:10.1111/pbi.13430.
55. Zhang, R.; Liu, J.; Chai, Z.; Chen, S.; Bai, Y.; Zong, Y.; Chen, K.; Li, J.; Jiang, L.; Gao, C. Generation of herbicide tolerance traits and a new selectable marker in wheat using base editing. *Nat. Plants* 2019, *5*, 480–485, doi:10.1038/s41477-019-0405-0.
56. Robertson, G.; Burger, J.; Campa, M. CRISPR/Cas-based tools for the targeted control of plant viruses. *Mol. Plant Pathol.* 2022, *23*, 1701–1718, doi:10.1111/mpp.13252.
57. Ali, Z.; Abul-Faraj, A.; Li, L.; Ghosh, N.; Piatek, M.; Mahjoub, A.; Aouida, M.; Piatek, A.; Baltes, N.J.; Voytas, D.F.; et al. Efficient virus-mediated genome editing in plants using the CRISPR/Cas9 system. *Mol. Plant* 2015, *8*, 1288–1291, doi:10.1016/j.molp.2015.02.011.
58. Baltes, N.J.; Hummel, A.W.; Konecna, E.; Cegan, R.; Bruns, A.N.; Bisaro, D.M.; Voytas, D.F. Conferring resistance to geminiviruses with the CRISPR-Cas prokaryotic immune system. *Nat. Plants* 2015, *1*, 4–7, doi:10.1038/NPLANTS.2015.145.
59. Ji, X.; Zhang, H.; Zhang, Y.; Wang, Y.; Gao, C. Establishing a CRISPR-Cas-like immune system conferring DNA virus resistance in plants. *Nat. Plants* 2015, *1*, 1–4, doi:10.1038/NPLANTS.2015.144.
60. Liu, H.; Soyars, C.L.; Li, J.; Fei, Q.; He, G.; Peterson, B.A.; Meyers, B.C.; Nimchuk, Z.L.; Wang, X. CRISPR/Cas9-mediated resistance to cauliflower mosaic virus. *Plant Direct* 2018, *2*, 1–9, doi:10.1002/pld3.47.
61. Tashkandi, M.; Ali, Z.; Aljedaani, F.; Shami, A.; Mahfouz, M.M. Engineering resistance against Tomato yellow leaf curl virus via the CRISPR/Cas9 system in tomato. *Plant Signal. Behav.* 2018, *13*, 1–7, doi:10.1080/15592324.2018.1525996.
62. Kis, A.; Hamar, É.; Tholt, G.; Bán, R.; Havelda, Z. Creating highly efficient resistance against wheat dwarf virus in barley by employing CRISPR/Cas9 system. *Plant Biotechnol. J.* 2019, *17*, 1004–1006, doi:10.1111/pbi.13077.
63. Zhang, T.; Zheng, Q.; Yi, X.; An, H.; Zhao, Y.; Ma, S.; Zhou, G. Establishing RNA virus resistance in plants by harnessing CRISPR immune system. *Plant Biotechnol J.* 2018, *16*, 1415–1423, doi:10.1111/pbi.12881.
64. Aman, R.; Mahas, A.; Butt, H.; Ali, Z.; Aljedaani, F.; Mahfouz, M. Engineering RNA virus interference via the CRISPR/Cas13 machinery in arabidopsis. *Viruses* 2018, *10*, doi:10.3390/v10120732.
65. Zhang, T.; Zhao, Y.; Ye, J.; Cao, X.; Xu, C.; Chen, B.; An, H.; Jiao, Y.; Zhang, F.; Yang, X.; et al. Establishing CRISPR/Cas13a immune system conferring RNA virus resistance in both dicot and monocot plants. *Plant Biotechnol. J.* 2019, *17*, 1185–1187, doi:10.1111/pbi.13095.
66. Zhan, X.; Zhang, F.; Zhong, Z.; Chen, R.; Wang, Y.; Chang, L.; Bock, R.; Nie, B.; Zhang, J. Generation of virus-resistant potato plants by RNA genome targeting. *Plant Biotechnol. J.* 2019, *17*, 1814–1822, doi:10.1111/pbi.13102.
67. Jiao, B.; Hao, X.; Liu, Z.; Liu, M.; Wang, J.; Liu, L.; Liu, N.; Song, R.; Zhang, J.; Fang, Y.; et al. Engineering CRISPR immune systems conferring GLRaV-3 resistance in grapevine. *Hortic. Res.* 2022, *9*, doi:10.1093/hr/uhab023.
68. Doebley, J.F.; Gaut, B.S.; Smith, B.D. The molecular genetics of crop domestication john. *Cell* 2006, *127*, 1309–1321, doi:10.1016.j.cell.2006.12.006.

69. Yang, X.P.; Yu, A.; Xu, C. [De novo domestication to create new crops]. *Yi Chuan* 2019, *41*, 827–835, doi:10.16288/j.yczz.19-151.
70. Li, T.; Yang, X.; Yu, Y.; Si, X.; Zhai, X.; Zhang, H.; Dong, W.; Gao, C.; Xu, C. Domestication of wild tomato is accelerated by genome editing. *Nat. Biotechnol.* 2018, *36*, 1160–1163, doi:10.1038/nbt.4273.
71. Zsögön, A.; Čermák, T.; Naves, E.R.; Notini, M.M.; Edel, K.H.; Weinl, S.; Freschi, L.; Voytas, D.F.; Kudla, J.; Peres, L.E.P. De novo domestication of wild tomato using genome editing. *Nat. Biotechnol.* 2018, *36*, 1211–1216, doi:10.1038/nbt.4272.
72. Ran, Y.; Liang, Z.; Gao, C. Current and future editing reagent delivery systems for plant genome editing. *Sci. China Life Sci.* 2017, *60*, 490–505, doi:10.1007/s11427-017-9022-1.
73. Tang, G.; Teotia, S.; Tang, X.; Singh, D. *RNA-Based Technologies for Functional Genomics in Plants*; 2021; Cham: Springer.
74. Abudayyeh, O.O.; Gootenberg, J.S.; Konermann, S.; Joung, J.; Slaymaker, I.M.; Cox, D.B.T.; Shmakov, S.; Makarova, K.S.; Semenova, E.; Minakhin, L.; et al. C2c2 is a single-component programmable RNA-guided RNA-targeting CRISPR effector. *Science* 2016, *353*, doi:10.1126/science.aaf5573.
75. East-Seletsky, A.; O'Connell, M.R.; Knight, S.C.; Burstein, D.; Cate, J.H.D.; Tjian, R.; Doudna, J.A. Two distinct RNase activities of CRISPR-C2c2 enable guide-RNA processing and RNA detection. *Nature* 2016, *538*, 270–273, doi:10.1038/nature19802.
76. Severinov, K.; Zhang, F.; Wolf, Y.I.; Shmakov, S.; Semenova, E.; Minakhin, L.; Makarova, K.S.; Koonin, E.; Konermann, S.; Joung, J.; et al. Novel CRISPR enzymes and systems. U.S. Patent Application 15/482,603. 2017.
77. Abudayyeh, O.O.; Gootenberg, J.S.; Essletzbichler, P.; Han, S.; Joung, J.; Belanto, J.J.; Verdine, V.; Cox, D.B.T.; Kellner, M.J.; Regev, A.; et al. RNA targeting with CRISPR-Cas13. *Nature* 2017, *550*, 280–284, doi:10.1038/nature24049.
78. Knott, G.J.; East-Seletsky, A.; Cofsky, J.C.; Holton, J.M.; Charles, E.; O'Connell, M.R.; Doudna, J.A. Guide-bound structures of an RNA-targeting A-cleaving CRISPR-Cas13a enzyme. *Nat. Struct. Mol. Biol.* 2017, *24*, 825–833, doi:10.1038/nsmb.3466.
79. Liu, L.; Li, X.; Ma, J.; Li, Z.; You, L.; Wang, J.; Wang, M.; Zhang, X.; Wang, Y. The molecular architecture for RNA-guided RNA cleavage by Cas13a. *Cell* 2017, *170*, 714–726.e10, doi:10.1016/j.cell.2017.06.050.
80. Xue, Y.; Chen, Z.; Zhang, W.; Zhang, J. Engineering CRISPR/Cas13 system against RNA viruses: From diagnostics to therapeutics. *Bioengineering* 2022, *9*, doi:10.3390/bioengineering9070291.
81. Burstein, D.; Harrington, L.B.; Strutt, S.C.; Probst, A.J.; Anantharaman, K.; Thomas, B.C.; Doudna, J.A.; Banfield, J.F. New CRISPR-Cas systems from uncultivated microbes. *Nature* 2017, *542*, 237–241, doi:10.1038/nature21059.
82. Smargon, A.A.; Cox, D.B.T.; Pyzocha, N.K.; Zheng, K.; Slaymaker, I.M.; Gootenberg, J.S.; Abudayyeh, O.A.; Essletzbichler, P.; Shmakov, S.; Makarova, K.S.; et al. Cas13b is a type VI-B CRISPR-associated RNA-guided RNase differentially regulated by accessory proteins Csx27 and Csx28. *Mol. Cell* 2017, *65*, 618–630.e7, doi:10.1016/j.molcel.2016.12.023.
83. Cox, D.B.; Gootenberg, J.S.; Abudayyeh, O.O.; Franklin, B.; Kellner, M.J.; Joung, J.; Zhang, F. RNA editing with CRISPR-Cas13. *Yearb. Paediatr. Endocrinol.* 2018, *15*, 14.11, doi:10.1530/ey.15.14.11.
84. Yan, W.X.; Chong, S.; Zhang, H.; Makarova, K.S.; Koonin, E. V.; Cheng, D.R.; Scott, D.A. Cas13d is a compact RNA-targeting type VI CRISPR effector positively modulated by a WYL-domain-containing accessory protein. *Mol. Cell* 2018, *70*, 327–339.e5, doi:10.1016/j.molcel.2018.02.028.
85. Xu, C.; Zhou, Y.; Xiao, Q.; He, B.; Geng, G.; Wang, Z.; Cao, B.; Dong, X.; Bai, W.; Wang, Y.; et al. Programmable RNA editing with compact CRISPR-Cas13 systems from uncultivated microbes. *Nat. Methods* 2021, *18*, 499–506, doi:10.1038/s41592-021-01124-4.
86. Freije, C.A.; Myhrvold, C.; Boehm, C.K.; Lin, A.E.; Welch, N.L.; Carter, A.; Metsky, H.C.; Luo, C.Y.; Abudayyeh, O.O.; Gootenberg, J.S.; et al. Programmable inhibition and detection of RNA viruses using Cas13. *Mol. Cell* 2019, *76*, 826–837.e11, doi:10.1016/j.molcel.2019.09.013.
87. Huynh, N.; Depner, N.; Larson, R.; King-Jones, K. A versatile toolkit for CRISPR-Cas13-based RNA manipulation in *Drosophila*. *Genome Biol.* 2020, *21*, 1–29, doi:10.1186/s13059-020-02193-y.
88. Wessels, H.H.; Méndez-Mancilla, A.; Guo, X.; Legut, M.; Daniloski, Z.; Sanjana, N.E. Massively parallel Cas13 screens reveal principles for guide RNA design. *Nat. Biotechnol.* 2020, *38*, 722–727, doi:10.1038/s41587-020-0456-9.
89. Kordyś, M.; Sen, R.; Warkocki, Z. Applications of the versatile CRISPR-Cas13 RNA targeting system. *Wiley Interdiscip. Rev. RNA* 2022, *13*, 1–30, doi:10.1002/wrna.1694.
90. Bot, J.F.; van der Oost, J.; Geijsen, N. The double life of CRISPR-Cas13. *Curr. Opin. Biotechnol.* 2022, *78*, 102789, doi:10.1016/j.copbio.2022.102789.

91. Aman, R.; Ali, Z.; Butt, H.; Mahas, A.; Aljedaani, F.; Khan, M.Z.; Ding, S.; Mahfouz, M. RNA virus interference via CRISPR/Cas13a system in plants. *Genome Biol.* 2018, *19*, 1–9, doi:10.1186/s13059-017-1381-1.
92. Mahas, A.; Aman, R.; Mahfouz, M. CRISPR-Cas13d mediates robust RNA virus interference in plants. *Genome Biol.* 2019, *20*, 1–16, doi:10.1186/s13059-019-1881-2.
93. Yu, Y.; Pan, Z.; Wang, X.; Bian, X.; Wang, W.; Liang, Q.; Kou, M.; Ji, H.; Li, Y.; Ma, D.; et al. Targeting of SPCSV-RNase3 via CRISPR-Cas13 confers resistance against sweet potato virus disease. *Mol. Plant Pathol.* 2022, *23*, 104–117, doi:10.1111/mpp.13146.

17 Genome Editing in Ornamental Plants

Current Findings and Future Perspectives

Kumaresan Kowsalya, Nandakumar Vidya, Packiaraj Gurusaravanan, Arumugam Vijaya Anand, Muthukrishnan Arun, and Bashyam Ramya

LIST OF ABBREVIATIONS

3GT	3-Glycosyl Transferase
AN4	Anthocyanin 4
CrRNA	CRISPR RNA
DFR-B	DihydroFlavonol-4-Reductase-B
DPL	Deep Purple
DSB	Double-Stranded Breaks
EFP	Enhancer of Flavonoid Production
GST1	Glutathione-S-Transferase
HDR	Homology-Directed Repair
InCCD4	Carotenoid Cleavage Dioxygenase
NHEJ	Non-homologous End Joining
PAM	Protospacer Adjacent Motif
SCF-SLF	Skp 1-Cullin-F-box-S-Locus F-box protein
TALEN	Transcription-Activator-Like Effector Nuclease
TracrRNA	Transactivating RNA
ZFN	Zinc Finger Nucleases
SgRNA	Single guide RNA
DSB	Double-stranded breaks

17.1 INTRODUCTION

Ornamental plants are highly desirable due to their attractive, vibrant, and colorful flowers, and they capture the attention of consumers in terms of pot plants as well as cut materials which are often propagated through seeds or vegetative cuttings. Among them, around 8,000 species of ornamentals exist worldwide in the ornamental flower industry. These attractive ornamental flowers have a dual commercial role as cut flowers for decorative purposes and fragrance raw material for the perfume industry. The demand for cut flowers has resulted in an escalation of global market value to 6.98 billion US dollars in 2017 and is estimated to reach nearly 11.68 billion US dollars by 2027 (Giovannini et al., 2021). Current ornamental plant breeding has evolved and progressed one step further to increase a broad spectrum of flower attributes such as innovative new colors, size, number

DOI: 10.1201/9781003387060-17

of flowers, shelf life, repeated blossoming, disease resistance, and enhanced nutrient uptake capability (De, 2017). To meet the industrial demand for ornamental flowers, several breeding strategies such as crossbreeding and mutation breeding have been employed to develop numerous cultivars. Unfortunately, the crossbreeding technique possesses certain limitations like breeding only with compatible species or closely related species resulting in disease susceptible hybrid plants. Also, these methods are laborious which consumes time and cost. For example, only 2%–3% of seedlings have the desired trait in 500,000 seedlings raised through crossbreeding, and proper selection of these plants in the upcoming 5–6 generations is a quite difficult process. Even, mutation breeding has several challenges which include introducing mutations in particular gene locations and resulting in chimeric mutants (Kishi-Kaboshi et al., 2018). On the other hand, mutation breeding also interchanges other unrelated genes during hybridization and requires the compulsory need of backcrossing to attain the closest form of a desirable trait. Moreover, transgenic technology surpasses the labor-intensive conventional breeding but possesses a few disadvantages like random integration of transferred genes in cells which affects and disrupts nearby functional neighbor genes. These shortcomings of conventional breeding and transgenic approaches can be addressed by using genome editing technologies (Khan et al., 2022). When compared to other genome editing tools like zinc finger nucleases (ZFN), transcription-activator like effector nuclease (TALEN), clustered regularly interspaced short palindromic repeats (CRISPR) technique outperforms as an effective choice of tool for targeted mutations which is cost-effective, less time consuming with reliability. Simultaneously, the CRISPR/Cas approach has attained numerous updated advances which involve CRISPRi, CRISPRa, prime editing, base editing, and several Cas enzyme variants like Cas12a, and dCas9-Fok I.

It has been suggested that CRISPR/Cas9 system is especially beneficial in ornamental plants. Moreover, this approach can be a promising platform for targeting unique sites that might provide specificity checks for a wide range of horticulture crops (Kaur et al., 2021). Lee et al. (2020) suggested that this technology can be effectively used to develop desirable traits (flower longevity, flowering time, color change, and fragrance improvement) by precisely editing a few bases. Hence, this chapter collectively summarizes the updates on applications of CRISPR/Cas-based genome editing in ornamental flowers.

17.2 MECHANISMS OF CRISPR/Cas TECHNOLOGY

CRISPR/Cas system utilizes the Cas9 nuclease enzyme that binds to CRISPR RNA (CrRNA) and trans-activating RNA (tracrRNA), collectively called single guide RNA, which is complementary to the distinct target sequences. This target region contains a short sequence known as the protospacer adjacent motif (PAM), next to the template DNA sequence in the downstream position. The CRISPR Cassette (Cas9/SgRNA) binds, cleaves, and processes the upstream portion of the PAM sequence (5′-NGG-3′ where N denotes nucleotide) (Corte et al., 2019). Simply, it helps the Cas9 enzyme to identify the target sequence in the genome of a cell. Another critical role of the PAM sequence is to avoid unnecessary endonuclease activity in DNA sequences other than the target sequence and is often referred to as a gatekeeper for the CRISPR cassette. In the target site, the endonuclease enzyme cleaves the DNA sequence causing double-stranded breaks (DSB). These double-stranded breaks in the target sequence are restored through DNA repair pathways such as non-homologous end joining (NHEJ) and homology-directed repair (HDR) (Figure 17.1) (Corte et al., 2019). NHEJ introduces the heterologous pool of insertion and deletion mutations (known as Indel mutations) in the broken ends of DNA in the specific target site. HDR repairs the target site with the help of template DNA using a homologous recombination strategy. In contrast to HDR, NHEJ repair occurs rapidly (Lee et al., 2020).

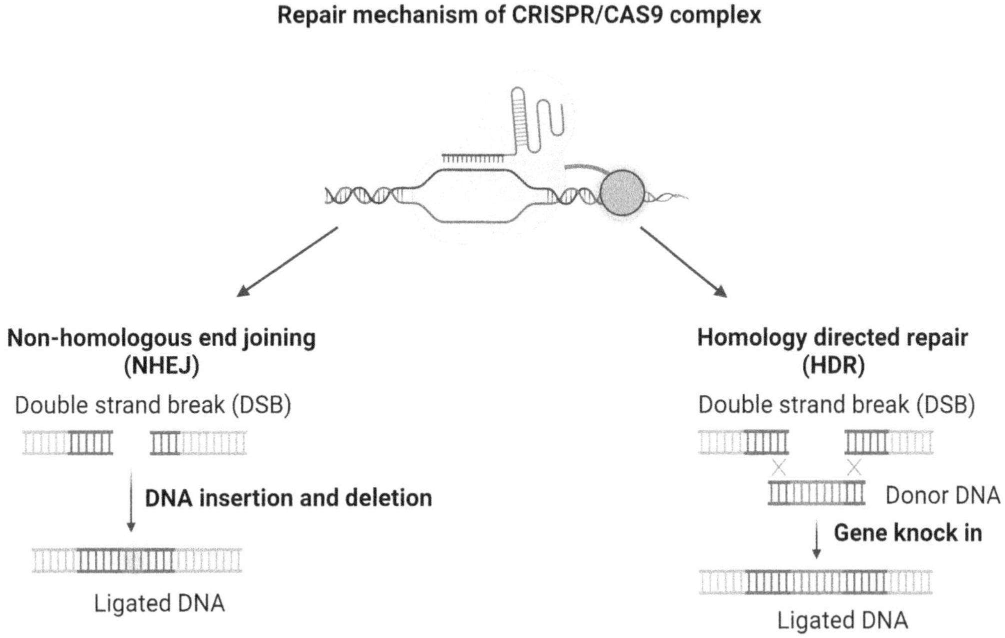

FIGURE 17.1 Overview of CRISPR/CAS technique repair mechanism (Non-homologous end joining and homology-directed repair).

17.3 CRISPR/Cas VARIANTS AND THEIR MECHANISMS

17.3.1 CRISPRi

CRISPRi refers to the type of CRISPR/Cas system that is used for gene silencing or repression. It inhibits the transcription process in which DNA is converted into RNA. Unlike RNA interference technology, which degrades mRNA in the cytoplasm, CRISPRi targets the DNA sequence of the target gene to prevent transcription. It has been proved that this method is more effective in gene silencing when compared to RNAi technology (Guo et al., 2019).

17.3.1.1 dCas9

Endonuclease-deactivated Cas9 (dCas9) is a Cas9 variant that can be used for gene regulation in the transcription process. It is produced by introducing point mutations in Cas9 nuclease domains, which inactivates its nuclease activity. The target gene can be silenced when dCas9 is used together with a single guide RNA without altering the gene sequence. Hence, it is a useful tool for CRISPR interference and transcriptional gene regulation (Hussain et al., 2018).

17.3.2 CRISPRa

CRISPRa is the CRISPR-based gene activation technique that is involved in fusing deactivated Cas9 protein to transcriptional activation domains such as VP64, VPR, p65AD, VP16, and VP160 to enhance transcription at target sites (Guo et al., 2019) It also works by using a modified version of CRISPR-Cas9 protein to target specific DNA sequences near the gene of interest. Upon binding to the target DNA, the CRISPRa complex triggers the recruitment of other proteins that facilitate gene expression resulting in elevated levels of protein produced by the gene (Koonin and Makarova, 2022).

17.3.3 Base Editing

Base editing is a CRISPR tool that can precisely convert one nucleotide to another nucleotide in a programmable manner. This technique does not generate DSB or NHEJ pathways or a homology-directed pathway for a template. It is mainly based on the fusion of catalytically impaired Cas9 nuclease and a single-stranded (SS) DNA deaminase. In this base editing, there are two main types of base editors such as cytosine base editors (CBE) and adenine base editors (ABE). This type of classification is based on the different deaminases involved in the fusion of Cas9 nuclease and ssDNA deaminase (Guo et al., 2019).

17.3.4 Prime Editing

Prime editing is primarily based on RNA template-based DNA modifications along with a few insertions and deletions (Vidya and Arun, 2023). It also utilizes a modified form of Cas9 protein and a reverse transcriptase enzyme to achieve precise modifications to DNA sequences through a two-step process. Firstly, the Cas9 protein is altered to include a reverse transcriptase (RT) domain, which is often called a "prime editor". It is guided to a specific location in the genome using a complementary guide RNA. Secondly, this prime editor creates a single-strand break in the DNA, allowing the RT to synthesize a new DNA strand that replaces the original sequence (Koonin and Makarova, 2022).

17.3.5 CRISPR/Cpf1

The mechanism of Cpf1 involves using a single short RNA guide molecule with 42 to 44 nucleotides (CrRNA). This guide molecule recognizes the PAM sequence and cleaves the DNA in a staggered manner producing a double-stranded break after the 18th base on the non-targeted strand and after the 23rd base on the targeted strand. This process results in indel mutations that are located far from the seed region (the first five nucleotides on the 5′ end of the spacer sequence). This seed region is preserved for subsequent cleavages and the enzyme generates 4- or 5-nucleotide long 5′ overhangs (Murovec et al., 2017).

17.3.6 SpCas9

The SpCas9 system derived from *Streptococcus pyogenes* has been utilized extensively as a genome editing tool for various purposes, including disrupting target genes, activating or repressing transcription, modifying epigenetic alterations, and converting single base pairs in different cell types and organisms. Like other variants, it also locates the target sequence with the PAM region which leads to the conformational change in SpCas9 activating endonuclease activity. The nuclease domains HNH (historical nuclease homing) and RuVC (Resolvase uVrC) present in SpCas9 create DSB that can be repaired by either HDR or NHEJ mechanism (Guo et al., 2019). To broaden the scope of application, SpCas9 has been developed with multiple variants like SpCas9NG, SpG, SpRY, and HscCas9, etc. (Li et al., 2023).

17.3.7 SaCas9

SaCas9 is derived from the bacterium, *Staphylococcus aureus*. This variant is a smaller version of Cas9 nuclease. It functions by attaching to the guide RNA that matches a particular target sequence and subsequently cutting both strands of the DNA at that location resulting in a double-strand break followed by regular repairing mechanisms (HDR or NHEJ) (Kumar et al., 2018).

17.3.8 nCas9

Nickase Cas9 (nCas9) can be produced from Cas9 by modifying the nuclear domains (RuvC and HNH) with single base mutation. This variant is capable of cleaving only one strand of the target DNA sequence (Guo et al., 2019).

17.3.9 CRISPR/Cas13 System

CRISPR/Cas13 system comprises of single RNA-guided Cas13 protein that can bind and cleave the single-stranded RNA. This system has many variants such as Cas13a (also called C2c2), Cas13b, Cas13c, and Cas13d. It has been successfully utilized in viral detection, labeling of transcripts, RNA knockdown, and regulation of splicing (Xu and Li, 2020).

17.4 ADVANCEMENTS OF CRISPR/Cas TECHNOLOGY IN ORNAMENTAL FLOWERS

17.4.1 Flower Color

The color of the ornamental flower is one of the most significant commercial traits in the flower industry. For decorative and aesthetic landscaping purposes, there is a constant demand for diversified color varieties that have a greater impact on the flower market. On the other hand, the major problem in ornamental flowers is the gradual fading of flower colors due to several external factors such as high temperature, soil, light, and diseases. In addition, raising plants that are resistant to these ecological attributes and seasonal fluctuations is quite challenging with conventional breeding. The aforesaid demand and environmental challenges could be addressed by CRISPR/Cas technology which holds an enormous potential to produce desirable hues ranging from almost all colors with desirable flower attributes and improved traits that sustain in changing climatic conditions (Wei et al., 2020, Zheng and Zhang, 2018) (Figure 17.2).

Watanabe et al. (2018) have done CRISPR/Cas9 mediated knock out of *carotenoid cleavage dioxygenase* (*InCCD4)* that resulted in a color shift from white petals to pale yellow in *Ipomea nil.*

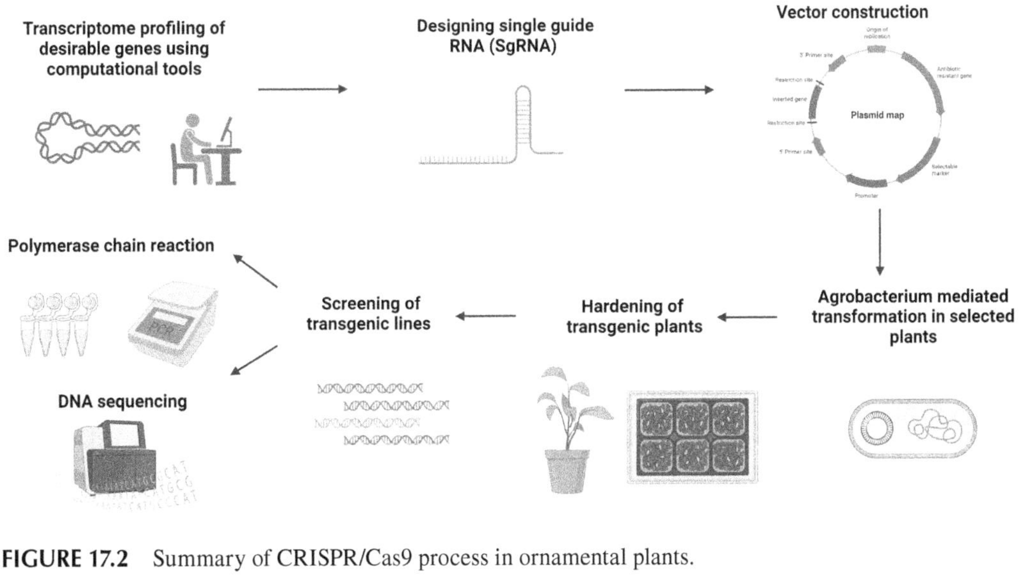

FIGURE 17.2 Summary of CRISPR/Cas9 process in ornamental plants.

The improvement in flower color was correlated with the elevated carotenoid production (20-fold) in *Ipomea* flowers compared to the control. This gene (*InCCD4*) is primarily responsible for the degradation of carotenoids by breaking the polyene chain of carotenoids. In *Brassica nappus*, gene silencing of zeaxanthin epoxidase candidate genes *BnaA09.ZEP* and *BnaC09.ZEP*in *OF1* (orange flower) and *OF2* loci respectively (which were regulated by an unidentified transcription factor) led to enhanced orange color flowers. The findings also revealed that editing these genes resulted in crosstalk between the carotenoid (activator) as well as flavonoid pathway (repressor) (Liu et al., 2020). Morita et al. (2018) have done CRISPR/Cas9-mediated mutagenesis of *DFR-B* (dihydroflavonol-4-reductase-B) in *Ipomea nil* which resulted in pale-colored flowers. The authors claimed that these gene catalyzes the dihydro flavonols to leucoanthocyanidins and are involved in the flavonoid as well as anthocyanin pathway. In addition, Morita et al. (2018) suggested a few potential candidates such as *EFP* (*enhancer of flavonoid production*) and *3-glycosyl transferase* (*3GT*) for the production of pale-colored flowers.

In *Gentiana scabra*, researchers have knocked out three genes *Gt5GT* (Anthocyanin 5-O-glycosyl transferase), *Gt3′GT* (anthocyanin 3′-O-glycosyl transferase), and *Gt5/3′AT* (Anthocyanin 3′-O-glycosyl transferase) which plays a major role in the modification of anthocyanin pathway. Among them, *Gt5GT, Gt3′GT* accumulated delphinidin 3G (pale red-violet flowers), and delphinidin 3G-5CafG (dull pink flowers) respectively. In addition to that, *Gt5/3′AT* mutated lines showed primary pigment delphinidin 3G-5G-3′Gand secondary pigment delphinidin 3G-5G (Pale mauve flowers). The overall mechanism is glycosylation and subsequent acylation of the 3′-hydroxy group of the B-ring in delphinidin aglycone in flowers (Tasaki et al., 2019). In the Japanese gentian, this research group has functionally characterized the glutathione-S-transferase (*GST1*) gene role of gentiodelphin anthocyanin pigmentation present in the petal vacuoles using *GST1*-CRISPR-edited mutant lines (white and pale blue flowers). This study also provides insights into the potential involvement of *GST1,* an anthocyanin transporter in response to sugar-induced stress conditions. Unfortunately, the underlying mechanism of anthocyanin transportation to petal vacuoles is still not known (Tasaki et al., 2020). Hence, the CRISPR/Cas technique allows precise and targeted modifications of specific genes involved in flower pigment pathways. By editing these genes, researchers can potentially enhance or alter the production of pigments responsible for flower color which are listed in Table 17.1.

17.4.2 Shelf Life

Ornamental flowers are live, metabolically active organs, and are very fragile. Ethylene, pathogens, water, and carbohydrate content, are the four main elements that affect the vase life in cultivation and postharvest (Vehniwal et al., 2019). *EPHEMERAL1 (EPH1),* a NAC transcription factor is a significant regulator of petal senescence in *Ipomea nil.* This study demonstrated that CRISPR/Cas9 technology successfully produced mutations in the target gene depicting that *EPH1* is a key player in petal senescence (Shibuya et al., 2018). In another study, CRISPR-edited mutant lines showed the role of *PhACO1 (1-aminocyclopropane-1-carboxylate oxidase 1)* genes responsible for flower longevity which resulted in reducing ethylenes in different flowering stages of *Petunia cv Mirage rose.* On the other hand, the edited plants have shown extended longevity when compared to control *Petunia* plants (Xu and Li, 2020). The *PhACO3* and *PhACO4*-edited lines demonstrated noticeably decreased ethylene production (2.8 to 3.0 fold in corollas and 1.5 fold in pistils) during flowering as well as increased flower lifespan (around 9.5 days). Therefore, editing the ethylene pathway-related genes is ideal for increasing flower longevity using CRISPR/Cas9 technology (Xu et al., 2021).

TABLE 17.1
Enhanced Flower Characteristics in Ornamental Plants through CRISPR/Cas9 Technology

S. No	Plant Species	Target Genes	Trait/Phenotype	Mode of Transformation	References
1	*Arabidopsis thaliana*	*PAP1*	Purple leaf	*Agrobacterium tumefaciens*	Liu et al. (2020)
2	*Chrysanthemum morifolium*	*CpYGFP*	GFP fluorescence	*Agrobacterium tumefaciens*	Kishi-Kaboshi et al. (2018)
3	*Dendrobium officinale*	*C3H, C4H, 4CL, CCR, IRX*	Ligno cellulose Biosynthesis is reduced	*Agrobacterium tumefaciens*	Kui et al. (2017)
4	*Ipomoea nil*	*InDFR-B*	Color of flower	*Agrobacterium tumefaciens*	Watanabe et al. (2017)
5	*I. nil 'AK77'*	*InCCD4*	Color of the flower	*Agrobacterium tumefaciens*	Watanabe et al. (2018)
6	*I. nil 'AK77'*	*EPH1*	Aging of flower	*Agrobacterium tumefaciens*	Shibuya et al. (2018)
7	*Lilium pumilum DC. Fisch*	*LpPDS*	Albino, pale yellow and albino–green chimeric mutants	*Agrobacterium tumefaciens*	Yan et al. (2019)
8	*L. longiflorum* 'White Heaven	*LlPDS*	Albino, pale yellow and albino–green chimeric mutants	*Agrobacterium tumefaciens*	Yan et al. (2019)
9	*P. hybrida inbred line* Mitchell Diploid	*PhPDS*	Albino and mosaic shoots	*Agrobacterium tumefaciens*	Zhang et al. (2016)
10	*Petunia 'Mirage Rose'*	*PhACO1*	Ethylene production reduced and flowerlongevity enhancement	*Agrobacterium tumefaciens*	Xu and Li (2020)
11	*Petunia hybrida*	*AN4, DPL*	Venation of corolla tube	*Agrobacterium tumefaciens*	Xu and Li (2020)
12	*P. inflata*	*PiSSK1*	Self-incompatibility	*Agrobacterium tumefaciens*	Sun and Kao (2018)
13	*Phalaenopsis equestris*	*MADS44, MADS36, MADS8*	MADS-null mutants	*Agrobacterium tumefaciens*	Tong et al. (2020)
14	*Torenia fournieri* 'Crown Violet'	*F3H*	Color of flower.	*Agrobacterium tumefaciens*	Nishihara et al. (2018)

ACO1, 1-Aminocyclopropane-1-carboxylic acid oxidase; *AN4*, Anthocyanin 4; *C3H*, Coumarate-3-hydroxylase; *C4H*, Coumarate-4-hydroxylase; *4CL4*, Coumarate CoA ligase; *CCD4*, Carotenoid cleavage dioxygenase4; *CCR*, Cinnamoyl CoA reductase; *CpYGFP*, Yellowish green fluorescent protein gene from *Chiridius poppei*; *DFR*, Bdihydroflavonol 4 reductase; *DPL*, Deep purple; *EPH1*, Ephemeral 1; *F3H*, Flavanone-3-hydroxylase; *IRX*, Irregular xylem; *MADS*44, *MADS*36, *MADS*8, Minichromosome maintenance factor agamous deficiens serum response factor; *PAP1*, Phenylalanine ammonia lyase; *PDS*, Phytoene desaturase; *PiSSK1*, *Petunia inflata* suppressor of sensor kinase;

17.4.3 Floral Development

A critical developmental phase in the life cycle of higher plants is floral development. This process of flowering is initiated in a plant by internal (age of plants, size) or external factors (photoperiodism, temperature, stress conditions). These external factors enable synchronized flowering within a population as well as provide the best timing of flowering throughout the year to ensure successful pollination and seed development. Any change in these factors can significantly impact the flowering pattern in ornamental flowers (Erwin, 2007). In *Petunia hybrida* flowers, the role of two genes, *DPL* (deep purple) and *AN4* (anthocyanin 4) in regulating floral development was investigated. In particular, *DPL* was found to play a critical role in determining vein-associated anthocyanin pattering on the abaxial epidermis of flower buds, while *AN4* is involved in determining corolla tube venation. By employing CRISPR/Cas9 mediated mutations, they were able to obtain *DPL* mutants, which allowed them to study the impact of *DPL* on venation patterning. These findings refine our understanding of the regulatory framework of anthocyanin patterns in *Petunia* providing new insights into floral development mechanisms (Zhang et al., 2021).

Another crucial characteristic feature of modified ornamental plants is flowering time (Boutigny et al., 2020). In *Fortunella hindsii*, they submitted the first genome sequence and the CRISPR-mediated mutagenesis of phytoene desaturase (*FhPDS*) exhibited global albino, mosaic albino, or no obvious abnormal phenotypes. By utilizing CRISPR/Cas genome editing technology to overcome the challenges of long juvenility and inherent apomixes in citrus, *F. hindsii* exists as a valuable tool for understanding and improving citrus crops. The early flowering characteristic of *F. hindsii* is vital because it allows for faster generation turnover and is important in studying and manipulating important genes in citrus. This can potentially lead to advancements in breeding, crop improvement, and the development of new ornamental citrus varieties with desirable traits (Zhu et al., 2019).

17.4.4 Self-Incompatibility

Self-incompatibility is a reproductive mechanism found in certain plant species that prevents self-fertilization and promotes outcrossing. It acts as an intraspecific barrier to ensure genetic diversity and prevents inbreeding. This recognition is often controlled by the S locus, a highly polymorphic genetic region that encodes proteins involved in the self-incompatibility response. The specific mechanisms of self-incompatibility can vary among plant species, but they generally involve molecular interactions between proteins in the pistil and pollen to determine compatibility or incompatibility (Sun and Kao, 2018). This characteristic feature of *Petunia inflata* related to the role of *PiSSK1*, a subunit of the SCF-SLF complex (Skp1-cullin1-F-box-S-locus F-box protein) was studied.

The researchers generated indel alleles (alleles with insertions or deletions) of *PiSSK1* using CRISPR/Cas9 technology to generate a T0 mutant plant containing *PiSSK1* with a specific indel allele in its pollen genome. By using bud selfing to bypass self-incompatibility, they obtained progeny plants that were homozygous for each indel allele independently. The mechanism behind this process is that SLF proteins collectively interact with non-self S-RNases, leading to their ubiquitination and subsequent degradation by the 26S proteasome which prevents inbreeding by allowing cross-compatible pollination. By understanding the molecular mechanisms involved in self-incompatibility, breeders can manipulate and control the breeding process to achieve desired phenotypes (Sun and Kao, 2018).

17.5 LIMITATIONS OF CRISPR/CAS TECHNOLOGY

The CRISPR/Cas technology has several limitations which include off-target effects, complexity of gene regulation in eukaryotes, large deletions and rearrangements of genomic sequences, and requirement of PAM (Vidya and Arun, 2023). Among them, off-targets are the base pair mismatches between single guide RNA (SgRNA) and non-target sequences resulting in one or multiple unknown mutations. Moreover, several sequencing methods such as whole genome sequencing and GUIDE seq have been developed to detect these off-targets. Furthermore, the complexity of gene regulation in eukaryotes (particularly epigenetic modifications) makes this technology more challenging as it leads to altered expression of multiple genes. On the other hand, massive deletions and chromosomal translocations in this method are not acceptable in clinical applications due to the potential risk of malignant diseases. Another limitation is the presence of PAM in the target sequence that is used as a recognition site by the Cas9 enzyme. Several Cas9 variants such as Cas12a, Cas13b, Cms1, and dCas9 which are not restricted to PAM sequence recognition have been recently developed (Li et al., 2023).

17.6 FUTURE PERSPECTIVES

CRISPR/Cas genome editing method seems to be essential to change the color and aroma of ornamental flowers with a suitable level of efficiency. Alterations made to ornamental flowers were frequently shown to be unstable and reversible using antisense, sense, or RNAi technology. Through genome editing techniques, a variety of strategies could be used to permanently alter ornamentals (Khan et al., 2022). However, the existing challenge is in the part of stable transformation as well as poor transformation efficiency in many ornamental plants which needs to be improved. Second, it is still necessary to identify the genes that control crucial flower attributes. Although recent developments in DNA sequencing have produced the complete genome sequences of numerous diploid ornamental plants, it is still difficult to identify the relevant genes for a lot of characteristics (Kishi-Kaboshi et al., 2018).On the other hand, pests such as aphids, bugs, whiteflies, sawflies, and moths are constantly a serious threat to ornamental plants. A study in *Arabidopsis* and *Brassica* revealed that the carotenoid pathway's *PDS* gene silencing reduced the caterpillar growth of *Pieris rapae* in the altered plants. Thus, future research can focus on employing genome editing methods to reduce insect infestations on ornamental plants (Khan et al., 2022). The effectiveness and applicability of genome editing techniques in a variety of ornamental species are constrained by highly variable, genotype-dependent responses, and a large number of recalcitrant cultivars. On the other hand, the public acceptance of genome editing technologies will depend on a balanced, ideal, and globally consistent regulatory framework that needs to be approved and channelized in society (Giovannini et al., 2021). In several significant horticultural crops, many successful instances using CRISPR/Cas technology have already been attained (Figure 17.3). Yet, some flower attributes like floral fragrance have an immense scope to be improved using the CRISPR/Cas technique as most of the genes related to scent and their pivotal roles have been already well studied. Moreover, the development of horticultural crops with improved agronomic traits will result from the expanding knowledge of CRISPR/Cas9-based tools. This will lead to the creation of innovative approaches for sustainable and competitive ornamental flower production (Corte et al., 2019).

ACKNOWLEDGMENT

The corresponding author is thankful to Bharathiar Cancer Theranostics Research Centre (BU/RUSA2.0/BCTRC/2020/BCTRC-CT08/14.12.2020), Bharathiar University for providing financial support.

DATA AVAILABILITY

The authors confirm that the data supporting the findings of this study are available within the book/chapter.

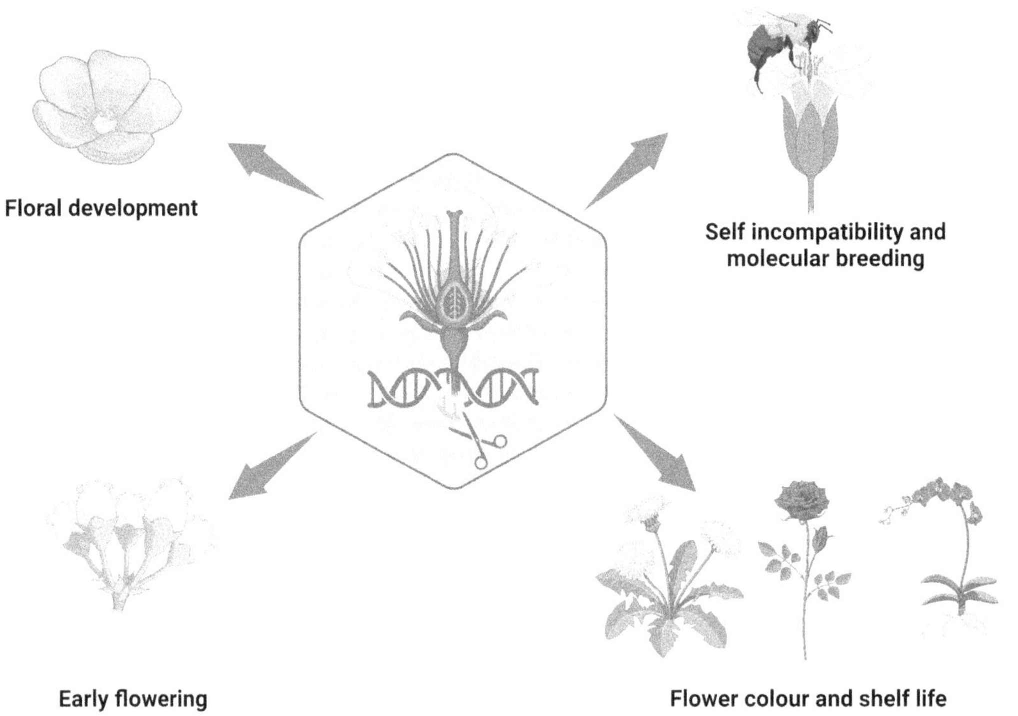

FIGURE 17.3 Overall flower attributes improved by using CRISPR/Cas9 technology.

REFERENCES

Boutigny, A. L., Dohin, N., Pornin, D., & Rolland, M. (2020). Overview and detectability of the genetic modifications in ornamental plants. *Horticulture Research, 7*(1). https://doi.org/10.1038/s41438-019-0232-5

Corte, L. E. D., Mahmoud, L. M., Moraes, T. S., Mou, Z., Grosser, J. W., & Dutt, M. (2019). Development of improved fruit, vegetable, and ornamental crops using the CRISPR/Cas9 genome editing technique. *Plants, 8*(12). https://doi.org/10.3390/plants8120601

De, L. (2017). Improvement of ornamental plants - a review. *International Journal of Horticulture*, (August). https://doi.org/10.5376/ijh.2017.07.0022

Erwin, J. (2007). Factors affecting flowering in ornamental plants. In: Anderson, N.O. (eds) *Flower Breeding and Genetics*. Springer, Dordrecht, pp. 7–48. https://doi.org/10.1007/978-1-4020-4428-1_1

Giovannini, A., Laura, M., Nesi, B., Savona, M., & Cardi, T. (2021). Genes and genome editing tools for breeding desirable phenotypes in ornamentals. *Plant Cell Reports*, *40*(3), 461–478. https://doi.org/10.1007/s00299-020-02632-x

Guo, M., Ren, K., Zhu, Y., Tang, Z., Wang, Y., Zhang, B., & Huang, Z. (2019). Structural insights into a high fi delity variant of SpCas9. *Cell Research*, (January). https://doi.org/10.1038/s41422-018-0131-6

Hussain, B., LuCas, S. J., & Budak, H. (2018). CRISPR/Cas9 in plants: At play in the genome and at work for crop improvement. *Briefings in Functional Genomics*, *17*(April), 319–328. https://doi.org/10.1093/bfgp/ely016

Kaur, H., Pandey, D. K., Goutam, U., & Kumar, V. (2021). CRISPR/Cas9-mediated genome editing is revolutionizing the improvement of horticultural crops: Recent advances and future prospects. *Scientia Horticulturae*, *289*(May), 110476. https://doi.org/10.1016/j.scienta.2021.110476

Khan, Z., Razzaq, A., Sattar, T., Ahmed, A., Habibullah Khan, S., & Zubair Ghouri, M. (2022). Understanding floral biology for CRISPR-based modification of color and fragrance in horticultural plants. *F1000Research*, *11*, 854. https://doi.org/10.12688/f1000research.122453.1

Kishi-Kaboshi, M., Aida, R., & Sasaki, K. (2018). Genome engineering in ornamental plants: Current status and future prospects. *Plant Physiology and Biochemistry*, *131*(December 2017), 47–52. https://doi.org/10.1016/j.plaphy.2018.03.015

Koonin, E. V, & Makarova, K. S. (2022). Evolutionary plasticity and functional versatility of CRISPR systems. *PLoS Biology*, *20*(1), e3001481. https://doi.org/10.1371/journal.pbio.3001481

Kui, L., Chen, H., Zhang, W., He, S., Xiong, Z., Zhang, Y., ... & Cai, J. (2017). Building a genetic manipulation tool box for orchid biology: identification of constitutive promoters and application of CRISPR/Cas9 in the orchid, Dendrobium officinale. *Frontiers in plant science*, *7*, 2036. https://doi.org/10.3389/fpls.2016.02036

Kumar, N., Stanford, W., Solis, C. De, Abraham, N. D., Dao, T. J., Thaseen, S., … Smith, G. (2018). The development of an AAV-based CRISPR SaCas9 genome editing system that can be delivered to neurons in vivo and regulated via doxycycline and cre-recombinase. *Frontiers in Molecular Neuroscience*, *11*(November), 1–14. https://doi.org/10.3389/fnmol.2018.00413

Lee, H., Subburaj, S., Tu, L., Lee, K.-Y., Park, G., & Lee, G.-J. (2020). Overview of CRISPR/Cas9: A chronicle of the CRISPR system and application to ornamental crops. *Korean Journal of Agricultural Science*, *47*(4), 903–920.

Li, T., Yang, Y., Qi, H., Cui, W., Zhang, L., Fu, X., … Li, P. (2023). CRISPR/Cas9 therapeutics: Progress and prospects. *Signal Transduction and Targeted Therapy*, *8*(1), 36. https://doi.org/10.1038/s41392-023-01309-7

Liu, Y., Ye, S., Yuan, G., Ma, X., Heng, S., Yi, B., … Wen, J. (2020). Gene silencing of *BnaA09.ZEP* and *BnaC09.ZEP* confers orange color in *Brassica napus* flowers. *Plant Journal*, *104*(4), 932–949. https://doi.org/10.1111/tpj.14970

Morita, Y., & Hoshino, A. (2018). Recent advances in flower color variation and patterning of Japanese morning glory and petunia. *Breeding Science*, *68*(1), 128–138. https://doi.org/10.1270/jsbbs.17107

Murovec, J., Pirc, Z., & Yang, B. (2017). New variants of CRISPR RNA-guided genome editing enzymes. *Plant Biotechnology Journal*, *15*(8), 917–926. https://doi.org/10.1111/pbi.12736

Nishihara, M., Higuchi, A., Watanabe, A., & Tasaki, K. (2018). Application of the CRISPR/Cas9 system for modification of flower color in Torenia fournieri. *BMC Plant Biology*, *18*(1), 1–9. https://doi.org/10.1186/s12870-018-1539-3

Shibuya, K., Watanabe, K., & Ono, M. (2018). CRISPR/Cas9-mediated mutagenesis of the EPHEMERAL1 locus that regulates petal senescence in Japanese morning glory. *Plant Physiology and Biochemistry*, *131*(February), 53–57. https://doi.org/10.1016/j.plaphy.2018.04.036

Sun, L., & Kao, T. H. (2018). CRISPR/Cas9-mediated knockout of PiSSK1 reveals essential role of S-locus F-box protein-containing SCF complexes in recognition of non-self S-RNases during cross-compatible pollination in self-incompatible *Petunia inflata. Plant Reproduction, 31*(2), 129–143. https://doi.org/10.1007/s00497-017-0314-1

Tasaki, K., Higuchi, A., Watanabe, A., Sasaki, N., & Nishihara, M. (2019). Effects of knocking out three anthocyanin modification genes on the blue pigmentation of gentian flowers. *Scientific Reports*, *9*(1), 1–10. https://doi.org/10.1038/s41598-019-51808-3

Tasaki, K., Yoshida, M., Nakajima, M., Higuchi, A., Watanabe, A., & Nishihara, M. (2020). Molecular characterization of an anthocyanin-related glutathione S- transferase gene in Japanese gentian with the CRISPR/Cas9 system. *BMC Plant Biology, 20*(1), 1–14. https://doi.org/10.1186/s12870-020-02565-3

Tong, C. G., Wu, F. H., Yuan, Y. H., Chen, Y. R., & Lin, C. S. (2020). High-efficiency CRISPR/Cas-based editing of Phalaenopsis orchid MADS genes. *Plant Biotechnology Journal*, *18*(4), 889. https://doi.org/10.1111/pbi.13264

Vehniwal, S. S., & Abbey, L. (2019). Cut flower vase life - influential factors, metabolism and organic formulation. *Horticulture International Journal*, *3*(6), 275–281. https://doi.org/10.15406/hij.2019.03.00142

Vidya, N., & Arun, M. (2023). Updates and applications of CRISPR/Cas technology in plants. *Journal of Plant Biology*, 1–20. https://doi.org/10.1007/s12374-023-09383-8

Watanabe, K., Kobayashi, A., Endo, M., Sage-Ono, K., Toki, S., & Ono, M. (2017). CRISPR/Cas9-mediated mutagenesis of the dihydroflavonol-4-reductase-B (DFR-B) locus in the Japanese morning glory Ipomoea (Pharbitis) nil. *Scientific reports*, *7*(1), 10028. https://doi.org/10.1038/s41598-017-10715-1

Watanabe, K., Oda-Yamamizo, C., Sage-Ono, K., Ohmiya, A., & Ono, M. (2018). Alteration of flower colour in Ipomoea nil through CRISPR/Cas9-mediated mutagenesis of carotenoid cleavage dioxygenase 4. *Transgenic research*, *27*, 25–38. https://doi.org/10.1007/s11248-017-0051-0

Wei, Z., Arazi, T., Hod, N., Zohar, M., Isaacson, T., Doron-Faigenboim, A., Yedidia, I. (2020). Transcriptome profiling of Ornithogalum dubium leaves and flowers to identify key carotenoid genes for CRISPR gene editing. *Plants*, *9*(4), 540. https://doi.org/10.3390/plants9040540

Xu, J., Naing, A. H., Bunch, H., Jeong, J., Kim, H., & Kim, C. K. (2021). Enhancement of the flower longevity of petunia by CRISPR/Cas9-mediated targeted editing of ethylene biosynthesis genes. *Postharvest Biology and Technology*, 174(December 2020), 111460. https://doi.org/10.1016/j.postharvbio.2020.111460

Xu, Y., & Li, Z. (2020). CRISPR-Cas systems : Overview, innovations and applications in human disease research and gene therapy. *Computational and Structural Biotechnology Journal*, *18*, 2401–2415. https://doi.org/10.1016/j.csbj.2020.08.031

Zhang, B., Xu, X., Huang, R., Yang, S., Li, M., & Guo, Y. (2021). CRISPR/Cas9-mediated targeted mutation reveals a role for *AN4* rather than *DPL* in regulating venation formation in the corolla tube of *Petunia hybrida. Horticulture Research, 8*(1). https://doi.org/10.1038/s41438-021-00555-6

Zhang, B., Yang, X., Yang, C., Li, M., & Guo, Y. (2016). Exploiting the CRISPR/Cas9 system for targeted genome mutagenesis in petunia. *Scientific Reports*, *6*(1), 20315. https://doi.org/10.1038/srep20315

Zheng, Y., & Zhang, H. (2018). The Method of Color Element Allocation of Ornamental Plants Considering Water Condition. *CCAMLR Science*, *25*(3), 261.

Zhu, C., Zheng, X., Huang, Y., Ye, J., Chen, P., Zhang, C., … Deng, X. (2019). Genome sequencing and CRISPR/Cas9 gene editing of an early flowering Mini-Citrus (*Fortunella hindsii*). *Plant Biotechnology Journal*, *17*(11), 2199–2210. https://doi.org/10.1111/pbi.13132

18 A Convenient CRISPR/Cas9 Mediated Plant Multiple Gene Editing Protocol by In-Fusion Technology

Jun-Li Wang, Luo-Yu Liang, and Lei Wu

LIST OF ABBREVIATIONS

CDS	coding DNA sequence
CRISPR/Cas9	clustered regularly interspaced short palindromic repeats-Cas9
DSBs	double-strand DNA breaks
IAA32	INDOLE-3-ACETIC ACID INDUCIBLE 32
IAA34	INDOLE-3-ACETIC ACID INDUCIBLE 34
MS	Murashige and Skoog
NHEJ	nonhomologous end-joining
PAM	protospacer adjacent motif
PCR	polymerase chain reaction
SgRNA	*single-guide RNA*
TALENs	TAL effector nucleases
TMK1	TRANSMEMBRANE PROTEIN KINASE 1
UTR	untranslated region
ZFNs	zinc finger nucleases

18.1 INTRODUCTION

Organisms have a variety of endogenous mechanisms to repair DNA double-strand breaks (DSBs) to ensure genomic integrity [1–4]. When the organism repairs the broken double-strand DNA and reconnects it, often causes base addition, deletion, or change [5]. Thus, the introduction of precise breaks at specific sites in the genome, and then depending on these DNA repair mechanisms can be targeted gene modification or editing. Given this, several gene editing tool enzymes that introduce DSB into target sites have emerged successively. In 2002, researchers proposed a strategy of DNA-targeted cleavage using artificially modified nucleases. The enzyme contains two domains: the zinc finger domain (recognizing target sites on DNA sequences) and the nuclease domain (cleaving DNA double strands). Therefore, the enzyme is also called zinc finger nucleases (ZFNs) [6,7]. In 2009, it was reported that transcription activator-like (TAL) effector nucleases (TALENs) were used for gene editing. TALENs consist of the TAL effector DNA-binding domain and the nuclease domain, which respectively perform the target sequences identification and cutting functions [8,9]. Although ZFNs and TALENs have been successfully used for gene editing in eukaryotes [10], ZFNs and TALENs are difficult to operate, low in accuracy and targeting efficiency, and the difficulties in synthesis, design, and validation of artificially modified nucleases required to restrict their widespread use in gene editing.

DOI: 10.1201/9781003387060-18

With the in-depth study of the CRISPR/Cas immune system from bacteria and archaea, it has been found that this immune mechanism can specifically remember, recognize, and clear foreign virus DNA sequences to avoid infection [11]. Based on this immune system of the prokaryotic organism can recognize and cut specific nucleic acid sequences, in 2014, a novel gene editing scheme using the CRISPR/Cas9 system was proposed by Doudna, J. A. and Charpentier, E. [12]. The artificially designed *sgRNA* consists of *crispr RNA* (*crRNA*) and *trans-activating crispr RNA* (*tracrRNA*), which are respectively responsible for specifically recognizing the DNA target and interacting with the Cas9 protein. The assembled CAS9-*sgRNA* complex can cleave the target site like a restriction endonuclease [12]. In addition, the target site to be recognized and cut also depends on an NGG (where N stands for any base) protospacer adjacent motif (PAM) sequence which is located downstream of the DNA target in genomic sequences [13,14]. Compared with ZFNs and TALENs, CRISPR/Cas9 technology is low-cost, has easy operation, and short production time.

The functional study of genes in plant science is important for modern agriculture to breed varieties with excellent agronomic characteristics, which depend on using mutants as materials. Traditional mutants come from chemical mutagens (such as intercalating agents or alkylating agents), physical mutagens (such as electromagnetic radiation or particle radiation), and biological mutagens (such as Transposons or insertion sequences) [15]. Due to the functional redundancy and complex interactions between genes, high-order mutants are often needed in scientific research and breeding. Although the construction of high-order mutants by using traditional hybridization techniques has been widely used, researchers have been plagued by problems such as heavy and complex screening and identification, and long construction cycles. Therefore, the establishment of a simple and rapid double/multiple gene editing system will be of great significance to the development of plant science.

Among the three tool enzymes capable of gene editing, ZFNs, and TALENs can specifically recognize target sequences for gene editing, depending on the zinc finger domain of the protein itself and the TAL effector DNA-binding domain [6,8,9]. So only there are more than two copies of the identified target sequence in the genome, or two or more ZFNs or TALENs are designed with the characteristics of recognizing different DNA target sites to achieve double/multiple gene editing. Unlike TALENs and ZFNs, CAS9 protein can interact with multiple *sgRNA*s to edit multiple target sites [13,16]. The size of each *sgRNA* is only about 100 bp, and the target of CRISPR/Cas9 can be changed only by changing the 20bp sequence of *sgRNA*, which means that multiple *sgRNA*s can be co-expressed easily and the design of *sgRNA* is very simple. Therefore, CRISPR/Cas9 is the most suitable enzyme for multiple gene editing among these three tool enzymes. In plant research, Cas9 and multiple *sgRNA*s are usually co-expressed in cells to achieve multi-gene editing and obtain high-order mutants. For example, Wang et al. (2020) constructed septuple and octuple mutants of *SAUR* family genes in the study [17]. Furthermore, in terms of obtaining and constructing multiple mutants, multiple gene editing also has an advantage that traditional hybridization technology does not have, that is, linkage genes are not easy to obtain double/multiple mutants through hybridization, while gene editing technology can be easily obtained.

To transcript multiple *sgRNA*s at the same time, tandem repeat expression cassettes of *sgRNA* were usually constructed, but in the research reported so far, their construction is a troublesome assembly process [16,18–22]. Here, A convenient tandem repeat expression cassette of *sgRNA* assembly strategy was reported by our laboratory [23]. Two cloning plasmids *pUC57-Amp-U6TU6P* and *pUC57-Kan-CRP* containing elements of *sgRNA* expression cassette (*U6 Promoter, sgRNAs Scaffold,* and *U6 Terminator*) were created as templates, and the 20bp *sgRNA* sequences from target genes added at the 5′-end of the fixed sequence were used as primers for PCR reactions to amplify fragments of tandem repeat expression cassettes of *sgRNA*. And then by In-Fusion technology, these fragments with homologous 20-base *sgRNA* sequences at both ends can be assembled into tandem repeat expression cassettes of *sgRNA*.

Moreover, we selected the fluorescence reporter system of Yu and Zhao 2019 [24] to isolate *Cas9*-free multiple gene editing homozygous mutants in T_2 generation, which depends on the reported gene *mCherry* specifically expressed in the seed coat [20,24]. Selecting gene editing mutants that do not contain exogenous *CAS9* and *sgRNA*s for the following studies has two advantages: Firstly, in studying mutants, genetic complementation experiments are usually carried out to verify whether phenotypes and genes are linked. If CAS9 and *sgRNA*s are still present in the plant, gene editing will continue to occur, resulting in the complemented wild-type genes failing their function. Secondly, although the homozygous mutants after gene editing have no *sgRNA* targets, there is still a small probability of off-target editing [25–27]. The gene editing plants with residual T-DNA containing exogenous genes produce CAS9 and *sgRNA*s in cells that will affect the stability of the genome.

Furthermore, to further improve the efficiency of gene editing, we changed the promoter of the *CAS9* gene in the CRISPR/Cas9 gene editing binary plasmid from the *35S* promoter to a more efficient constitutive UBQ10 promoter, which improved the editing efficiency by increasing the abundance of Cas9 protein. We named this new binary vector *pHDE-UBQ10-Cas9-mCherry*, and our existing experiments have confirmed that this improvement can significantly improve editing efficiency.

In this protocol, the construction of tandem repeat *sgRNA* expression cassettes with three *sgRNA*s (the target sequences of them respectively come from gene *IAA32* (*INDOLE-3-ACETIC ACID INDUCIBLE 32*), *IAA34*, and *TMK1* (*TRANSMEMBRANE PROTEIN KINASE 1*)) is taken as an example. The sequences and location of the targets of *sgRNA*s on genes *IAA32*, *IAA34*, and *TMK1* are shown in Figure 18.1.

18.2 MATERIALS

18.2.1 Growth of Plants

1. The wild-type Columbia-0 *Arabidopsis thaliana* was used in this chapter.
2. Half MS medium plates: 2.22 g/L Murashige and Skoog (MS) Salts [*Phyto* Technology Laboratories, M404, http://www.phytotechlab.com/], 1% Sucrose (weight/volume, w/v), 1% Agar powder (w/v), pH = 5.75–5.80.
3. Plants were grown in a greenhouse at 22°C, Medium light intensity (80 μmol/m²·s), and a long day condition (16 h light/8 h dark).

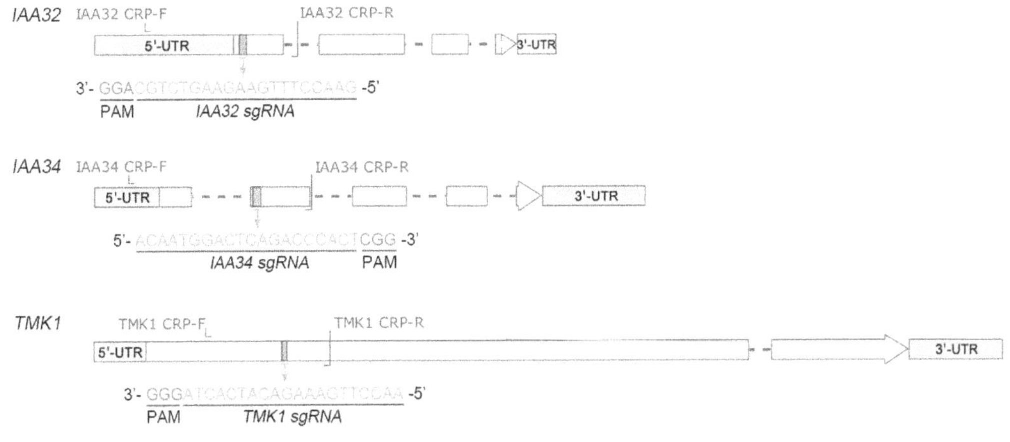

FIGURE 18.1 Gene structure, location and sequence of sgRNAs, and genotyping primers. Using white squares to represent CDS (Coding DNA Sequence), and cadet gray squares to represent UTR (Untranslated Region). Purple lines indicate the location of genotyping primers.

18.2.2 Vector Construction

1. Cloning plasmids *pUC57-Amp-U6TU6P* and *pUC57-Kan-CRP* [23], CRISPR/Cas9 gene editing binary plasmid *pHDE-35S-Cas9-mCherry* and *pHDE-35S-Cas9-mCherry-UBQ* [20] [Addgene, Plasmid No. #78931 and #78932, http://www.addgene.org/].
2. Restriction endonuclease: Fast Digest *EcoR I,* Fast Digest *Mss I,* and Fast Digest *Mfe I* for linearizing plasmids [Thermo Scientific™, FD0274, FD1344 and FD0754, https://www.thermofisher.cn/].
3. Plasmid extraction kit: The plasmids were obtained by plasmid extraction kit [Axygen, AP-MN-P-50G, www.axygen.com/].
4. Prime STAR® GXL DNA Polymerase for phusion PCR reaction [TaKaRa, R050Q, https://www.takarabio.com/].
5. 1% Agarose gel (w/v): Agarose, TBE buffer, and 0.5–1.0 μg/mL Ethidium Bromide (EB).
6. DNA Gel Purification Kit: For purifying DNA fragments [Axygen, AP-GX-50G].
7. PCR cleaning Kit: For purifying DNA fragments [Axygen, AP-PCR-50G].
8. In-Fusion® HD Cloning Kit for assembly of DNA fragments [Clontech, 639648, https://www.takarabio.com/].
9. LB medium: LB agar, powder [Solarbio, L1015, https://www.solarbio.com/].
10. Kanamycin, Ampicillin, Spectinomycin, Rifampicin, and Gentamycin were used for the identification of positive colonies of *Escherichia coli* and *Agrobacterium tumefaciens.*
11. *Escherichia coli* DH5α competent cells stored at −80°C.
12. Vector construction and sequencing primers.

18.2.3 Genetic Transformation of Gene Editing

1. *Agrobacterium tumefaciens* GV3101 competent cells stored at −80°C.
2. 5% sucrose solution: Dissolve 5 g sucrose into 100 mL deionized water.
3. Silwet® L-77 [*Phyto* Technology Laboratories, CAS NO. 27306-78-1].

18.2.4 Selection of *Cas9*-Free Homozygous Mutants

1. Fluorescence microscopic imaging system with mCherry ULTRA Widefield Fluorescence Filter.
2. The primers for genotyping.

18.3 METHODS

18.3.1 Growth of Plants

1. *Arabidopsis* Seeds were sterilized with 0.1% $HgCl_2$, thoroughly washed with ddH_2O three times, and placed on MS plates.
2. After keeping at 4°C for 48 hours in darkness, the plates with seeds were transferred to the greenhouse and cultured for 7 days.
3. Transfer 7-day-old seedlings to the soil and continue cultivating them in the greenhouse until they bloom.

18.3.2 Vector Construction

1. Tandem repeat sgRNA expression cassette design (see Note 1).
   1.1. As shown in Figure 18.2, the essential elements of tandem repeat *sgRNA* expression cassettes: *U6 Promoter, sgRNA Scaffold,* and *U6 Terminator* were contained in cloning plasmids *pUC57-Amp-U6TU6P* and *pUC57-Kan-CRP* [23]. Using these

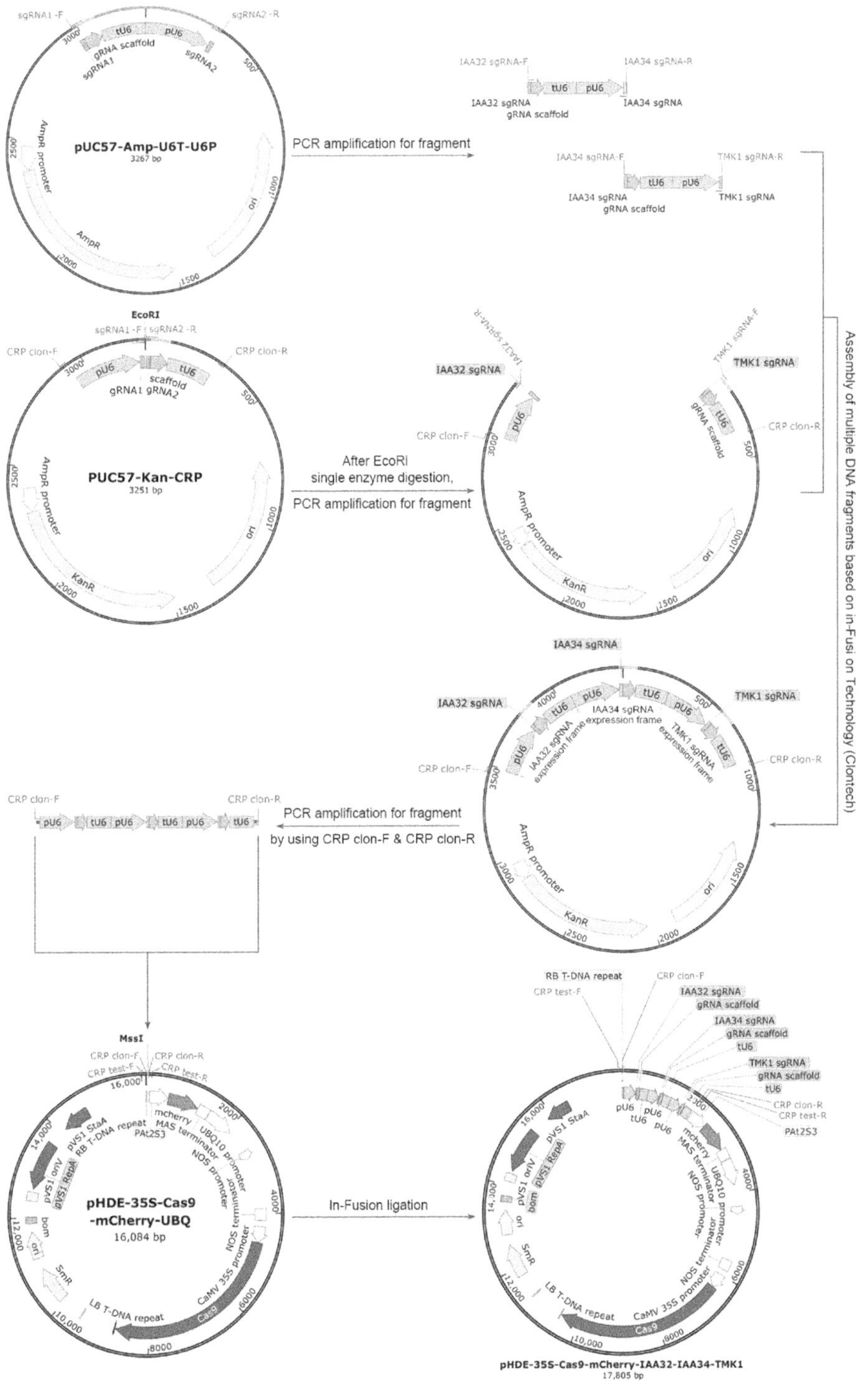

FIGURE 18.2 The workflow for constructing tandem repeat sgRNA expression cassettes. Three steps to complete the assembly of the CRISPR/Cas9 gene editing binary plasmid. (1) Two cloning plasmids, pUC57-Amp-U6TU6P and pUC57-Kan-CRP, were created as templates for amplifying DNA fragments of tandem repeat sgRNA expression cassettes. (2) All the fragments were joined to each other to assemble into tandem repeat sgRNA expression cassettes. (3) The tandem repeat expression cassettes of sgRNA were re-amplified and inserted into the *Mss I* site of the CRISPR/Cas9 gene editing binary plasmid.

two cloning plasmids as PCR reaction templates and primers containing 20bp *sgRNA* target sequence at 5′-end amplify DNA fragments, at both ends of which are 20bp *sgRNA* target modules (see Note 2). Primers are listed in Table 18.3.

1.2. All PCR fragments are assembled by In-Fusion technology so that the *pUC57-Kan* plasmid backbone contains several tandem repeat expression cassettes of *sgRNA* (Figure 18.2) (see Note 3).

1.3. The Primers *CRP clon F* and *CRP clon R* were used to amplify tandem repeat expression cassettes of *sgRNA* (Table 18.3) (see Note 4), and then PCR product was inserted into the *Mss I* site of CRISPR/Cas9 gene editing binary plasmid by In-Fusion technology (Figure 18.2).

2. A simplified protocol of Vector Construction.

2.1. Amplified DNA fragments (Table 18.1):

PCR condition was 35 cycles of 98°C for 10 sec, 60°C for 15 sec, and 68°C for 1 min/kb.

2.2. DNA fragments purification: Agarose gel electrophoresis was used for the separation of DNA fragments. The specific bands were cut and purified by a DNA gel purification kit (see Note 5).

2.3. Using In-Fusion cloning reaction assemble DNA fragments (Table 18.2):

Place the In-Fusion reaction mixture in a water bath at 50°C for 15 minutes, then ice bath until used (see Note 8).

2.4. Transformation Procedure: 2.5 μL In-Fusion reaction mixture was sucked into a tube containing 50 μL *Escherichia coli* DH5α competent cells and then mixed gently. Followed by an ice bath for 30 minutes, heat shock at 42°C water bath for 45 sec, and then an ice bath until used.

2.5. 500 μL liquid medium without antibiotics was added to a tube containing competent cells and then cultivated at 37°C shaker 200 rpm for 1 hour. Spread on an LB plate containing antibiotics (determined according to plasmid screening markers), overnight culture at 37°C (see Note 9).

TABLE 18.1
PCR Reaction Component

Composition of PCR Reaction Mixture	
5×Prime STAR GXL Buffer	10.0 μL
dNTP Mixture (2.5 mM each)	4.0 μL
Forward primer	2.5 μL
Reverse primer	2.5 μL
PrimeSTAR GXL DNA Polymerase	1.25 μL
Sterilized distilled water	to final reaction volume of 50 μL

TABLE 18.2
In-Fusion Reaction Component

In-Fusion Cloning Reaction Mixture (see Note 6)	
5×In-Fusion HD Enzyme Premix	2.0 μL
Purified Linearized vector (see Note 7)	× μL
Each Insert (see Note 7)	× μL
Sterilized distilled water	to final reaction volume of 10 μL

TABLE 18.3
The Primers Used in this Chapter [23]

Primer Name	Sequence
IAA32 sgRNA F	5′-GAACCTTTGAAGAAGTCTGCGTTTTAGAGCTAGAAATAGCAAGTT-3′
IAA32 sgRNA R	3′-TAACCATCTCAGCTGATTCTCTAACCTTGGAAACTTCTTCAGACG-5′
IAA34 sgRNA F	5′-ACAATGGACTCAGACCCACTGTTTTAGAGCTAGAAATAGCAAGTT-3′
IAA34 sgRNA R	3′-TAACCATCTCAGCTGATTCTCTAACTGTTACCTGAGTCTGGGTGA-5′
TMK1 sgRNA F	5’-AACCTTGAAAGACATCACTAGTTTTAGAGCTAGAAATAGCAAGTT-3′
TMK1 sgRNA R	3′-TAACCATCTCAGCTGATTCTCTAACTTGGAACTTTCTGTAGTGAT-5′
CRP clon F	5′-ATATCCTGTCAAACACTGATAGTTT-3′
CRP clon R	5′-AGACCAAGCTTCACTTCACTAGTTT-3′
CRP test F	5′-GGTTGGCATGCACATACAAATGGAC-3′
CRP test R	5′-CAATCTTTTGATGGTAGCTAGTGATGTTGG-3′
IAA32 CRP F	5′-AAGCATTGGCAGACCTGGTTGG-3′
IAA32 CRP R	5′-ATGAGGTGAAAGCACGGCTATTC-3′
IAA34 CRP F	5′-GAGATGTGCTTGCTGTGATTAG-3′
IAA34 CRP R	5′-GCTTCCATGATCAAGAACACAG-3′
TMK1 CRP F	5′-GGTACACTTTCTCCTGATCTACG-3′
TMK1 CRP F	5′-TTAAACCAGTCATGTTCTGAAGA-3′
pHDE F	5′-CGCAATTGATGGACAAGAAGTACTCCATTGGGCTCGATAT-3′
pHDE R	5′-TGTCCATCAATTGCGAGCTGTTAATCAGAAAAACTCAGAT-3′

2.6. Check positive colonies: To determine the presence of an insert, Firstly, a portion of several individual isolated colonies were picked from the plate and transferred to an independent tube containing 10 μL PCR premixture and performed colony PCR program. The primers (*CRP test F* & *CRP test R*) used are listed in Table 18.3. Secondly, agarose gel electrophoresis analysis to identify positive colonies. A positive colony was selected and inoculated in an LB liquid medium containing antibiotics overnight for culture at 37°C shaker at 200 rpm. Finally, extract plasmid DNA using a Plasmid extraction kit, and analyze the DNA via sequencing. (see Note 10).

18.3.3 Genetic Transformation of Gene Editing

After the successful sequencing identification, the recombinant CRISPR/Cas9 gene editing binary plasmid was transferred into *Agrobacterium tumefaciens* GV3101 and then was used for genetic transformation using the floral dipping method [28].

1. The *Agrobacterium tumefaciens* GV3101 competent cells stored at −80°C were kept at room temperature or the palm for a moment to be partially melted, and inserted into the ice when it was mixed with ice water.
2. 1–2 μL plasmid DNA was sucked into a tube containing 50 μL competent cells, and mixed gently. The following processes are carried out in sequence: ice bath for 5 minutes, liquid nitrogen for 5 minutes, water bath at 37°C for 5 minutes, and ice bath for 5 minutes.
3. 700 μL LB liquid medium without antibiotics was added to a tube containing competent cells, then cultivated at 28°C shaker 160 rpm for 2 hours.
4. Take about 300 μL bacterial liquid coated on an LB plate containing spectinomycin, rifampicin, and gentamycin, then cultivate at 28°C for 48–72 hours.

5. Check positive colonies: Pick up several individual isolated colonies, transfer them to an independent tube containing 10 μL PCR premixture, and perform colony PCR program. Agarose gel electrophoresis analysis to identify positive colonies.
6. Select a positive colony of *Agrobacterium* to inoculate into a 15 mL LB liquid medium containing spectinomycin, rifampicin, and gentamycin, then cultivate at a 28°C shaker at 160 rpm for 30 hours.
7. Suction 5 mL *Agrobacterium* solution 3,900 rpm centrifugation, discard the supernatant, and re-suspend with 5% sucrose solution to OD_{600}=0.8–1.0. Add Silwet® L-77 (3 μL/1 mL) and mix.
8. Drop the solution onto the unopened buds with a pipettor. Place the plants in a dark environment for 1 day.
9. Infect every 7 days until there are no flower buds.

18.3.4 Selection of *Cas9*-Free Homozygous Mutants

1. T_1 seeds were selected with red fluorescence by observing the mCherry reporter (Figure 18.3).
2. Planting *Arabidopsis* T_1 lines with red fluorescence, and taking 2-week-old leaves to extract DNA.
3. The sequences near editing target sites were amplified (The names and positions of primers are shown in Figure 18.1, and the sequences of primers are shown in Table 18.3).
4. Sequencing the PCR product. Determine gene editing by comparing it with the wild-type sequence.
5. T_1 plants with gene disruption detected at all three targets were retained and used to produce T_2 seeds (see Note 11).
6. T_2 seeds were selected without red fluorescence by observing the mCherry reporter (Figure 18.3).
7. Planting *Arabidopsis* T_2 lines without red fluorescence, and taking 2-week-old leaves to extract DNA. Identification is similar to that of the T_1 generation.
8. Sequencing all editing target sites to isolate *Cas9*-free homozygous mutants (see Note 12).

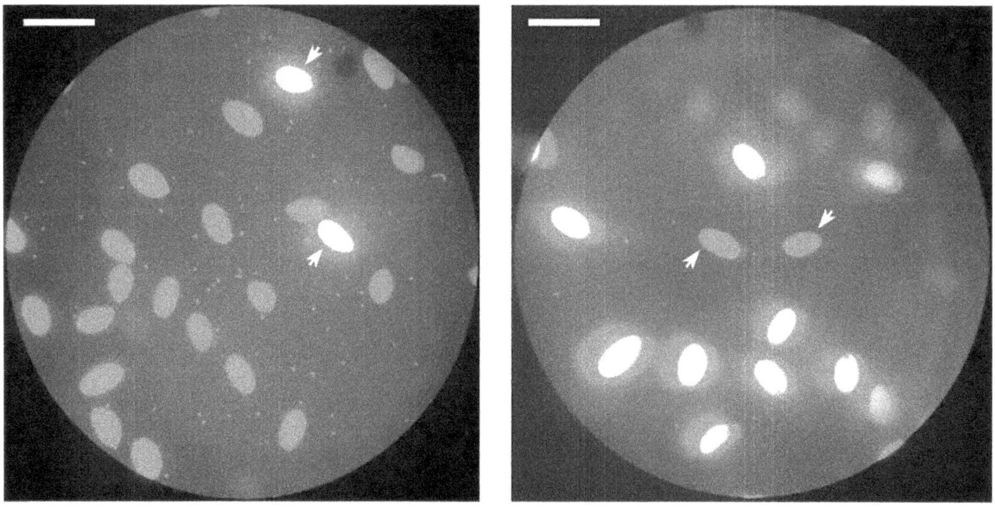

FIGURE 18.3 The seeds of T1 and T2 generation were isolated by red fluorescence in the seed coat. The white arrows respectively represent the seeds with (left) and without (right) red fluorescence that were isolated in T1 (left) and T2 (right). The scale bar represents 1 mm.

18.4 NOTES

1. Assembly of tandem repeat *sgRNA* expression cassettes relies on Clontech's In-Fusion technology. By using that, two or more DNA fragments are joined together depending on the 15–20 bp homologous sequences at their ends. In our protocol, there are the same *U6 promoter*, *gRNA scaffold,* and *U6 terminator* sequences in each of the assembled tandem repeat *sgRNA* expression cassettes, with the only difference between them being the 20bp *sgRNA*. In view of this, two cloning plasmids, *pUC57-Amp-U6TU6P* and *pUC57-Kan-CRP*, were created to be used as templates for amplifying fragments of tandem repeat *sgRNA* expression cassettes [23] (Figure 18.2). The 20bp *sgRNA* sequences were present at both sides of every PCR fragment and were used as the homologous sequence for joining other fragments by In-Fusion cloning.
2. The number of fragments (amplified by using *pUC57-Amp-U6TU6P* as the template) can be determined according to the needs of the research. In our example, two fragments were amplified for subsequent gene editing experiments (Figure 18.2). Moreover, to improve editing efficiency, multiple *sgRNA* sequences can be selected from one gene to construct tandem repeat *sgRNA* expression cassettes. One *sgRNA* target site was selected from each of the three genes as an example of vector construction (Figure 18.1).
3. The recombinant *pUC57-Kan* plasmid containing tandem repeat *sgRNA* expression cassettes can be directly used as the template for the next PCR reaction, or it can be used to transform *Escherichia coli* DH5α competent cells and then extract the plasmid as the template. No matter what scheme is adopted, cloning vector *pUC57-Kan-CRP* was best linearized by endonuclease *EcoR I* before being used as a PCR template in protocol III.B.1.1. Because this can avoid producing non-specific amplification without tandem repeat *sgRNA* expression cassettes, or avoid false positive clones (empty vector *pUC57-Kan-CRP*) during DH5α transformation of assembled *pUC57-Kan* containing tandem repeat *sgRNA* expression cassettes.
4. The homologous sequences of 15 bases on both sides of *Mss I* in CRISPR/Cas9 gene editing binary plasmid were present in the 5′-end of primers *CRP clon F* and *CRP clon R* for In-Fusion cloning, respectively.
5. If electrophoresis shows a single band without other non-specific bands, PCR products can also be purified with a PCR cleaning Kit.
6. To reduce the cost, the volume of the In-Fusion reaction can be 5 μL.
7. In protocol III.B.1.2, the linearized vector is a PCR product amplified by using *pUC57-Kan-CRP* as a template, and inserts are PCR products amplified by using *pUC57-Amp-U6TU6P* as a template. In protocol III.B.1.3, the linearized vector is the CRISPR/Cas9 gene editing binary plasmid linearized by *Mss I*, and the insert is a PCR product amplified by using *pUC57-Kan* containing tandem repeat *sgRNA* expression cassettes as a template. In the In-Fusion reaction, by using a 50–200 ng linearized vector and one or multiple inserts respectively, good cloning efficiency is achieved, regardless of their lengths.
8. In-Fusion Cloning Procedure for PCR-Linearized vector and multiple PCR fragments: place the In-Fusin reaction mixture in a water bath at 37°C for 15 minutes, water bath at 50°C for 15 minutes, then ice bath until use.
9. If the In-Fusion reaction mixture transformed to competent cells was more than 5 μL, the efficiency of transformation will be reduced.
10. If there are too many tandem repeat *sgRNA* expression cassettes, an additional sequencing primer of the *sgRNA* sequence can be used.
11. Gene editing in T_1 somatic cells cannot be passed on to offspring, so T_2 generation may not be able to detect editing at target sites.
12. If two copies of the target site were edited in the T_1 plant germ cells, two different homozygous mutants would be isolated in the T_2 plants.

13. *UBQ10* promoter can be used to replace the *35S* promoter to initiate *Cas9* expression, which can effectively improve the efficiency of gene editing by increasing the cas9 protein in cells. The primers (*pHDE F* & *pHDE R*, Table 18.3) were designed to amplify plasmid sequences except the red region using *pHDE-35S-Cas9-mCherry-UBQ* plasmid (linearized with *Mfe I*) as a template, and then the gained fragment was head-to-tail ligation by In-Fusion technology. This newly constructed vector was named *pHDE-UBQ10-Cas9-mCherry* (Figure 18.4).
14. Xie et al. (2015) found that if the number of *sgRNA*s expressed at the same time exceeded 6 by using CRISPR/Cas9 mediated multiple gene editing, the efficiency of editing would decrease significantly, thus it was speculated that too many *sgRNA*s compete with each other, reducing editing efficiency [16]. In a study by Wang et al. (2020), Only 10 of the 12 *sgRNA*s expressed simultaneously participated in gene editing [22]. Therefore, the construction of high-order mutants also faces the risk that gene editing cannot be detected at some target sites, and the acquisition of some high-order mutants may require secondary gene editing.
15. When the genes corresponding to the high-order mutant to be constructed are sequence-conserved homologous genes, Li et al. (2013) proposed that editing targets could be selected in the sequence-conserved region so that one *sgRNA* could target multiple editing sites in the genome [29], thus optimizing the experimental scheme to the greatest extent and obtaining high-order mutants.

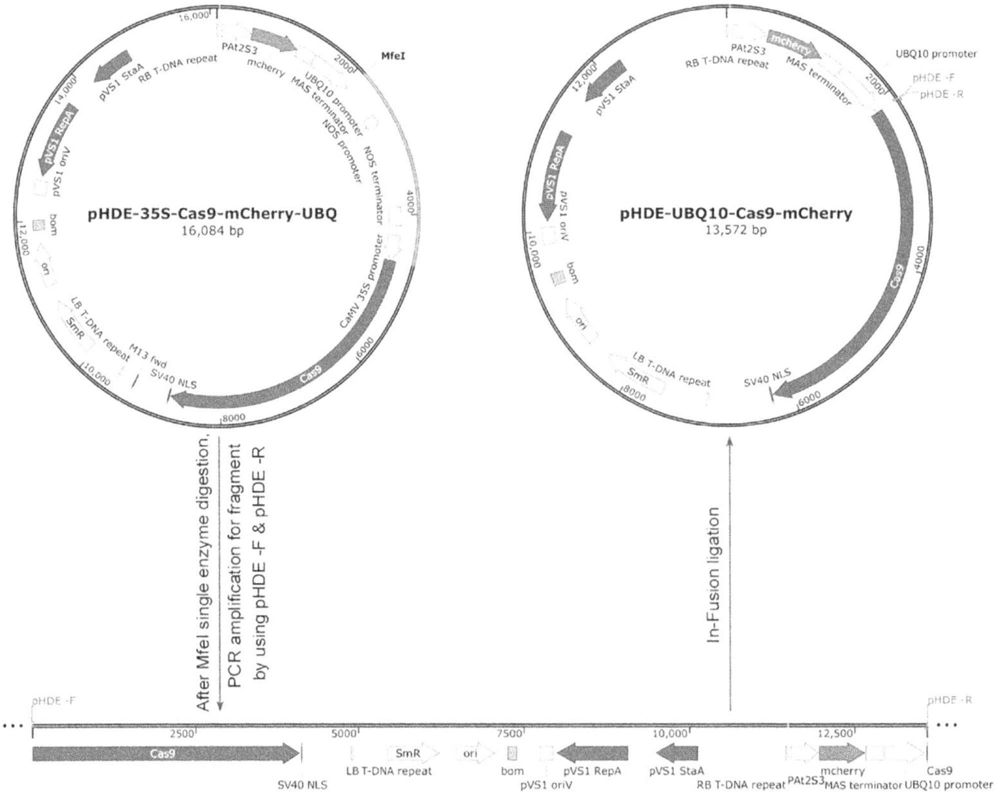

FIGURE 18.4 CRISPR/Cas9 gene editing binary plasmid structure. The workflow for constructing **plasmid** pHDE-UBQ10-Cas9-mCherry from the **plasmid** pHDE-35S-Cas9-mCherry-UBQ.

REFERENCES

1. Rudin N, Sugarman E, Haber JE: Genetic and physical analysis of double-strand break repair and recombination in Saccharomyces cerevisiae. *Genetics* 1989, **122**(3):519–534.
2. Plessis A, Perrin A, Haber JE, Dujon B: Site-specific recombination determined by I-SceI, a mitochondrial group I intron-encoded endonuclease expressed in the yeast nucleus. *Genetics* 1992, **130**(3):451–460.
3. Rouet P, Smih F, Jasin M: Introduction of double-strand breaks into the genome of mouse cells by expression of a rare-cutting endonuclease. *Mol Cell Biol* 1994, **14**(12):8096–8106.
4. Choulika A, Perrin A, Dujon B, Nicolas JF: Induction of homologous recombination in mammalian chromosomes by using the I-SceI system of Saccharomyces cerevisiae. *Mol Cell Biol* 1995, **15**(4):1968–1973.
5. Kim D, Kim J, Hur JK, Been KW, Yoon SH, Kim JS: Genome-wide analysis reveals specificities of Cpf1 endonucleases in human cells. *Nat Biotechnol* 2016, **34**(8):863–868.
6. Bibikova M, Golic M, Golic KG, Carroll D: Targeted chromosomal cleavage and mutagenesis in Drosophila using zinc-finger nucleases. *Genetics* 2002, **161**(3):1169–1175.
7. Lloyd A, Plaisier CL, Carroll D, Drews GN: Targeted mutagenesis using zinc-finger nucleases in *Arabidopsis. Proc Natl Acad Sci U S A* 2005, **102**(6):2232–2237.
8. Boch J, Scholze H, Schornack S, Landgraf A, Hahn S, Kay S, Lahaye T, Nickstadt A, Bonas U: Breaking the code of DNA binding specificity of TAL-type III effectors. *Science* 2009, **326**(5959):1509–1512.
9. Christian M, Cermak T, Doyle EL, Schmidt C, Zhang F, Hummel A, Bogdanove AJ, Voytas DF: Targeting DNA double-strand breaks with TAL effector nucleases. *Genetics* 2010, **186**(2):757–761.
10. Gaj T, Gersbach CA, Barbas CF, 3rd: ZFN, TALEN, and CRISPR/Cas-based methods for genome engineering. *Trends Biotechnol* 2013, **31**(7):397–405.
11. Barrangou R, Fremaux C, Deveau H, Richards M, Boyaval P, Moineau S, Romero DA, Horvath P: CRISPR provides acquired resistance against viruses in prokaryotes. *Science* 2007, **315**(5819):1709–1712.
12. Doudna JA, Charpentier E: Genome editing. The new frontier of genome engineering with CRISPR-Cas9. *Science* 2014, **346**(6213):1258096.
13. Cong L, Ran FA, Cox D, Lin S, Barretto R, Habib N, Hsu PD, Wu X, Jiang W, Marraffini LA et al: Multiplex genome engineering using CRISPR/Cas systems. *Science* 2013, **339**(6121):819–823.
14. Mali P, Yang L, Esvelt KM, Aach J, Guell M, DiCarlo JE, Norville JE, Church GM: RNA-guided human genome engineering via Cas9. *Science* 2013, **339**(6121):823–826.
15. Zhang Y, Ma X, Xie X, Liu YG: CRISPR/Cas9-based genome editing in plants. *Prog Mol Biol Transl Sci* 2017, **149**:133–150.
16. Xie K, Minkenberg B, Yang Y: Boosting CRISPR/Cas9 multiplex editing capability with the endogenous tRNA-processing system. *Proc Natl Acad Sci U S A* 2015, **112**(11):3570–3575.
17. Wang J, Sun N, Zhang F, Yu R, Chen H, Deng XW, Wei N: SAUR17 and SAUR50 differentially regulate PP2C-D1 during apical hook development and cotyledon opening in Arabidopsis. *Plant Cell* 2020, **32**(12):3792–3811.
18. Ma X, Zhang Q, Zhu Q, Liu W, Chen Y, Qiu R, Wang B, Yang Z, Li H, Lin Y et al: A robust CRISPR/Cas9 system for convenient, high-efficiency multiplex genome editing in monocot and dicot plants. *Mol Plant* 2015, **8**(8):1274–1284.
19. Wang C, Shen L, Fu Y, Yan C, Wang K: A Simple CRISPR/Cas9 system for multiplex genome editing in rice. *J Genet Genomics* 2015, **42**(12):703–706.
20. Gao X, Chen J, Dai X, Zhang D, Zhao Y: An effective strategy for reliably isolating heritable and Cas9-free Arabidopsis mutants generated by CRISPR/Cas9-mediated genome editing. *Plant Physiol* 2016, **171**(3):1794–1800.
21. Zhang Z, Mao Y, Ha S, Liu W, Botella JR, Zhu JK: A multiplex CRISPR/Cas9 platform for fast and efficient editing of multiple genes in Arabidopsis. *Plant Cell Rep* 2016, **35**(7):1519–1533.
22. Wang J, Chen H: A novel CRISPR/Cas9 system for efficiently generating Cas9-free multiplex mutants in *Arabidopsis. aBIOTECH* 2020, **1**(1):6–14.
23. Wang J-L, Liang L-Y, Lang N, Li Y-Y, Guo W, Cui J, Zou Y-J, Liu H-Q, Fei Q-H, Li X-F et al: An improved plant CRISPR-Cas9 system for generating Cas9-free multiplex mutants. *Biotechnology & Biotechnological Equipment* 2022, **35**(1):1850–1857.
24. Yu H, Zhao Y: Fluorescence marker-assisted isolation of Cas9-free and CRISPR-edited Arabidopsis plants. *Methods Mol Biol* 2019, **1917**:147–154.
25. Pattanayak V, Lin S, Guilinger JP, Ma E, Doudna JA, Liu DR: High-throughput profiling of off-target DNA cleavage reveals RNA-programmed Cas9 nuclease specificity. *Nat Biotechnol* 2013, **31**(9):839–843.

26. Cho SW, Kim S, Kim Y, Kweon J, Kim HS, Bae S, Kim JS: Analysis of off-target effects of CRISPR/Cas-derived RNA-guided endonucleases and nickases. *Genome Res* 2014, **24**(1):132–141.
27. Kuscu C, Arslan S, Singh R, Thorpe J, Adli M: Genome-wide analysis reveals characteristics of off-target sites bound by the Cas9 endonuclease. *Nat Biotechnol* 2014, **32**(7):677–683.
28. Weigel D, Glazebrook J: In planta transformation of Arabidopsis. *CSH Protoc* 2006, **2006**(7):pdb.prot4668.
29. Li JF, Norville JE, Aach J, McCormack M, Zhang D, Bush J, Church GM, Sheen J: Multiplex and homologous recombination-mediated genome editing in Arabidopsis and Nicotiana benthamiana using guide RNA and Cas9. *Nat Biotechnol* 2013, **31**(8):688–691.

19 The Application of CRISPR Technology for Functional Genomics in Oil Palm and Coconut

Siti Nor Akmar Abdullah and Muhammad Asyraf Md Hatta

19.1 INTRODUCTION

Crops belonging to the Arecaceae family, including oil palm (*Elaeis guineensis* L.) and coconut (*Cocos nucifera* L.), possess considerable economic importance in many developing countries, particularly in the South East Asian regions. According to FAOSTAT (2021), the world's annual production of oil palm and coconut reached 75 million tons and 63 million tons, respectively. Notably, Asia accounted for over 80% of the total production. Similar to other crops, oil palm and coconut production are continuously threatened by both abiotic and biotic stresses. Climate change and global warming have led to unpredictable and severe weather patterns in certain regions, resulting in substantial decreases in crop productivity. Furthermore, the prevailing circumstances will be further aggravated owing to the persistent presence of plant pathogens, including bacteria, viruses, fungi, and oomycetes, as well as pests such as insects, nematodes, mites, and vertebrates. Thus, it is crucial to increase the production of oil palm and coconut by developing novel varieties/hybrids that exhibit enhanced resistance to drought, heat, floods, pests, and diseases.

Marker-assisted selection has been employed to improve the selection process in the breeding program over the past decade. Rather than assessing traits solely based on a plant's physical characteristics, it is possible to conduct selection at the DNA level. This approach is based on the understanding that certain genes are responsible for encoding proteins and harboring the instructions for specific traits. The identification of a particular DNA fragment associated with resistance or tolerance to certain stress traits allows for a more efficient evaluation of the offspring at an early developmental stage (Herzog & Frisch, 2011). The utilization of marker-assisted selection has been extensively adopted by breeders due to its significant impact on reducing selection cost and time (Hasan et al., 2021). However, it is important to note that crossbreeding encounters a further constraint, which is the fundamental requirement of the crosses between individual plants of the same species (Wulff et al., 2011).

The transfer of desired genes as transgenes into a single plant, even across different species has been possible with the development of genetically modified (GM) technology in the late 1970s. Consequently, it has been argued that GM technology exhibits a higher degree of precision, predictability, and controllability in contrast to the conventional method of crossbreeding. However, this technology continues to face challenges pertaining to public acceptance and political concern. To address these issues, new methods with greater precision and efficacy to enable the changes of DNA in a more targeted manner are required.

In recent times, the emergence of clustered regularly interspaced short palindromic repeats (CRISPR)-based genome editing technology has enabled precise DNA manipulation in various plant species (Zaidi et al., 2020; Zhu et al., 2020). CRISPR technology has demonstrated its potential in

DOI: 10.1201/9781003387060-19

the field of genetic engineering by allowing precise modifications to the DNA sequence, thereby altering the phenotypic traits of interest. Since its initial discovery, the CRISPR toolbox has been continuously expanded and multiple engineered variants that possess specific functions have been developed including CRISPRi, CRISPRa, (Lowder et al., 2015; Piatek et al., 2015) base editors (Lin et al., 2020), and prime editors (Kim et al., 2017).

Due to numerous advantages of CRISPR, such as its adaptability and ability to yield reliable and rapid results, several key research institutes and universities have started employing this technology to accelerate the genetic improvement of oil palm and coconut. The continuous decline in sequencing costs has allowed access to more genomic resources of oil palm and coconut, including high-quality reference genomes, transcriptomes, and proteomes that provide novel insights into the genetic basis of various agriculturally important traits. Following the identification of candidate genes linked to the desired trait, CRISPR-based approaches can be utilized to characterize their function. Although this strategy has been hampered by inadequate advancements in tissue culture and genetic transformation methodologies, significant progress has been made in recent years, particularly for oil palm, allowing for a wider application of CRISPR-based approaches for functional genomics.

This chapter outlines the potential of CRISPR-based techniques for improving the traits of oil palm and coconut. The primary focus will be on the existing techniques that have been established on these crops and how they can be effectively used to support the applications of the CRISPR technology for a better understanding of the physiological, biochemical, and molecular mechanisms governing the diverse key traits. The application of GM technology and DNA-free methods, as well as techniques involving stable and transient transformation, will be discussed.

19.2 TECHNIQUES AND RESOURCES FOR FUNCTIONAL GENOMICS USING CRISPR

The advantage of the CRISPR system is that it allows functional characterization in the homologous system, and changes to the gene sequences are made at the exact chromosomal location of the gene in the oil palm or coconut genome. This is achieved through the recruitment of the sgRNA sequence, which consists of a complementary sequence to the target site in the genome and a sequence that binds the Cas enzyme with endonuclease activity (Jinek et al., 2012; Sander & Joung, 2014). This enables the evaluation of gene function in the natural biological system without having to take into consideration the position effects of the inserted genes. As multiple genes may be influencing a particular trait, the different genes can be simultaneously studied through a multiplexing approach in the CRISPR system (McCarty et al., 2020). Whereas stacking of genes in transgenic systems typically requires each gene to be controlled independently by separate promoters resulting in large gene constructs which significantly reduce the transformation efficiency. CRISPR knockout (CRISPR-KO) is a powerful tool to investigate the function of single or multiple genes together. Repair of the double-stranded DNA break created by Cas9 through non-homologous end joining (NHEJ) is an efficient way of introducing small random DNA insertions and deletions that disrupt gene functions through frameshift mutations or introduction of premature stop codons affecting the open reading frame (ORF) of the target genes (Symington & Gautier, 2011; Zhang et al., 2018). These strategies can be usefully applied to study the different pathways influencing important traits such as oil quality and response to biotic and abiotic factors in oil palm and coconut. Dead Cas9 (dCas9) which has lost its endonuclease activity while still retaining its function in delivering sgRNA to specific genomic sites proves to be a versatile tool for functional genomics. CRISPR/dCas9 can be used for activating (CRISPRa) or suppressing (CRISPRi) the expression of the target genes through the recruitment of trans-activators or suppressors, respectively, to the proximal promoter region of these genes. Furthermore, the non-functional dCas9 can be fused with various modifying enzymes to achieve specific research interests. Fusion of the dCas9 with methylase enzyme can have valuable applications in epigenetic studies in oil palm and coconut (Jogam et al., 2022; Lee et al., 2019).

Like in other eukaryotes, most agronomically important genetic variations in oil palm and coconut are single-nucleotide polymorphisms requiring precision genome editing tools for characterizing their functions. Application of CRISPR/Cas9 in precise gene editing (PGE) still suffers problems associated with double-stranded break and the inefficiency of homology-directed repair (HDR) of the introduced donor DNA template and its potential in introducing high levels of indels and unintended products (Puchta, 2005; Van Vu et al., 2019). CRISPR/dCas9 which offers a viable alternative strategy for PGE through fusion with adenine and cytidine deaminases is currently widely applied for functional studies in animals and plants and can have useful applications in oil palm and coconut. Even though it is highly efficient, there is a limitation to the type of substitution that can be performed by the deaminases confining to four transitions, C → T, G → A, A → G and T → C (Kim et al., 2017; Mishra et al., 2020). The more recent development to single base editing strategy is through prime editing (PE) which is highly versatile as it can perform all possible base conversions, small indels, and their combinations at target sites (Anzalone et al., 2019) and PE has shown great promise when tested in different plant species such as rice (Lin et al., 2020), wheat (Lin et al., 2020), tomato (Lu et al., 2021), and maize (Jiang et al., 2020). In PE, the prime editing guide RNA (pegRNA) not only specifies the target site but also contains the primer binding site (PBS) fused to a reverse transcription template for introducing nucleotide substitution. The nCas introduces a nick at the non-target strand, which anneals with the PBS, enabling reverse transcription to be performed, generating copies of the edited sequence which is incorporated into the target site following a complex DNA repair process.

Even without having 100% homology, the sgRNA could still bind target DNA and such off-target binding is of great concern among researchers. Therefore, it is critical to use a sgRNA designing platform that is equipped with features that can evaluate sgRNA characteristics particularly the position where it binds in the genome, its GC content, and secondary structure, which are important properties influencing the sgRNA functionality. Most of the tools are equipped with on and off-target scores that enable sgRNA to be ranked for selecting the one that gives the highest efficiency. Cas-Designer (http://www.rgenome.net/cas-designer/) enables sgRNA to be designed for more than 30 PAM types, thus not limited to PAM from *Streptococcus pyogenes* (Park et al., 2015). CRISPOR (http://crispor.org) provides a complete solution for gRNA selection, cloning, and expression, also primers for testing guide activity and potential off-targets (Concordet & Haeussler, 2018). An important consideration by biologists is the user-friendliness of the online tools for expediting the process of designing efficient sgRNA with minimum off-targets as demonstrated by CRISPR-P 2.0 (http://crispr.hzau.edu.cn/CRISPR2) (Liu et al., 2017).

19.3 TARGET GENES FROM OIL PALM FOR FUNCTIONAL CHARACTERIZATION BY CRISPR SYSTEMS

Oil palm is the source of two types of oil with commercial values obtained from two different tissues of the oil palm fruits, which are palm oil from the mesocarp and palm kernel oil. Palm oil has a balanced composition of saturated and monounsaturated fatty acids suitable for various food and non-food applications and rich in vitamin E and beta-carotene. The kernel oil mainly consists of medium-chain saturated fatty acid, which meets the requirement as feedstock for the oleochemical industry (Murphy et al., 2021). Besides studies associated with oil production and composition and its nutritional value, there has been active research on improving the tissue culture process through somatic embryogenesis. Tissue culture can produce high-quality planting materials genetically identical to the source of explants (ortets) and yield improvement as high as 30% has been achieved in the clonal palms. Even though the oil palm micropropagation technique has been established since the 1980s, the low somatic embryogenesis rate and the risk of somaclonal epigenetic alterations are major concerns restricting the commercialization of oil palm tissue culture clones (Ong-Abdullah et al., 2015; Sahara et al., 2023). Oil palm is also susceptible to diseases, and there are many reports on different strategies to overcome the devastating basal stem rot caused by the pathogenic fungal pathogen, *Ganoderma boninense* (Bharudin et al., 2022; Siddiqui et al., 2021). The El Niño phenomenon due to the effects of climate change is causing unpredictable weather conditions resulting in prolonged drought, which significantly affected oil palm productivity in recent years (Khor et al., 2023). The high price of fertilizer and its detrimental effects on the environment is also a major concern to be addressed through enhancing nutrient uptake and nutrient use efficiency for the sustainability of the oil palm industry (Arifin et al., 2022).

Potential candidate genes for oil palm improvement have been identified from various single (Hamzah et al., 2021; Kong et al., 2021) and high throughput (RNA sequencing) gene discovery studies (Dussert et al., 2013; Ho et al., 2016). Functional characterization is a requisite to unravel their functions and associated mechanisms before using them in genetic engineering or molecular breeding programs and even in molecular diagnostics. The genes of interest that were identified include genes that are involved in the biosynthesis of fatty acids and high-value minor components (vitamin E and beta-carotene), transcription factors as key regulators of the developmental process and responses to biotic and abiotic stresses, nutrient transporters, and many others.

19.3.1 Tissue Culture for Production of Elite Planting Materials

RNA-seq approach was used to unravel the molecular mechanism affecting the oil palm tissue culture process in order to overcome the low somatic embryogenesis rate of less than 6% (Sahara et al., 2023). The RNA-seq data revealed that physiological differences between high- and low-embryogenic ortets led to the different capacities for somatic embryogenesis. Up-regulation of abscisic acid signaling-related genes including LEA, DDX28, and vicilin-like protein as well as other hormone signaling genes such as HD-ZIP (brassinosteroids) and NPF (auxin) was observed in high-embryogenic ortets. Previously, based on the expression patterns under auxin treatment, dehydration, osmotic stresses, and light exposure, the involvement of a novel oil palm HD-ZIP II gene during the acquisition of embryogenic competency in early somatic embryogenesis and subsequent vascular development was suggested (Ooi et al., 2016). This concurs with the expected role of HD-ZIP genes based on the reports in other species (Li et al., 2022). The differentially expressed genes (DEGs) discovered through RNA-seq can be evaluated through the CRISPRa system, either singly or through the multiplexing approach to understand gene networks. These will assess their roles and potential use as biomarkers for improving the rate of somatic embryogenesis. By studying a transcription factor like HD-ZIP, the effects on downstream genes regulated by this transcription factor can also be evaluated.

There is a potential risk in increasing the frequency of somaclonal variation as the embryogenic structures are propagated extensively for increasing the efficiency of somatic embryogenesis in oil palm (Rival et al., 2013). Differential accumulation of transcripts for *EGAD1* which encodes a putative oil palm defensin gene was observed at the callus stage of tissue-cultured oil palm, which correlated with the occurrence of mantled flowering abnormality (Tregear et al., 2002). The upstream regulatory region of the *EGAD1* contained regulatory elements that are normally found in defense-related genes explaining the expression observed in tissue culture. Through an epigenome-wide association study, Ong-Abdullah et al. (2015) discovered that the oil palm fruit abnormality is associated with methylation events near the intron splice site of the homeotic gene, *DEFICIENS* which determines alternate splicing and premature termination of this gene. Functional characterization through fusing dCas9 with methylase enzyme targeting selected sites in the oil palm *EGAD1* and *DEFICIENS* may provide more in-depth information on epigenetic mechanisms associated with clonal abnormality to support their development and utility as biomarkers for predicting somaclonal variation.

19.3.2 Overcoming Loss Due to Biotic and Abiotic Stresses

Oil palm requires high fertilizer input to sustain high yield and phosphorus is the major limiting macronutrient in acidic tropical soil. P deficiency leads to poor oil palm growth and productivity

(Arifin et al., 2022). RT-qPCR analysis (Hamzah et al., 2021) showed upregulated expression of oil palm phosphate starvation response factor *EgPHR2* in response to Pi deficiency and the transcriptional response correlated with the expression of the high-affinity phosphate transporters *EgPHT1;4* and *EgPHT1;7* harboring the P1BS motif. High throughput studies using RNA-seq suggested a potentially critical role played by oil palm *PHL7* and vacuolar influx transporter, *VPT* in Pi homeostasis under Pi deficiency (Kong et al., 2021). In rice, it was demonstrated by Ruan et al. (2017) that alteration of nucleotide in the P1BS motif can enhance the transcriptional response of a phosphate starvation inducible gene, *PHOSPHATE TRANSPORTER TRAFFIC FACILITATOR 1 (PHF1)*, and this can be explored in oil palm through PE using CRISPR-dCas9. Furthermore, the complex mechanism in regulating Pi homeostasis involving *EgPHR2*, *EgPHL7*, and the *EgVPT* can potentially be unraveled through CRISPRa and CRISPRi.

There have been several reports on high throughput gene expression analysis on molecular response mechanisms of oil palm to *G. boninense* infection generating a lot of useful data (Bahari et al., 2018; Othman et al., 2019; Zuhar et al., 2021). Comparison with the response of oil palm to the biocontrol agent (BCA) such as *Trichoderma harzianum* was also of interest since these provide an understanding of how the plant defense against *G. boninense* is enhanced or made effective by the BCA (Ho et al., 2016; 2018). Several candidate genes have been identified for further characterization from the different RNA-seq data including genes involved in phytohormones ethylene and jasmonic acid and Ca^{2+}signaling, reactive oxygen species production and scavenging, cell wall modification and antifungal proteins (pathogenesis-related proteins, THAUMATIN, and fungal protease inhibitor) which can be carried out through functional studies using CRISPR-mediated systems. Recently, Bahari et al. (2018) and Sakeh et al. (2021) identified transcription factors that may potentially be the key regulators of early-stage infection at the biotrophic and necrotrophic phases, respectively. The ability to delay or prevent the fungal switching from the initial biotrophic mode of infection for establishing colonization of root cells to the cell-damaging necrotrophic mode could potentially save the oil palm plant. Several downstream genes co-expressing with these transcription factors were determined. Through PE or recruitment of deaminase by CRISPR/dCas9, valuable information on transcriptional regulation can be obtained from editing the nucleotide in the functional domain particularly the DNA binding domain of these transcription factors or the promoter elements, which serves as their binding sites. The promoter elements that bind to *JUNGBRUNNEN 1* and *Ethylene Responsive Factor 113* have been validated through *in vitro* and *in vivo* assays by Sakeh et al. (2021).

19.3.3 Oil Yield and Quality Improvement

Earlier studies for functional characterization of isolated fatty acid biosynthetic gene promoters and oil palm tissue-specific promoters were performed in model plants such as *Arabidopsis thaliana* and tomato (Kamaladini et al., 2013; Saed Taha et al., 2012; Zhu et al., 2018). The studies aimed to evaluate the function of genes and genetic engineering tools for modifying the oil composition of palm oil. Even though the copy number and position of oil palm genes being inserted have an important influence on the results, the researchers had no control in determining them in these approaches. Furthermore, the results obtained using the heterologous transgenic systems may not accurately reflect the function of the structural genes or the regulatory regions in oil palm.

Oil palm germplasm resources can be tapped to identify functional variants associated with traits of economic importance. *Elaeis oleifera* which is the oil palm species from South America typically contains higher levels of unsaturated fatty acids (Montoya et al., 2014). The *Elaeis guineensis* germplasm materials from Africa exhibit a wide range of variation in the content of beneficial vitamin E. RT-qPCR analysis showed a high level of expression of homogentisate geranylgeranyl transferase (HGGT) in the mesocarp tissues during the period of oil synthesis in *E. guineensis* while homogentisate phytyltransferase (HPT) demonstrated high upregulated expression in the mesocarp of *E. oleifera* during oil synthesis period (Kong et al., 2016). HGGT and HPT are involved in the production of tocotrienol and tocopherol isomers of vitamin E in plants, respectively.

It was demonstrated that one nucleotide change, which altered the sequence of the proximal promoter element, CAAT-box influenced the expression of *HGGT* in oil palm. This is a natural nucleotide variant observed in the oil palm germplasm materials associated with high tocotrienol content (Karim et al., 2021).

The different RNA-seq efforts (Dussert et al., 2013; Morris et al., 2020) have identified *WRINKLED1* (*WRI1*) as the key transcription factor regulating palm oil synthesis. The gene paralogs, *EgWRI1-1* and *EgWRI1–2* were very highly transcribed during the period of oil accumulation in the two oil-bearing tissues of the oil palm. Furthermore, the expression level of fatty acid biosynthesis genes correlated with WRI1 transcripts level as well as oil content. In addition, Yeap et al. (2017) reported that EgNF-YA3 has direct interaction with EgWRI1-1, forming a transcription complex with EgNF-YC2 and EgABI5 for modulating the transcription of palm oil biosynthetic genes. With the availability of the oil palm genome sequence (Mohd Sanusi et al., 2023; Singh et al., 2013; Wang et al., 2022), whole genome resequencing or targeted sequencing can be used for the discovery of natural variants in the regulatory or coding regions of the vitamin E biosynthetic gene and key regulatory transcription factors of oil biosynthesis. The functional importance of the variants can be further validated using CRISPR PE. This will help in providing information for future knowledge-guided strategies for improving the oil and nutritional content of palm oil. Table 19.1 provides a list of potential candidate genes from oil palm for functional analysis through CRISPR technology that has been discussed.

TABLE 19.1
List of Potential Targets for Functional Characterization through CRISPR-Mediated Technology in Oil Palm

Target Gene	Type of Editing	Trait	Reference
LEA, DDX28, NPF, HD-ZIP	CRISPRa and CRISPR-KO	The efficiency of somatic embryogenesis	Sahara et al. (2023) Ooi et al. (2016)
EGAD1 DEFICIENS	CRISPR/dCas 9 fused to methylase for epigenetic analysis	Tissue culture abnormality	Tregear et al. (2002) Ong-Abdullah et al. (2015)
EgPHR2, EgPHT1;4 and EgPHT1;7 EgPHL7, EgVPT	Prime editing or dCas9 fused to deaminases	Phosphate deficiency response	Hamzah et al. (2021) Kong et al. (2021)
Phytohormones ethylene and jasmonic acid and Ca^{2+} signaling, reactive oxygen species production and scavenging, cell wall modification and antifungal proteins (pathogenesis-related proteins, THAUMATIN and fungal protease inhibitor) *JUNGBRUNNEN 1* and *Ethylene Responsive Factor 113*	CRISPRa or CRISPR- KO Prime editing or dCas9 fused to deaminases	Response to *Ganoderma boninense* infection Biotrophy to necrotrophy switch	Bahari et al. (2018) Othman et al. (2019) Zuhar et al. (2021) Sakeh et al. (2021)
HPT and HGGT	CRISPRa and prime editing targeting to modify promoter or enzyme activity	Vitamin E biosynthesis	Kong et al., (2016) Karim et al., (2021)
WRINKLED1 EgNF-YA3, EgNF-YC2 and EgABI5	CRISPRa and CRISPR-KO	Oil synthesis	Dussert et al. (2013) Morris et al. (2020) Yeap et al. (2017)

19.4 RESOURCES AND ESTABLISHED TOOLS FOR FUNCTIONAL STUDIES THROUGH TRANSIENT EXPRESSION AND STABLE TRANSFORMATION IN OIL PALM

19.4.1 Oil Palm Genome Sequence

The high-quality genome sequence is essential for designing gRNA for targeting the genomic site for functional characterization through CRISPR-based technology. Oil palm is a diploid organism ($2n=32$) and the estimated genome size is 1.8 Gb. The AVROS *Pisifera* is used as the male parent in commercial crossing with *Dura* as the female parent to produce high-yielding *Tenera* progenies. Singh et al. (2013) reported on the first draft of the oil palm genome sequence of AVROS *Pisifera* consisting of 83% (1.535 Gb) of the total genome size from 16 genetic scaffolds (one per chromosome). A total of 34,802 identified candidate genes showed similarity at the peptide level with known proteins with 96% of them found in transcriptome data from 30 tissue types. Among them, the genes of importance for improving key traits have been identified including genes involved in fatty acid biosynthesis, response to biotic and abiotic stimuli, regulation of gene expression, signal transduction, embryo development, flower development, and ripening. Besides the assembly of contigs into scaffolds, a good genome assembly requires the formation of the physical map through the placement and contiguity of scaffolds on chromosomes. Recently Mohd Sanusi et al. (2023) reported an improvement to 54% from 43% of the first reported genome assembly (Singh et al., 2013), achieved through integrating a number of SNP and SSR-based genetic maps to produce a consensus map-generating scaffolds anchored to the 16 pseudochromosome. In addition, Wang et al. (2022) reported on a chromosome-level reference genome that is of high quality from a *Dura* palm. The genome was assembled using a long reads sequencing platform with ~ 150× genome coverage resulting in 1.7 Gb of the assembled genome. This constitutes 94.5% of the estimated genome size, where 91.6% were assigned to 16 pseudochromosomes.

19.4.2 Gene Functional Studies through Transient Expression in Protoplasts

Transient expression analysis in transfected oil palm protoplast is suggested to be the most efficient approach for the functional study of oil palm genes. Protoplast is a living plant cell that does not possess a cell wall and can be transfected by exogenous macromolecules including DNA, RNA, and proteins (Yue et al., 2021). It provides a versatile single-cell-based system for studying functions of macromolecules through gene activation and silencing and genome editing and the GM protoplast can be regenerated to produce a whole plant. Using the protoplast system different cellular activities and properties of proteins can be studied including their regulatory roles (such as transcription factors and protein kinases), enzymatic activities, subcellular localization and trafficking, interaction with other proteins, and role in signal transduction pathway (Fraiture et al., 2014).

Fizree et al. (2021) reported on optimized protocols and cell wall degradation enzyme combinations for protoplast isolation from oil palm leaf and mesocarp tissues. Chloroplast yield from leaf tissues of up to 2.5×10^6 protoplasts g^{-1} fresh weight $(FW)^{-1}$ with 95% viability and from mesocarp of 3.98×10^6 protoplasts g^{-1} FW^{-1} with 85% viability was achieved. Introduction of plasmid DNA into oil palm protoplasts through microinjection and polyethylene glycol (PEG)-mediated transfection has been reported (Masani et al., 2014). PEG-mediated protoplast transfection is preferred for routine application as the technique is simple and robust without requiring special or expensive equipment and suitable for high-throughput analysis (Ghose et al., 2022). A recent report showed an improvement in the oil palm protoplast transfection efficiency to about 56%. Fizree et al. (2021) used PEG-mediated transfection compared to a previous report of only 5% (Masani et al., 2014). This improvement was due to the optimization of several parameters including the incubation period as well as increasing the concentration of PEG and oil palm DNA. It was found that the amount of DNA taken up by the protoplasts is directly proportional to the DNA concentration. There is an optimal concentration for PEG to produce the highest transfection efficiency and prolonged PEG exposure should be avoided to reduce DNA degradation by nucleases released from the protoplasts.

Protoplast can be transfected with plasmid DNA or ribonucleoproteins (RNPs) to introduce the gene transcriptional modification or editing agents Transfection with RNP complex provides a platform for DNA-free gene editing without the concern associated with GM issues if the plants are regenerated (Moradpour & Abdulah, 2020). Oil palm is mainly used in food products where the application of GM technology is sensitive and subjected to public scrutiny. The RNP complex consists of sgRNA and dCas9 for recruiting various transcriptional activating domains or modifying enzymes such as nucleotide deaminases or methylase. The edited plants can immediately be evaluated following the transfection process through sequencing and gene expression analysis (Tuncel & Qi, 2022). Several reports showed that editing at off-target sites is significantly reduced, particularly with the RNP approach (Farboud et al., 2018; Zhang et al., 2021). A fluorescent protein can be used to aid with the selection of the transfected protoplast before proceeding with regeneration.

19.4.3 Stable Oil Palm Transformation for Gene Functional Evaluation in Transgenic System

Stable oil palm transformation can be achieved through biolistic transformation and *Agrobacterium*-mediated transformation methods (Masani et al., 2018). The oil palm biolistic technique shows a very low transformation efficiency rate of 1.5% or less and insertion of multiple genes that could potentially cause gene silencing is a major concern. The standard and reproducible protocol for *Agrobacterium*-mediated transformation of oil palm has been developed albeit at very low transformation efficiency (Izawati et al., 2015; Izawati et al., 2012). The study performed by (Promchan & Te-chato, 2013) showed that the *Agrobacterium* strain EHA101 was found to be better at transforming oil palm cells compared to AGL-1. Since oil palm has one meristem, its genetic transformation relies on callus as target tissue. Embryogenic calli from suspension culture are the preferred target tissue for both oil palm transformation techniques as they provide a higher regeneration rate compared to transformed calli that originated from solid media (Abang Masli et al., 2009). The continuous availability of oil palm suspension cultures that produce a large number of embryogenic calli is an advantage as numerous transformation events are needed to produce one successfully transformed plant.

The established transformation and regeneration protocols of oil palm embryogenic calli could be utilized for functional evaluation of single or multiple oil palm genes through CRISPR gene editing techniques. The vector constructs harboring the editing agents can be introduced through biolistic or *Agrobacterium*-mediated transformation techniques for the production of transgenic plants that can be used in evaluating the biochemical and physiological functions of these genes and the interaction between them (Masani et al., 2018). In addition, the transgenic oil palm seedlings can be subjected to various abiotic and biotic stress treatments to provide a more comprehensive evaluation of their functions. This approach is more feasible for studies involving vegetative tissues at the seedling stage. As oil palm is already a tall tree when it starts fruiting after 2 and ½ to 3 years (Herdiansyah et al., 2020) of planting, phenotypic, physiological, and molecular analysis on fruit tissues require a large greenhouse or isolated areas for containment that can prevent the spread of pollens (Krishan Kumar et al., 2020) from the transgenic oil palm plants due to their GM status with inserted gene construct in the genome. Figure 19.1 illustrates the route for CRISPR-mediated functional studies using stable transformation or transient assay in protoplast that have been discussed.

19.4.4 Reports of CRISPR/Cas9 Application in Oil Palm

Knockout of two fatty acid biosynthetic genes, *palmitoyl-acyl carrier protein thioesterase* and *fatty acid desaturase 2* in oil palm by multiplex genome editing using CRISPR/Cas9 was reported (Bahariah et al., 2023) in order to reduce the level of saturation for a healthier oil. The multiple

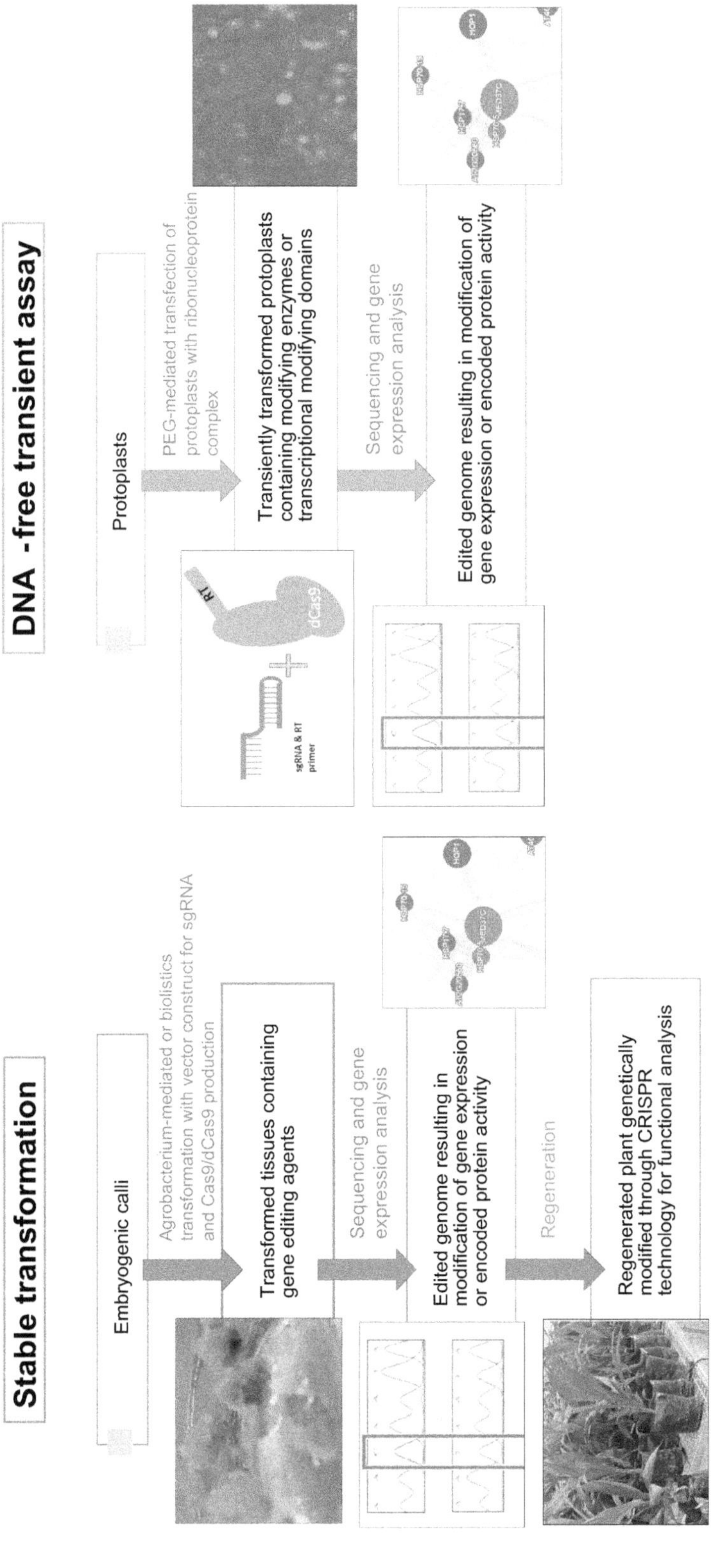

FIGURE 19.1 CRISPR-mediated functional studies using stable transformation or transient assay in protoplasts. Oil palm has established resources and tools for using these two routes for gene functional studies. RT represents reverse transcriptase enzyme for prime editing using CRISPR/dCas9.

sgRNAs of the target genes were introduced into oil palm embryogenic calli and protoplasts. The different transformation methods including *Agrobacterium*-mediated, PEG-mediated (for protoplasts), and biolistics were attempted. The samples were co-bombarded with the pAMDsRED plasmid to aid in the selection of successfully transformed tissues. Subsequent PCR analysis and sequencing showed successful production of single and double knockout mutants of both genes and small and large deletions were found within the targeted sites of transformed tissues. Expression of Cas9 was also detected.

Earlier, Yeap et al. (2021) reported on the introduction of *phytoene desaturase* (*EgPDS*) and *brassinosteroid-insensitve 1* (*EgBRI1*) separately into oil palm protoplasts for targeted mutagenesis using CRISPR/Cas9. The protoplasts were used for a rapid transient assay of cleavage frequency and biolistic particle bombardment of immature embryos was subsequently carried out to produce stably transformed plants. For *EgPDS*, 25.49% cleavage efficiency was achieved, however, the transgenic oil palm shoots exhibited abnormal chimeric albino phenotype due to the genome alteration, with high mutation efficiency between 60% and 80%. Genome alterations were also successfully achieved for *EgBRI1*, however abnormal stunted phenotypes were reported due to DNA InDels.

19.5 COCONUT'S GENETIC RESOURCES AND POTENTIAL TARGETS FOR CRISPR-MEDIATED GENE EDITING

The coconut tree is a perennial plant that grows naturally along tropical islands and coastlines. It is a diploid organism containing 32 chromosomes ($2n = 16$) (Nasir et al., 2016). Despite its economic significance, there has been little progress on coconut's genetic improvement via biotechnological approaches. The majority of prior research has primarily relied on conventional breeding methods. Due to its long generation time and perennial nature, this approach presents a major challenge (Lineesha & Antony, 2021). Hence, fundamental and applied research on this crop is limited.

19.5.1 Coconut Genome Sequencing Initiatives

The availability of a high-quality reference genome is crucial for accelerating the genetic improvement of coconut. This is essential as it enables the identification of novel genes that contribute to economically valuable characteristics such as resistance to pests and diseases, tolerance to drought, and high productivity, among other traits. A recent publication by (Xiao et al., 2017) unveiled the initial version of a complete coconut genome, which was generated through the Illumina HiSeq 2000 platform. The sequencing of the Hainan Tall cultivar's genome has resulted in the production of a total of 419.67 gigabases (Gb) of high-quality reads, which encompasses approximately 90.91% of the estimated genome size of *Cocos nucifera* (2.42 Gb).

Following that, a group from the Philippines released the first draft of a dwarf coconut, cv. Catigan Green Dwarf (CATD) using Illumina Miseq, PacBio Single Molecule Real Time (SMRT), and Dovetail Chicago sequencing technologies (Lantican et al., 2019). The estimated genome size is 2.15 Gb, 280 Mbp lower than the Hainan Tall variety (Xiao et al., 2017). In terms of sequencing coverage, a higher percentage was achieved as compared to the previous effort with 97.6% of the estimated genome size. The study revealed 34,958 protein-coding genes, which play crucial roles in various significant traits such as pest and disease resistance, drought tolerance, and the biosynthesis of coconut oil. One of the potential target genes for CRISPR-mediated gene editing is 1-acyl-sn-glycerol-3-phosphate acyltransferase (LPAAT), which plays a crucial role in the significant accumulation of lauric acid (Liang et al., 2014). The potential improvement of copra oil quality in coconut may be achieved through the overexpression of this gene. Notably, the gene's ortholog is also present in African oil palm, although with a shorter length of 28,344 bp (Singh et al., 2013).

More recently, the utilization of Nanopore single-molecule sequencing and Hi-C technology has generated *de novo* assembly of each palm of tall and dwarf variety. The study presented significant findings regarding the essential genes that play a role in the divergence of tall and dwarf height traits in coconut (Wang et al., 2021). In the same study, transcriptome data was also generated to compare the gene expression levels of certain key genes. One of the important traits in coconut is plant height as it is significantly related to both yield and harvesting efficiency. The gibberellin (GA) biosynthetic enzyme GA-20 oxidase (*GA20ox*) gene was found to be more highly expressed in the tall variety, making it a promising target for CRISPR to develop dwarf palms. It has been previously reported that the targeted mutagenesis of the *GA20ox* gene has generated a semi-dwarf phenotype in rice (Santoso et al., 2020) and maize (Zhang et al., 2020).

The tall variety was also discovered to have a higher number of NAC and WRKY family transcription factors, which govern salt tolerance, as well as genes associated with lignin and lipid synthesis than the dwarf (Wang et al., 2021). Lignin is one of the main components of coconut fiber. Among potential targets for lignin synthesis are cinnamoyl-CoA reductase (CCR) and Ferulate 5-hydroxylase (F5H) enzymes, as well as the transcription factors MYB85, MYB58, and MYB46. Since MYB46 was the most highly expressed transcription factor in the transcriptome, it may be the primary target for overexpression via CRISPR/Cas9-based activation to increase the fiber content for biocomposite production. The cellulose synthase (CESA) gene involved in the biosynthesis and regulation of cellulose in coconut could also be one of the target genes to enhance fiber content.

19.5.2 Identification of Candidate Genes through Transcriptomics

Various transcriptomics studies have been reported in coconut, which may provide insights into the discovery of novel functional genes associated with economically important traits. To enhance plant pathogen proliferation, they often exploit genes in the host plant known as susceptibility (*S*) genes. The pathogen's ability to cause disease could be reduced by inactivating these genes, potentially leading to more durable and broad-spectrum resistance (Bozkurt et al., 2015; Koseoglou et al., 2022; Lapin & Van den Ackerveken, 2013). Comparative transcriptome profiling of healthy and afflicted cv. Chowghat Green Dwarf upon root wilt disease (RWD) infection has revealed some candidate *S* genes for CRISPR-mediated gene knockout (Rajesh et al., 2018). One of them includes a gene that encodes for endo-1, 4-β-mannosidase, an enzyme involved in the degradation of hemicelluloses present in the cell wall. The impairment of this gene may block the degradation of host cell walls during the pathogen attack, rendering the coconut plant resistant to the RWD. Another potential target is asparagine synthetase, which regulates the amino acid biosynthesis in response to the attack. A study in the tomato-*Botrytis cinerea* pathosystem has proposed its role in providing a nitrogen source for facilitating the pathogen's growth and subsequent colonization (Seifi et al., 2013). Apart from the discovery of genes linked to susceptible response, the transcriptomics data also uncovered genes that are involved in resistance mechanisms such as 3-ketoacyl-CoA synthase and Ca2+ and calmodulin-dependent protein kinase.

Transcriptome profiling of naturally infected leaves of cv. Malayan Red Dwarf upon phytoplasma infection has revealed genes that encode pathogenesis-related proteins (PRs) such as thaumatin-like proteins (TLPs) (PR-5), chitinase (PR-3), glucan endo-1,3-beta-d-glucosidase (PR-2), peroxidase (PR-9) and Bet v 1-related proteins (PR-10) (Nejat et al., 2015). Numerous other transcripts that are linked to resistance response have also been highly expressed during pathogen attacks including genes of WRKY transcription factors, receptor-like kinase (RLK), reactive oxygen species (ROS), phytoalexin biosynthesis, hormone biosynthesis, and lignin biosynthesis.

To elucidate the signaling pathways involved in salt stress, a transcriptomics study was performed on the leaves of cv. Hainan Tall and Aromatic Dwarf seedlings (Yang et al., 2021). Sixty-five DEGs were identified during the salt stress response. An interesting one is a gene that encodes for chloroplastic CU-Zn superoxide dismutase (SODCP). This gene has been identified in Hainan Tall but is absent in the Aromatic Dwarf. Dwarf, which exhibits signs of human domestication may have undergone a genetic evolution resulting in the absence of the SODCP gene. Conversely, the Tall variety, which primarily remains in coastal regions, tends to retain this salt-stress tolerance trait. However, the protein phosphatase 2C (PP2C) gene of the ABA-dependent pathway emerges as the most promising target for gene editing. The repression or deletion of this gene has the potential to enhance ABA signaling, resulting in the closure of stomata and the biosynthesis of protective molecules that contribute to osmotic homeostasis (Kundan Kumar et al., 2013), especially in dwarf varieties (da Silva et al., 2017).

19.5.3 Limitations of CRISPR Applications in Coconut

The significant advancements in the generation of genomic and transcriptomic data have revealed various candidate genes associated with agriculturally important traits in coconut (Table 19.2). CRISPR-mediated functional characterization of these genes will accelerate the breeding of superior coconut cultivars/hybrids (Figure 19.2). Despite its widespread use in other crops, genetic improvement of coconut continues to pose difficulties in using this method. As of present, there has been no successful application of genome editing techniques in the coconut plant reported. The limited advancement in this approach can be attributed to the lack of well-established methodologies for tissue culture and genetic transformation. So far, there is only a report on the transient genetic transformation in which embryogenic callus has been used as explants (Andrade-Torres et al., 2011). The functional characterization of genes using CRISPR should ideally be carried out in a stable coconut transformation. However, due to its extended life cycle, this may take a long time to observe the phenotypic traits of interest. Alternatively, previous studies used *in vitro* biochemical tests (Davies et al., 1995; Knutzon et al., 1999) and gene transformation in model plants such as *Arabidopsis thaliana* and rice to characterize the genes (Reynolds et al., 2015; Sun et al., 2017). The efficacy of genome editing technology in crops is primarily dependent on highly efficient transformation techniques. Therefore, further research is necessary to develop such procedures in coconut.

19.6 CONCLUSION

To effectively address the challenges posed by the constantly evolved pests and diseases, as well as the impact of climate change, breeders of oil palm and coconut must generate superior varieties/hybrids. Over the past decade, the genetic improvement of crops has been revolutionized by the advent of genome editing techniques, most notably CRISPR. This approach has been successfully employed in various important crops including wheat, rice, tomato, and maize, among others.

TABLE 19.2
List of Potential Targets for Functional Characterization through CRISPR-Mediated Technology in Coconut

Target Gene	Type of Editing	Trait	Reference
LPAAT	CRISPRa	Improvement of copra oil quality	Lantican et al., (2019) Liang et al., (2014)
GA20ox	CRISPR-KO	Plant height (dwarf phenotype)	Wang et al. (2021) Santoso et al. (2020) Zhang et al. (2020)
MYB46	CRISPRa	Fiber content	Wang et al. (2021)
CESA	CRISPRa	Fiber content	Wang et al. (2022)
endo-1, 4-β-mannosidase	CRISPR-KO	Root wild disease resistance	Rajesh et al. (2018)
asparagine synthetase	CRISPR-KO	Root wild disease resistance	Rajesh et al. (2018)
PP2C	CRISPRi or CRISPR-KO	Salt stress tolerance	Yang et al. (2021) Kumar et al. (2013) da Silva et al. (2017)

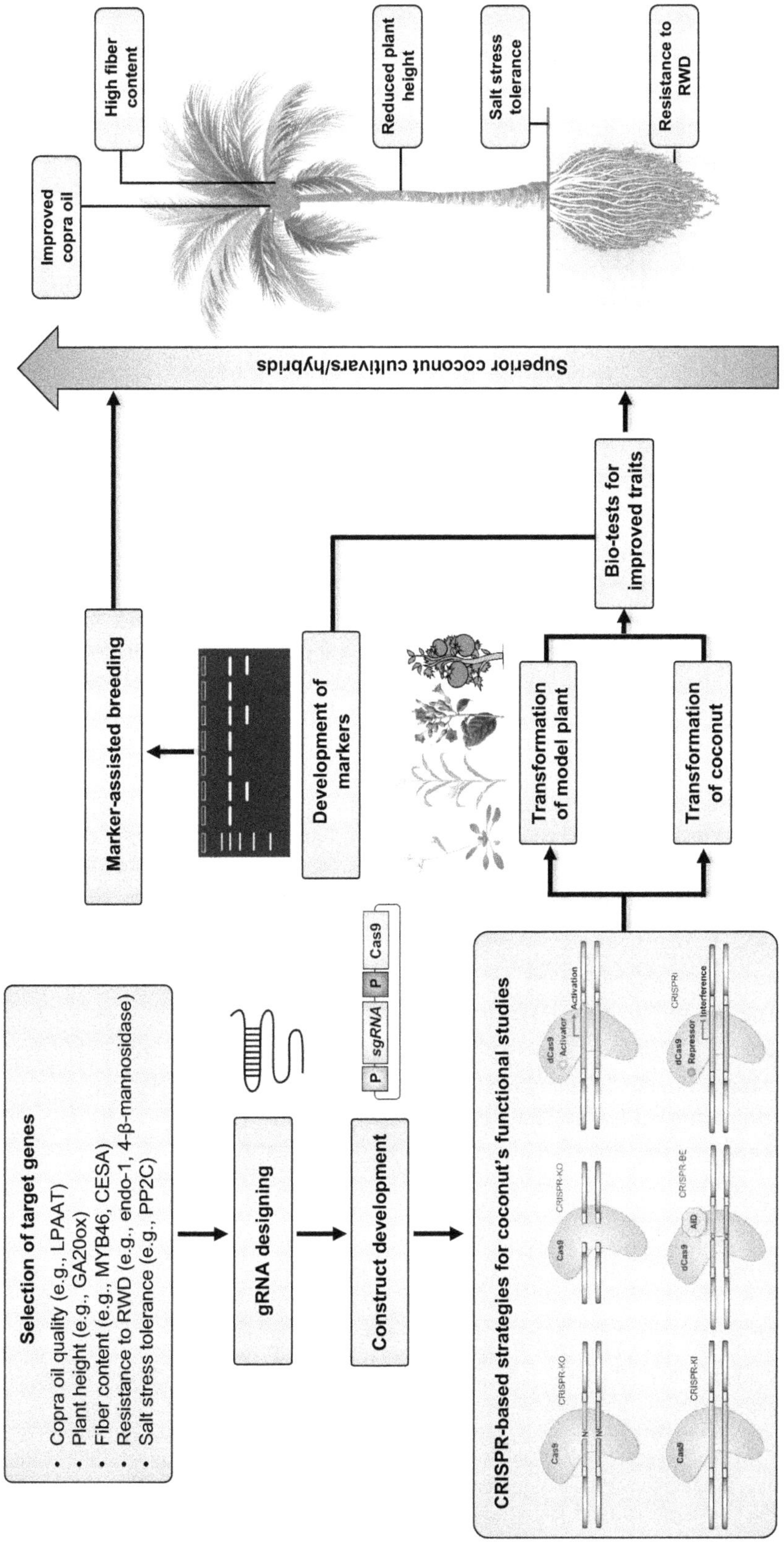

FIGURE 19.2 CRISPR-based strategies for genetic improvement of coconut.

The CRISPR technology exhibits remarkable versatility and flexibility, providing robust methods for the functional characterization of specific genes. These methods include targeted gene knockout, gene knockin, base editing, PE, transcriptional activation, and transcriptional repression. Yet, the wide adoption of this technology in perennial crops such as oil palm and coconut has been hindered by the complexity and heterozygosity of their genomes, as well as their poor regeneration efficiency in tissue culture. The identification of potential candidate genes for the genetic improvement of these crops has been extensively conducted using genomics and transcriptomics resources. However, it is crucial to elucidate their specific roles and underlying mechanisms prior to incorporating them into the breeding program. In oil palm, desirable traits include improved oil quality, resistance to basal stem rot, tolerance to drought, and enhanced phosphate uptake. While reduced plant height, better copra oil quality, salt-stress tolerance, and resistance to RWD are among the traits in coconut that can be targeted using CRISPR. With the ongoing advancements in stable transformation protocols for oil palm and coconut, alternative methods for CRISPR-mediated gene editing can be explored through the utilization of model plants and the transient expression of protoplasts.

19.7 FUTURE PERSPECTIVES

With the anticipated decline in sequencing costs, there is an opportunity to re-sequence a greater number of oil palm and coconut genotypes in order to establish a comprehensive "pan-genome". This effort would yield valuable insights into the underlying genetic mechanisms governing complex traits like yield and drought tolerance. Despite the availability of various genetic resources, the application of CRISPR-based strategies to understand the underlying mechanisms of important traits in oil palm and coconut is hindered by the absence of well-established tissue culture protocols and efficient genetic transformation methods. Therefore, the primary focus for the scientific communities of these crops in the coming years should be to enhance the plantlet regeneration system. This improvement is necessary to fully exploit the potential of CRISPR as a tool for characterizing candidate genes. Nonetheless, the phenotypic characterization of the mutants may be time-consuming owing to the long generation time observed in perennial crops. To accelerate this process, the CRISPR techniques can be combined with speed breeding technology. This method has been successful in reducing the generation time of diverse crops. We envision that a similar advancement in oil palm and coconut would be a game changer in the genetic improvement of these crops via CRISPR-mediated gene editing.

FUNDING

This work was funded by the Ministry of Higher Education Malaysia (MoHE) through Long-Term Research Grant Scheme (LRGS/1/2020/UPM/01/2/2).

REFERENCES

Abang Masli, D. I., Ghulam Kadir, A. P., & Mat Yunus, A. M. (2009). Transformation of oil palm using *Agrobacterium tumefaciens*. *Journal of Oil Palm Research*, 21, 643–652.

Andrade-Torres, A., Oropeza, C., Sáenz, L., González-Estrada, T., Ramírez-Benítez, J. E., et al. (2011). Transient genetic transformation of embryogenic callus of *Cocos nucifera*. *Biologia,* 66(5), 790–800. https://doi.org/10.2478/s11756-011-0104-4

Anzalone, A. V., Randolph, P. B., Davis, J. R., Sousa, A. A., Koblan, L. W., et al. (2019). Search-and-replace genome editing without double-strand breaks or donor DNA. *Nature,* 576(7785), 149–157. https://doi.org/10.1038/s41586-019-1711-4

Arifin, I., Hanafi, M. M., Roslan, I., Ubaydah, M. U., Karim, Y. A., et al. (2022). Responses of irrigated oil palm to nitrogen, phosphorus and potassium fertilizers on clayey soil. *Agricultural Water Management*, 274, 107922. https://doi.org/10.1016/j.agwat.2022.107922

Bahari, M. N. A., Sakeh, N. M., Abdullah, S. N. A., Ramli, R. R., & Kadkhodaei, S. (2018). Transciptome profiling at early infection of *Elaeis guineensis* by *Ganoderma boninense* provides novel insights on fungal transition from biotrophic to necrotrophic phase. *BMC Plant Biology*, 18(1), 377. https://doi.org/10.1186/s12870-018-1594-9

Bahariah, B., Masani, M. Y. A., Fizree, M., Rasid, O. A., & Parveez, G. K. A. (2023). Multiplex CRISPR/Cas9 gene-editing platform in oil palm targeting mutations in *EgFAD2* and *EgPAT* genes. *Journal of Genetic Engineering and Biotechnology*, 21(1), 3. https://doi.org/10.1186/s43141-022-00459-5

Bharudin, I., Ab Wahab, A. F. F., Abd Samad, M. A., Xin Yie, N., Zairun, M. A., et al. (2022). Review update on the life cycle, plant-microbe interaction, genomics, detection and control strategies of the oil palm pathogen *Ganoderma boninense*. *Biology*, 11(2). https://doi.org/10.3390/biology11020251

Bozkurt, T. O., Belhaj, K., Dagdas, Y. F., Chaparro-Garcia, A., Wu, C. H., et al. (2015). Rerouting of plant late endocytic trafficking toward a pathogen interface. *Traffic*, 16(2), 204–226. https://doi.org/10.1111/tra.12245

Concordet, J.-P., & Haeussler, M. (2018). CRISPOR: Intuitive guide selection for CRISPR/Cas9 genome editing experiments and screens. *Nucleic Acids Research*, 46. https://doi.org/10.1093/nar/gky354

da Silva, A. R. A., Bezerra, F. M. L., de Lacerda, C. F., de Sousa, C. H. C., & Bezerra, M. A. (2017). Physiological responses of dwarf coconut plants under water deficit in salt-affected soils. *Revista Caatinga*, 30, 447–457.

Davies, H. M., Hawkins, D. J., & Nelsen, J. S. (1995). Lysophosphatidic acid acyltransferase from immature coconut endosperm having medium chain length substrate specificity. *Phytochemistry*, 39(5), 989–996. https://doi.org/10.1016/0031-9422(95)00046-A

Dussert, S., Guerin, C., Andersson, M., Joët, T., Tranbarger, T. J., et al. (2013). Comparative transcriptome analysis of three oil palm fruit and seed tissues that differ in oil content and fatty acid composition. *Plant Physiology*, 162(3), 1337–1358. https://doi.org/10.1104/pp.113.220525

FAOSTAT. (2021). Crops and livestock products. https://www.fao.org/faostat/en/#data/QCL/visualize. from Accessed 16 June 2023.

Farboud, B., Jarvis, E., Roth, T. L., Shin, J., Corn, J. E., et al. (2018). Enhanced genome editing with Cas9 ribonucleoprotein in diverse cells and organisms. *JoVE (Journal of Visualized Experiments)* (135). https://doi.org/10.3791/57350

Fizree, M. D. P. M. A. A., Shaharuddin, N. A., Ho, C.-L., Manaf, M. A. A., Parveez, G. K. A., et al. (2021). Efficient protocol improved the yield and viability of oil palm protoplasts isolated from *in vitro* leaf and mesocarp. *Scientia Horticulturae*, 290, 110522. https://doi.org/10.1016/j.scienta.2021.110522

Fraiture, M., Zheng, X., & Brunner, F. (2014). An *Arabidopsis* and tomato mesophyll protoplast system for fast identification of early MAMP-triggered immunity-suppressing effectors. In P. Birch, J. Jones, J. Bos, (Eds.), *Plant-Pathogen Interactions. Methods in Molecular Biology* (Vol. 1127, .pp. 213–230). Totowa, NJ: Humana Press. https://doi.org/10.1007/978-1-62703-986-4_17

Ghose, A. K., Abdullah, S. N. A., Md Hatta, M. A., & Megat Wahab, P. E. (2022). DNA free CRISPR/dCas9 based transcriptional activation system for UGT76G1 gene in *Stevia rebaudiana* bertoni protoplasts. *Plants*, 11(18), 2393. https://doi.org/10.3390/plants11182393

Hamzah, M. L., Abdullah, S. N. A., & Azzeme, A. M. (2021). Genome-wide molecular characterization of *Phosphate Transporter 1* and *Phosphate Starvation Response* gene families in *Elaeis guineensis* Jacq. and their transcriptional response under different levels of phosphate starvation. *Acta Physiologiae Plantarum*, 43(8), 113. https://doi.org/10.1007/s11738-021-03282-6

Hasan, N., Choudhary, S., Naaz, N., Sharma, N., & Laskar, R. A. (2021). Recent advancements in molecular marker-assisted selection and applications in plant breeding programmes. *Journal of Genetic Engineering and Biotechnology*, 19(1), 128. https://doi.org/10.1186/s43141-021-00231-1

Herdiansyah, H., Negoro, H., Rusdayanti, N., & Shara, S. (2020). Palm oil plantation and cultivation: Prosperity and productivity of smallholders. *Open Agriculture*, 5, 617–630. https://doi.org/10.1515/opag-2020-0063

Herzog, E., & Frisch, M. (2011). Selection strategies for marker-assisted backcrossing with high-throughput marker systems. *Theoretical and Applied Genetics*, 123(2), 251–260. https://doi.org/10.1007/s00122-011-1581-0

Ho, C. L., Tan, Y. C., Yeoh, K. A., Ghazali, A. K., Yee, W. Y., et al. (2016). *De novo* transcriptome analyses of host-fungal interactions in oil palm (*Elaeis guineensis* Jacq.). *BMC Genomics*, 17, 66. https://doi.org/10.1186/s12864-016-2368-0

Ho, C.-L., Tan, Y.-C., Yeoh, K.-A., Lee, W.-K., Ghazali, A.-K., et al. (2018). Transcriptional response of oil palm (*Elaeis guineensis* Jacq.) inoculated simultaneously with both *Ganoderma boninense* and *Trichoderma harzianum*. *Plant Gene*, 13, 56–63. https://doi.org/10.1016/j.plgene.2018.01.003

Izawati, A. M., Masani, M. Y., Ismanizan, I., & Parveez, G. K. (2015). Evaluation on the effectiveness of 2-deoxyglucose-6-phosphate phosphatase (*DOGR1*) gene as a selectable marker for oil palm (*Elaeis guineensis* Jacq.) embryogenic calli transformation mediated by *Agrobacterium tumefaciens*. *Frontiers in Plant Science*, 6, 727. https://doi.org/10.3389/fpls.2015.00727

Izawati, A. M., Parveez, G. K., & Masani, M. Y. (2012). Transformation of oil palm using *Agrobacterium tumefaciens*. In J. Dunwell, A. Wetten (Eds.), *Transgenic Plants. Methods in Molecular Biology* (Vol 847, pp. 177–188). Humana Press. https://doi.org/10.1007/978-1-61779-558-9_15

Jiang, Y.-Y., Chai, Y.-P., Lu, M.-H., Han, X.-L., Lin, Q., et al. (2020). Prime editing efficiently generates W542L and S621I double mutations in two ALS genes in maize. *Genome Biology*, 21(1), 257. https://doi.org/10.1186/s13059-020-02170-5

Jinek, M., Chylinski, K., Fonfara, I., Hauer, M., Doudna, J. A., et al. (2012). A programmable dual-RNA-guided DNA endonuclease in adaptive bacterial immunity. *Science*, 337(6096), 816. https://doi.org/10.1126/science.1225829

Jogam, P., Sandhya, D., Alok, A., Peddaboina, V., Allini, V. R., et al. (2022). A review on CRISPR/Cas-based epigenetic regulation in plants. *International Journal of Biological Macromolecules*, 219, 1261–1271. https://doi.org/10.1016/j.ijbiomac.2022.08.182

Kamaladini, H., Nor Akmar Abdullah, S., Aziz, M. A., Ismail, I. B., & Haddadi, F. (2013). Breaking-off tissue specific activity of the oil palm metallothionein-like gene promoter in T(1) seedlings of tomato exposed to metal ions. *Journal of Plant Physiology,* 170(3), 346–354. https://doi.org/10.1016/j.jplph.2012.10.017

Karim, M. S. N., Abdullah, S. N. A., Nadzir, M. M. M., Moradpour, M., Shaharuddin, N. A., et al. (2021). Single nucleotide polymorphisms in oil palm *HOMOGENTISATE GERANYL-GERANYL TRANSFERASE* promoter for species differentiation and TOCOTRIENOL improvement. *Meta Gene,* 27, 100818. https://doi.org/10.1016/j.mgene.2020.100818

Khor, J. F., Ling, L., Yusop, Z., Chin, R. J., Lai, S. H., et al. (2023). Impact comparison of El Niño and ageing crops on Malaysian oil palm yield. *Plants,* 12(3). https://doi.org/10.3390/plants12030424

Kim, Y. B., Komor, A. C., Levy, J. M., Packer, M. S., Zhao, K. T., et al. (2017). Increasing the genome-targeting scope and precision of base editing with engineered Cas9-cytidine deaminase fusions. *Nature Biotechnology,* 35(4), 371–376. https://doi.org/10.1038/nbt.3803

Knutzon, D. S., Hayes, T. R., Wyrick, A., Xiong, H., Maelor Davies, H., et al. (1999). Lysophosphatidic acid acyltransferase from coconut endosperm mediates the insertion of laurate at the *sn-2* position of triacylglycerols in lauric rapeseed oil and can increase total laurate levels. *Plant Physiology,* 120(3), 739–746. https://doi.org/10.1104/pp.120.3.739

Kong, S.-L., Abdullah, S. N. A., Ho, C. L., & Amiruddin, M. D. (2016). Molecular cloning, gene expression profiling and in silico sequence analysis of vitamin E biosynthetic genes from the oil palm. *Plant Gene,* 5, 100–108. https://doi.org/10.1016/j.plgene.2016.01.003

Kong, S.-L., Abdullah, S. N. A., Ho, C.-L., Musa, M. H. b., & Yeap, W.-C. (2021). Comparative transcriptome analysis reveals novel insights into transcriptional responses to phosphorus starvation in oil palm (*Elaeis guineensis*) root. *BMC Genomic Data,* 22(1), 6. https://doi.org/10.1186/s12863-021-00962-7

Koseoglou, E., van der Wolf, J. M., Visser, R. G. F., & Bai, Y. (2022). Susceptibility reversed: modified plant susceptibility genes for resistance to bacteria. *Trends in Plant Science*, 27(1), 69–79. https://doi.org/10.1016/j.tplants.2021.07.018

Kumar, K., Gambhir, G., Dass, A., Tripathi, A. K., Singh, A., et al. (2020). Genetically modified crops: current status and future prospects. *Planta*, 251(4), 91. https://doi.org/10.1007/s00425-020-03372-8

Kumar, K., Kumar, M., Kim, S.-R., Ryu, H., & Cho, Y.-G. (2013). Insights into genomics of salt stress response in rice. *Rice,* 6(1), 27. https://doi.org/10.1186/1939-8433-6-27

Lantican, D. V., Strickler, S. R., Canama, A. O., Gardoce, R. R., Mueller, L. A., et al. (2019). *De Novo* genome sequence assembly of dwarf coconut (*Cocos nucifera* L. 'Catigan Green Dwarf') provides insights into genomic variation between coconut types and related palm species. *G3: Genes, Genomes, Genetics*, 9(8), 2377–2393. https://doi.org/10.1534/g3.119.400215

Lapin, D., & Van den Ackerveken, G. (2013). Susceptibility to plant disease: more than a failure of host immunity. *Trends in Plant Science*, 18(10), 546–554. https://doi.org/10.1016/j.tplants.2013.05.005

Lee, J. E., Neumann, M., Duro, D. I., & Schmid, M. (2019). CRISPR-based tools for targeted transcriptional and epigenetic regulation in plants. *PLoS One*, 14(9), e0222778. https://doi.org/10.1371/journal.pone.0222778

Li, Y., Yang, Z., Zhang, Y., Guo, J., Liu, L., et al. (2022). The roles of HD-ZIP proteins in plant abiotic stress tolerance. *Frontiers in Plant Science*, 13, 1027071. https://doi.org/10.3389/fpls.2022.1027071

Liang, Y., Yuan, Y., Liu, T., Mao, W., Zheng, Y., et al. (2014). Identification and computational annotation of genes differentially expressed in pulp development of *Cocos nucifera* L. by suppression subtractive hybridization. *BMC Plant Biology*, 14, 205. https://doi.org/10.1186/s12870-014-0205-7

Lin, Q., Zong, Y., Xue, C., Wang, S., Jin, S., et al. (2020). Prime genome editing in rice and wheat. *Nature Biotechnology*, 38(5), 582–585. https://doi.org/10.1038/s41587-020-0455-x

Lineesha, K. A., & Antony, G. (2021). Genome editing: Prospects and challenges. In M. K. Rajesh, S. V. Ramesh, L. Perera, & C. Kole (Eds.), *The Coconut Genome* (pp. 191–203). Cham: Springer International Publishing. https://doi.org/10.1007/978-3-030-76649-8_14

Liu, H., Ding, Y., Zhou, Y., Jin, W., Xie, K., et al. (2017). CRISPR-P 2.0: An improved CRISPR-Cas9 tool for genome editing in plants. *Molecular Plant*, 10(3), 530–532. https://doi.org/10.1016/j.molp.2017.01.003

Lowder, L. G., Zhang, D., Baltes, N. J., Paul, J. W., 3rd, Tang, X., et al. (2015). A CRISPR/Cas9 Toolbox for multiplexed plant genome editing and transcriptional regulation. *Plant Physiology*, 169(2), 971–985. https://doi.org/10.1104/pp.15.00636

Lu, Y., Tian, Y., Shen, R., Yao, Q., Zhong, D., et al. (2021). Precise genome modification in tomato using an improved prime editing system. *Plant Biotechnology Journal*, 19(3), 415–417. https://doi.org/10.1111/pbi.13497

Masani, M. Y. A., Izawati, A. M. D., Rasid, O. A., & Parveez, G. K. A. (2018). Biotechnology of oil palm: Current status of oil palm genetic transformation. *Biocatalysis and Agricultural Biotechnology*, 15, 335–347. https://doi.org/10.1016/j.bcab.2018.07.008

Masani, M. Y., Noll, G. A., Parveez, G. K., Sambanthamurthi, R., & Prüfer, D. (2014). Efficient transformation of oil palm protoplasts by PEG-mediated transfection and DNA microinjection. *PLoS One*, 9(5), e96831. https://doi.org/10.1371/journal.pone.0096831

McCarty, N. S., Graham, A. E., Studená, L., & Ledesma-Amaro, R. (2020). Multiplexed CRISPR technologies for gene editing and transcriptional regulation. *Nature Communications*, 11(1), 1281. https://doi.org/10.1038/s41467-020-15053-x

Mishra, R., Joshi, R. K., & Zhao, K. (2020). Base editing in crops: current advances, limitations and future implications. *Plant Biotechnology Journal*, 18(1), 20–31. https://doi.org/10.1111/pbi.13225

Mohd Sanusi, N. S. N., Rosli, R., Chan, K. L., Halim, M. A. A., Ting, N. C., et al. (2023). Integrated consensus genetic map and genomic scaffold re-ordering of oil palm (*Elaeis guineensis*) genome. *Computational Biology and Chemistry*, 102, 107801. https://doi.org/10.1016/j.compbiolchem.2022.107801

Montoya, C., Cochard, B., Flori, A., Cros, D., Lopes, R., et al. (2014). Genetic architecture of palm oil fatty acid composition in cultivated oil palm (Elaeis guineensis Jacq.) compared to its wild relative E. oleifera (H.B.K) Cortés. PLoS One, 9(5), e95412. https://doi.org/10.1371/journal.pone.0095412

Moradpour, M., & Abdulah, S. N. A. (2020). CRISPR/dCas9 platforms in plants: strategies and applications beyond genome editing. *Plant Biotechnology Journal*, 18(1), 32–44. https://doi.org/10.1111/pbi.13232

Morris, P. E., Chan, P.-L., Cheng-Li, L. O., Ong, P.-W., Kamaruddin, K., et al. (2020). Transcriptome analysis across various mesocarp developmental stages of MPOB-Angola *Dura*. *Journal of Oil Palm Research*, 32(3), 406–419. https://doi.org/10.21894/jopr.2020.0034

Murphy, D. J., Goggin, K., & Paterson, R. R. M. (2021). Oil palm in the 2020s and beyond: challenges and solutions. *CABI Agriculture and Bioscience*, 2(1), 39. https://doi.org/10.1186/s43170-021-00058-3

Nasir, K. H., Azwan, J., Shahril, A. R., Sentoor, G. K., N., S., et al. (2016). Genetic diversity evaluation of MARDI's coconut (*Cocos nucifer*a L.) germplasm using simple sequence repeats. *International Journal on Coconut R & D*, 32, 37–45. https://doi.org/10.37833/cord.v32i2.34

Nejat, N., Cahill, D. M., Vadamalai, G., Ziemann, M., Rookes, J., et al. (2015). Transcriptomics-based analysis using RNA-Seq of the coconut (*Cocos nucifera*) leaf in response to yellow decline phytoplasma infection. *Molecular Genetics and Genomics*, 290(5), 1899–1910. https://doi.org/10.1007/s00438-015-1046-2

Ong-Abdullah, M., Ordway, J. M., Jiang, N., Ooi, S. E., Kok, S. Y., et al. (2015). Loss of *Karma* transposon methylation underlies the mantled somaclonal variant of oil palm. *Nature*, 525(7570), 533–537. https://doi.org/10.1038/nature15365

Ooi, S.-E., Ramli, Z., Syed Alwee, S. S. R., Kulaveerasingam, H., & Ong-Abdullah, M. (2016). *EgHOX1*, a HD-Zip II gene, is highly expressed during early oil palm (*Elaeis guineensis* Jacq.) somatic embryogenesis. *Plant Gene*, 8, 16–25. https://doi.org/10.1016/j.plgene.2016.09.006

Othman, N. Q., Sulaiman, S., Lee, Y. P., & Tan, J. S. (2019). Transcriptomic data of mature oil palm basal trunk tissue infected with *Ganoderma boninense*. *Data Brief*, 25, 104288. https://doi.org/10.1016/j.dib.2019.104288

Park, J., Bae, S., & Kim, J. S. (2015). Cas-Designer: A web-based tool for choice of CRISPR-Cas9 target sites. *Bioinformatics*, 31(24), 4014–4016. https://doi.org/10.1093/bioinformatics/btv537

Piatek, A., Ali, Z., Baazim, H., Li, L., Abulfaraj, A., et al. (2015). RNA-guided transcriptional regulation in planta via synthetic dCas9-based transcription factors. *Plant Biotechnology Journal*, 13(4), 578–589. https://doi.org/10.1111/pbi.12284

Promchan, T., & Te-chato, S. (2013). Strains of *Agrobacterium* affecting gene transformation through embryogenic cell suspension of hybrid tenera oil palm *International Journal of Agricultutral Technology*, 9(3), 669–679.

Puchta, H. (2005). The repair of double-strand breaks in plants: Mechanisms and consequences for genome evolution. *Journal of Experimental Botany*, 56(409), 1–14. https://doi.org/10.1093/jxb/eri025

Rajesh, M. K., Rachana, K. E., Kulkarni, K., Sahu, B. B., Thomas, R. J., et al. (2018). Comparative transcriptome profiling of healthy and diseased Chowghat Green Dwarf coconut palms from root (wilt) disease hot spots. *European Journal of Plant Pathology,* 151(1), 173–193. https://doi.org/10.1007/s10658-017-1365-8

Reynolds, K. B., Taylor, M. C., Zhou, X. R., Vanhercke, T., Wood, C. C., et al. (2015). Metabolic engineering of medium-chain fatty acid biosynthesis in *Nicotiana benthamiana* plant leaf lipids. *Frontiers in Plant Science*, 6, 164. https://doi.org/10.3389/fpls.2015.00164

Rival, A., Ilbert, P., Labeyrie, A., Torres, E., Doulbeau, S., et al. (2013). Variations in genomic DNA methylation during the long-term in vitro proliferation of oil palm embryogenic suspension cultures. *Plant Cell Reports*, 32(3), 359–368. https://doi.org/10.1007/s00299-012-1369-y

Ruan, W., Guo, M., Wu, P., & Yi, K. (2017). Phosphate starvation induced OsPHR4 mediates Pi-signaling and homeostasis in rice. *Plant Molecular Biology*, 93(3), 327–340. https://doi.org/10.1007/s11103-016-0564-6

Saed Taha, R., Ismail, I., Zainal, Z., & Abdullah, S. N. (2012). The stearoyl-acyl-carrier-protein desaturase promoter (Des) from oil palm confers fruit-specific GUS expression in transgenic tomato. *Journal of Plant Physiology*, 169(13), 1290–1300. https://doi.org/10.1016/j.jplph.2012.05.001

Sahara, A., Roberdi, R., Wiendi, N. M. A., & Liwang, T. (2023). Transcriptome profiling of high and low somatic embryogenesis rate of oil palm (*Elaeis guineensis* Jacq. var. Tenera). *Frontiers in Plant Science*, 14, 1142868. https://doi.org/10.3389/fpls.2023.1142868

Sakeh, N. M., Abdullah, S. N. A., Bahari, M. N. A., Azzeme, A. M., Shaharuddin, N. A., et al. (2021). EgJUB1 and EgERF113 transcription factors as potential master regulators of defense response in *Elaeis guineensis* against the hemibiotrophic *Ganoderma boninense*. *BMC Plant Biology*, 21(1), 59. https://doi.org/10.1186/s12870-020-02812-7

Sander, J. D., & Joung, J. K. (2014). CRISPR-Cas systems for editing, regulating and targeting genomes. *Nature Biotechnology*, 32(4), 347–355. https://doi.org/10.1038/nbt.2842

Santoso, T. J., Trijatmiko, K. R., Char, S. N., Yang, B., & Wang, K. (2020). Targeted mutation of *GA20ox-2* gene using CRISPR/Cas9 system generated semi-dwarf phenotype in rice. In *IOP Conference Series: Earth and Environmental Science* (Vol. 482, No.1, p. 012027). https://doi.org/10.1088/1755-1315/482/1/012027

Seifi, H. S., Van Bockhaven, J., Angenon, G., & Höfte, M. (2013). Glutamate metabolism in plant disease and defense: friend or foe? *Molecular Plant-Microbe Interactions*, 26(5), 475–485. https://doi.org/10.1094/mpmi-07-12-0176-cr

Siddiqui, Y., Surendran, A., Paterson, R. R. M., Ali, A., & Ahmad, K. (2021). Current strategies and perspectives in detection and control of basal stem rot of oil palm. *Saudi Journal of Biological Sciences*, 28(5), 2840–2849. https://doi.org/10.1016/j.sjbs.2021.02.016

Singh, R., Ong-Abdullah, M., Low, E.-T. L., Manaf, M. A. A., Rosli, R., et al. (2013). Oil palm genome sequence reveals divergence of interfertile species in Old and New worlds. *Nature*, 500(7462), 335–339. https://doi.org/10.1038/nature12309

Sun, R., Ye, R., Gao, L., Zhang, L., Wang, R., et al. (2017). Characterization and ectopic expression of CoWRI1, an AP2/EREBP domain-containing transcription factor from coconut (*Cocos nucifera* L.) endosperm, changes the seeds oil content in transgenic *Arabidopsis thaliana* and rice (*Oryza sativa* L.). *Frontiers in Plant Science*, 8, 63. https://doi.org/10.3389/fpls.2017.00063

Symington, L. S., & Gautier, J. (2011). Double-strand break end resection and repair pathway choice. *Annual Review of Genetics*, 45, 247–271. https://doi.org/10.1146/annurev-genet-110410-132435

Tregear, J. W., Morcillo, F., Richaud, F., Berger, A., Singh, R., et al. (2002). Characterization of a defensin gene expressed in oil palm inflorescences: Induction during tissue culture and possible association with epigenetic somaclonal variation events. *Journal of Experimental Botany*, 53(373), 1387–1396.

Tuncel, A., & Qi, Y. (2022). CRISPR/Cas mediated genome editing in potato: Past achievements and future directions. *Plant Science*, 325, 111474. https://doi.org/10.1016/j.plantsci.2022.111474

Van Vu, T., Sung, Y. W., Kim, J., Doan, D. T. H., Tran, M. T., et al. (2019). Challenges and perspectives in homology-directed gene targeting in monocot plants. *Rice*, 12(1), 95. https://doi.org/10.1186/s12284-019-0355-1

Wang, L., Lee, M., Yi Wan, Z., Bai, B., Ye, B., et al. (2022). Chromosome-level reference genome provides insights into divergence and stress adaptation of the african oil palm. *Genomics, Proteomics & Bioinformatics*. https://doi.org/10.1016/j.gpb.2022.11.002

Wang, S., Xiao, Y., Zhou, Z. W., Yuan, J., Guo, H., et al. (2021). High-quality reference genome sequences of two coconut cultivars provide insights into evolution of monocot chromosomes and differentiation of fiber content and plant height. *Genome Biology*, 22(1), 304. https://doi.org/10.1186/s13059-021-02522-9

Wulff, B. B., Horvath, D. M., & Ward, E. R. (2011). Improving immunity in crops: new tactics in an old game. *Current Opinion in Plant Biology,* 14(4), 468–476. https://doi.org/10.1016/j.pbi.2011.04.002

Xiao, Y., Xu, P., Fan, H., Baudouin, L., Xia, W., et al. (2017). The genome draft of coconut (*Cocos nucifera*). *Gigascience*, 6(11), 1–11. https://doi.org/10.1093/gigascience/gix095

Yang, Y., Bocs, S., Fan, H., Armero, A., Baudouin, L., et al. (2021). Coconut genome assembly enables evolutionary analysis of palms and highlights signaling pathways involved in salt tolerance. *Communications Biology*, 4(1), 105. https://doi.org/10.1038/s42003-020-01593-x

Yeap, W. C., Lee, F. C., Shabari Shan, D. K., Musa, H., Appleton, D. R., et al. (2017). WRI1-1, ABI5, NF-YA3 and NF-YC2 increase oil biosynthesis in coordination with hormonal signaling during fruit development in oil palm. *The Plant Journal*, 91(1), 97–113. https://doi.org/10.1111/tpj.13549

Yeap, W. C., Norkhairunnisa Che Mohd, K., Norfadzilah, J., Muad, M. R., Appleton, D. R., et al. (2021). An efficient clustered regularly interspaced short palindromic repeat (CRISPR)/CRISPR-associated protein 9 mutagenesis system for oil palm (*Elaeis guineensis*). *Frontiers in Plant Science*, 12, 773656. https://doi.org/10.3389/fpls.2021.773656

Yue, J. J., Yuan, J. L., Wu, F. H., Yuan, Y. H., Cheng, Q. W., et al. (2021). Protoplasts: From isolation to CRISPR/Cas genome editing application. *Frontiers in Genome Editing*, 3, 717017. https://doi.org/10.3389/fgeed.2021.717017

Zaidi, S. S.-E.-A., Mahas, A., Vanderschuren, H., & Mahfouz, M. M. (2020). Engineering crops of the future: CRISPR approaches to develop climate-resilient and disease-resistant plants. *Genome Biology*, 21(1), 289. https://doi.org/10.1186/s13059-020-02204-y

Zhang, J., Zhang, X., Chen, R., Yang, L., Fan, K., et al. (2020). Generation of transgene-free semidwarf maize plants by gene editing of *Gibberellin-Oxid*ase20-3 using CRISPR/Cas9. *Frontiers in Plant Science*, 11, 1048. https://doi.org/10.3389/fpls.2020.01048

Zhang, S., Shen, J., Li, D., & Cheng, Y. (2021). Strategies in the delivery of Cas9 ribonucleoprotein for CRISPR/Cas9 genome editing. *Theranostics*, 11(2), 614–648. https://doi.org/10.7150/thno.47007

Zhang, Y., Massel, K., Godwin, I. D., & Gao, C. (2018). Applications and potential of genome editing in crop improvement. *Genome Biology*, 19(1), 210. https://doi.org/10.1186/s13059-018-1586-y

Zhu, H., Li, C., & Gao, C. (2020). Applications of CRISPR-Cas in agriculture and plant biotechnology. *Nature Reviews Molecular Cell Biology*, 21(11), 661–677. https://doi.org/10.1038/s41580-020-00288-9

Zhu, S., Zhu, Z., Wang, H., Wang, L., Cheng, L., et al. (2018). Characterization and functional analysis of a plastidial *FAD6* gene and its promoter in the mesocarp of oil palm (*Elaeis guineensis*). *Scientia Horticulturae*, 239, 163–170. https://doi.org/10.1016/j.scienta.2018.05.042

Zuhar, L. M., Madihah, A. Z., Ahmad, S. A., Zainal, Z., Idris, A. S., et al. (2021). Identification of oil palm's consistently upregulated genes during early infections of *Ganoderma boninense* via RNA-Seq technology and real-time quantitative PCR. *Plants*, 10(10). https://doi.org/10.3390/plants10102026

20 Advances and Perspectives in Genetic Engineering and the CRISPR/Cas-Based Technology for Oil Crop Enhancement

Kattilaparambil Roshna, Muthukrishnan Arun, Arumugam Vijaya Anand, Packiaraj Gurusaravanan, Sathasivam Vinoth, Bashyam Ramya, and Annamalai Sivaranjini

LIST OF ABBREVIATIONS

ALA	α-Linoleic Acid
CIPKs	Calcineurin B-like Protein Kinases
DRE	Dehydration Response Factor
DREBs	Dehydration-Responsive Element Binding Proteins
HKT1	High Affinity K^+ Transporter
hpRNA	Hairpin RNA
LEA	Late Embryogenesis Abundance
NGS	Next Generation Sequencing
PEPCase	Phosphoenolpyruvate Carboxylase
PLCPs	Plant Papain-like Cysteine Proteases
ROS	Reactive Oxygen Species
sgRNA	Single Guide RNA
SOS1	Salt Overly Sensitive 1

20.1 INTRODUCTION

India is the world's fourth most considerable producer of oilseed crops and the second most significant factor for agricultural economizing. Soybean, peanut, cotton, sunflower, and brassica are oil-producing crops that are high in dietary fibers, proteins, fatty acids, minerals, and vitamin B (Desmae et al., 2018). Oil seeds valued at US$ 28.2 million would be sold to various countries in 2021, with production estimated to be over 31.5 million/kg. Research in oilseed crops has recently received increased attention attributable to the increased need in the oleochemical, biofuels, animal feed, and medicinal sectors. For a long period, the plants were predisposed to different agricultural diseases as a result of climatic circumstances, resulting in a decline in crop output and, eventually, a major loss in the Indian economy. As a result, farmers are having difficulty growing plants with significant yields, putting a burden on agricultural marketing. Drought, salinity, flood, and infections are the key climatic conditions that cause crop diseases.

Among all yield-limiting factors, salinity is one of the most persistent, impacting 20%–33% of farmed and cultivated land worldwide (Machado and Serralheiro, 2017). The increased accumulation of K Na^+/Na^+ proportion, as well as the soil concentration of Na^+ and Cl^-, may cause saltiness

 DOI: 10.1201/9781003387060-20

caused by a lack of water. Salinity can stress plants by raising relative electrolyte leakage, which affects the plants, and even induces the creation of reactive oxygen species (ROS) and it can harm the photosynthetic machinery (Cui et al., 2018; Zhu et al., 2021).

Through co-transport, the high affinity K^+ transporter (HKT1) maintains Na^+ homeostasis. Under K^+ deficiency conditions, HKT allows for greater Na^+ absorption, which increases the plant's tolerance to salinity (Figure 20.1). Salt excessively sensitive 1 (SOS1) transports Na^+ intracellular components to the area outside the cell via Na^+/H^+ antiporter in semi-permeable membranes. To reduce the high level of Na^+, plasma lemma-attached membrane protein Na^+/proton exchangers preferentially interchange Na^+ protons from the cytoplasm to the extracellular space and vacuole. Surplus H^+ is transmitted from the cytoplasm to the extracellular space via ATP, resulting in a balance of sodium/hydrogen ions (Shi et al., 2022).

This type of stress has an impact on mineral concentrations, the development of pods, and crop output (Taffouo et al., 2010; Meena et al., 2016). Another strategy for overcoming these challenges is to grow crop plants under a pure water supply. In biotic stress, approximately 67,000 bug species generate a decline in the economic value of oil-producing plants (Kumar et al., 2020). Insects suck the sap and devour plant parts like leaves and roots, transferring pathogens to plants. Farmers have utilized chemically synthesized fertilizers or insecticides to control pests and insects. It has major environmental consequences and has an impact on human and animal livelihoods. Crop plants are mostly self-pollinated and sensitive to a variety of stressors. Traditional plant breeding methods are limited and time-consuming. As a result, scientists are concentrating on genetic engineering strategies for boosting crops under a variety of environmental challenges. Genetic engineering is one of the ways that can be used to create improved kinds by modifying the host DNA. This strategy improves a species' existing feature by presenting a unique gene that does not ordinarily reside in the plant.

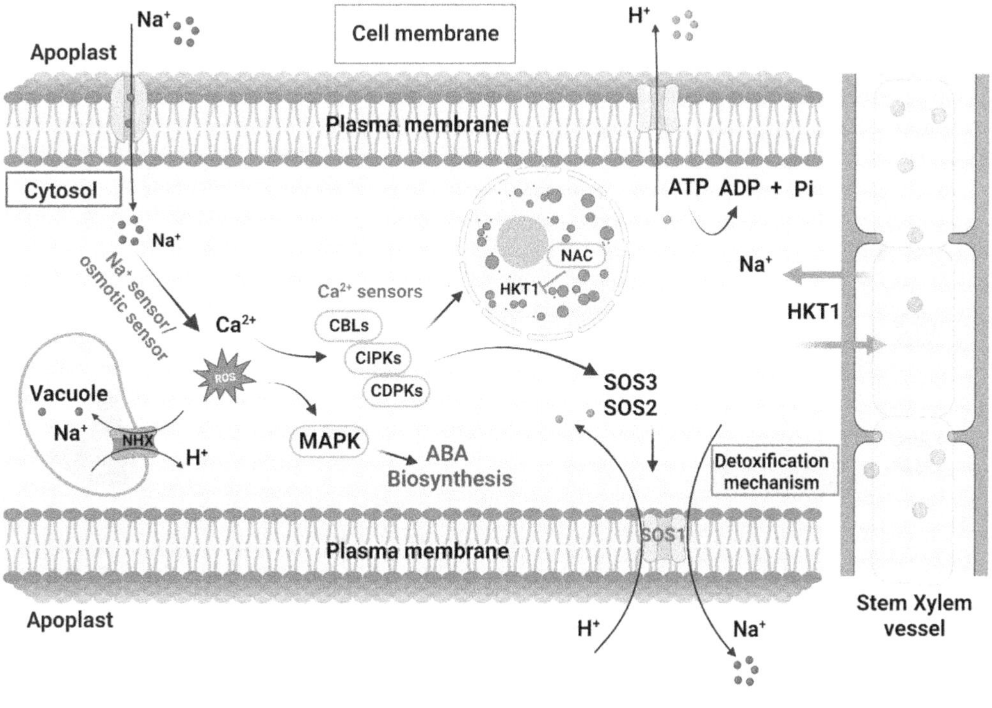

FIGURE 20.1 Diagrammatic presentation of the Na^+ transport mechanism in plant cells under salt stress.

Transgenic plants such as Agrobacterium-mediated transformation procedures have incorporated specified parts of foreign DNA or sequences of genes into their genomes. The approach, which was developed in the year 1977, involves introducing Ti plasmid DNA into the genome of the host plant. CRISPR/Cas techniques can currently modify the plant genome to develop unique traits in agricultural plants. Several efforts are undertaken on oil kernel crops to develop the best varieties in enhanced quality, which can grow in the face of abiotic and biotic pressures. In 1977, Ti plasmid DNA was introduced into the genome of the host plant's cell. CRISPR/Cas technology can now modify the plant genome to develop unique traits in crop plants. Several attempts are undertaken on oil seed crops to generate the greatest cultivars with enhanced quality, which can flourish under abiotic and biotic challenges.

20.2 PRELIMINARY STRATEGIES FOR THE OIL SEED CROP ADVANCEMENT

The need for vegetable oil has increased rapidly during the last few decades. The global population expansion, the need for dietary shifts, and the demand for sustainability all highlight the relevance of plant-based goods such as oil crops. Palm is the world's most extensive vegetable oil source, followed by soybean, cotton, peanut, sunflower, and rapeseed. These oil crops supply better than half of the needs of the world (United States Department of Agriculture Foreign Agricultural Service, 2020). Nevertheless, numerous uncontrollable environmental conditions, such as rising temperatures, global warming, flooding, freezing, drought, and salt, harm oil seed yield.

Apart from the climatic effects of climate change, there is also an increase in disease, which is caused by bacteria, viruses, and fungus diseases) and pest infection and these are all predicted to evolve more complicated in the forthcoming eras (Jaradat, 2016; Raman et al., 2019). As a result, different technologies can be used to achieve enhanced yield in oil seeds, like raising oil content in seeds, increasing the yield of the seeds, or lowering seed losses rendered by breaking of pod/silique, overgrowth of the weeds, and resilience to biotic/abiotic stress (Raman et al., 2019).

Before producing oilseed crop cultivars, conventional breeding procedures such as hybridization, artificial selection, and the use of chemical mutagens inducing mutagenesis were used. Because of the polyploidy of most oilseed crops, these procedures are difficult, requiring considerable effort and time to produce advancements (Yang et al., 2017). The crops generated from transgenic techniques have been confirmed challenging to market due to rejection from the public and demanding regulatory frameworks. Many key crop genomes have been sequenced in a concise period employing current and developed sequencing approaches like Next generation sequencing (NGS). These data were investigated, and the genomic data collected was used to boost crop yield. The metabolomic genetic investigation was used to specify the genes that govern critical agronomic features, and optimum breeding cultivars were generated using these genetic facts. Customized algorithm outcomes via bioinformatics, combined with distinct tools to edit or delete or alter target genes, will be the most effective crop enhancement strategy.

20.3 GENETIC ENGINEERING FOR ENHANCING OIL CROP TRAITS

The production of genetically modified crops has expanded obviously from 1.7 million ha in the year 1996 to 191.7 million ha in the year 2018. As yet, 525 transgenic occurrences in 32 crops have been exploited (ISAAA, 2018, ISAAA database, 2019). The foremost transgenic plants were antibiotic-resistant petunia and tobacco, which were produced in 1983 (Fraley et al., 1983). Through genetic engineering, crop properties were improved under biotic and abiotic pressures. The tag of stress genes implicated in plant processes is critical for the result of genetically altered plants. Genetically engineered crops express certain features at the upright level and boost the quality of crops (Gurusaravanan et al., 2020). Distinct transcription factors are concerned with the excitant of particular active genes at the time of stress and the regulation of distinct signal transduction paths.

Furthermore, transcriptional factors can translate signals due to stress into defensive cellular reactions (Century et al., 2008). Furthermore, transgenic plants containing these characteristics were better functioning than usual plants in a variety of stress conditions (Ribichich et al., 2014).

20.4 CROP IMPROVEMENT IN SOYBEAN

Soybean is a nutritionally and commercially significant crop all over the world. Brazil and the USA are the most considerable cultivators of soybeans. It has a high concentration of saturated, monosaturated, and polyunsaturated fatty acids, as well as valuable secondary active metabolites (USDA, 2018). Many microbial diseases, insect pests, and other environmental stressors have a substantial impact on soybean productivity. The Indian soybean has been discovered to be resistant to tissue culture and gene transfer techniques. As a result, a dependable transformation method is necessary for better Indian soybean properties. Several investigations on soybeans demonstrated the utility of known gene transformation procedures for crop development and studies about its genome in soybeans (Tiwari et al., 2023). Resistance to diseases, tolerance for herbicides and salt, and improvement of nutritional status are all achieved through genetic engineering in various nations (Pérez-Massot et al., 2013; Siehl et al., 2014).

Genetic transformation has been used in recent years to create salt, herbicide, and drought-tolerant soybean breeders. Researchers are attempting to create superior cultivars with improved nutritional quality through gene alteration. Agrobacterium-mediated gene technology of transformation can be used to transfer certain genes into the genome of soybean plants. This crop has great genotypic specificity in tissue culture-mediated transgenic recovery, making genetic transformation difficult. Furthermore, the responsive genotypes of soybeans have resulted in a relatively low proportion of transgenic plant recovery. Salinity has an important effect on the productivity of soybeans and its yield. Salt stress is a key issue that reduces the quality and yield of soybeans (Zhang et al., 2020; Li et al., 2021).

For environmental adaptation, the plant produced many protection systems. There are several methods to protect them from stress, such as maintaining ion homeostasis and restoring osmotic equilibrium, and calcineurin B-like protein kinases (CIPKs) recreate an important role in structurally adjusting the stress due to salt (Phang et al., 2008; Athar et al., 2022). Besides, the Perilla frutescens gene FAD3a has excessive expression in soybeans caused by CaMV P35S to improve tolerance to salinity. Over-expression of PfFAD3a in the seeds and leaves of transgenic soybeans boosted FAD3 expression and improved the content of α-linolenic acid (Li et al., 2023).

NHX gene over-expression increases salt toleration in a variety of plants, implying that TaNHX2, a member of the NHX family, may play a considerable role in the transgenic soybean (Cao et al., 2011). The NAC gene is another significant protein family. 152 NAC genes are identified in soybean and Arabidopsis thaliana consists of 117 NAC genes (Nuruzzaman et al., 2010; Le et al., 2011). Wang et al. (2015) found that over-expression of the gene GmWRKY27 in soybean hairy roots relates with GmMYB174 to diminish the expression of GmNAC29 and enhance drought and salinity toleration. PLCPs (Plant papain-like cysteine proteases) from the cysprot family, according to Yuan et al. (2020), aid in soybean nodule development and minimize leaf senescence due to salt stress. Cys Protgenes serve to up-regulate proteolytic enzyme activity, which leads to protein complex remodeling and reduced cell damage and nutrition mobilization. Table 20.1 lists some transgenic oil crops and their improved properties. Chen et al. (2022) highlighted that Dehydration-responsive element binding proteins (DREBs) are one of the transcription factors that belong to the AP2 family and contain domains made up of amino acids, which are connected to the dehydration response factor (DRE). Transcription factors attach to the target gene promoter region, triggering the expression of abiotic stress alerts from the surroundings. Marinho et al. (2016) created a transgenic soybean with the transgene AtDREB1A that controls the rd29A proponent and over-expression of the gene AREB1 subsequently created in improved pod and seed number and a greater survival rate.

TABLE 20.1
Information about Numerous Oil-Producing Transgenic Crops

Transgenic Plant	Transgene	Character Development	Reference
Soybean	GmWRKY27	Increase the salt and drought tolerance of soybeans	Wang et al. (2015)
	FAD3A and FAD3B	High oleic acid content	Haun et al. (2014); Waltz (2018)
	GmDREB6	Increasing drought and salt tolerance	Nguyen et al. (2019)
	FAD2-1A and FAD2-1B	High oleic acid content	Waltz (2019)
	GmNAC06	Proline and glycine accumulation for salinity tolerance	Li et al. (2021)
Ground nut	SbASR1	Tolerance to salinity and drought stress	Tiwari et al. (2015)
	AtNHX1 gene	Tolerance to salinity	Gantait and Mondal (2018)
Cotton	GhCKX	Improving fruiting branches and bolls, as well as seed size	Zhao et al. (2015)
	AtNHX1 cross with AVP1	Drought resistance and salt tolerance	Shen et al. (2015)
	GhRaf19	Tolerant to cold stress	Jia et al. (2016)
	GhACCase	Increased oil content	Cui et al. (2017)
Castor	SbNHX1	Tolerant to salt stress	Patel et al. (2015)
	fae1 RcFAH12	Increase oil content and fatty acid synthesis	Adhikari et al. (2016)
Rapeseed	Gly I (*glyoxalase* I)	Tolerant to abiotic stress	Rajwanshi et al. (2016)
	BnSIP1-1	Tolerance to abiotic stress and ABA signaling	Luo et al. (2017)
Sunflower	TaNHX2	Tolerance to salinity	Mushke et al. (2019)

20.5 GENETIC ENGINEERING FOR COTTON CROP IMPROVEMENT

The green revolution is the enhancement of farming to boost the production of food and other products. Factors including salt, cold, heat, and drought reduce crop productivity, putting the agricultural economy at risk (Altaf et al., 2023). However, adopting contemporary farming technologies can aid in the production of a large amount of plant material and food. There are numerous strategies for producing biotic and abiotic stress-tolerant crops. Biotechnology approaches have played a significant role in agricultural sectors, particularly in the generation of transgenic plants with covetable features. Cotton, which is the basis of cloth fiber and plays a vital role in farming economizing, is an important cash crop (Abdelraheem et al., 2019). It is mostly grown in subtropical and tropical climates around the world. Cotton (Gossypium hirsutum) is the world's fifth-largest oilseed cash crop (Chen et al., 2021). Crops provide a living for around 8 million farmers in India.

Cottonseed oil has a moisture content of 28.24%–44.05% and is rich in unsaturated fatty acids (Shang et al., 2016). Cotton cultivation is employed for gossypol production in addition to fiber and oil production because of its broad spectrum of biological qualities, which include treating parasitic infections, infertility, and cancer (Wang et al., 2009; Ninkuu et al., 2023). Environmental factors harm the maturation and growth of the plant, and also its productivity. Biotic and abiotic challenges can considerably lower plant output worldwide, and some scientists are working hard to enhance agricultural yield and characteristics resistance to biotic stresses by using genetic tools (Rivero et al., 2022). Because there are only a few studies accessible to reduce cell damage; even if favorable qualities have not yet evolved under diverse conditions.

Abiotic stress usually harms plants by changing physical processes, metabolic reactions, and molecular strategies (Zhang et al., 2021). These environmental conditions may have a significant impact on the concentration of fatty acids and lipid metabolism in plant cells. In general, transcription factors and associated genes protect against abiotic stressors. It has several functions in increasing proline levels and producing stress-associated defensive enzymes (Guo et al., 2022). The transcription factors (TF) genes were discovered and evaluated on several crop cultivars due to abiotic stressors. Genetic engineering techniques are a valuable approach for enhancing the quality of cotton fiber and output under varying circumstances. As a result, investigating the over-expression of stress-associated genes and the modification of preferred genes can boost the output of cotton and the quality of oil due to abiotic stress. In recent years, the stress-associated gene GhBCCP1 was inserted into the cotton genome and significantly raised the oil levels to 21.9% in transgenic cotton plants by GhBCCP1 gene over-expression (Cui et al., 2017; Zhao et al., 2018). Phosphoenolpyruvate carboxylase (PEPCase) plays an important part in the biosynthesis of protein as well as fatty acid pathways. It enhances the cotton seed quality of the oil by triggering PEPCase in the GhACCase transgenic plant (Xu et al., 2016). Cotton over-expression of GhACCase, which encodes PEPCase, may improve the toleration for salinity and drought in transgenic plants.

Cotton bollworm is a prominent herbivorous pest bug that generates considerable harvest losses in crops such as cotton, maize, and others. For plant-mediated RNAi studies, the dsRNA-HaHR3 component suppressed HaHR3 in cotton bollworms. The overall number of 19 transgenic cotton lines expressing HaHR3 were effectively cultivated, with seven of the lines employed in feeding bioassays (Han et al., 2017).

Bollgard 1 was released in 2002 and is India's foremost genetically altered crop for the control of insects and pests, particularly the pink bollworm. It quickly acquired popularity among farmers due to its excellent efficiency against pests (Raman, 2017). Bollgard-1 has the gene cry1Ac obtained from *Bacillus thuringiensis*, which boosts crop output. Invading pathogens are detected in the host plant by active plant recognition receptors such as protein/kinase. If the plant's basal and specialized defense mechanisms are not engaged, it may be vulnerable to pathogen attack, which causes cell damage and nucleus fragmentation. The synthesis of defensive proteins and associated cry gene products by the *Bacillus thuringiensis* transgenic plant prevents pathogen entrance. Cry gene products are poisonous to insects, and crystalline inclusions and ROS that injure insect cells are formed in bacterial spores. The harmful substance binds to distinct receptors and enters the midgut of the insect via the epithelial cell lining the plasma membrane. The domain I attaches to the receptor embeds into the membrane, and causes holes to form in the insect stomach. Because of the active endotoxin, the insect suffers from malnutrition, which eventually leads to paralysis and death, exhibiting insecticidal qualities (Figure 20.2).

20.6 GENE TRANSFORMATION IN SUNFLOWER FOR STRESS TOLERANCE

Sunflower (*Helianthus annuus*) is an attractive plant whose seed oil is highly valued in the oil industry. With a global production of 47.9 million tons, it ranks fourth behind brassica, soybean, and groundnut. Russia and Ukraine are the biggest sunflower producers (FAOSTAT, 2021). Its oil is employed in the production of vegetable oil, cosmetics, and pharmaceuticals. Sunflower oil is regarded as superior to other oils since it includes a high content of polyunsaturated linoleic acid, which reduces the pathogenesis of vascular complications (Seiler et al., 2018). Sunflower is sensitive at various phases of growth due to abiotic and biotic stress. It has an impact on oil production in the crucial sunflower oil-producing region, particularly in China and India (Chappa et al., 2022). Plants developed a unique system of cat-ion exchange channels (NHXs) around the vacuolar and cell membranes to control the shifting of Na^+ and K^+. Ion balance in the cytoplasm is maintained with higher K^+ and lower Na^+ under salinity stress conditions. So far, three wheat vacuolar NHX genes (TaNHA1, TaNHX2, and TaNHX3) that encode Na^+/H^+ antiporters have been functionally studied.

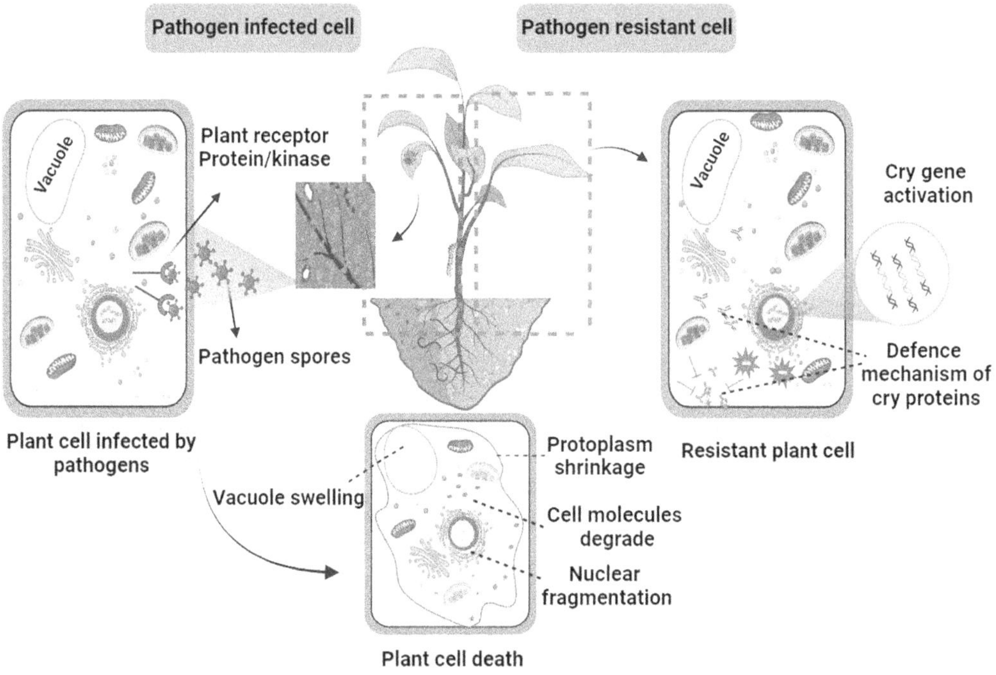

FIGURE 20.2 Diagrammatic presentation of defense responses by plant cells against pathogens.

When compared to non-transformed plants, wheat promoter TaNHX2 gene transfer in sunflower utilizing Agrobacterium tumefaciens boosted development and ionic balance in roots and leaves. This finding suggests that TaNHX2 participates in Na^+ and K^+ ion transportation. In the face of salinity, increased chlorophyll content and proline concentration is an essential process for plant adaptability and healthy maturation. Under salt conditions, transgenic sunflowers express more TaNHX2 than non-transformed plants (Mushke et al., 2019). Transgenic plant cell membrane integrity is compromised under saline stress conditions due to the low MDA level. It promotes the homeostasis of the plasma membrane (Ebrahimian and Bybordi, 2012; Chakraborty et al., 2018). WRI1 and AtWRI1 orthologue expression improved the amount of oil in the plant seed and tissue of the vegetable. WRINKLED (WRI1) encodes an APETALA (AP2) transcription factor that facilitates sucrose-to-oil transformation by activating genes encoding enzymes, which regulates the process of glycolysis and the synthesis of fatty acid (Lim et al., 2022).

20.7 GENETIC ENGINEERING FOR RAPESEED CROP DEVELOPMENT

Rapeseed is one of the third most significant oil-producing crops after soybean and palm oil (Ahmar et al., 2019). Brassica crops are widely cultivated in the world's dry and semi-arid regions, where they meet different kinds of biotic and abiotic stresses. Because of the salinity of the soil conditions, crop productivity is significantly reduced (Sastre-Conde et al., 2015). Late embryogenesis abundance (LEA) is protein in nature, which protects plants from several abiotic stressors including drought, salt, high temperature, flooding, freezing, and deposition of heavy metal. The Arabidopsis AtLEA4-1 gene was over-expressed in mustard to reduce salinity stress, resulting in higher levels of antioxidative enzyme activity, relative water content, chlorophyll content, and low concentrations of lipid peroxidation (Saha et al., 2016).

α-Linoleic acid (ALA) is an essential fatty acid found in extremely low amounts in common dietary oils. The seed of the chia (Salvia hispanica) plant has a lot of ALA (Muñoz et al., 2013). The FAD2 and FAD3 genes from the seeds of the chia plant were produced in rapeseed using an Agrobacterium tumefaciens-mediated gene delivery system. ShFAD2 and ShFAD3 over-expression increased genes in the fatty acid pathway, resulting in increased crop oil production (Xue et al., 2023). This transgenic technology has aided in the cultivation of oil-producing plants, benefiting both farmers and consumers.

20.8 CROP IMPROVEMENT IN PEANUT BY GENETIC ENGINEERING METHOD

Worldwide, 29.5 million peanuts (*Arachis hypogaea*) were grown over a real planted region of 29.5 million acres (FAOSTAT, 2020; Zhang et al., 2022). India, China, the USA, and Nigeria are the top groundnut producers (FAOSTAT 2019). India is the world's second most considerable producer of peanuts behind China. In 2017, 5.3 million ha of peanuts were produced in India, with a whole yield of 9.179 million tons (FAOSTAT, 2019). The annual output of peanut seeds in India's southern and eastern regions has increased by 1,000 quintals to 17,500 quintals (Motagi et al., 2022). The consequence of necrosis diseases on ground nuts caused by peanut stem necrosis disease caused by the tobacco streak virus has dramatically reduced production rates in India. Transgenic peanut-containing the desired gene construct encoding tobacco streak virus-coat protein hairpin RNA (hpRNA). Using Agrobacterium tumefaciens-mediated transformation, this gene sequence was exploited to provide TSC-based RNA interference opposition upon necrotic illness (Shomo et al., 2016).

Both kinds of stresses reduce ground nut yield. Drought and salt stress, in particular, harm groundnut growth, water relations, photosynthesis, and yield. Numerous functional genes implicated in tolerance mechanisms are regulated by TFs. WRKY, NAC, AP2/EREBP, MYB, and bZIP TFs control genes implicated in drought stress. The WRKY gene is one of these TFs that responds to stress. Macrotyloma uniflorum genes (MuWRKY3, MuNAC4, and MuMYB96) have been linked to increased drought tolerance in groundnut plants. These Agrobacterium-transformed genes produced less ROS, caused less damage to the membrane, and boosted the action of ascorbate peroxidase and superoxide dismutase, suggesting enhanced antioxidative defense systems due to drought stress. It increases the development of the root, the formation of secondary metabolite, and the postponement of the senescence of the leaf (Kiranmai et al., 2018; Zhu et al., 2021; Venkatesh et al., 2022; Dwivedi et al., 2023).

Transgenic peanuts containing the AhMYB30 protein, which has been linked to the activation of MYB TFs in response to salinity, increased toleration by up-regulating stress-responsive genes and enzymes of antioxidants. This also enhanced osmolytes and proline levels in roots and stems, decreased the density of stomatal, and increased chlorophyll concentration (Amareshwari 2017; Banavath et al., 2018; Chen et al., 2023).

20.9 CRISPR TECHNOLOGY AND CROP IMPROVEMENT

CRISPR/Cas (Alamillo et al., 2023) have widespread applications in frontier fields of plant science. CRISPR necessitates the nuclease protein Cas, which causes a double-stranded DNA break, and a 20 bp sequence short single guide RNA (sgRNA) that directs Cas upstream to the precise locus of a Cas-dependent protospacer-adjacent motif (Mohanta et al., 2017). Cas nuclease specificity has increased due to changes in PAM recognition domains (Raitskin et al., 2019). Nature Biotechnology released the first paper on genome editing of agricultural species using CRISPR/Cas9 techniques in 2013. The effectiveness and amounts of guide RNA expression and nuclease Cas9 are

TABLE 20.2
Represent the Vectors and Genes Employed for Crop Improvement in Various Oilseed Crops

Plant	Construct	Gene	Reference
Arachis hypogaea	pDW3873/PmCDA1 and pDW3876/rAPOBEC-1	*AhFAD2*	Neelakandan et al. (2022a)
Arachis hypogaea	BGK41-Cas9 recombinant vector FAD2B-1	*FAD2B*	Han et al. (2023)
Arachis hypogaea	pDW3877 & pDW3872	*AhFAD2A* and *AhFAD2B*	Neelakandan et al. (2022b)
Arachis hypogaea	p201G/Cas9: sgRNAs	*AhNFR1* and *AhNFR5*	Shu et al. (2020)
Arachis hypogaea	p201B-Cas9	*AhFAD2*	Yuan et al. (2019)
Arachis hypogaea	pDW3872	*AhFAD2*	Neelakandan et al. (2022c)
Brassica napus	Binary pYLCRISPR/Cas9	*BnFAD2*	Liu et al. (2022)
Brassica napus	pCas9-TPC	*BnITPK*	Sashidhar et al. (2020)
Brassica napus	pLYCRISPR/Cas9P35S-H	*BnFAD2* and *BnFAE1*	Shi et al. (2022)
Brassica napus	pKSE401-sgRNA	BnaRGA	Yang et al. (2017)
Brassica napus	pMD18-T vector	Lysophosphatidic acid acyltransferase	Zhang et al. (2019)
Glycine max	pCas9-AtU6-sgRNA vector	*Glyma06g14180, Glyma08g02290* and *Glyma12g37050*	Sun et al. (2015)
Glycine max	*pBGK041*	*GmHdz4*	Zhong et al. (2022)
Glycine max	*pFGC5941*-Cas9-GmTAP1	*GmTAP1*	Liu et al. (2023)

influenced by several variables. It precisely includes sgRNA and target DNA secondary structure, Cas9 nuclease codons, and target DNA GC content. The researchers then utilized modified Cas nucleases (Cas9, Cas12a, and Cas12b) for crop enhancement (Kaya et al., 2016; Zaidi et al., 2017; Li Hao et al., 2018; Ming et al., 2020).

Several endeavors have been undertaken to effectively modify the genome sequence of agricultural oil seed plants, including soybean (Li et al., 2015; Cai et al., 2015), camelina (Jiang et al., 2017), Brassica spp. (Yang et al., 2017), and cotton (Zhang Ge, et al., 2018). The nuclease proteins utilized for editing the plant genome are zinc finger nucleases, transcription activator-like effector nucleases, and the CRISPR/ Cas9 technique (Osakabe et al., 2010; Miller et al., 2011; Kumar and Jain, 2015). Because of its simplicity and efficiency, CRISPR Cas9 nucleases are generally used for editing the plant genome. Table 20.2 lists the many vectors and genes employed in the enhancement of oil seed crops.

20.10 RAPESEED GENE EDITING

Rapeseed (*Brassica napus* (L.)) is the third most significant vegetable oil source and the most important oil-producing crop with good nutritious content. The nutritive quality of oilseed is governed by the fatty acid proportion, and it has widespread usage in the rape seed oil industries and is a more valuable source of biodiesel fuels (Nesi et al., 2008). Rapeseed oil exhibited worldwide in a year was 27.98 million tons (Ohlrogge, 1994; Thelen and Ohlrogge, 2002). In addition to traditional breeding procedures, gene editing techniques were used to enhance the nutritive content of the oil seed plant. The bnFAD2 gene was altered, and new variants containing high amounts of oleic acid (Okuzaki et al., 2018; Huang et al., 2020).

In recent years, two sgRNAs were created for editing the genes BnFAD2 and BnFAE1, and the editing effectiveness for the two genes. The gene BnFAD2 has modified by CRISPR/Cas9 system to increase the fatty acid content in *Brassica napus*, subsequently increasing oleic acid and a normal phenotype of the plant (Okuzaki et al., 2018; Huang et al., 2020). All of these investigations show that by strengthening the genetic material, the content of oleic acid can be raised while simultaneously maintaining an acceptable PUFA concentration in fatty acid profiles.

20.11 CONCLUSION

Oils derived from plants are essential in the human health, biofuel, and pharmaceutical industries. Oil seed crop development is a pressing demand in frontier research, and many scientists have used conventional and genetic engineering techniques. Several gene delivery strategies have been developed for oil seed crops including soybean, brassica, cotton, sunflower, and peanut, with only a few being particularly successful. To combat the extreme environmental stressors, technologies that bridge the gap between farmer and consumer are required. CRISPR/Cas methods must be improved to overcome severe challenges in plant genetic engineering. Basic plant tissue culture techniques, along with precise genome editing tools, appear to be viable crop improvement strategies.

ACKNOWLEDGMENT

All of the authors like to state their gratitude to their various universities and institutions for their invaluable assistance.

DATA AVAILABILITY

The authors confirm that the data supporting the study's findings are incorporated in the book/chapter.

REFERENCES

Abdelraheem, A., Esmaeili, N., O'Connell, M., & Zhang, J. (2019). Progress and perspective on drought and salt stress tolerance in cotton. *Industrial Crops and Products*, *130*, 118–129. https://doi.org/10.1016/j.indcrop.2018.12.070

Adhikari, N. D., Bates, P. D., & Browse, J. (2016). WRINKLED1 rescues feedback inhibition of fatty acid synthesis in hydroxylase-expressing seeds. *Plant Physiology*, *171*(1), 179–191. https://doi.org/10.1104/pp.15.01906.

Ahmar, S., Liaqat, N., Hussain, M., Salim, M. A., Shabbir, M. A., Ali, M. Y., & Rizwan, M. (2019). Effect of abiotic stresses on Brassica species and role of transgenic breeding for adaptation. *Asian Journal of Research in Crop Science*, *3*(1), 1–10. https://doi.org/10.9734/AJRCS/2019/v3i130037.

Alamillo, J. M., López, C. M., Rivas, F. J. M., Torralbo, F., Bulut, M., & Alseekh, S. (2023). Clustered regularly interspaced short palindromic repeats/CRISPR-associated protein and hairy roots: A perfect match for gene functional analysis and crop improvement. *Current Opinion in Biotechnology*, *79*, 102876. https://doi.org/10.1016/j.copbio.2022.102876.

Altaf, M. T., Liaqat, W., Nadeem, M. A., & Baloch, F. S. (2023). Recent trends and applications of omics-based knowledge to end global food hunger. In *Sustainable Agriculture in the Era of the OMICs Revolution*, Springer International Publishing: Cham, pp. 381–397. https://doi.org/10.1007/978-3-031-15568-0_18.

Amareshwari, P. (2017). Overexpression of Sorghum bicolor vacuolar H+ pyrophosphatase (SbVPPase) to improve salt and drought stress tolerance in peanut (Arachis hypogaea L.) (Doctoral dissertation, Thesis submitted to the Osmania University, Hyderabad for the award of Ph. D. Degree).

Athar, H. U. R., Zulfiqar, F., Moosa, A., Ashraf, M., Zafar, Z. U., Zhang, L., & Siddique, K. H. (2022). Salt stress proteins in plants: An overview. *Frontiers in Plant Science*, *13*, 999058. https://doi.org/10.3389/fpls.2022.999058.

Banavath, J. N., Chakradhar, T., Pandit, V., Konduru, S., Guduru, K. K., Akila, C. S., & Puli, C. O. (2018). Stress inducible overexpression of *AtHDG11* leads to improved drought and salt stress tolerance in peanut (*Arachis hypogaea* L.). *Frontiers in Chemistry*, *6*, 34. https://doi.org/10.3389/fchem.2018.00034.

Cai, Y., Chen, L., Liu, X., Sun, S., Wu, C., Jiang, B., & Hou, W. (2015). CRISPR/Cas9-mediated genome editing in soybean hairy roots. *PLoS One*, *10*(8), e0136064. https://doi.org/10.1371/journal.pone.0136064.

Cao, D., Hou, W., Liu, W., Yao, W., Wu, C., Liu, X., & Han, T. (2011). Overexpression of *TaNHX2* enhances salt tolerance of 'composite' and whole transgenic soybean plants. *Plant Cell, Tissue and Organ Culture (PCTOC), 107*, 541–552. https://doi.org/10.1007/s11240-011-0005-9.

Century, K., Reuber, T. L., & Ratcliffe, O. J. (2008). Regulating the regulators: The future prospects for transcription-factor-based agricultural biotechnology products. *Plant Physiology, 147*(1), 20–29. https://doi.org/10.1104/pp.108.117887.

Chakraborty, K., Basak, N., Bhaduri, D., Ray, S., Vijayan, J., Chattopadhyay, K., & Sarkar, R. K. (2018). Ionic basis of salt tolerance in plants: Nutrient homeostasis and oxidative stress tolerance. In Plant Nutrients and Abiotic Stress Tolerance, pp. 325–362. https://doi.org/10.1007/978-981-10-9044-8_14.

Chappa, L. R., Mugwe, J., Maitra, S., & Gitari, H. (2022). Current status and prospects of improving sunflower production in Tanzania through intercropping with Sunn hemp. *International Journal of Bioresource Science, 9*, 1–8. https://doi.org/10.30954/2347-9655.01.2022.1

Chen, K., Tang, W., Zhou, Y., Chen, J., Xu, Z., Ma, R., & Chen, M. (2022). AP2/ERF transcription factor GmDREB1 confers drought tolerance in transgenic soybean by interacting with GmERFs. *Plant Physiology and Biochemistry, 170*, 287–295. https://doi.org/10.1016/j.plaphy.2021.12.014.

Chen, N., Pan, L., Yang, Z., Su, M., Xu, J., Jiang, X., & Chi, X. (2023). A MYB-related transcription factor from peanut, AhMYB30, improves freezing and salt stress tolerance in transgenic *Arabidopsis* through both DREB/CBF and ABA-signaling pathways. *Frontiers in Plant Science, 14*, 11–10. https://doi.org/10.3389/fpls.2023.1136626.

Chen, Y., Fu, M., Li, H., Wang, L., Liu, R., Liu, Z., & Jin, S. (2021). High-oleic acid content, nontransgenicallotetraploid cotton (*Gossypium hirsutum* L.) generated by knockout of *GhFAD2* genes with CRISPR/Cas9 system. *Plant Biotechnology Journal, 19*(3), 424–426.https://doi.org/10.3389/fpls.2023.1136626.

Cui, F., Sui, N., Duan, G., Liu, Y., Han, Y., Liu, S., & Li, G. (2018). Identification of metabolites and transcripts involved in salt stress and recovery in peanut. *Frontiers in Plant Science, 9*, 217. https://doi.org/10.3389/fpls.2018.00217

Cui, Y., Liu, Z., Zhao, Y., Wang, Y., Huang, Y., Li, L., & Hua, J. (2017). Overexpression of heteromericGhACCase subunits enhanced oil accumulation in upland cotton. *Plant Molecular Biology Reporter, 35*, 287–297. https://doi.org/10.1007/s11105-016-1022-y.

Desmae, H., Janila, P., Okori, P., Pandey, M. K., Motagi, B. N., Monyo, E., & Varshney, R. K. (2018). Genetics, genomics and breeding of groundnut (*Arachis hypogaea* L.). *Plant Breeding, 138*(4), 425–444. https://doi.org/10.1111/pbr.12645.

Dwivedi, S. L., Chapman, M. A., Abberton, M. T., Akpojotor, U. L., & Ortiz, R. (2023). Exploiting genetic and genomic resources to enhance productivity and abiotic stress adaptation of underutilized pulses. *Frontiers in Genetics, 14*, 1193780. https://doi.org/10.3389/fgene.2023.1193780.

Ebrahimian, E., & Bybordi, A. (2012). Effect of salinity, salicylic acid, silicium and ascorbic acid on lipid peroxidation, antioxidant enzyme activity and fatty acid content of sunflower. *African Journal of Agricultural Research, 7*(25), 3685–3694. https://doi.org/10.5897//AJAR11.799.

FAOSTAT, 2021. https://www.fao.org/3/cb4477en/cb4477en.pdf

Food and Agriculture Organization Statistical Databases (FAOSTAT) 2019. Available online: http://faostat.fao.org/.

Food and Agriculture Organization Statistical Databases (FAOSTAT) 2020. Available online: http://faostat.fao.org/.

Food and Agriculture Organization Statistical Databases (FAOSTAT) 2021. Available online: http://faostat.fao.org/.

Fraley, R. T., Rogers, S. G., Horsch, R. B., Sanders, P. R., Flick, J. S., Adams, S. P., & Woo, S. C. (1983). Expression of bacterial genes in plant cells. Proceedings of the National Academy of Sciences of the United States of America, *80*(15), 4803–4807. https://doi.org/10.1073/pnas.80.15.4803.

Gantait, S., & Mondal, S. (2018). Transgenic approaches for genetic improvement in groundnut (*Arachis hypogaea* L.) against major biotic and abiotic stress factors. *Journal of Genetic Engineering and Biotechnology, 16*(2), 537–544. https://doi.org/10.1016/j.jgeb.2018.08.005.

Guo, X., Ullah, A., Siuta, D., Kukfisz, B., & Iqbal, S. (2022). Role of WRKY transcription factors in regulation of abiotic stress responses in cotton. *Life, 12*(9), 1410. https://doi.org/10.3390/life12091410.

Gurusaravanan, P., Vinoth, S., & Jayabalan, N. (2020). An improved *Agrobacterium*-mediated transformation method for cotton (*Gossypium hirsutum* L. 'KC3') assisted by microinjection and sonication. *In vitro Cellular & Developmental Biology-Plant, 56*, 111–121. https://doi.org/10.1007/s11627-019-10030-6.

Han, H. W., Yu, S. T., Wang, Z. W., Yang, Z., Jiang, C. J., Wang, X. Z., & Wang, C. T. (2023). In planta genetic transformation to produce CRISPRed high-oleic peanut. Plant Growth Regulation, *101*, 443–451. https://doi.org/10.1007/s10725-023-01031-y.

Han, Q., Wang, Z., He, Y., Xiong, Y., Lv, S., Li, S., & Zeng, H. (2017). Transgenic cotton plants expressing the *HaHR3* gene conferred enhanced resistance to *Helicoverpa armigera* and improved cotton yield. *International Journal of Molecular Sciences*, *18*(9), 1874. https://doi.org/10.3390/ijms18091874.

Haun, W., Coffman, A., Clasen, B. M., Demorest, Z. L., Lowy, A., Ray, E., Retterath, A., Stoddard, T., Juillerat, A., Cedrone, F., Mathis, L., Voytas, D., Zhang, F. (2014). Improved soybean oil quality by targeted mutagenesis of the fatty acid desaturase 2 gene family. *Plant biotechnology journal*, *12*(7), 934–940. https://doi.org/10.1111/pbi.12201

Huang, H., Cui, T., Zhang, L., Yang, Q., Yang, Y., Xie, K., … & Zhou, Y. (2020). Modifications of fatty acid profile through targeted mutation at *BnaFAD2* gene with CRISPR/Cas9-mediated gene editing in *Brassica napus*. *Theoretical and Applied Genetics*, *133*, 2401–2411. https://doi.org/10.1007/s00122-020-03607-y.

ISAAA (2018). Global Status of Commercialized Biotech/GM Crops in 2018: Biotech Crops Continue to Help Meet the Challenges of Increased Population and Climate Change. ISAAA Brief No. 54. ISAAA: Ithaca, NY. https://www.isaaa.org/resources/publi cations/briefs/54/executivesummary/pdf/B54-ExecSum-Engli sh.pdf

ISAAA database (2019). GM Approval Database. Retrieved on 17 Nov 2019. https://www.isaaa.org/gmapprovaldatabase/default.asp

Jaradat, A. A. (2016). Breeding oilseed crops for climate change. In *Breeding Oilseed Crops for Sustainable Production*, Academic Press, pp. 421–472. https://doi.org/10.1016/B978-0-12-801309-0.00018-5.

Jia, H., Hao, L., Guo, X., Liu, S., Yan, Y., & Guo, X. (2016). A Raf-like MAPKKK gene, GhRaf19, negatively regulates tolerance to drought and salt and positively regulates resistance to cold stress by modulating reactive oxygen species in cotton. *Plant Science*, *252*, 267–281. https://doi.org/10.1016/j.plantsci.2016.07.014.

Jiang, W. Z., Henry, I. M., Lynagh, P. G., Comai, L., Cahoon, E. B., & Weeks, D. P. (2017). Significant enhancement of fatty acid composition in seeds of the allohexaploid, Camelina sativa, using CRISPR/Cas9 gene editing. *Plant Biotechnology Journal*, *15*(5), 648–657. https://doi.org/10.1111/pbi.12663.

Kaya, H., Mikami, M., Endo, A., Endo, M., & Toki, S. (2016). Highly specific targeted mutagenesis in plants using Staphylococcus aureus Cas9. *Scientific Reports*, *6*(1), 26871. https://doi.org/10.1038/srep26871.

Kiranmai, K., LokanadhaRao, G., Pandurangaiah, M., Nareshkumar, A., Amaranatha Reddy, V., Lokesh, U., & Sudhakar, C. (2018). A novel WRKY transcription factor, *MuWRKY3* (*Macrotyloma uniflorum* Lam. Verdc.) enhances drought stress tolerance in transgenic groundnut (*Arachis hypogaea* L.) plants. *Frontiers in Plant Science*, *9*, 346. https://doi.org/10.3389/fpls.2018.00346.

Kumar, K., Gambhir, G., Dass, A., Tripathi, A. K., Singh, A., Jha, A. K., & Rakshit, S. (2020). Genetically modified crops: Current status and future prospects. *Planta*, *251*, 1–27. https://doi.org/10.1007/s00425-020-03372-8.

Kumar, V., & Jain, M. (2015). The CRISPR-Cas system for plant genome editing: Advances and opportunities. *Journal of Experimental Botany*, *66*(1), 47–57.https://doi.org/10.1093/jxb/eru429.

Le, D. T., Nishiyama, R. I. E., Watanabe, Y., Mochida, K., Yamaguchi-Shinozaki, K., Shinozaki, K., & Tran, L. S. P. (2011). Genome-wide survey and expression analysis of the plant-specific NAC transcription factor family in soybean during development and dehydration stress. *DNA Research*, *18*(4), 263–276.https://doi.org/10.1093/dnares/dsr015.

Li Z, Wang, Y., Yu, L., Gu, Y., Zhang, L., Wang, J., & Qiu, L. (2023). Overexpression of the Purple Perilla (*Perilla frutescens* (L.)) *FAD3a* gene enhances salt tolerance in soybean. *International Journal of Molecular Sciences*, *24*(13), 10533. https://doi.org/10.3390/ijms241310533.

Li, C., Hao, M., Wang, W., Wang, H., Chen, F., Chu, W., & Hu, Q. (2018). An efficient CRISPR/Cas9 platform for rapidly generating simultaneous mutagenesis of multiple gene homoeologs in allotetraploid oilseed rape. *Frontiers in Plant Science*, *9*, 442.https://doi.org/10.3389/fpls.2018.00442.

Li, M., Chen, R., Jiang, Q., Sun, X., Zhang, H., & Hu, Z. (2021). GmNAC06, a NAC domain transcription factor enhances salt stress tolerance in soybean. *Plant Molecular Biology*, *105*, 333–345. https://doi.org/10.1007/s11103-020-01091-y.

Li, Z., Liu, Z. B., Xing, A., Moon, B. P., Koellhoffer, J. P., Huang, L., & Cigan, A. M. (2015). Cas9-guide RNA directed genome editing in soybean. Plant Physiology, *169*(2), 960–970. https://doi.org/10.1104/pp.15.00783.

Liu, H., Lin, B., Ren, Y., Hao, P., Huang, L., Xue, B., Jiang, L., Zhu, Y., & Hua, S. (2022). CRISPR/Cas9-mediated editing of double loci of BnFAD2 increased the seed oleic acid content of rapeseed (Brassica napus L.). *Frontiers in Plant Science*, *13*, 1034215. https://doi.org/10.3389/fpls.2022.1034215.

Lim, A. R., Kong, Q., Singh, S. K., Guo, L., Yuan, L., & Ma, W. (2022). Sunflower WRINKLED1 plays a key role in transcriptional regulation of oil biosynthesis. *International Journal of Molecular Sciences*, *23*(6), 3054. https://doi.org/10.3390/ijms23063054.

Liu, T., Ji, J., Cheng, Y., Zhang, S., Wang, Z., Duan, K., & Wang, Y. (2023) CRISPR/Cas9-mediated editing of *GmTAP1* confers enhanced resistance to *Phytophthora sojae* in soybean. *Journal of Integrative Plant Biology*, *65*(7), 1609–1612. https://doi.org/10.1111/jipb.13476.

Luo, J., Tang, S., Mei, F., Peng, X., Li, J., Li, X., … & Wu, G. (2017). *BnSIP1-1*, a trihelix family gene, mediates abiotic stress tolerance and ABA signaling in *Brassica napus*. *Frontiers in Plant Science*, *8*, 44. https://doi.org/10.3389/fpls.2017.00044.

Machado, R. M. A., & Serralheiro, R. P. (2017). Soil salinity: Effect on vegetable crop growth. Management practices to prevent and mitigate soil salinization. *Horticulturae*, *3*(2), 30. https://doi.org/10.3390/horticulturae3020030.

Marinho, J. P., Kanamori, N., Ferreira, L. C., Fuganti-Pagliarini, R., CorrêaCarvalho, J. D. F., Freitas, R. A., & Nepomuceno, A. L. (2016). Characterization of molecular and physiological responses under water deficit of genetically modified soybean plants overexpressing the AtAREB1 transcription factor. *Plant Molecular Biology Reporter, 34*, 410–426. https://doi.org/10.1007/s11105-015-0928-0.

Meena, H. N., Meena, M., & Yadav, R. S. (2016). Comparative performance of seed types on yield potential of peanut (*Arachis hypogaea* L.) under saline irrigation. *Field Crops Research*, *196*, 305–310. https://doi.org/10.1016/j.fcr.2016.06.006.

Miller, J. C., Tan, S., Qiao, G., Barlow, K. A., Wang, J., Xia, D. F., & Rebar, E. J. (2011). A TALE nuclease architecture for efficient genome editing. *Nature Biotechnology*, *29*(2), 143–148. https://doi.org/10.1038/nbt.1755.

Ming, M., Ren, Q., Pan, C., He, Y., Zhang, Y., Liu, S., & Qi, Y. (2020). CRISPR-Cas12b enables efficient plant genome engineering. *Nature Plants, 6*(3), 202–208. https://doi.org/10.1038/s41477-020-0614-6.

Mohanta, T. K., Bashir, T., Hashem, A., Abd_Allah, E. F., & Bae, H. (2017). Genome editing tools in plants. *Genes*, *8*(12), 399. https://doi.org/10.3390/genes8120399.

Motagi, B. N., Bhat, R. S., Pujer, S., Nayak, S. N., Pasupaleti, J., Pandey, M. K., & Gowda, M. C. (2022). Genetic enhancement of groundnut: Current status and future prospects. In *Accelerated Plant Breeding, Oil Crops*, Vol. 4, pp. 63–110. https://doi.org/10.1007/978-3-030-81107-5_3.

Muñoz, L. A., Cobos, A., Diaz, O., & Aguilera, J. M. (2013). Chia seed (*Salvia hispanica*): An ancient grain and a new functional food. *Food Reviews International*, *29*(4), 394–408. https://doi.org/10.1080/87559129.2013.818014.

Mushke, R., Yarra, R., & Kirti, P. B. (2019). Improved salinity tolerance and growth performance in transgenic sunflower plants via ectopic expression of a wheat antiporter gene (*TaNHX2*). *Molecular Biology Reports*, *46*(6), 5941–5953. https://doi.org/10.1007/s11033-019-05028-7.

Neelakandan, A. K., Subedi, B., Traore, S. M., Binagwa, P., Wright, D. A., & He, G. (2022)a. Base editing in peanut using CRISPR/nCas9. *Frontiers in Genome Editing*, *4*, 901444. https://doi.org/10.3389/fgeed.2022.901444.

Neelakandan, A. K., Wright, D. A., Traore, S. M., Chen, X., Spalding, M. H., & He, G. (2022b). CRISPR/Cas9 based site-specific modification of FAD2 cis-regulatory motifs in peanut (*Arachis hypogaea* L). *Frontiers in Genetics*, *13*, 849961. https://doi.org/10.3389/fgene.2022.849961.

Neelakandan, A. K., Wright, D. A., Traore, S. M., Ma, X., Subedi, B., Veeramasu, S., & He, G. (2022c). Application of CRISPR/Cas9 system for efficient gene editing in peanut. *Plants*, 11(10), 1361. https://doi.org/10.3390/plants11101361.

Nesi, N., Delourme, R., Brégeon, M., Falentin, C., & Renard, M. (2008). Genetic and molecular approaches to improve nutritional value of *Brassica napus* L. seed. *Comptesrendusbiologies*, *331*(10), 763–771. https://doi.org/10.1016/j.crvi.2008.07.018.

Nguyen, Q. H., Vu, L. T. K., Nguyen, L. T. N., Pham, N. T. T., Nguyen, Y. T. H., Le, S. V., & Chu, M. H. (2019). Overexpression of the GmDREB6 gene enhances proline accumulation and salt tolerance in genetically modified soybean plants. *Scientific Reports*, 9(1), 19663. https://doi.org/10.1038/s41598-019-55895-0.

Ninkuu, V., Liu, Z., Zhou, Y., & Sun, X. (2023). The nutritional and industrial significance of cottonseeds and genetic techniques in gossypol detoxification. *Plants, People, Plane,* 2023, 1–16. https://doi.org/10.1002/ppp3.10433.

Nuruzzaman, M., Manimekalai, R., Sharoni, A. M., Satoh, K., Kondoh, H., Ooka, H., & Kikuchi, S. (2010). Genome-wide analysis of NAC transcription factor family in rice. *Gene*, *465*(1–2), 30–44. https://doi.org/10.1016/j.gene.2010.06.008.

Ohlrogge, J. B. (1994). Design of new plant products: Engineering of fatty acid metabolism. *Plant Physiology*, *104*(3), 821. https://doi.org/10.1104/pp.104.3.821.

Okuzaki, A., Ogawa, T., Koizuka, C., Kaneko, K., Inaba, M., Imamura, J., & Koizuka, N. (2018). CRISPR/Cas9-mediated genome editing of the fatty acid desaturase 2 gene in *Brassica napus*. *Plant Physiology and Biochemistry*, *131*, 63–69. https://doi.org/10.1016/j.plaphy.2018.04.025.

Osakabe, K., Osakabe, Y., & Toki, S. (2010). Site-directed mutagenesis in *Arabidopsis* using custom-designed zinc finger nucleases. *Proceedings of the National Academy of Sciences of the United States of America*, *107*(26), 12034–12039. https://doi.org/10.1073/pnas.1000234107.

Patel, M. K., Joshi, M., Mishra, A., & Jha, B. (2015). Ectopic expression of *SbNHX1* gene in transgenic castor (*Ricinus communis* L.) enhances salt stress by modulating physiological process. *Plant Cell, Tissue and Organ Culture (PCTOC)*, *122*, 477–490. https://doi.org/10.1007/s11240-015-0785-4.

Pérez-Massot, E., Banakar, R., Gómez-Galera, S., Zorrilla-López, U., Sanahuja, G., Arjó, G., & Zhu, C. (2013). The contribution of transgenic plants to better health through improved nutrition: Opportunities and constraints. *Genes & Nutrition*, *8*, 29–41. https://doi.org/10.1007/s12263-012-0315-5.

Phang, T. H., Shao, G., & Lam, H. M. (2008). Salt tolerance in soybean. *Journal of Integrative Plant Biology*, *50*(10), 1196–1212. https://doi.org/10.1111/j.1744-7909.2008.00760.x.

Raitskin, O., Schudoma, C., West, A., & Patron, N. J. (2019). Comparison of efficiency and specificity of CRISPR-associated (Cas) nucleases in plants: An expanded toolkit for precision genome engineering. *PLoS One*, *14*(2), e0211598. https://doi.org/10.1371/journal.pone.0211598.

Rajwanshi, R., Kumar, D., Yusuf, M. A., DebRoy, S., & Sarin, N. B. (2016). Stress-inducible overexpression of *glyoxalase I* is preferable to its constitutive overexpression for abiotic stress tolerance in transgenic *Brassica juncea*. *Molecular Breeding*, *36*, 1–15. https://doi.org/10.1007/s11032-016-0495-6.

Raman, H., Uppal, R. K., & Raman, R. (2019). Genetic solutions to improve resilience of canola to climate change. *In Genomic Designing of Climate-Smart Oilseed Crops*, pp. 75–131. https://doi.org/10.1007/978-3-319-93536-2_2.

Raman, R. (2017). The impact of Genetically Modified (GM) crops in modern agriculture: A review. *GM Crops & Food*, *8*(4), 195–208. https://doi.org/10.1080.21645698.2017.1413522.

Ribichich, K. F., Arce, A. L., & Chan, R. L. (2014). *Climate Change and Plant Abiotic Stress Tolerance*, John Wiley & Sons. https://doi.org/10.1002/9783527675265.ch24.

Rivero, R. M., Mittler, R., Blumwald, E., & Zandalinas, S. I. (2022). Developing climate resilient crops: Improving plant tolerance to stress combination. *The Plant Journal*, *109*(2), 373–389. https://doi.org/10.1111/tpj.15483.

Saha, B., Mishra, S., Awasthi, J. P., Sahoo, L., & Panda, S. K. (2016). Enhanced drought and salinity tolerance in transgenic mustard [*Brassica juncea* (L.) Czern& Coss.] overexpressing *Arabidopsis* group 4 late embryogenesis abundant gene (*AtLEA4-1*). *Environmental and Experimental Botany*, *128*, 99–111. https://doi.org/10.1016/j.envexpbot.2016.04.010.

Sashidhar, N., Harloff, H. J., Potgieter, L., & Jung, C. (2020). Gene editing of three BnITPK genes in tetraploid oilseed rape leads to significant reduction of phytic acid in seeds. *Plant Biotechnology Journal*, *18*(11), 2241–2250. https://doi.org/10.1111/pbi.13380.

Sastre-Conde, I., Lobo, M. C., Beltrán-Hernández, R. I., Poggi-Varaldo, H. M. (2015) Remediation of saline soils by a two-step process: Washing and amendment with sludge. *Geoderma*, *247*(248), 140–150. https://doi.org/10.1016/j.geoderma.2014.12.002.

Seiler, G. J., Gulya, T., Kong, G., Thompson, S., & Mitchell, J. (2018). Oil concentration and fatty-acid profile of naturalized *Helianthus annuus* populations from Australia. *Genetic Resources and Crop Evolution*, *65*, 2215–2229. https://doi.org/10.1007/s10722-018-0686-6.

Shang, L., Abduweli, A., Wang, Y., & Hua, J. (2016). Genetic analysis and QTL mapping of oil content and seed index using two recombinant inbred lines and two backcross populations in Upland cotton. *Plant Breeding*, *135*(2), 224–231. https://doi.org/10.1111/pbr.12352.

Shen, G., Wei, J., Qiu, X., Hu, R., Kuppu, S., Auld, D., & Zhang, H. (2015). Co-overexpression of AVP1 and AtNHX1 in cotton further improves drought and salt tolerance in transgenic cotton plants. *Plant Molecular Biology Reporter*, *33*, 167–177. https://doi.org/10.1007/s11105-014-0739-8.

Shi, J., Ni, X., Huang, J., Fu, Y., Wang, T., Yu, H., & Zhang, Y. (2022). CRISPR/Cas9-mediated gene editing of *BnFAD2* and *BnFAE1* modifies fatty acid profiles in *Brassica napus*. *Genes*, 13(10), 1681. https://doi.org/10.3390/genes13101681.

Shomo, S., Senthilraja, C., & Velazhahan, R. (2016). Growth performance of transgenic groundnut (*Arachis hypogaea* L.) plants engineered for resistance against Tobacco streak virus. *Biochemical and Cellular Archives*, *16*(2), 267–270.

Shu, H., Luo, Z., Peng, Z., & Wang, J. (2020). The application of CRISPR/Cas9 in hairy roots to explore the functions of AhNFR1 and AhNFR5 genes during peanut nodulation. *BMC Plant Biology*, *20*(1), 1–15. https://doi.org/10.1186/s12870-020-02614-x.

Siehl, D.L., Tao, Y., Albert, H., Dong, Y., Heckert, M., Madrigal, A., Lincoln-Cabatu, B. (2014). Broad 4-hydroxyphenylpyruvate dioxygenase inhibitor herbicide tolerance in soybean with an optimized enzyme and expression cassette. *Plant Physiology*, 166, 1162–1176. https://doi.org/10.1104/pp.114.247205.

Sun, X., Hu, Z., Chen, R., Jiang, Q., Song, G., Zhang, H., & Xi, Y. (2015). Targeted mutagenesis in soybean using the CRISPR-Cas9 system. *Scientific Reports*, *5*(1), 10342. https://doi.org/10.1038/srep10342.

Taffouo, V. D., Wamba, O. F., Youmbi, E., Nono, G. V., & Akoa, A. (2010). Growth, yield, water status and ionic distribution response of three bambara groundnut (*Vigna subterranea* (L.) Verdc.) landraces grown under saline conditions. *International Journal of Botany*, *6*(1), 53–58. https://doi.org/10.3923/ijb.2010.53.58.

Thelen, J. J., & Ohlrogge, J. B. (2002). Metabolic engineering of fatty acid biosynthesis in plants. *Metabolic Engineering*, *4*(1), 12–21. https://doi.org/10.1006/mben.2001.0204.

Tiwari, R., Singh, A.K. & Rajam, M.V. (2023). Improved and reliable plant regeneration and *Agrobacterium*-mediated genetic transformation in soybean (*Glycine max* L.). *Journal of Crop Science and Biotechnology*, *26*, 275–284. https://doi.org/10.1007/s12892-022-00179-9.

Tiwari, V., Chaturvedi, A. K., Mishra, A., & Jha, B. (2015). Introgression of the SbASR-1 gene cloned from a halophyte *Salicornia brachiata* enhances salinity and drought endurance in transgenic groundnut (*Arachis hypogaea*) and acts as a transcription factor. *PLoS One*, 10(7), e0131567. https://doi.org/10.1371/journal.pone.0131567

United States Department of Agriculture Foreign Agricultural Service. (2020). Oilseeds: World markets and trade. https://apps.fas.usda.gov/psdonline/circulars/oilseeds.pdf

USDA (United States Department of Agriculture) (2018) Basic Report: 11450, Soybeans, green, raw. Agricultural Research Service. National Nutrient Database for Standard Reference Legacy Release. https://ndb.nal.usda.gov/ndb/foods/show/11450 (Last accessed: 8 June 2018).

USDA, U. (2019). Food composition databases. https://www.nal.usda.gov/human-nutrition-and-food-safety/food-composition. (Last accessed: 8 June 2018).

Venkatesh, B., Vennapusa, A. R., Kumar, N. J., Jayamma, N., Reddy, B. M., Johnson, A. M., & Sudhakar, C. (2022). Co-expression of stress-responsive regulatory genes, *MuNAC4*, *MuWRKY3* and *MuMYB96* associated with resistant-traits improves drought adaptation in transgenic groundnut (*Arachis hypogaea* L.) plants. *Frontiers in Plant Science, 13*, 4642. https://doi.org/10.3389/fpls.2022.1055851.

Waltz, E. (2018). With a free pass, CRISPR-edited plants reach market in record time. *Nature Biotechnology*, *36*(1), 6–8. https://doi.org/10.1038/nbt0118-6b.

Waltz, E. (2019). Appetite grows for biotech foods with health benefits. *Nature Biotechnology*. https://doi.org/10.1038/d41587-019-00012-9.

Wang, F., Chen, H. W., Li, Q. T., Wei, W., Li, W., Zhang, W. K., & Chen, S. Y. (2015). Gm WRKY 27 interacts with Gm MYB 174 to reduce expression of Gm NAC 29 for stress tolerance in soybean plants. *The Plant Journal*, *83*(2), 224–236. https://doi.org/10.1111/tpj.12879.

Wang, X., Howell, C. P., Chen, F., Yin, J., & Jiang, Y. (2009). Gossypol-a polyphenolic compound from cotton plant. *Advances in Food and Nutrition Research*, *58*, 215–263. https://doi.org/10.1016/S1043-4526(09)58006-0.

Xu, Z., Li, J., Guo, X., Jin, S., & Zhang, X. (2016). Metabolic engineering of cottonseed oil biosynthesis pathway via RNA interference. *Scientific Reports*, *6*(1), 33342. https://doi.org/10.1038/srep33342.

Xue, Y. F., Tseke, A. I., Yin, N. W., Jiang, J. Y., Zhao, Y. P., Kun, L. U., & Chai, Y. R. (2023). Biotechnology of α-linolenic acid in oilseed rape (*Brassica napus*) using *FAD2* and *FAD3* from chia (*Salvia hispanica*). *Journal of Integrative Agriculture*. https://doi.org/10.1016/j.jia.2023.05.018.

Yang, H., Wu, J. J., Tang, T., Liu, K. D., & Dai, C. (2017). CRISPR/Cas9-mediated genome editing efficiently creates specific mutations at multiple loci using one sgRNA in Brassica napus. Scientific Reports, *7*(1), 7489. https://doi.org/10.1038/s41598-017-07871-9.

Yuan, M., Zhu, J., Gong, L., He, L., Lee, C., Han, S., Chen, C., & He, G. (2019). Mutagenesis of FAD2 genes in peanut with CRISPR/Cas9 based gene editing. *BMC Biotechnology*, *19*(1), 1–7. https://doi.org/10.1186/s12896-019-0516-8.

Yuan, S., Ke, D., Li, R., Li, X., Wang, L., Chen, H., & Zhou, X. (2020). Genome-wide survey of soybean papain-like cysteine proteases and their expression analysis in root nodule symbiosis. *BMC Plant Biology*, *20*, 1–16. https://doi.org/10.1186/s12870-020-02725-5.

Zaidi, S. S. E. A., Mahfouz, M. M., & Mansoor, S. (2017). CRISPR-Cpf1: A new tool for plant genome editing. *Trends in Plant Science*, *22*(7), 550–553. https://doi.org/10.1016/j.tplants.2017.05.001.

Zhang, K., Nie, L., Cheng, Q., Yin, Y., Chen, K., Qi, F., & Li, M. (2019). Effective editing for lysophosphatidic acid acyltransferase 2/5 in allotetraploid rapeseed (*Brassica napus* L.) using CRISPR-Cas9 system. *Biotechnology for Biofuels*, *12*(1), 1–18. https://doi.org/10.1186/s13068-019-1567-8.

Zhang, W., Wang, N., Yang, J., Guo, H., Liu, Z., Zheng, X., & Xiang, F. (2020). The salt-induced transcription factor GmMYB84 confers salinity tolerance in soybean. *Plant Science*, *291*, 110326. https://doi.org/10.1016/j.plantsci.2019.110326.

Zhang, Z., Gangurde, S. S., Chen, S., Mandlik, R., Liu, H., Deshmukh, R., & Li, Y. (2022). Overexpression of peanut (*Arachis hypogaea* L.)AhGRFi gene in *Arabidopsis thaliana* enhanced root growth inhibition under exogenous NAA treatment. https://doi.org/10.21203/rs.3.rs-2227535/v1.

Zhang, Z., Ge, X., Luo, X., Wang, P., Fan, Q., Hu, G., & Wu, J. (2018). Simultaneous editing of two copies of Gh14-3-3d confers enhanced transgene-clean plant defense against *Verticillium dahliae* in allotetraploid upland cotton. *Frontiers in Plant Science*, *9*, 842. https://doi.org/10.3389/fpls.2018.00842.

Zhang, Z., Zhu, L., Li, D., Wang, N., Sun, H., Zhang, Y., & Liu, L. (2021). In situ root phenotypes of cotton seedlings under phosphorus stress revealed through RhizoPot. *Frontiers in Plant Science*, *12*, 716691. https://doi.org/10.3389/fpls.2021.716691.

Zhao, J., Bai, W., Zeng, Q., Song, S., Zhang, M., Li, X., & Pei, Y. (2015). Moderately enhancing cytokinin level by down-regulation of GhCKX expression in cotton concurrently increases fiber and seed yield. *Molecular Breeding*, *35*, 1–11. https://doi.org/10.1007/s11032-015-0232-6.

Zhao, Y., Huang, Y., Wang, Y., Cui, Y., Liu, Z., & Hua, J. (2018). RNA interference of GhPEPC2 enhanced seed oil accumulation and salt tolerance in Upland cotton. *Plant Science*, *271*, 52–61. https://doi.org/10.1016/j.plantsci.2018.03.015.

Zhong, X., Hong, W., Shu, Y., Li, J., Liu, L., Chen, X., & Tang, G. (2022). CRISPR/Cas9 mediated gene-editing of GmHdz4 transcription factor enhances drought tolerance in soybean (Glycine max [L.] Merr.). *Frontiers in Plant Science, 13*, 988505. https://doi.org/10.3389/fpls.2022.988505.

Zhu, H., Jiang, Y., Guo, Y., Huang, J., Zhou, M., ang, Y., & Qiao, L. (2021). A novel salt inducible WRKY transcription factor gene, AhWRKY75, confers salt tolerance in transgenic peanut. *Plant Physiology and Biochemistry*, *160*, 175–183. https://doi.org/10.1016/j.plaphy.2021.01.014.

21 CRISPR/Cas9-Based Technology for Functional Genomics and Crop Improvement in Soybean

Cuong Xuan Nguyen and Phat Tien Do

21.1 INTRODUCTION

Soybean (*Glycine max* (L.) Merr.) is a valuable crop grown worldwide for seed protein and oil production. Moreover, the crop has a unique mutualistic relationship- nitrogen-fixing symbiosis with rhizobia; thus, it becomes a suitable crop for sustainable agriculture. Soybean is the first legume species whose genome was completely sequenced (Schmutz et al., 2010) and several new assemblies of both cultivated and wild soybeans have been released (Liu et al., 2020b; Valliyodan et al., 2019, 2021). These genome sequence databases are fundamental genetic resources for functional genomic studies and crop improvements in soybeans. In addition to genome sequences, transcriptomics, translatomics, proteomics, and metabolomics databases offer valuable resources for soybean genetic research and have made soybean an emerging model plant beside Arabidopsis or Medicago (Brechenmacher et al., 2012; Libault et al., 2010; Song et al., 2022; Sreedasyam et al., 2023; Valdés-López et al., 2016; Zhou et al., 2022). However, large genome sizes and highly duplicated sequences would be a challenge for gene function studies and crop improvement in soybeans.

Introducing new genetic variations into the populations is the main driving force of crop breeding. However, the relatively low genetic diversity in soybeans limits the available phenotypic variations that can be exploited for crop improvement (Hyten et al., 2006). To overcome the limitation, physical and chemical-induced mutagenesis are common approaches to introduce random genetic variations in plant genomes. In soybeans, mutant populations have been generated and utilized in breeding and gene function studies; however, developing the mutant population and screening for the trait of interest is time-consuming and labor-intensive (Campbell and Stupar, 2016).

Gene editing technologies are powerful molecular genetic tools that have revolutionized and accelerated gene discovery and breeding applications in plants. Generating large mutant populations, lowering the viability of the mutants, and inducing random gene mutations are challenges that the technologies overcome some challenges associated with conventional mutageneses such as generation of large mutant populations, lower viability of the mutants, and random gene mutations. Compared to other technologies such as TALEN, and ZFN, the CRISPR/Cas9 technology is widely applied in a variety of plant species for crop improvement because it is easier to construct, higher in efficiency, and more precise (Jinek et al., 2012; Luo et al., 2016; Zhu et al., 2020). In line with other species, the CRISPR/Cas9 system has been successfully utilized in soybeans to improve traits of interest such as yield components, seed quality, pathogenic resistance, and so on. Beyond that, the technology is becoming a main genetic tool for gene function studies in combination with mutant screening (Cai et al., 2015; Do et al., 2019; Jacobs et al., 2015; Michno et al., 2015; Nguyen et al., 2023). Therefore, the CRISPR/Cas9 system is a powerful tool for studying gene functions and crop improvement in soybean.

 DOI: 10.1201/9781003387060-21

21.2 RECENT APPLICATIONS OF CRISPR/Cas TECHNOLOGIES IN SOYBEANS

21.2.1 Application of CRISPR/Cas9 System to Improve Traits Related to Soybean Yield

Utilizing the CRISPR/Cas9 system to alter different agronomic and physiological traits, especially soybean seed yield, has tremendously gained success and demonstrated in recent reports. Genes that are involved in regulating plant morphology and architecture, and flowering time have been focused on modulating and showing some potential results.

Cheng and colleagues designed a CRISPR/Cas9 system with four guide RNAs in order to simultaneously induce mutations of *GmLHY* genes (*GmLHY1a*, *GmLHY1b*, *GmLHY2a,* and *GmLHY2b*) in soybean for altering the internodal length as well as the plant height (Cheng et al., 2019). The T2 mutant lines carrying quadruple mutations of the *GmLHY* genes showed shorter internodes and dwarfed plants. This phenotype phenomenon resulted from reductions in the bioactive GA3 content due to the lower expression of genes related to the gibberellins biosynthesis pathway in soybeans. The molecular markers for identifying potential mutations were developed and validated for further soybean breeding programs. In the same research strategy to change soybean plant morphology, Bao and co-workers created mutations of four genes encoding transcriptional factors (SQUAMOSA PROMOTER BINDING PROTEIN-LIKE,) of the SPL9 gene family (Bao et al., 2019). At the T4 generation, combinations of CRISPR/Cas9-induced mutations of the four tested genes resulted in increases in the internode number of the main stem as well as the branches. The report also indicated the contribution of *GmSPL9a, GmSPL9b, GmSPL9c*, and *GmSPL9d* genes in the regulation of the soybean plant architecture.

Flowering time is a critical factor affecting plant growth, development, and yield performance of different seed crops. Recently, CRISPR/Cas9 has been applied to induce targeted mutations in several genes involved in the flowering time to promote the adaptability of legume plants to different photoperiod and temperature conditions. Cai and colleagues generated CRISPR/Cas9-targeted mutations of *GmFT2a* and *GmFT5a* genes and indicated the function and impact of these genes on soybean flowering under diverse photoperiods (short days and long days) (Cai et al., 2020). Particularly, under short-day conditions, soybean lines containing double mutations of *GmFT2a* and *GmFT5a* genes showed late flowering, and higher pod and seed numbers per plant as compared to wild-type. In addition, CRISPR/Cas9-induced *GmFT* mutations enhanced soybean adaptability to diverse latitudes and farming systems, indicating the potential for expanding soybean cultivated area and production. In another study for functional analysis of the APETALA1 (AP1) gene family that contributed to the formation and development of flower organs, Chen and colleagues successfully created mutations in four *GmAP1* genes (*GmAP1a*, *GmAP1b*, *GmAP1c*, and *GmAP1d*) by the CRISPR/Cas9 system (Chen et al., 2020). Under short-day conditions, the quadruple mutant lines exhibited changes in flower morphology, delays in flowering, and increases in node number, internode length, and plant height compared to wild-type plants. This result provides the potential approach to improve soybean yield by modifying the expression of these candidate genes.

Seed size is one of the major components that directly impacts the yield performance of soybean crops. Targeted mutations of *GmSWEET10a* and *GmSWEET10b* genes were created in the Williams 82 cultivar using the CRISPR/Cas9 system for functional analysis of the *GmSWEET10* gene family in soybeans (Wang et al., 2020). The reduction in seed size was observed from soybean lines carrying single or double mutations of *GmSWEET10a* and *GmSWEET10b* genes. In contrast, CRISPR/Cas9-induced mutations of the *GmKIX8-1* gene resulted in big leaves and big seed phenotypes (Nguyen et al., 2021). *GmKIX8-1* encodes for KINASE-INDUCIBLE DOMAIN INTERACTION 8, an essential component participating in the PPD/KIX/TPL protein complex that regulates cell division. The seed weight of *GmKIX8-1* mutant lines increased up to 30% compared to the wild-type plants. Recently, the loss-of-function mutations of *GmST1*, a gene encoding the UDP-D-glucuronate

4-epimerase enzyme, were also generated by the CRISPR/Cas9 system. The *GmST1* mutants showed decreases in all seed size parameters such as length, width, and thickness (Li et al., 2022). In addition to seed size, increasing the number of seeds per pod per plant might also improve seed yield. In soybean, the *Ln* locus, which encodes GmJAGGED1 (GmJAG1), was shown to be a key regulator of seed number per pod (Jeong et al., 2012). Loss of function of *gmjag1* significantly increases the frequency of four-seeded pods (Jeong et al., 2011). Introducing the *ln* locus using the conventional crossing method into the cultivar Kedou1 has generated a new high-yield variety, Kedou 17 with 8%–10% higher in yield than the original cultivar (Liu et al., 2020a). Consistent with cultivars carrying *ln* locus, the CRISPR/Cas9 *gmjag* mutant line significantly increased the number of four-seeded pods, and ~8.67% of seed yield (Cai et al., 2021, 6).

21.2.2 Utilization of CRISPR/Cas9 for Improvement of Soybean Seed Components

Fatty acid components and contents have been seen as crucial criteria in determining soybean seed quality. Due to the low level of monounsaturated fatty acids in soybean oil, additional steps have to be performed during production processes which add up to the cost of expenses as well as generate unbeneficial compounds to human health. As a consequence, soybean oil is less competitive compared to other vegetable oils of other crops such as olive, canola, sunflower, etc. Therefore, enhancing monounsaturated fatty acid contents, especially oleic acid, is considered an important soybean breeding strategy to improve soybean oil quality and its adaptation in the market. In addition to conventional breeding methods, scientists have successfully applied advanced biotechnology approaches including CRISPR/Cas9 system to induce mutations in genes involved in fatty acid biosynthesis to increase the oleic acid content in soybean seed, recently.

The CRISPR/Cas9 system with two guide RNAs (gRNAs) was effectively utilized to cause loss-of-function mutations of two important genes *GmFAD2-1A* and *GmFAD2-1B*) in the soybean Maverick cultivar (Do et al., 2019). According to this report, 40% of the T0 transgenic plants carried the null mutations of both tested genes. The oleic acid content in mature seeds of T1 soybean lines harbored homozygous mutations in both *GmFAD2-1A* and *GmFAD2-1B* genes increased by up to 80%, while the linoleic acid content declined by below 2%. Importantly, no change in the total protein and fatty acid contents in soybean seeds was found between mutant lines and wild-type plants. In another report, the oleic acid content increased to 65.58%, whereas the linoleic acid content reduced to 16.08% in *GmFAD2-2* mutant lines created by the CRISPR/Cas9 system (Al Amin et al., 2019). However, the total fatty acid content was changed in CRISPR/Cas-mutant lines as compared to the wild-type. With the same strategy to improve the soybean oil quality, Wu and colleagues used CRISPR/Cas9 to induce mutations in *GmFAD2-1A* and *GmFAD2-2A* in a local soybean cultivar JN38 (Wu *et al.*, 2020). The seed component analysis was performed for both T2 and T3 mutant lines and showed the enhanced acid oleic amount (from 32.11% to 73.5%) and the decreased linoleic acid content (from 62.9% to 12.23%). Moreover, the increased protein content was also observed in the seeds of all mutant lines. Meanwhile, no difference in plant phenotype and yield performance was observed between mutant lines of the parent cultivar. Recently, CRISPR/Cas9 knock-out mutations of *GmFATB* genes (*GmFATB1a* and *GmFATB1b*) reduced palmitic and stearic contents but declined the amount of oleic and linoleic acids in soybean seeds (Ma et al., 2021). However, mutant lines had short stems, slow plant growth, as well as male sterility. In addition, decreases in the total seed oil content were found in all mutant lines as compared to the wild-type plants.

Besides changing fatty acid composition and contents, the CRISPR/Cas9 system was also utilized to alter other soybean seed components. For example, lipoxygenase-free soybean lines were generated by creating simultaneous mutations of multiple *GmLox* genes (*GmLox1*, *GmLox2*, and *GmLox3*) via the CRISPR/Cas9 system (Wang et al., 2019). In addition, two potentially allergenic proteins (Gly mBd 28 K or Gly m Bd 30 K) in the soybean seed of two local Japanese cultivars were eliminated through CRISPR/Cas9-induced mutations of their encoding genes (Sugano et al., 2020).

Together with fatty acid and protein composition, reducing raffinose family oligosaccharides in seed components is also a critical strategy in soybean breeding. CRISPR/Cas9- knock-out mutations of two genes (*GmGOLS1A* and *GmGOLS1B*) encoding for galactinol synthase, a key enzyme that catalyzed the initial reaction of the raffinose biosynthesis pathway, enhanced the soybean seed quality (Le et al., 2020). Particularly, the total RFO content was decreased by 35.2% in mutant lines carrying homozygous mutations of both *GmGOLS1A* and *GmGOLS1B*. Some mutant lines also had higher total protein and fatty acid contents. Interestingly, mutant lines were identical in terms of plant phenotype, growth and development, seed weight, and seed germination with non-mutant control plants. Recently, soybean lines carrying single or simultaneous mutations of raffinose synthase genes (*GmRS2* and *GmRS3*) were conducted using a multiplex CRISPR/Cas9 system (Cao et al., 2022). The mutant lines showed significant decreases in both stachyose and raffinose contents and increases in the sucrose concentration.

21.2.3 Enhancing Resistance to Biotic and Abiotic Stresses

Abiotic stress including drought, salt, temperature stress, and flooding causes significant loss in soybean yield. Abiotic-tolerant cultivars are mostly developed from the conventional breeding program; however, the approach is labor and cost-intensive. CRISPR/Cas9 genome editing is a key technology to generating new stress-tolerant soybeans, though the use of CRISPR/Cas9 to enhance abiotic tolerance in soybeans is rarely reported. Salinity stress is one of the major abiotic stresses that decrease soybean yields and growth. ABA-induced transcription repressors (AITRs) are involved in the feedback regulation of ABA signaling in plants. Loss of function of *aitrs* in Arabidopsis enhanced tolerance to salt and drought stresses (Chen et al., 2021; Tian et al., 2017). Translating the results into soybean, the knock-out of *GmAITR* genes by CRISPR/Cas9, the doube and quintuple mutants *gmaitr36* and *gmaitr23456*, respectively, increased salinity tolerance without fitness cost in soybean (Wang et al., 2021). In addition, modulating the ABA signaling by knock-out soybean circadian oscillator genes (*GmLCLs*) decreased water loss under dehydrated stress conditions (Yuan et al., 2021). Beyond abiotic stress, pests and pathogens including fungi, bacteria, viruses, insects, and nematodes are projected as a major cause of 40% yield losses in soybeans. Simultaneously knocking out the expression of three genes *GmF3H1*, *GmF3H2*, and *GmFNSII-1* that are involved in the phenylpropanoid pathway resulted in increased isoflavone content and enhanced resistance to soybean mosaic virus (SMV) in soybean plants (Zhang et al., 2020). These successful studies showed the potential application of the CRISPR/Cas9 system to improve stress tolerance in soybeans, although more efforts are needed to adopt a more useful way to generate tolerant cultivars.

21.2.4 Application of Hairy Root Systems for Validating CRISPR/Cas9 System Activities and Gene Function Studies

Rhizobium rhizogenes-mediated systems for hairy root induction and transformation have been established in soybeans via both *in vivo* and *in vitro* methods. The *in vitro* procedure using cotyledons as explants for bacterial infection is considered the most effective method for soybean hairy root induction (Niazian et al., 2022). Recently, the *R. rhizogenes*-mediated transformation has been successfully utilized for assessing, validating, and optimizing the performance of CRISPR/Cas9 gene editing constructs in different important crops. This system was conducted for the first time to assess the efficiency of the CRISPR/Cas9 vectors for inducing mutations of both exogenous and endogenous genes in soybeans (Jacobs et al., 2015). The induced mutant efficiency of the CRISPR/Cas9 system was recorded by over 95% of transgenic hairy root lines. After that, various studies have used the *R. rhizogenes*-mediated hairy root transformation system to evaluate the activity and efficiency of the designed CRISPR/Cas9 constructs (Cai et al., 2015; Cheng et al., 2021; Do et al., 2019, 2; Kong et al., 2023; Le et al., 2020; Li et al., 2019, 9; Sun et al., 2015). In addition, Trinh and colleagues also utilized the soybean hairy root system to confirm the sequential editing probability of the CRISPR/Cas9 system in soybean (Trinh et al., 2022). Particularly, soybean lines carrying a CRISPR/Cas9 expression cassette were used as a fast and sufficient system to assess and compare the gene editing efficiency of gRNAs. The report indicated that the sequential method has the potential to screen and select optimized gRNA sequences for the CRISPR/Cas9 vector construction before conducting soybean stable transformation.

Furthermore, recently, the hairy root induction and transformation systems have been used in combination with CRISPR/Cas9 systems for gene functional analysis. Typically, the role of the *Rfg1* allele in inhibiting soybean nodule formation under the infection of *S. fredii* USDA193 strain was analyzed and confirmed using hairy root lines carrying targeted mutations generated by the CRISPR/Cas9 system (Fan et al., 2017, 193). Similarly, the role of *GmROP9* genes in controlling symbiosis between rhizobium bacteria and soybean roots was performed and verified through the CRISPR/Cas9 hairy root transformation system (Gao et al., 2021). In addition, Nguyen and co-workers utilized the CRISPR/Cas9 constructs for functional analysis of two genes (*GmUOX* and *GmXDH)* encoding Uricase and Xanthine Dehydrogenase, two important enzymes for nitrogen fixation and nodule development in soybean (Nguyen et al., 2023). The loss-of-function mutations of two miRNAs (miR156b and miR156f) on soybean hairy roots were generated by the CRISPR/Cas9 system to access the interaction and contribution of these two factors on the regulation of different genes involved in the nodule formation (Yun et al., 2022).

21.3 CONCLUSION AND FUTURE PERSPECTIVES

In recent years, some achievements have been made in soybean improvement using the system based on CRISPR/Cas9 (Figure 21.1). Additionally, the CRISPR/Cas system now plays a tremendous role in precision plant breeding in which seed yield and quality, and stress-resistant traits of several crop species such as rice, maize, tomato, wheat, and soybean have been significantly improved

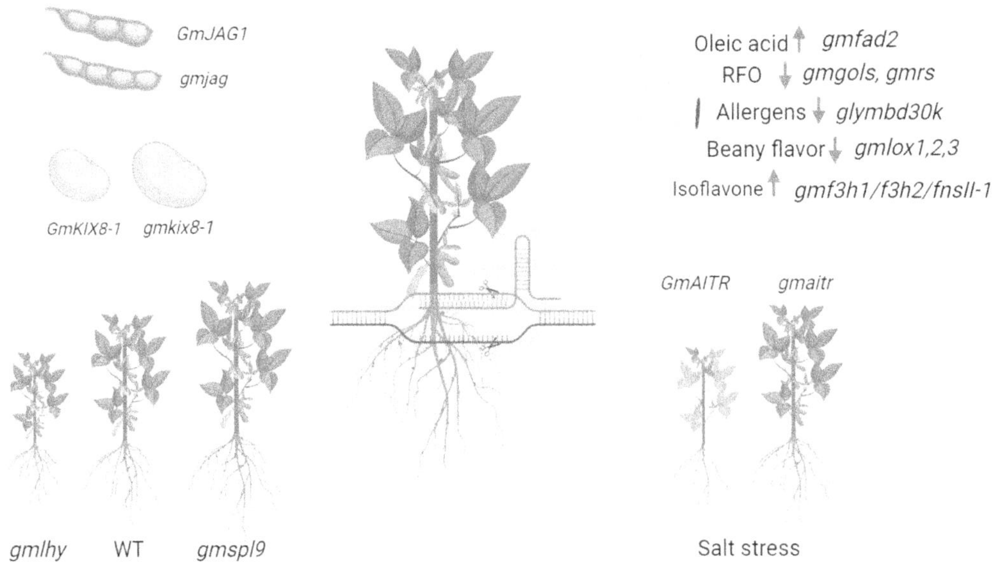

FIGURE 21.1 Achievements of using CRISPR/Cas9 system for trait improvement in soybean. Soybean seed yield could be improved by modulating genes involved in regulating seed size (*GmKIX8-1*), number of seed per pod (*GmJAG1*) or plant architectures (*GmSPL9, GmLHY*). Seed quality traits such as high oleic acid, isoflavone contents or low allergic proteins could be archived by knocking-out genes involved in the biosynthesis or metabolism of the target traits using the CRISPR/Cas9 approach. Salt tolerance could be improved in soybean by precisely mutate *gmaitr* that is involved in ABA signaling in plant. The figure was created with biorender.com.

(Chen et al., 2019). However, these traits are mainly gained from knock-out mediated CRISPR/Cas approaches. Precise gene targeting is still challenging in the plant research community due to the low efficiency of HDR or prime editing technology, unintended In-Dels, and high off-target of the base editing system, as well as to limitations of gene delivery system to specific plant species, genotypes, and tissues (Altpeter et al., 2016; Chen et al., 2019; Kumlehn et al., 2018). Therefore, there is still a need for the development of a comprehensive CRISPR/Cas toolkit for plants.

FUNDING

This work was supported by the National Foundation for Science and Technology of Vietnam (106.03-2019.11).

REFERENCES

Al Amin, N., Ahmad, N., Wu, N., Pu, X., Ma, T., Du, Y., et al. (2019) CRISPR-Cas9 mediated targeted disruption of FAD2-2 microsomal omega-6 desaturase in soybean (*Glycine max.* L). *BMC Biotechnology*, **19**, 1–10.

Altpeter, F., Springer, N.M., Bartley, L.E., Blechl, A.E., Brutnell, T.P., Citovsky, V., et al. (2016) Advancing crop transformation in the era of genome editing. *Plant Cell*, 28, 1510.

Bao, A., Chen, H., Chen, L., Chen, S., Hao, Q., Guo, W., et al. (2019) CRISPR/Cas9-mediated targeted mutagenesis of GmSPL9 genes alters plant architecture in soybean. *BMC Plant Biology*, **19**, 1–12.

Brechenmacher, L., Nguyen, T.H.N., Hixson, K., Libault, M., Aldrich, J., Pasa-Tolic, L., and Stacey, G. (2012) Identification of soybean proteins from a single cell type: The root hair. *Proteomics*, **12**, 3365–3373.

Cai, Y., Chen, L., Liu, X., Sun, S., Wu, C., Jiang, B., et al. (2015) CRISPR/Cas9-mediated genome editing in soybean hairy roots. *PLoS One*, 10, e0136064.

Cai, Y., Wang, L., Chen, L., Wu, T., Liu, L., Sun, S., et al. (2020) Mutagenesis of GmFT2a and GmFT5a mediated by CRISPR/Cas9 contributes for expanding the regional adaptability of soybean. *Plant Biotechnology Journal*, **18**, 298–309.

Cai, Z., Xian, P., Cheng, Y., Ma, Q., Lian, T., Nian, H., and Ge, L. (2021) CRISPR/Cas9-mediated gene editing of GmJAGGED1 increased yield in the low-latitude soybean variety Huachun 6. *Plant Biotechnology Journal*, **19**, 1898–1900.

Campbell, B.W. and Stupar, R.M. (2016) Soybean (*Glycine max*) mutant and germplasm resources: Current status and future prospects. *Current Protocols in Plant Biology*, **1**, 307–327.

Cao, L., Wang, Z., Ma, H., Liu, T., Ji, J., and Duan, K. (2022) Multiplex CRISPR/Cas9-mediated raffinose synthase gene editing reduces raffinose family oligosaccharides in soybean. *Frontiers in Plant Science*, **13**, 1048967.

Chen, K., Wang, Y., Zhang, R., Zhang, H., and Gao, C. (2019) CRISPR/Cas genome editing and precision plant breeding in agriculture. *Annual Review of Plant Biology*, **70**, 667–697.

Chen, L., Nan, H., Kong, L., Yue, L., Yang, H., Zhao, Q., et al. (2020) Soybean AP1 homologs control flowering time and plant height. *Journal of Integrative Plant Biology*, **62**, 1868–1879.

Chen, S., Zhang, N., Zhou, G., Hussain, S., Ahmed, S., Tian, H., and Wang, S. (2021) Knockout of the entire family of AITR genes in Arabidopsis leads to enhanced drought and salinity tolerance without fitness costs. *BMC Plant Biology*, **21**, 137.

Cheng, Q., Dong, L., Su, T., Li, T., Gan, Z., Nan, H., et al. (2019) CRISPR/Cas9-mediated targeted mutagenesis of GmLHY genes alters plant height and internode length in soybean. *BMC Plant Biology*, **19**, 1–11.

Cheng, Y., Wang, X., Cao, L., Ji, J., Liu, T., and Duan, K. (2021) Highly efficient *Agrobacterium rhizogenes*-mediated hairy root transformation for gene functional and gene editing analysis in soybean. *Plant Methods*, **17**, 1–12.

Do, P.T., Nguyen, C.X., Bui, H.T., Tran, L.T., Stacey, G., Gillman, J.D., et al. (2019) Demonstration of highly efficient dual gRNA CRISPR/Cas9 editing of the homeologous GmFAD2-1A and GmFAD2-1B genes to yield a high oleic, low linoleic and α-linolenic acid phenotype in soybean. *BMC Plant Biology*, **19**, 1–14.

Fan, Y., Liu, J., Lyu, S., Wang, Q., Yang, S., and Zhu, H. (2017) The soybean Rfg1 gene restricts nodulation by Sinorhizobium fredii USDA193. *Frontiers in Plant Science*, **8**, 1548.

Gao, J.-P., Xu, P., Wang, M., Zhang, X., Yang, J., Zhou, Y., et al. (2021) Nod factor receptor complex phosphorylates GmGEF2 to stimulate ROP signaling during nodulation. *Current Biology*, **31**, 3538–3550.

Hyten, D.L., Song, Q., Zhu, Y., Choi, I.-Y., Nelson, R.L., Costa, J.M., et al. (2006) Impacts of genetic bottlenecks on soybean genome diversity *Proceedings of the National Academy of Sciences of the United States of America*, **103**, 16666.

Jacobs, T.B., LaFayette, P.R., Schmitz, R.J., and Parrott, W.A. (2015) Targeted genome modifications in soybean with CRISPR/Cas9. *BMC Biotechnology*, **15**, 16.

Jeong, N., Moon, J.-K., Kim, H.S., Kim, C.-G., and Jeong, S.-C. (2011) Fine genetic mapping of the genomic region controlling leaflet shape and number of seeds per pod in the soybean. *Theoretical and Applied Genetics*, **122**, 865–874.

Jeong, N., Suh, S.J., Kim, M.-H., Lee, S., Moon, J.-K., Kim, H.S., and Jeong, S.-C. (2012) Ln is a key regulator of leaflet shape and number of seeds per pod in soybean. *The Plant Cell*, **24**, 4807–4818.

Jinek, M., Chylinski, K., Fonfara, I., Hauer, M., Doudna, J.A., and Charpentier, E. (2012) A programmable dual-RNA-guided DNA endonuclease in adaptive bacterial immunity. *Science*, **337**, 816.

Kong, Q., Li, J., Wang, S., Feng, X., and Shou, H. (2023) Combination of hairy root and whole-plant transformation protocols to achieve efficient CRISPR/Cas9 genome editing in soybean. *Plants*, 12, 1017.

Kumlehn, J., Pietralla, J., Hensel, G., Pacher, M., and Puchta, H. (2018) The CRISPR/Cas revolution continues: From efficient gene editing for crop breeding to plant synthetic biology. *Journal of Integrative Plant Biology*, **60**, 1127–1153.

Le, H., Nguyen, N.H., Ta, D.T., Le, T.N.T., Bui, T.P., Le, N.T., et al. (2020) CRISPR/Cas9-mediated knockout of galactinol synthase-encoding genes reduces raffinose family oligosaccharide levels in soybean seeds. *Frontiers in Plant Science*, **11**, 2033.

Li, C., Nguyen, V., Liu, J., Fu, W., Chen, C., Yu, K., and Cui, Y. (2019) Mutagenesis of seed storage protein genes in Soybean using CRISPR/Cas9. *BMC Research Notes*, **12**, 1–7.

Li, J., Zhang, Y., Ma, R., Huang, W., Hou, J., Fang, C., et al. (2022) Identification of ST1 reveals a selection involving hitchhiking of seed morphology and oil content during soybean domestication. *Plant Biotechnology Journal*, **20**, 1110–1121.

Libault, M., Farmer, A., Joshi, T., Takahashi, K., Langley, R.J., Franklin, L.D., et al. (2010) An integrated transcriptome atlas of the crop model *Glycine max*, and its use in comparative analyses in plants. *The Plant Journal*, **63**, 86–99.

Liu, S., Zhang, M., Feng, F., and Tian, Z. (2020a) Toward a "green revolution" for soybean. *Molecular Plant*, **13**, 688–697.

Liu, Y., Du, H., Li, P., Shen, Y., Peng, H., Liu, S., et al. (2020b) Pan-genome of wild and cultivated soybeans. *Cell*, **182**, 162–176.e13.

Luo, M., Gilbert, B., and Ayliffe, M. (2016) Applications of CRISPR/Cas9 technology for targeted mutagenesis, gene replacement and stacking of genes in higher plants. *Plant Cell Reports*, **35**, 1439–1450.

Ma, J., Sun, S., Whelan, J., and Shou, H. (2021) CRISPR/Cas9-mediated knockout of GmFATB1 significantly reduced the amount of saturated fatty acids in soybean seeds. *International Journal of Molecular Sciences*, **22**, 3877.

Michno, J.-M., Wang, X., Liu, J., Curtin, S.J., Kono, T.J., and Stupar, R.M. (2015) CRISPR/Cas mutagenesis of soybean and Medicago truncatula using a new web-tool and a modified Cas9 enzyme. *GM Crops & Food*, **6**, 243–252.

Nguyen, C.X., Dohnalkova, A., Hancock, C.N., Kirk, K.R., Stacey, G., and Stacey, M.G. (2023) Critical role for uricase and xanthine dehydrogenase in soybean nitrogen fixation and nodule development. *The Plant Genome*, **16**, e20171.

Nguyen, C.X., Paddock, K.J., Zhang, Z., and Stacey, M.G. (2021) GmKIX8-1 regulates organ size in soybean and is the causative gene for the major seed weight QTL qSw17-1. *New Phytologist*, **229**, 920–934.

Niazian, M., Belzile, F., and Torkamaneh, D. (2022) CRISPR/Cas9 in planta hairy root transformation: A powerful platform for functional analysis of root traits in soybean. *Plants*, **11**, 1044.

Schmutz, J., Cannon, S.B., Schlueter, J., Ma, J., Mitros, T., Nelson, W., et al. (2010) Genome sequence of the palaeopolyploid soybean. *Nature*, **463**, 178–183.

Song, J.H., Montes-Luz, B., Tadra-Sfeir, M.Z., Cui, Y., Su, L., Xu, D., and Stacey, G. (2022) High-resolution translatome analysis reveals cortical cell programs during early soybean nodulation. *Frontiers in Plant Science*, **13**, 820348.

Sreedasyam, A., Plott, C., Hossain, M.S., Lovell, J.T., Grimwood, J., Jenkins, J.W., et al. (2023) JGI Plant Gene Atlas: An updateable transcriptome resource to improve functional gene descriptions across the plant kingdom. *Nucleic Acids Research*, **51**(16), 8383–8401.

Sugano, S., Hirose, A., Kanazashi, Y., Adachi, K., Hibara, M., Itoh, T., et al. (2020) Simultaneous induction of mutant alleles of two allergenic genes in soybean by using site-directed mutagenesis. *BMC Plant Biology*, **20**, 1–15.

Sun, X., Hu, Z., Chen, R., Jiang, Q., Song, G., Zhang, H., and Xi, Y. (2015) Targeted mutagenesis in soybean using the CRISPR-Cas9 system. *Scientific Reports*, **5**, 10342.

Tian, H., Chen, S., Yang, W., Wang, T., Zheng, K., Wang, Y., et al. (2017) A novel family of transcription factors conserved in angiosperms is required for ABA signalling. *Plant, Cell & Environment*, **40**, 2958–2971.

Trinh, D.D., Le, N.T., Bui, T.P., Le, T.N.T., Nguyen, C.X., Chu, H.H., and Do, P.T. (2022) A sequential transformation method for validating soybean genome editing by CRISPR/Cas9 system. *Saudi Journal of Biological Sciences*, **29**, 103420.

Valdés-López, O., Batek, J., Gomez-Hernandez, N., Nguyen, C.T., Isidra-Arellano, M.C., Zhang, N., et al. (2016) Soybean roots grown under heat stress show global changes in their transcriptional and proteomic profiles. *Frontiers in Plant Science*, **7**, 517–517.

Valliyodan, B., Brown, A.V., Wang, J., Patil, G., Liu, Y., Otyama, P.I., et al. (2021) Genetic variation among 481 diverse soybean accessions, inferred from genomic re-sequencing. *Scientific Data*, **8**, 50.

Valliyodan, B., Cannon, S.B., Bayer, P.E., Shu, S., Brown, A.V., and Ren, L., et al. (2019) Construction and comparison of three reference-quality genome assemblies for soybean. *The Plant Journal*, **100**, 1066–1082.

Wang, J., Kuang, H., Zhang, Z., Yang, Y., Yan, L., Zhang, M., et al. (2019) Generation of seed lipoxygenase-free soybean using CRISPR-Cas9. *The Crop Journal*, **8**(3), 432–439.

Wang, L., Sun, S., Wu, T., Liu, L., Sun, X., Cai, Y., et al. (2020) Natural variation and CRISPR/Cas9-mediated mutation in GmPRR37 affect photoperiodic flowering and contribute to regional adaptation of soybean. *Plant Biotechnology Journal*, **18**, 1869–1881.

Wang, T., Xun, H., Wang, W., Ding, X., Tian, H., Hussain, S., et al. (2021) Mutation of GmAITR genes by CRISPR/Cas9 genome editing results in enhanced salinity stress tolerance in soybean. *Frontiers in Plant Science*, **12**, 779598.

Wu, N., Lu, Q., Wang, P., Zhang, Q., Zhang, J., Qu, J., and Wang, N. (2020) Construction and analysis of GmFAD2-1A and GmFAD2-2A soybean fatty acid desaturase mutants based on CRISPR/Cas9 technology. *International Journal of Molecular Sciences*, **21**, 1104.

Yuan, L., Xie, G.Z., Zhang, S., Li, B., Wang, X., Li, Y., et al. (2021) GmLCLs negatively regulate ABA perception and signalling genes in soybean leaf dehydration response. Plant, *Cell & Environment*, **44**, 412–424.

Yun, J., Sun, Z., Jiang, Q., Wang, Y., Wang, C., Luo, Y., et al. (2022) The miR156b-GmSPL9d module modulates nodulation by targeting multiple core nodulation genes in soybean. *New Phytologist*, **233**, 1881–1899.

Zhang, P., Du, H., Wang, J., Pu, Y., Yang, C., Yan, R., et al. (2020) Multiplex CRISPR/Cas9-mediated metabolic engineering increases soya bean isoflavone content and resistance to soya bean mosaic virus. *Plant Biotechnology Journal*, **18**, 1384–1395.

Zhou, M., Fulcher, J.M., Zemaitis, K.J., Degnan, D.J., Liao, Y.-C., Veličković, M., et al. (2022) Discovery top-down proteomics in symbiotic soybean root nodules. *Frontiers in Analytical Science*, **2**, 1012707.

Zhu, H., Li, C., and Gao, C. (2020) Applications of CRISPR-Cas in agriculture and plant biotechnology. *Nature Reviews Molecular Cell Biology*, **21**, 661–677.

Index

Note: **Bold** page numbers refer to tables and *italic* page numbers refer to figures.

Milton Keynes UK
Ingram Content Group UK Ltd.
UKHW010351101224
451979UK00006BA/100

9 781032 469492

Milton Keynes UK
Ingram Content Group UK Ltd.
UKHW010351101224
451979UK00006BA/101

9 781032 259970